# Quantum Mechanics and Path Integrals

**Book 4 of Physics from Maximal Information Emanation, a seven-book physics series.**

**ISBN 979-8-89487-011-3**

# Quantum Mechanics and Path Integrals

**by**

**Stephen Winters-Hilt**

**ISBN 979-8-89487-011-3**

**Golden Tao Publishing**
**Angel Fire, NM**
**USA**

**Dedication**

This book is dedicated to my family that helped on this lengthy road of discovery: Cindy, Nathaniel, Zachary, Sybil, Eric, Joshua, Teresa, Steffen, Hannah, Anders, Angelo, John and Susan.

# Contents

*Preface to Physics Series on:*

## Physics from Maximal Information Emanation

The Road goes ever on and on
Down from the door where it began.
Now far ahead the Road has gone,
And I must follow, if I can,
Pursuing it with eager feet,
Until it joins some larger way
Where many paths and errands meet.
And whither then? I cannot say
**— J.R.R. Tolkien, The Fellowship of the Ring**

### Variation, Propagation, and Emanation

This is a seven book Physics Series that starts with Classical Mechanics (Book 1 [16]), then Classical Field Theory, such as electromagnetism (Book 2 [17]), then Manifold Dynamics, such a General Relativity (Book 3 [18]). The switch to a quantum mechanics description is given in Book 4 [19], and to a quantum field theory, Quantum Electrodynamics in particular, in Book 5 [15]. A 'quantum manifold theory' would be the obvious next step except it cannot be done (there is not a renormalizable Field theory for Gravitation). Instead a thermal quantum manifold theory is considered, as well as Black Hole thermodynamics in general, in Book 6 [20]. Book 7 [21] describes a new theory, Emanator Theory, that provides a deeper mathematical construct that undergirds Quantum theory, much like quantum theory can be shown to provide a deeper (complexified) mathematical construct based on the classical theory.

This is a modern exposition where subtleties of chaos theory are described in Book 1, of Lorentz Invariance in Book 2, of Covariant Derivatives (General Relativity) and Gauge Covariant Derivatives (Yang-Mills Field Theory) in Book3. Book 4 on Quantum Mechanics provides an extensive review of quantum mechanics, then considers a full self-adjoint analysis on the full general relativistic solution to the spherical shell in-fall system (a result carried over from Book 3). Book 5 considers quantum field theory basics in detail, along with alternate vacua in specific scenarios. Book 6 considers thermodynamics from the basics to the Hamiltonian thermodynamics of some Black Hole systems. Throughout, the odd recurrence of the alpha parameter is noted. In Book 7 we look to a deeper mathematical formulation from which the Quantum Path Integral formulation would result, as well as explaining the odd parameters and structures that have been discovered (such as alpha and Lorentz Invariance).

The physical description starts with the classic formulations of point particle motion. The first approach to doing this is using differential equations (Newton's $1^{st}$ and $2^{nd}$ Law); the second is using a variational function formulation to select the differential equation (Lagrangian variation); the third is using a variational functional formulation (Action formulation) to select the variational function formulation. Historically, it wasn't realized until much later that there are two domains for motion in many systems: non-chaotic; and chaotic.

In a description of particle motion, assuming not in a parameter domain with chaotic motion, several important limits are found to exist. Examples include: the universal constants from the aforementioned chaos phenomenon, that are still encountered in non-chaos regimes if driven "to the edge of chaos". Limits are found where scattering is defined in the asymptotic limit and perturbation theory is well-defined in the sense that it is convergent. Overall, if the evolution is described as a 'process' it is often a Martingale process, which has well-defined limits. So, we have descriptions for motion, typically reducible to an ordinary differential equation (ODE), and for which solutions (requiring limit-definitions) are typically found to exist.

The physical description then contends with field dynamics in 2D, 3D, and 4D (in Book 3 [18]). Two-dimensional ("2D") field dynamics can be described as a complex function (that maps complex numbers to complex numbers). A novelty of the 2D complex function is it also shows how to handle many types of singularities (the residue theorem), thus provides important information about fundamental structures in physics as well as fundamental mathematical techniques for solving many integrals. For the 3D field dynamics we do an analysis of the electromagnetic field in 3D. The level of coverage begins at an overview of electrostatics at the level of the graduate text Jackson [190]. Some problems from Jackson Ch's 1-3 are examined closely in developing the theory itself. For some this material (in Book 2 [17]) might provide a useful accompaniment to Jackson's text in a full course on Electromagnetism (based from

Jackson's text). A quick review of electrodynamics and electromagnetic wave phenomena is then given. In essence, we see many more examples of ODE problems with solutions, such as for the 3D Laplacian, usually involving separation of variables. We then review the famous transform, discovered by Lorentz in 1899 [191], that relates the electromagnetic field as seen by two observers differing by a relative velocity. With the existence of this transform, that brings in the time dimension along with the relative velocity, we effectively have a 4D theory.

From Lorentz Invariance we have, as a point transformation, rotational invariance under SO(3) or SU(2). If Lorentz Invariance is fundamental, then we should see both forms of rotation invariance, one of vector/tensor type from SO(3), and one of spinorial type from SU(2). This is the case, as gauge fields are vectorial and matter fields are spinorial. From Lorenz Invariance as a local invariance we have the Minkowski (flat) spacetime metric, which then generalizes to the Riemannian metric (in General Relativity).

As with the point particle dynamics, for the field dynamics we have three ways to formulate the behavior: (1) differential equation; (2) function variation (on Lagrangian); and (3) functional variation (on the Action). We will see similar limit phenomena as before, but also new phenomena, including (i) inevitable Black Hole singularity formation (the Penrose singularity theorem); (ii) FRW Universe formation (from homogeneity and isotropy); (iii) the Black Hole collapse singularity; (iv) the atomic collapse radiative 'singularity'.

Classical dynamics, thus, has two field-like formulations to describe the world: field and manifold. Such formulations can be interrelated mathematically, so what is happening is more a matter of physics emphasis and convenience. The emphasis on this difference, that appears to be no difference (mathematically), is that different physical phenomenologies are at play. Field descriptions appear to work for 'matter', where the fundamental elements are spinorial. Manifold descriptions appear to work best for geometrodynamics (General Relativity), where the fundamental elements are vectorial (or tensorial, such as the metric). Matter fields are renormalizable, thus quantizable in the standard quantum field theory formulation (to be described in Book 5 [15]), while gravitational manifolds are not renormalizable, and have constraints (weak energy condition and positive energy condition given the existence of spinor fields on the manifold).

The presentation in Books 1-3 [16-18], on 'classical' physics, is partly done to make the transition to quantum physics simple, obvious, and in some cases, trivial. Consider the functional variation (Action) formulation of the behavior (whether point-particle or field), this can be captured in integral form, as was done by D'Alembert very early [22] (then by Laplace [23]). Note the use of a large constant to effect a 'highly damped' integral for selection purposes (on variational extremum of the action). To transition to the quantum theory we also have the large constant from 1/h, and so the only difference is the introduction of a factor of 'i', to effect a 'highly oscillatory' integral for selection purposes.

After the transition to a quantum theory, for the point-particle descriptions, the classical collapse problem for atomic nuclei is eliminated. The spectral predictions have excellent agreement with theory, but there is still fine-structure in the spectra not fully explained. The theory is not relativistic and some initial corrections for this are possible (without going to a field-theory) and these indicate closer agreement and explain most of the fine-structure constant discrepancy (and reveal alpha in another place in the theory). It is shown in Book 3 [18] and Book 4 [19], that the General Relativity singularity problem, however, remains unresolved (for the test case of spherical dust shell collapse, done in a full General Relativity analysis, then quantized in a full self-adjoint quantization analysis).

In Book 5 [15], the transition to quantum theory is continued to the field theory descriptions. A precise description/agreement of atomic nuclei is now possible with Quantum Electrodynamics, and within the nuclei themselves (quark confinement) with Quantum Chromodynamics. The field theories have a small set of bothersome infinities, however, which is eventually solved by renormalization [15]. As mentioned, the quantization of manifold theories, such as General Relativity, does not appear to be possible due to non-renormalizability. Not to be deterred, in Book 6 [20] we consider a Hamiltonian description of a General Relativity system whose quantization would involve an energy spectrum based on that Hamiltonian, if we then use analytic continuation to take us to the thermal ensemble theory based on the partition function that results, we can consider the thermal quantum gravity of such systems.

This last example (from Book 6), showing a consistent thermal quantum gravity theory if we use analyticity, is part of a long sequence of successful maneuvers involving analytic continuations in different settings. What is indicated is the presence of an actual complex structure to the stated theory. There is the trivial complex structure extension mentioned above that brought us from the standard classical physics theory to the standard path integral quantum theory. But we also see actual complex structure at the component level with time complexation (that ties to thermal version of the theory by defining the partition function), and we have complex structure as the dimension-level in the form of the successfully applied dimensional regularization procedure used in the renormalization program.

As well as covering the breadth of core physics topics at both undergraduate and graduate level (for courses taken at Caltech and Oxford), including extensive presentation of problems and their solutions, the Series also examines, in specific cases, the boundaries of the physical world "from the inside" (and then later "from the outside"). To this end exploration of spherical dust collapse to form a singularity is examined in a fully general relativistic formalism, and then carried-over to a quantum minisuperspace (quantum gravity) analysis (in Books 3 and 4 [18,19]). Also examined in-depth are the topics of black hole thermodynamics and quantum field theory with alternate vacua (part of Books 5 and 6 [15,20]). The in-depth material comprises the topics covered in my PhD dissertation [147], portions of which are published [125,126,172,173].

In recent work on machine learning, that includes statistical learning on neuromanifolds [170,21], we find a possible new source for a foundational element for statistical mechanics (entropy) via seeking a minimal learning process/path on a neuromanifold [170,21]. By the time the Series reaches thermodynamics in Book 6 [20], therefore, the foundational thermodynamics elements have all been established from the physical descriptions discovered in Books 1-5 [15-19], they just haven't been put together in a comprehensive analysis that gives us the fundamental constructs of thermodynamics and statistical mechanics. That said, it would seem that thermodynamics is, thus, entirely derivative from other, truly fundamental theories. Not so, in the joining of the parts to make thermodynamics we have something greater than the sum of the parts. In the 'system' descriptions we find that emergent phenomena exist. This, at least, is unique to thermodynamics, so it is fundamental in this "sum greater than the parts' aspect.

In Book 7 (the last) of the Series [21], we consider the standard physical world, described by modern physics, "from the outside." In doing this we've already eliminated part of the mystery of entropy by the geometric 'neuromanifold' description. If we can understand other oddities of the standard theory, and arrive at them naturally, then we might have an even deeper dive into modern physics, testing the limits of what is possible, and see possible future developments and unifications of the theory. This is what is described in papers [104,168-170,192-196], and organized along with current results into the final Book of the series.

Efforts in the last book of the Series involve choices and concepts identified in the prior six books of the Series, and theoretical maneuvers gleaned from the most advanced courses in physics and mathematical physics taken while at Caltech (as an undergraduate and then as a graduate) and the Oxford Mathematics Institute (as a graduate), and the University of Wisconsin at Milwaukee (as a graduate).

The broad range of topics covered in the Series is, initially, similar to the Landau & Lifshitz graduate textbook series (see [197]), with a similar exposition on classical mechanics at the start of Book 1. Even with well-established classical mechanics, however, there are significant, modern, updates, such as (modern) chaos theory. In the final two books of the Series (Books 6 and 7 [20,21]) we arrive at statistical mechanics and thermodynamics, together with modern topics such as black hole thermodynamics, thermal quantum gravity, and emanator theory.

Key constants and structures of physics, their discovery from the experimental data, and their theoretical placement in the "Grand Scheme," are emphasized throughout the Series. The constant alpha, a.k.a. the fine structure constant, appears in numerous settings so special note of the occurrence of alpha will be made in each chapter. This is the case even at the outset with Book 1, due to fundamental numerical constants appearing from chaos theory. In Book 7 we see the origin of alpha, as a maximal perturbation amount, appears naturally in a formalism for maximal information 'emanation'. But maximal perturbation in what space and in what manner? In Book 7 of the series [21] we will see a possible representation of such an information entity, and its space of existence, in terms of chiral trigintaduonions.

Thus, in the end, this is an effort to tell of a journey to a special place, where many paths and errands meet, giving rise to emanator theory and an answer to the mystery of alpha. Part of this journey is equivalent to finding the alpha similarly to the arkenstone in Tolkien's story, in the most unlikely of places, the trigintaduonion emanation mathematics underpinning the emanator formalism (e.g., Smaug's Lair, described in Book 7 [21]). Why I should have wandered into such an odd place (mathematically speaking), and why I should posit a deeper form of quantum propagation using hypercomplex trigintaduonions, here called emanation, is why there is such extensive background on standard topics. This extensive background even impacts the classical mechanics description via its modern chaos theory material (due to a possible relation between $C_\infty$ and alpha). The critical role of emergent phenomena is only understood at the end, including for manifolds in geometry and neuromanifolds in statistical mechanics, and leads to a Book 6 that goes from very basic (introductory thermodynamics) to very advanced (emergent phenomena). Much is made clear with emanator theory, including how reality is both fractal and emergent. At this point in the journey, as with Tolkien, this much I can say to repeat Tolkien: The Road goes ever on and on … And whither then? I cannot say.

The seven books in the Series are as follows:

- Book 1. Classical Mechanics and Chaos
- Book 2. Classical Field Theory
- Book 3. Classical Manifold Theory
- Book 4. Quantum Mechanics and the Path Integral Foundation
- Book 5. Quantum Field Theory and the Standard Model
- Book 6. Thermal & Statistical Mechanics, and Black Hole Thermodynamics
- Book 7. Maximum Information Emanation and Emanator Theory

***Overview of Book 1***

Book 1 is a modern exposition of classical mechanics, including chaos theory, and including ties to later theoretical developments as well. The exposition consists, throughout, of the presentation of interesting problems with many solved, the others left for the reader. The problems are drawn from classical mechanics and mathematics courses taken at Caltech, Oxford, and the University of Wisconsin. The courses range from undergraduate level to advanced graduate level. The courses had a rich and sophisticated selection of textbook and reference material, as you might expect, and those reference texts are, similarly, drawn on here. Those classical mechanics texts, listed by author, include: Landau and

Lifshitz [197]; Goldstein [198]; Fetter & Walecka [199]; Percival & Richards [200]; Arnold (ODE) [201]; Arnold (CM) [202]; Woodhouse [203]; and Bender & Orszag [204]. Notice how the first Arnold reference and the Bender and Orszag reference involve textbooks focused on ordinary differential equations (ODEs). Likewise, an analysis of the excellent, and rapid, exposition by Landau and Lifshitz, reveals that it partly progresses through the material by going through ODEs of increasing complexity (corresponding to more complicated pendulum motion, for example, such as by adding a frictional force). This strong alignment with the underlying mathematics of ODEs is continued in this exposition, so much so that an appendix is provided for a quick review of ODEs from the applied mathematics perspective.

Particle dynamics, with and without forces, are described, with all arriving at descriptions with chaotic motion, with chaos described in the latter half of Book 1 [16]. Universally it is found that systems transitioning to chaotic behavior do so with a remarkable period-doubling process and this will be described both mathematically and with computer results. In the analysis of such dynamical systems we will find that periodic physical systems can be described in terms of repeated "mappings", e.g., classic dynamic mappings [205], and when described in this way the transition to chaos is made much more mathematically evident (as will be shown). The familiar Mandelbrot set is generated by such a repeated mapping, where it's "edge of chaos" is defined by the fractal boundary of the classic Mandelbrot image.

Properties of the classic Mandelbrot set will be relevant to the physics discussed in Book 1 and Book 7, including the property that the fractal boundary has a fractal dimension of 2 (the fractal dimension of the boundary can be between 1 and 2, to get equal to 2 is special). With the Mandelbrot set we also recover the well-studied constants associated with the universal Feigenbaum constants [15]. In the Mandelbrot set we can clearly see the fundamental constant for maximum perturbation that is at maximum antiphase (negative) with magnitude $C_\infty$, where the same results hold for a family of basic formulations (for a variety of Lagrangian formulations, for example).

From the Lagrangian variational formulation of 'action' for particle motion we will eventually define the path integral functional variational formulation involving that same Lagrangian to arrive at a quantum description for the non-relativistic quantum particle motion (described in detail in Book 4 [19], and relativistic in Book 5 [15]). From the quantum description we arrive at the propagator formalism for describing dynamics (this exists in the classical formulation too, but typically is not used much in that context). Complex propagators will then be found to have ties to statistical mechanics and thermodynamics properties (Book 6 [20]). The ties to statistical mechanics are further emphasized when at the "edge of chaos" but with the orbit motion still confined. This may be associated with an ergodic regime, thus an equilibrium and martingale regime, the existence of which can then be used at the start of Book 6 [20] statistical mechanics and thermodynamics derivations with the existence of equilibria established at the outset. The existence of the familiar entropy measures are already indicated in the neuromanifold description (Book 3 [18]), thus, together with equilibria, the Book 6 thermodynamics description is able to begin with a well-established foundation that is not claimed by fiat, rather claimed as a direct result of what has already been determined in the theory/experiment described in the previous books of the Series.

***Overview of Books 2 & 3***

When moving from a theory of point particles to a theory of fields, there's not much discussion in the core physics books on fields in a general sense, it usually just directly jumps to the main field of relevance, Electromagnetism. If advanced, it may also cover General Relativity, as with [206]. In what follows we will cover these topics, but we will also cover the more basic fields in 1, 2, and 3D (including fluid dynamics), as well as 4D Lorentzian Field formulations (for Special Relativity), the Gauge Field formulation (thus Yang Mills covered in a classical context), and the General Relativity geometric and gauge formulations. This establishes the foundation for the standard forces, and upon quantization (Books 4 and 5 in the Series), lays the foundation for the standard renormalizable forces (all but gravitation).

The gravitational coupling constant 'G' is a dimensionful coupling (not like with alpha in electromagnetism), and gravitation with manifold construct can be described as a gauge field construct, although not renormalizable. Gravitation, and associated geometry/manifolds, appears to relate to its own emergent structure, as will be discussed in Book 6. From the local Lorentzian geometry and Lorentzian field descriptions we also see the first of many examples where there is system information in the complexification of some parameter, here the time component. If the Lorentzian is shifted to complex time, this shifts it to being a Euclidean field, with formally well-defined convergence properties (as occurs in statistical mechanics). Complex time also shows deep connections between classical motion and associated Brownian motion (where random walk reveals pi). Thus, it should not be surprising that an emergent manifold may have complex structure such that there is also an emergent 'thermal' manifold, possibly the neuromanifold described in Book 3 and the related partition functions examined in Book 6. Just like locally flat space-time is a natural construct in General Relativity, so too are optimization "learning" steps on a neuromanifold such that relative entropy is selected as a preferred measure, and from it Shannon entropy and Boltzmann's statistical entropy. Thus, the manifold construct appearing at Book 3 has far reaching impact into the foundations of the thermodynamic and statistical mechanical theory described in Book 6.

Before we even get to the manifold/geometry complexities of General Relativity, however, we have already established much with the electromagnetic field part of the theory: (i) from 'free' electromagnetism without matter we get the speed of light c, Lorentz invariance, and from that special relativity and locally flat space-time; (ii) from electromagnetism with matter we get the dimensionless coupling constant alpha.

In going over field theories to describe matter, force fields, and radiation we first describe the classical field theories of fluid mechanics, electromagnetism, and General Relativity, with many examples shown. This is then carried over to the quantum field theory description in Book 5. A review of the core mathematical constructs employed in classical field theory and Quantum Field Theory is given in the Appendix. Even as the mathematical physics approach grows in sophistication, we still obtain solutions via variational extrema. Thus, determining the evolution of the system from its variational optimum now becomes the focus of the effort. System 'propagation' from one time to a later time can be described by a propagator. Although a 'propagator' formulation is possible mathematically in classical mechanics and classical field theory, which are shown, this is usually not done, in favor of simpler representations for the experimental application at hand. As we move to descriptions in the quantum realm, however, the use of the propagator formalism becomes typical, and when used in the path integral formulations we arrive at a compact formulation describing both the evolution and stationary-phase solution at once.

In Book 2 the focus is on classical field theory in a fixed geometry, the main physical example is electromagnetism. In this setting alpha appears, for example, in the description of an electron-positron pair: $F = e^2/(4\pi\varepsilon a^2)$ for electron-positron distance 'a' apart, where alpha appears as the coupling constant. Later, in quantum mechanics, both modern and in the early Bohr model, we have that alpha = $[e^2/(4\pi\varepsilon)]/(c\hbar)$. The appearance of alpha in these situations is occurring in bound systems. If we examine electromagnetic interactions that are unbound, on the other hand, such as with the Lorentz Force $F = q(E \times v)$, here there arises no alpha parameter, nor with the early quantum mechanical analysis of such systems such as with Compton scattering. Thus, we see an early role for alpha, but only in bound systems, thus only in systems with (convergent) perturbative expansions in system variables.

In Book 3, classical field theory with *dynamic* geometry, i.e. General Relativity, we don't see alpha at all. Instead we see manifold constructs and the mathematics of differential geometry (and to some extent differential topology and algebraic topology). Manifold constructs are entirely encapsulated in the math background given in Book 3 and the Appendix there. An application in the area of neuromanifolds (see [179]), shows the equivalent of a geodesic path in this setting is evolution involving minimum relative entropy steps. Similar to the description of a locally flat space-time we now have a description of 'entropy' increasing/evolving according to minimum relative entropy.

General relativity stands apart from the other force fields. All the other force fields are part of an adjoint representation of the standard model vis-à-vis the stability subgroup $U(1)xSU(2)_LxSU(3)$. The form of which is derivable from the chiral T one-sided products described in Book 7. The standard model is uniquely obtained in this process, and with no mention of General Relativity. Keep in mind, however, that the adjoint representation has operation on some space (hyperspinorial in case of simple octonion right-products, for example). The 'force' due to gravity is that due to manifold curvature, where the manifold construct is possibly emergent on the space of operation. Thus, the origin of the General Relativity force is entirely different, and it will not allow quantization like the other forces, nor will its singular solutions be resolvable via quantum physics alone, as with electromagnetism in Books 4&5, but will also need thermal physics (as will be described in Book 6).

The existence of singular General Relativity solutions, outside of specially symmetric cases (the classic Black hole solutions), wasn't firmly established until the Penrose singularity theorem [207] (awarded Nobel prize in Physics for this in 2020). Some of this material is covered in Book 3 to show how the mathematical formalism shifts to differential topology methods to describe the singularities, with examples referencing the Hawking and Ellis classic [208] and using Penrose diagrams. This, in turn, will come in handy when describing the classic FRW cosmologies with radiation and matter dominated phases (using notes from Peebles [209], Peebles won the Nobel in Physics in 2019).

The General Relativity development would be remiss if it didn't briefly delve into cosmological models, the classic FRW cosmologies in particular. With the General Relativity tools developed, cosmological results are examined, starting with the entry of the cosmological constant into the formalism (a candidate for Dark energy). Various observational data on galaxy rotations and universe simulations of galaxy cluster formation both indicate the existence of Dark matter. This, then, means we have new matter, non-interacting except gravitationally, and this is actually consistent with the latest observational data on the muon g-2 value [210], where the discrepancy between theory and experiment has grown to 4.2 standard deviations, where an extension in the Standard Model appears to be in the works. This is convenient as Emanator theory (Book 7 [21]), predicts such an extension.

We can thus arrive at field equations for electromagnetism, General Relativity, and Yang-Mills Gauge Fields (Strong and weak). We can obtain wave and vortex phenomena (as hinted in fluid dynamics). We show the classical instability for atomic matter (classical electromagnetism instability) and classical gravitational instability (leading to black hole formation with singularity). From Lagrangian formulations we can then arrive at a Quantum Field Theory formulation (Book 5). The Quantum Field Theory formulation completes the quantum mechanics (Book 4) cure of "non-relativistic atomic instability" with the cure of the fully relativistic atomic description of the radiative-collapse instability. Introduction of Quantum Field Theory also leads to new instability or infinities, but these can be eliminated by renormalization for the electromagnetic and electroweak formulations, and the Yang-Mills strong formulation, but not the General Relativity (gauge) formulation. The current theoretical formulation in modern physics has one glaring gap, therefore: a quantum theory of gravitation. Perhaps this is not a missing element, however, if geometry/General Relativity is a derivative phenomenon, like the field of statistical mechanics and thermodynamics appeared as derivative phenomenon when the complexified quantum propagator gives rise to a real (quantum) partition function. The hint of a deeper emanator theory suggests emergent structures of geometry and thermodynamics are arrived at in the process of emanation, with the information emanated being that of the renormalizable quantum matter fields. In Book 7 [21] a precise mathematical meaning will be found for describing maximal information emanation.

***Overview of Book 4***

By 1834, with Hamilton's Principle, there was a strong foundation for what is now called classical mechanics. By 1905, with Einstein's publication on the photoelectric effect [51], the rules of classical mechanics were being superseded by the new rules of quantum mechanics. The earliest appearance of quantum mechanics, however, began with the various observations of quantization of light, starting with the strange occurrence of spectral lines for hydrogen. The hydrogen spectrum was made even stranger by a precise fit to a succinct empirical formula by Balmer in 1885 [31]. This is the beginning of an amazing period of discovery. The developments of quantum mechanics from introductory to advanced roughly follows that history.

The early phase of discovery for quantum mechanics moved into the modern quantum mechanics formalism with the discovery of Heisenberg of the successful application of matrix mechanics and the resultant uncertainty principle (1925) [25]. In 1926, Schrodinger showed that the problem of finding a diagonal Hamiltonian matrix in the Heisenberg's mechanics is equivalent to finding wavefunction solutions to his wave equation [26]. An interpretation of the wavefunction was then clarified in 1927 by Born [28]. Dirac developed a manifestly relativistic formalism for the wavefunction and wave-equation for fermionic matter (1928) [29]. An axiomatic reformulation of quantum mechanics was then given by Dirac (1930) [5], laying the foundation for much of modern quantum notation and for critical issues such as self-adjointness. Dirac then described a formulation of a quantum propagation path, with quantum propagator having the familiar phase factor involving the action, in his paper "The Lagrangian in Quantum Mechanics" in 1933 [60]. In essence, Dirac had obtained a single path, in what would eventually be generalized by Feynman to all paths with the invention of the path integral formalism (1942 & 1948) [61,62]. The equivalence of a quantum mechanical formulation in terms of path integrals and the Schrodinger formalism was shown by Feynman in 1948 [62].

In a path integral description, the quantum mixture state, semiclassical physics, and classical trajectories are all given by the stationary phase dominated component. A stationary phase solution that is dominated by a single path is typical for a classical system. Thus, variational methods are fundamental to analysis of physical systems, whether it be in the form of Lagrangian and Hamiltonian analysis, or in various equivalent integral formulations.

Feynman's discovery of the path integral formalism wasn't solely based on the prior work of Dirac (1933) [60], although by appending that paper to his PhD thesis (1946) its importance was clearly emphasized. Feynman also benefited from work going as far back as Laplace [23] for selection process based on highly oscillatory integral constructions that self-select for their stationary phase component. This branch of mathematics eventually became associated with Laplace's method of steepest descents, then to the work of Stokes and Lord Kelvin, then to the work of Erdelyi (1953) [112-114].

Feynman and others then invented quantum field theory for electromagnetism (QED) during 1946-1949 (more on this later). Extension to electroweak occurred in 1959, and to Quantum Chromodynamics (QCD) in 1973, and to the "Standard Model" in 1973-1975. Thus, the impact of the path integral revolution in quantum physics was felt well into the 1970's, but this was only the beginning. At their inception path integrals were examined by Norbert Wiener, with the introduction of the Wiener Integral, for solving problems in statistical mechanics in diffusion and Brownian motion. In the 1970's this led to what is now known as "the grand synthesis" which unified quantum field theory and statistical field theory of a fluctuating field near a second-order phase transition, and where use of renormalization group methods enabled significant advances from Quantum Field Theory to be carried over to statistical field theory.

The grand synthesis is one of many instances to come where we see analytic continuation of a constant or a parameter giving rise to familiar physics in the thermodynamic and statistical mechanics domains, showing a deeper connection (still not fully understood, see Book 7). The Schrödinger equation, for example, can be seen to be a diffusion equation with an imaginary diffusion constant. Likewise, the path integral can be seen to be an analytic continuation of the method for summing up all possible random walks.

In Book 4 we also carefully examine the closest gravitational equivalent to the hydrogenic atom (dust shell collapse). What results is an incomplete formulation due to boundary conditions, where to get the time choice you must input that time choice. No specific choice of time is indicated to avoid infall-collapse. The results, however, can show stability and consistency in a "full" thermal quantum gravity description where analyticity is employed. Success in this way, and not others, suggests possible fundamental role of analyticity and thermality (Books 6&7) and also suggests that thermal quantum gravity may 'exist' or be well-formulate-able, while quantum gravity generally might not 'exist'. These results, shown in Book 6, provide the lead-in to the Book 7 discussion on Emanator theory, where core concepts in Books 1-6 that tie to emanator theory are brought together in a new theoretical synthesis.

***Overview of Book 5***

In Book 5 we show quantum field theories in the gauge field representation, which clearly relates the choice of field theory to a choice of Lie algebra, which, in turn, can be related to a choice of group theory (such as U(1) and SU(3)). From this we can see that non-classical algebraic constructs are ubiquitous in quantum mechanics and quantum field theory. We find that the Schrodinger and Heisenberg formulations, however, often provide the only tractable way to get a solution for bound systems. In other theoretical considerations, however, the path integral approach is best, as will be shown, with the renormalization examples.

In Book 5 we get the highest precision result for the value of alpha, in its role as perturbation parameter. If a calculation of the electron magnetic moment parameter g-2 is performed, with all of the Feynman diagrams appropriate to expansions up to $5^{th}$ order, we get a determination of alpha up to 14 digits, where 1/alpha=137.05999….. . This gives us one of the most precise measurements of alpha known. When a similar analysis is done for the muon g-2, given the much larger muon mass, particle production pairs of other particles have a measurable effect, and we are able to probe the lower masses of the standard model that are present. In doing this, in preliminary experiments, there is a discrepancy indicating more particles, e.g. the Standard Model will need to be extended (possibly with a type of 'sterile' neutrino). These missing particles could be the missing “Dark Matter”. The prediction of such in Emanator Theory, and why there should be an imbalance between the left and right neutrinos (hint: maximum information transmission) is described in Book 7, with a summary in App. D.

Part of the description of quantum field theory entails use of analyticity and other complex structures to encapsulate more of the physics in a complex extension to the space (or dimension). This often leads to formulations in terms of complex integration, with the choice of complex contour specified, such as with the Feynman propagator. One of the main renormalization methods, for example, is to use dimensional regularization, which entails analytically continuing expressions with dimensionality to dimensionality as a complex parameter. There is also the aforementioned shift to complex and to “Wick rotate” expressions with real time to expressions with pure complex time. In doing this the statistical mechanical partition function for the system is obtained, with well-defined summation. Thus, a connection between ‘thermality’ and complex structure, in the time dimension at least, is indicated.

The second part of Book 5 describes quantum field theory on curved space-time, where we arrive at an early analysis of Black Hole thermodynamics. Here we find that space-time curvature gives rise to thermality (a thermal vacuum) and particle production effects. Black Hole thermality was revealed in Hawking radiation [211], due to the causal boundary at the horizon. Such thermality is even seen in flat space-time (Book 5) if causal boundaries are induced, such as in the case of an accelerated observer [212].

Quantum field theory on curved space-time has one further gift, critical to the statistical mechanics formalism to follow in Book 6, and that’s the spin-statistics relation. The spin-statistics relation is kinematically determined in flat spacetime quantum field theory, but it is separately determined to also be enforced dynamically in curved spacetime quantum field theory [15].

The choice of time is related to choice of vacuum, which is related to choice of field geometry or observer motion (such as constant acceleration or expansion). If you have flat spacetime quantum field theory with a boundary, then you have thermodynamic effects (e.g., the Rindler observer). In this setting we can compare the Hawking derivation of Hawking Radiation using the Euclideanization ‘trick’ vs the

Bogoliubov transformations of the field to the Rindler geometry from the Minkowski geometry (if chosen as the asymptotic vacuum reference). With quantum field theory on curved space-time we also arrive at spin-statistics as mentioned, and get the final extension of the theory by way of Grassman algebras, to arrive at thermodynamically consistent Bose and Fermi statistical descriptions on quantum matter.

***Overview of Book 6***

Thermodynamics is the oldest of the physics disciplines (fire), with unapologetic use of phenomenological arguments and mysterious thermodynamic potentials (entropy). Obviously, thermodynamics is still prevalent today, including in its more quantified form statistical mechanics. How is this not a failure of the mechanistic description of the universe indicated by classical mechanics and even quantum mechanics? Concepts that appeared in quantum mechanics, such as probability, are now occurring again. Other new concepts appear as well, including: approximate statistical laws; equations of state; heat as a form of energy; entropy as a variable of state; existence of equilibria; ensembles and distributions; and existence of the partition function. Many of these concepts appear in the path integral descriptions with the analyticity methods and extensions mentioned previously, so there are hints of a deeper theory that arrives at much of thermodynamics/Statistical mechanics foundation from the existing quantum theory.

Book 6 has been placed after the other chapters to await identification of entropy as fundamental in that it can be identified as an intrinsic system function even before getting to thermodynamics. We also already have experience with many particle systems, via quantum field theory (especially in curved space-time where particle creation is almost unavoidable), without directly tackling that scenario (due to quantum field theory effectively already being many-particle, with analytic determination of many-particle system functions, such as entropy). With entropy presented at the outset as an important system variable, the derivation of thermodynamic potentials is then a straightforward process, as will be shown. The standard statistical mechanics connections to thermodynamics can then be given. Thus, in covering Thermodynamics and Statistical Mechanics we start with the foundations of the theory mostly established, such as entropy (also with equipartition equivalent to sum on paths with no weightings, etc.), with no assumptions. Everything follows directly from the theoretical discoveries outlined in the preceding books in the Series. We don't see new connections to alpha, but we do see new structures/effects, especially manifold constructs (as with General Relativity, where we also saw no role for alpha).

The close ties between quantum mechanics complexified giving rise to a particle ensemble partition function, and quantum field theory complexified and field ensemble partition function, is now simply a derivative aspect of the fundamental complexation posited. This complexation will be posed in Book 7 with emanation in a complexified perturbation space.

From Atomic Physics we obtain the standard rules on electron shell completion (that is encoded in the periodic table). Similarly, we can also understand the origins of the intermolecular quantum chemistry rules. When taken to the statistical mechanics extreme we have thermodynamic equilibrium emergent from the Law of Large Numbers and reverse Martingale convergence. With completion of application to chemical processes we have clear phase-transition effects, as well as equilibrium and near-equilibrium effects. The familiar chemistry results, with phases of matter.

From chemical equilibrium and near-equilibrium, with $10^{23}$ elements that interact weakly or not at all, we have two generalizations. The first is to consider chemical near-equilibrium and directly obtain an emergent process at this level, this is the branch that gives us biology/life at its most primitive level. The second is to consider equilibrium and near-equilibrium in general when the elements interact strongly (with$10^{10}$elements, say), this is the branch that describes biology/life at its most advanced social level and economics. In classic shot noise, the granularity of low-current flow (due to discreteness off electron charge) leads to a noise effect. Thus, as we consider situations with fewer elements, there are more complications, not less, due to granularity noise effects, and we enter the realm of machine learning with sparse data. Noise effects can be significant in complex systems, especially in biology where it is part of what is selected (such as in hearing, for background noise cancellation).

The second part of Book 6 explores the role of thermodynamics in efforts to extend to thermal quantum field theory and thermal quantum gravity. This is done by exploring Black Hole settings. The recognition of a role for complex structure on system variables becomes apparent in this process (on top of the generalization to non-trivial algebras as already revealed).

In Book 6, part 2, we examine the Hamiltonian thermodynamics of some black hole geometries with stabilizing boundary conditions. In this foray into directly exploring a thermal quantum gravity solution we assume a path integral form for the General Relativity problem and shift directly to a partition function (by 'Wick rotation'). We see that thermal quantum gravity is possible, where positive heat capacity shows stability. Another encouraging result as to an eventual unifying theory comes from String theory via its explanation of Black Hole thermodynamics and Black Hole horizon effects with the Black Hole fuzz solution (via use of the holographic hypothesis and the related AdS-CFT relation [213,214]).

In Book 6, part 2, we also examine the propagator to partition-function transformation upon complexation, which leads to a thermodynamic theory for some equilibrium formulation, with certain parameter settings required for stability (positive heat capacity). This is doable in a variety of settings, suggesting how such thermodynamically consistent boundary conditions may be what constrains the classical motion and Black Hole singularity formulation by the effect of this stabilization manifesting for certain internal geometries. Successful thermal quantum gravity formulations, such as for RNadS and Lovelock spacetimes shown in Book 6, via reformulation using analyticity, and not via non-analytic approaches, suggests a possible fundamental role of analyticity once again and also suggest that thermal quantum gravity may 'exist' or be well-formulate-able, while quantum gravity generally might not 'exist'. These results, together with core concepts from Books 1-6 that tie to emanator theory, are brought together in a new theoretical synthesis in Book 7.

***Overview of Book 7***

In Books 4,5, and 6 of the Series, we explored examples of quantum mechanics with imaginary time, quantum field theory in curved space-time, thermal quantum field theory, minisuperspace quantum gravity, and thermal quantum gravity. In this effort we find the path integral, and PI propagator, to provide the most general representation. In seeking a deeper theory in Book 7 we build on the sum-on-paths with propagator formulation to arrive at a sum-on-emanations with emanator formulation.

Propagation in a complex Hilbert space, in a standard quantum mechanics or quantum field theory formulation, requires the propagator function to be a complex number (not real or quaternionic, etc., [99]). This prohibits what would otherwise be an obvious generalization to hypercomplex algebras. In order to achieve this generalization, we have to introduce a new layer to the theory, one with universal emanation involving hypercomplex algebras (trigintaduonions) that is hypothesized to project to the familiar complex Hilbert space propagation with associated fixed elements (e.g., the emanator formalism projects out the observed constants and group structure of the standard model). The 'projection' is an induced mathematical construct, like having SU(3) on products of octonions, but here it we be the standard model U(1)xSU(2)xSU(3) on products of emanator trigintaduonions. Thus, in Book 7 a unified variational formulation is posed, one that arrives at alpha as a natural structural element, among other things, uniquely specified by the condition of maximal information emanation.

In Book 7 we also make note of the implications of a fundamental mathematical operation on a space that is repeated or added. The non-General Relativity forces are given by the form of the operation (the sequence forming an associative algebra), the General Relativity forces are given indirectly by the form of the space, this leaves the aspect "repeated or added" to be considered with care. If a purely 'repeated' operation, or mapping, occurs we can return to the dynamical mapping discussion of Book 1, where chaos can occur and is ubiquitous. There, the primal 'phase transition', the transition to chaos, is evident. If an operation with addition is involved (in the statistical sense of multiple elements), along with repeated overall steps, we arrive at the general framework of statistical mechanics with effects from the Law of Large Numbers and reverse Martingale convergence, among other things (Book 6). Most notable, however, is the prevalence of a new effect, that of phase transitions and the emergence of new structure (order from disorder), including the remarkable structures of chemistry and biology.

Why the recurring 'Cabbalistic formula'? was a question even in the time of Sommerfeld [36]. Now, the numerological parallel is more exact than realized at that time, so is too much a coincidence to be by chance. The non-coincidence appears to be due to the maximal nature of information transmission in a variety of circumstances (in physics, biology, and even human communication with sufficient optimization) as well as with the fractal-like repetition of key parameter sets that occurs in these different settings {10,22,78,137≅1/alpha}. We see that 10 expresses the dimensionality of propagation (or nodes of connectivity), while 22 corresponds to the number of fixed parameters in the propagation (in Book 7 we explore propagation in a 10 dimensional subspace of the 32 dimensional trigintaduonion space, leaving 22 dimensions at fixed values that appear as parameters in the theory). We will see the number 78 relates to generators of the motion, and that there are 4 chiralities of motion ('doubly chiral'). We will also see that 137 is simply the number of independent tri-octonionic product terms in the general chiral trigintaduonion 'emanation'.

***Synopsis – Frodo Lives***

Tolkien wrote of eucatastrophes [215], perhaps he anticipated the constructive role of emergent phenomena in maximum information transmission.

*Preface to Physics Series, Book #4, on:*

## Quantum Mechanics and Path Integrals

This is a book on Quantum Mechanics. It starts with a description of the pre-Quantum physics developments that led to Quantum Theory. A thorough description of the modern Schrodinger and Heisenberg theories and applications is then given. After the standard, undergraduate-level, quantum mechanics description, a more mathematically formal construction is made with Dirac's axiomatic description, and more advanced quantum mechanical issues are addressed, such as relativistic quantum mechanics. At this point we've reached the normal extent of a first-year undergraduate course on quantum mechanics. In coving the undergraduate quantum topics (introductory and advanced) a very large number of worked problems are considered in detail.

Part of the selection on quantum mechanics topics (or emphasis) is meant to connect with the next book in the Physics Series [15] where Quantum Field Theory is described. Thus, examples and background development relevant to establishing a quantum field theory motivate part of the selection of advanced topics. In the Quantum Theory described above the quantization approach is equivalent to canonical quantization, and it will do a great job giving the spectral lines of hydrogen and other atoms (as will be discussed) but there are areas where canonical quantization, directly, is not an optimal approach to quantization (and getting results). Instead, there is the Feynman path integral formulation of quantum mechanics, so this will be described in detail as well.

Other than path integrals, the list of advanced topics that are covered includes relativistic quantum mechanics and exploration of theoretical foundations starting with deconstructing the propagator, Lorentz Invariance Representations, Spin-Statistics, and the Euclideanized Propagator. Lastly, a direct attempt is made at quantizing gravity by examination of the fully general relativistic shell collapse system and its quantization. Thus, the book covers undergraduate-level material and graduate-level material preparatory to a quantum field theory course (with a large number of carefully worked problems). But the book will also discuss quantum measurement theory, the nature of time, and explore quantum gravity, so may be of interest to advanced graduate students as well.

# Chapter 1. Introduction

This is a book on Quantum Mechanics. It starts with the pre-Quantum physics history that leads into Quantum Theory. A thorough description of the modern Schrodinger and Heisenberg theories and applications is then given. Material used in this endeavor draws from lecture notes and material from over a dozen excellent quantum mechanics textbooks [1-14]. For further detail on any given topic, simply explore the textbook cited in that section. After the standard, undergraduate-level, quantum mechanics description, a more mathematically formal construction is made with Dirac's axiomatic description, and more advanced quantum mechanical issues are addressed, such as relativistic quantum mechanics. At this point we've reached the normal extent of a first-year undergraduate course on quantum mechanics. In this book, however, part of the selection on quantum mechanics topics (or emphasis) is meant to connect with Book 5 of the Series on Quantum Field Theory [15]. Thus, examples and background development relevant to establishing a quantum field theory motivate part of the selection of advanced topics. In the Quantum Theory described above the quantization approach is equivalent to canonical quantization, and it will do a great job giving the spectral lines of hydrogen and other atoms (as will be discussed) but there are areas where canonical quantization, directly, is not an optimal approach to quantization (and getting results). Instead, there is the Feynman path integral formulation of quantum mechanics, so this will be described in detail as well. The path integral approach is not optional in quantum field theory, so it is good to understand it in the "first quantization" context anyway, as will be seen.

So far, the book description covers undergraduate-level material and graduate-level material preparatory to a quantum field theory course. Part of the objectives of the Series of Seven [15-21] is to cover the fundamental physics topics and to tie them into the most modern physics understanding, with direct connections given in Book 7 [21] to Emanator Theory. Thus, this book will also discuss quantum measurement theory (with objective reduction due to objective emanation described in App. D and Book7 [21]), and the nature of time. In this book an attempt is made to tackle the problem of time, and to see if a direct quantum gravity solution can be obtained in a highly symmetric ('simple') situation: the complete analysis of the quantization of a gravitational dust shell collapse (Chapter 7). The mathematics is for a fully general relativistic dust shell collapse, which is lengthy, so is partly placed in an appendix.. Then a choice of time reparameterization will be given using a new type of time reparameterization, and using this to full advantage, a quantization of the dust shell collapse is successfully completed in a fully self-adjoint formulation. In the process of doing the quantization, a greater understanding of the subtleties of time is obtained. Not surprisingly, the nature of time is too large an issue to be resolved in the context of the minisuperspace problem mentioned above alone, instead, further developments in Thermodynamics and Black Hole Thermodynamics, will be needed first (explored in Book 6 [20] and Book 7 [21]). Thus, in this Book, #4 of the Maximum Information Emanation Series [15-21], we begin the transition from a classical formulation of 'reality' to a quantum formulation.

In the classical formulation we often had system descriptions involving position and momentum that described a (classical) 'trajectory', e.g., a solution to a differential equation on real functions. We saw in Book 1 of the Series [16] the expression of the classical solution to the (real) differential equation could also be expressed as an integral expression involving the Lagrangian with a large (real) exponential factor (proposed by D'Alembert 1744 [22], and explored by Laplace [23]). The simplest path to understanding Quantum Mechanics is to understand both the differential and integrodifferential classical implementations, and generalize them to involve complex functions. Starting with the integrodifferential formulation, the generalization to complex theory is trivial with the aforementioned large exponential factor now pure imaginary (a phase), again indicating an equal weighting (***superposition principle***), and thereby directly arriving at the path integral formulation (interpreted as a ***sum over all paths***), where the action of each path contributes its own phase argument.

The Path Integral formulation to Quantum Mechanics will be seen to be most fundamental, but not always the easiest to work with. As with the classical theory, the integrodifferential formulation gives rise to a differential formulation that is often more useful. For Quantum Mechanics this differential theory is often expressed in the (non-relativistic) formulation of Schrodinger's Equation. If expressed in terms of position and momentum interpreted as 'quantum observables' we find that they obey a non-trivial constraint, the Heisenberg Uncertainty relation, and such a formulation is often represented in terms of Heisenberg's 'Matrix Mechanics'. Note that, given the assumption of an underlying wave theory, we could have presumed an uncertainty principle would exist on that basis alone, as established in the mathematics of wave theory (Benedict's Theorem [24]). From that it is required that there be an uncertainty relation on the 'operators' of the quantized theory, therefore, we could have guessed non-commutation on canonical variables and shown consistency on such solutions, thereby arriving at the canonical commutation relations for operators of the theory merely from it being a wave theory.

An example of where the Path Integral formulation fails to solve an important task is in solving the hydrogen atom and obtaining the precisely known spectral results that have been observed (as can be done using the Schrodinger, Heisenberg, or other operator approaches). Initial efforts to solve the Hydrogen quantization with Path Integrals failed. The 'simplest' approach to solving the problem is to generalize to a Path Integral formulation for curved spacetime and making use of analytic time parameter [9,12]. This will be the first of many examples where analytic time, or a shift to pure imaginary time, will be shown to regularize the problem and arrive at a solution. Understanding time, and how it has an analytic aspect, is beyond the scope of this text (see [20] and [21]), buts hints of such analytic time properties will appear nonetheless, and be noted when they occur.

According to classical mechanics, the atom should be unstable and collapse. There is no collapse, instead there are discrete atomic states, including a 'ground' state, as well as 'allowed' transitions between atomic 'states' giving rise to atomic spectra obeying quantization rules (a brief history of early Quantum Mechanics theory, such as the Bohr Model, is given in Section 2.5.2 to follow). These bound state solutions are generally the simplest solutions to the Schrodinger equation (a second-order differential equation, so many easy examples will be given). In a description of atomic matter, as far back as the Early Quantum Mechanics models and their spectral fits, there appears the fit parameter 'alpha'. In the more advanced quantum field theory (Book 5 [15]) this fit parameter is retained as a fundamental perturbation expansion parameter on electron-photon systems. Thus, we see the fundamental parameter alpha whenever we examine systems of bound (atomic) matter. Likewise, both the Schrodinger and Heisenberg formulations show a new, very small, constant, Planck's constant (the inverse of which provides the large constant for the aforementioned integrodifferential formulation). The Planck constant first appears as a (quantization) fit parameter in an analysis of the Black Body radiation (the history of the discovery of quantization and of the Planck parameter is given in the next section.)

The shift from the real-function formulation of reality (classical mechanics) to an inherently complex function formulation in quantum mechanics was a major paradigm shift, requiring re-interpretation of key variables and processes. A new vocabulary had to be invented having to do with quantum observables and classical apparatus. This process began with the mathematical formalisms of Quantum Mechanics formulated by Heisenberg (1925) [25] and Schrodinger (1926) [26]. Physical repercussions, such as the Heisenberg Uncertainty Principle (1927) [27], began to give a better sense of the new theory, as well as interpretation of the wave-function (Born 1927 [28]). Even so, there were still major hurdles in the theory due to the observed point like nature matter/charge for the fundamental particles. This requires admission of Dirac delta functions into the formalism, e.g., we need to extend from integration over functions to integration over L2-distibutions. A description of early distribution theory based functional analysis (e.g., as described in Richtmyer [10]) as given by Dirac (1930 [5]), is also given.

The relativistic formulation was also done by Dirac (1934) [29], and will be described in Section 6.2, to complete the formal underpinnings of the single-particle quantum theory. Interpretation problems begin to occur in many applications of Dirac's equation, however, and indicates the need for a many-particle theory that is best described in terms of a field theory – and this is the content, e.g. Quantum Field Theory, described in Book 5 [15] of the Series [15-21].

Lastly, we turn to an advanced application involving analysis of a fully generally relativistic dust shell collapse. Upon phase-space reduction this will be what's known as a minisuperspace model, and its quantization can then be directly performed using standard (single-particle) methods already ascertained. This is where the axiomatic Quantum Mechanics theory of Dirac helps to focus on the need to obtain a physical observable that is mathematically strictly associated with a self-adjoint operator. Put another way, the parameter-domain over which an operator is self-adjoint describes the domain over which a physical observable exists, and a description is possible. So, the dust shell analysis reduces to a self-adjoint Hamiltonian operator analysis that is given in Section 8.4. A minisuperspace solution is obtained but it has key parameters specified by boundary conditions that make the answer, effectively, whatever you put in, with no quantized spectra, thus disappointing in that it seems to resolve nothing vis-à-vis the quantum gravity problem. However, the tools used in the perturbative and self-adjoint analysis, especially invoking the Kato-Rellich Theorem (Section 8.4), are critical methods in performing the emanator perturbation analysis in Book 7 [21] in the derivation of the constant alpha (the fine structure constant – a shortened derivation is given in the appendix here as well), and the indications of analytic time and thermality in obtaining the dust shell solution will help to motivate the Black Hole Thermodynamic analysis done in Books 5 and 6 [15,20].

# Chapter 2. Early Atomic Theory and Early Quantum Theory

## 2.1 Introduction

The review of early atomic theory will start with the limits of the classic theory as regards the atom, where classical theory predicts a (rapid) radiative decay and collapse of all atoms. Since that didn't happen, a new theory was indicated. Better than a predictive failure of the old model is new data to support the new model, and that is what occurred with the observation of spectral lines in hydrogen by Balmer in 1885 [31].

Balmer's didn't just see spectral lines, he identified a precise mathematical relation describing their positions that required only one 'universal' constant as fit parameter. This aspect astonished the scientists of the day, for good reason, and led to efforts by Rydberg and Sommerfeld to look at spectral lines and 'fine-splitting' in those lines. What is revealed from the analysis of spectral lines are universal constants, alpha directly (aka, the fine structure constant), and Planck's constant indirectly (grouped with alpha in Rydberg's universal spectral constant). Also, in the fine-structure splitting, there appears to be quantization according to angular momentum…

The history of the quantum and appearance of Planck's constant is probably more familiar:
Quantization on energy change – Planck (1901)
Quantization on energy (photon) – Einstein (1905)
Bohr Model – Bohr (1913) and Sommerfeld-Wilson (1919)
Stern-Gerlach experiment – intrinsic spin is revealed (1922)
Wave-particle duality – deBroglie relation (1923) → Quantization on matter
Wave-particle duality – Compton effect (1924) → photon-electron scattering

A synthesis of the quantization and wave-particle duality concepts was occurring with the Bohr Model in 1913, with refinements in 1919 , and with Bohr's recovery of the Rydberg equation (and explanation of the

Rydberg constant) in 1923 there was then proposed the correspondence principle [ref]. Wave-particle duality was revealed in the results of 1923 and 1924, and the transition to modern quantum theory was about to occur (see Chapter 3 for further details).

## 2.2 Classical Atomic Theory

The brief history of classical atomic theory starts with Maxwell's unification of the radiation picture (1855). In 1887 there were three major discoveries (to be discussed next), but only the Michelson-Morley 'ether' experiments (showing no ether for electromagnetic wave propagation) are commonly known. With the discovery of the electron in 1897 by J.J. Thomson, there then followed a discovery of he structure of the atom. Thomson's original "plum pudding' model (1904) would be superseded by the Rutherford model with use of scattering experimental results (1911). In this process it was only becoming clearer, however, that there was a fundamental instability in atomic model due to radiative collapse. A simple approximation using the Larmor formula (Section 4.2.3) will show that this atomic collapse should happen faster than a nanosecond – so there's clearly a problem with the classical theory.

### *Maxwell unifies Radiation picture (1855)*

In 1855 Maxwell read a paper to the Cambridge Philosophical society titled "On Faraday's lines of force" [32], this was his first presentation of a simplified model of Faraday's work. His work unified electricity and magnetism via a linked set of differential equations. Eventually the work was published more completely in 1861 [33].

### *Hertz experiments on electromagnetic waves indicates photoelectric effect (1887)*

In 1887, while engaged in his famous experiments on electromagnetic waves, Hertz made the initial observation of the photoelectric effect [34]. He found that the length of spark induced in a secondary circuit was reduced when the terminals of the spark gap were shielded from the ultraviolet light emitted from the sparking on the primary circuit. This led the Hertz to perform further experiments and establish:
(i) only electrons or negative ions appear to emitted from polished metal plates upon illumination;
(ii) the electrons (or negative ions) are only emitted if the light is above a threshold frequency;

(iii) when the current is produced, it's magnitude depends on the intensity of the illumination; and
(iv) The energy of the emitted photoelectrons is independent of the intensity of the light and varied linearly with the frequency of the light.

***Michelson-Morley experiment on existence of an "ether" for electromagnetic wave propagation (1887)***

Maxwell's theory had indicated propagation of electromagnetic waves with a velocity that of the speed of light as early as 1862 (in comments made by Maxwell) and in explorations along these lines in 1887 the experimental question had arisen about whether a medium was necessary or implied by the existence of such waves. If there were such a medium, or "ether," it was proposed that the Earth's motion would reveal that ether with any interferometer dialed to cancel waves (maximum interference at detector), and simply wait for motion in that ether to result in the interference being undone, revealing the effect of such an ether. But this ether effect was not observed, ergo no ether.

***Michelson-Morley observation of the fine structure in spectral lines (1887)***

The fine structure in spectral lines was first observed by Michelson and Morley in 1887 [35]. The early Bohy model could not explain these line splittings. It wasn't until Sommerfeld modified the Bohr relation for relativistic effects, similar to those indicated by Einstein's Mercury precession results, that a prediction for the splittings within the Bohr-Sommerfeld model became possible [36]. This is where the constant alpha came into the form that it is known today (alpha~1/137).

***Electron Discovered (J.J. Thomson, 1897)***

Electrons were first revealed in the form of a cathode ray composed of negatively charged particles that had very small masses, as indicated by a charge-to-mass ratio that was very large [37]. This would be the first sub-atomic particle discovered.

***Thomson Model (plum pudding model, with electrons as 'plums': 1904)***

In 1904 Thomson put forward his famous "plum pudding" model, where the atom was like a plum pudding, with small electrons like plums, stuck (mostly immobilized) in that pudding, consisting of a positive background pudding or cloud. We now know the roles are switched, with electrons moving, and instead of a diffuse counter-charge background, it is a highly concentrated positive charge, at the 'nucleus' [38].

***Geiger and Marsden (1908-1909)***

Before the complete Rutherford-Geiger-Marsden Model would be described, a number of scattering experiments would be performed. From this early work the power and utility of scattering experiments would be revealed, the workhorse of the particle physics experimental process. In these efforts the term 'nucleus' was used for the first time [39,40].

***Rutherford Model (1911)***

Having identified the constituent positive and negative charged elements, and identifying a positive nucleus together with negative small electron in a form of orbital model, the Rutherford model was obtained [41].

## 2.3 The observation of atomic spectra and the beginning of the quantum revolution

The first signs of a break from the classical theory was in the observation of spectral lines from heated gases or vapors. There was nothing in the classical theory to explain this 'discrete' like phenomenon of emission only at particular spectral 'lines'. It wasn't until Balmer showed an almost miraculous fit on these spectral lines, with only a single fit parameter, that the scientific community became interested.

***Balmer (1885)***

In 1885 Balmer observed the spectral emission lines for hydrogen and was able to describe the collection of lines with a simple equation involving only one constant [31]. Here is a modern picture of the spectral lines:

Figure 2.1. Visible spectrum of hydrogen (from the Wikimedia commons).

Here is the modern form of Balmer's equation:

$$\lambda = B\left(\frac{m^2}{m^2 - 2^2}\right)$$

where $\lambda$ is the observed wavelength, $B = 364.50682$ nanometers, $m > 2$ is the initial state. The red line in the image above has the $m = 3$ initial state and produces light at wavelength 656 nm. This result, with its simple description with one parameter, did not fit in the classical mechanical description known at that time, yet it obviously was tied to something fundamental in that only one fit parameter was involved, thus the nature of the constant $B$ became of great interest.

***Rydberg (1888)***

In 1888 Rydberg clarified matters further by providing an equation that would describe all the "main" spectral lines for hydrogen [42]:

$$\lambda^{-1} = R_H\left(\frac{1}{2^2} - \frac{1}{n^2}\right), \qquad n = 3,4, \ldots$$

where $R_H = 1.09677583 \times 10^7 m^{-1}$ is the Rydberg constant for hydrogen. In 1890 [43] Rydberg extended his result to all hydrogen-like atoms (having only one electron), with different constant for different atom by replacing $R_H \rightarrow Z^2 R_H$, where $Z$ is the atomic number of the hydrogen-like atom. Looking ahead for a moment, with the simple Bohr model (1913) and with the modern Schrodinger Equation solution we get the same result, describing energy levels that depend on the $n$ appearing before (now considered a principle quantum number) as $1/n^2$ and the results are now seen as an energy difference between energy levels that corresponds to the emitted light. The more modern theories then give the same formula as Rydberg, except it is revealed what comprises the Rydberg constant:

$$R_H = \frac{\alpha^2 m_e c}{2h} \quad , \quad \alpha = \frac{e^2}{2\varepsilon_0 hc},$$

where $\alpha$ is alpha, the fine structure constant, $h$ is Planck's constant, $m_e$ is the mass of the electron, and $c$ is the speed of light. Thus, the Rydberg constant is a surprising collection of the two main quantum constants as well as the speed of light and the mass of electron. Although Planck's constant would not be seen until 1901, the constant alpha had already been seen.

As mentioned, in 1887 Michelson and Morley [35] observed the 'fine structure' in the spectral lines for Hydrogen, where each line had an energy-level splitting according to the level's energy with a factor of $(Z\alpha)^2$, where $\alpha \cong 1/137$ was an experimentally observed value. The appearance of the number 137 was not given much heed at the time.

***Sommerfeld (1919) – the first designation of the constant alpha as it is now known***

In the modern quantum theory the line splitting on energy levels will have precisely the leading $(Z\alpha)^2$ form indicated by Michelson and Morley. If taking the Schrodinger approach and adding relativistic corrections to lowest order (1927), there is obtained [44]:

$$\Delta E = \frac{(Z\alpha)^2}{n}\left(\frac{1}{j+\frac{1}{2}} - \frac{3}{4n}\right).$$

Note, however, that the above result was already obtained in 1919 by Sommerfeld using Bohr theory. It is from Sommerfeld that we have the notation and definition of alpha that we use today.

**2.4 The Mystery of Alpha (see [45] for extensive details)**

Sommerfeld noted the almost cabbalistic underpinnings of the mathematics describing spectral lines, referring to the Rydberg inverse square equation as a 'cabbalistic' formula in his 1919 paper [36]. In that effort Sommerfeld updated the Bohr model of hydrogen to explain the relativistic fine-splitting in the spectral lines of hydrogen. He obtained a result with a single "fit parameter", now called alpha, or written $\alpha$. The value alpha introduced quantified the gap "fine structure" between spectral lines and, as mentioned, is also known as the fine structure constant for that reason. Oddly, there was nothing cabbalistic about Rydberg's constant or equation, but there certainly was for the result obtained by Sommerfeld. As noted already, Sommerfeld adopted the notation $\alpha \cong 1/137$ for the fine structure constant, where 137 is an actual Kabbalistic number. So, the mystery deepens. This mystery is explained in [45] where it is seen that a maximum information propagation construct, whether 'reality' through the emanation process (Book 7 [21]), or simply some form of communication optimization like written language with gematria, will often give rise to the appearance of alpha as a maximum perturbation parameter in some sense.

The obsession with alpha did not stop with Sommerfeld. His famous student, Wolfgang Pauli, was very interested in the origins of alpha as well. In Pauli's Nobel Prize Lecture [46] he writes (italics mine): "From the view of logic my report on 'Exclusion principle and quantum mechanics' has no conclusion. I believe it will only be possible to write the conclusion if *a theory will be established which will determine the value of the fine structure constant* and will thus explain the atomistic of electric fields actually occurring in nature." Pauli was obsessed with the origins of alpha, so much so that he sought psychological help from the famed psychoanalyst Carl Jung, with whom he eventually partnered to try to solve the mystery of $\alpha$ (the madness is contagious) [47]. Carl Jung went on to write about archetypes (which were curiously similar to the Major Arcana as outlined in the Tarot). A popular history book describing this 'odd couple' interaction can be found in Arthur Miller's "137" [48].

The obsession with $\alpha$ continued with the next generation of great Physicists as well, particularly Feynman, who said [49]:

> There is a most profound and beautiful question associated with the observed coupling constant, *e* – the amplitude for a real electron to emit or absorb a real photon. It is a simple number that has been experimentally determined to be close to 0.08542455. (My physicist friends won't recognize this number, because they like to remember it as the inverse of its square: about 137.03597 with about an uncertainty of about 2 in the last decimal place. It has been a mystery ever since it was discovered more than fifty years ago, and all good theoretical physicists put this number up on their wall and worry about it.) Immediately you would like to know where this number for a coupling comes from: is it related to pi or perhaps to the base of natural logarithms? Nobody knows. It's one of the greatest damn mysteries of physics: a magic number that comes to us with no understanding by man. You might say the "hand of God" wrote that number, and "we don't know how He pushed his pencil." We know what kind of a dance to do experimentally to measure this number very accurately, but we don't know what kind of dance to do on the computer to make this number come out, without putting it in secretly!

## 2.5 Early Quantum Theory

***Planck (1901)***

Max Planck was performing black body experiments in 1900 and he couldn't get agreement between experiment and theory for the black body radiation produced. He could get good agreement, however, if he constrained his oscillator model to only allow changes in energy of minimal increment that was proportional to the frequency of the associated electromagnetic wave (the constant in this relation would become known as Planck's constant). We consider the Black Body experiment in detail in the next section.

### 2.5.1 Black Body Model

The emissive power from a heated (radiating) body depends on the wavelength observed and the temperature of the body, where $E(\lambda, T)$ is the energy emitted per unit time. In work in 1859 Kirchhoff [50] defined absorptivity, $A(\lambda)$, as the fraction of incident radiation of particular wavelength that is absorbed by the body, such that

$$\frac{E(\lambda, T)}{A(\lambda)} = C$$

where $C$ is the same for all bodies. For the theoretical extreme of a "black body" we will have $A(\lambda) = 1$, and the discussion turns to determination of $E(\lambda, T)$ for this instance as a universal function (for all bodies as indicated). According the Kirchhoff, the implication of the 2$^{nd}$ law of Thermodynamics in such a circumstance is that the radiation from the body be isotropic, homogenous, and the same in all cavities at the same temperature, and also holding for each wavelength.

***Relation between energy density in a cavity and emissive power***

Suppose we have emissive area $dA$ and a volume element a distance $r$ from the area, at an angle $\theta$ from the normal to the area. In spherical coordinates the volume element is

$$dV = dr\,(r\sin\theta\,d\theta)\,(rd\varphi) = r^2 dr \sin\theta\,d\theta d\varphi$$

The emitted energy in the volume element has differential:

$$d\{emerging\ energy\} = \left(\frac{\{projected\ area\ of\ hole\}}{\{surface\ area\ of\ shell\ at\ r\}}\right)(energy\ in\ Volume)$$

$$d\{emerging\ energy\} = \left(\frac{dA\cos\theta}{4\pi r^2}\right)u(\lambda, T)dV$$

To get the flow of radiation in time interval $\Delta t$ we must integrate the volume element with $dr$ going from 0 to $c\Delta t$. Thus:

$$d\{E(\lambda, T)\} = \frac{d\{emerging\ energy\}}{d(Area)\Delta t} \quad \rightarrow \quad E(\lambda, T) = \frac{1}{4}c\,u(\lambda, T)$$

So, emitted energy density is:

$$u(\lambda, T) = \frac{4}{c}E(\lambda, T).$$

By considering the special case of a perfectly reflecting spherical cavity contracting adiabatically, where the redistribution of energy as a function of wavelength is caused by the Doppler shift on reflection, Wien is able to show in general there must be the form:

$$u(\lambda, T) = \lambda^{-5} f(\lambda T) \quad \rightarrow \quad u(\nu, T) = \nu^3 g(\nu/T).$$

The Wien result fits well for the high frequency data observations with the empirical formula:

$$g(\nu/T) = Ce^{-\beta\nu/T}.$$

At low frequencies, the theoretical form indicated by Wien is completely wrong and a new model for the physical behavior is needed. In 1900 Rayleigh derives what is now known as the Rayleigh-Jeans Law:

$$u(\nu, T) = \frac{8\pi\nu^2}{c^3} kT.$$

The model giving rise to this result is built on the model of a simple harmonic oscillator. A simple harmonic oscillator that is in equilibrium with a thermal reservoir at temperature $T$ will have average kinetic energy $kT/2$ (see Book 6 [20] for thermodynamics details). A modal decomposition of the electromagnetic field in the classical theory reveals that each mode has the mathematical form of a simple harmonic oscillator. If the classical thermodynamic law of equipartition of energy is applied, each mode having energy $kT/2$, we can describe the energy density if we know the number of modes for an electromagnetic field confined to a cavity. A calculation for the latter premised on a differential formulation, and including a factor of 2 for the two polarizations possible in each mode, gives the above formula.

Faced with two mathematical forms, one good at high frequencies and one good at low frequencies, Planck in 1900 came up with the Black body energy density by use of an interpolating function with an adjustable constant ($h$):

$$u(\nu, T) = \frac{8\pi h\nu^3}{c^3} \frac{1}{(e^{h\nu/kT} - 1)}$$

When the function is made to fit the low and high frequency regimes (matching the functional forms above), the adjustable constant is fixed at the value now known as Planck's Constant:

$$h = 6.63 \times 10^{-27} \, erg \, sec.$$

The total energy in a cavity of unit volume is then

$$U(T) = \frac{8\pi h}{c^3} \int_0^\infty \frac{\nu^3 d\nu}{(e^{h\nu/kT} - 1)} = aT^4,$$

where

$$a = 7.56 \times 10^{-15}\, erg/(cm^3K^4)$$

is the Stefan-Boltzmann constant. This result for total energy is computed on purely thermodynamic grounds in Book 6 [20], aside from the derivation for the Stefan-Boltzmann constant, which from the quantum mechanic derivation above must have the form:

$$a = \frac{8\pi k^4}{(\hbar c)^3}\left(\frac{\pi^4}{15}\right) T^4,$$

where $k$ is Boltzmann's constant and $\hbar = 2\pi h$.

Note that Planck's expression for energy density can be written:

$$u(\nu, T) = \frac{8\pi\nu^2}{c^3}\frac{h\nu}{(e^{h\nu/kT} - 1)},$$

which has the modal density term with an average energy per mode term:

$$\frac{h\nu}{(e^{h\nu/kT} - 1)}.$$

Planck found that if the energy per mode was set at $\varepsilon = h\nu$ and he considered a Boltzmann probability distribution for system at equilibrium temperature $T$:

$$P(E) = \frac{e^{-\frac{\varepsilon}{kT}}}{\sum e^{-\frac{\varepsilon}{kT}}},$$

then the resulting average energy density for the system is:

$$\langle E\rangle = \sum E\, P(E) = \frac{\varepsilon}{(e^{\varepsilon/kT} - 1)} = \frac{h\nu}{(e^{h\nu/kT} - 1)}$$

as indicated. The implications of this are that the energy per mode is quantized and is directly related to the frequency of the mode. It wasn't until 1905 (Einstein photoelectric effect analysis [51]), however, that it would be understood that this also related the energy of a 'quantum' or corpuscle (photon) of light at a particular frequency by the same relation in general.

***Lenard (1902)***

Lenard's work with cathode rays began in 1888. He was the first show that the energy of rays in the photoelectric effect was independent of light intensity [52].

***Einstein Photoelectric Effect (1905) [51]***
Consider a work function for a given material given by 'W', with the experimental relation at maximal observed electron emission kinetic energy given by conservation of energy as:

(energy of light quantum) =
(observed electron emission kinetic energy)
+(empirically determined electron escape 'Work function')
+(thermal/inelastic energy loss)

This relation simplifies further when written for the special case of maximal electron recoil energy, where a mostly inelastic process has occurred, thus eliminating the thermal/inelastic term, to give the relation (for a given material under photon excitation):

$$E_{max} = (\text{energy of light quantum}) - W.$$

We also have the linear thresholding behavior in terms of the light quantum's frequency, which can be described with the equation for the photoelectric effect in the classic form:

$$E_{max} = h\nu - W,$$

where $h$ is Planck's constant, and the result that light has energy 'quantized' according to frequency ($E = h\nu$). Planck found a quantization on light according to $E = h\nu$ allowed the Black Body radiation profile to be explained, where $h$ was a new (quantum) constant. In 1905, Einstein adopts the idea of a quantum of light (motivated by Planck's studies) to conjecture the relation now known as the photoelectric effect:

$$\frac{1}{2}mv^2 = h\nu - W,$$

where $\frac{1}{2}mv^2$ is the kinetic energy of the electron, $\nu$ is the frequency of the illumination. and where $W$ is the "work function" of the metal, and is of the order of several eV (where 1eV=1.6x$10^{-12}$ erg). There results the Planck-Einstein relations: E=hv, $\vec{p} = \hbar\vec{k}$ (since E=pc), h=6.62 x$10^{-34}$J sec and $\hbar = h / 2\pi$.

***DeBroglie Relation (1923) [53]***
In 1923 DeBroglie considers scattering of electrons off of a crystal lattice. and observes preferred scattering directions at odds with classical theory. As with Einstein's indication of a wave-particle duality, here he attributes a wavelength to the electron according to a similar relationship to Einstein's relation for photon energy ($E = h\nu$), where here we have the wavelength relation for (massive) particulate matter: $\lambda = h/p$. Scattering off a lattice (crystal) with spacing between scattering sites 'a'. For constructive interference on the scattering, the phase shift on the scattered

wave must match the added distance phase shift. The extra distance traversed on scattering a wavefront goes as $2a\sin\theta$, where the added phase for a particular wavelength is:

$$\Delta\varphi = \frac{2\pi}{\lambda}(2a\sin\theta).$$

This added phase, for constructive interference, must equal $2\pi n$. We thus have preferred scattering directions as observed:

$$\lambda = \frac{(2a\sin\theta)}{n}.$$

***Existence of a four-vector description on energy-momentum***

Let's now recover the special relativity results of rest mass energy = $\boldsymbol{mc^2}$ and the existence of a four-vector description on energy-momentum. We will use an anachronistic argument by making use of DeBroglie's matter-wavefunction relation between wavelength and momentum:

$$\frac{h}{\lambda} = |\boldsymbol{p}|.$$

(recall, $\boldsymbol{p} = \hbar\boldsymbol{k}$, where $\lambda = 2\pi/|\boldsymbol{k}|$). For light we thus have:

$$E = h\nu = \frac{h}{\lambda}c = pc.$$

For light, since we don't have $\nu = 0$, we don't have $p = 0$, so the above equation is applicable to describe any energy state of light. For matter (excluding massless matter, if it exists) we have the momentum given by $p = mv$, where $v$ is velocity and this can most certainly be zero (in a particular frame of reference) and thus have zero momentum. Thus, to say that

$$E = pc = mvc,$$

for matter is clearly missing a rest-energy term. Let's have such a term consisting only of mass parameter $m$ and constants – thus dimensional analysis suggests

$$E_{rest} = mc^2.$$

Let's now guess a four-vector formulation on energy momentum such that the energy scalar can be written as (adding component-wise in quadrature):

$$E = \sqrt{(mvc)^2 + (mc^2)^2}.$$

Is this guess consistent with the standard low-velocity classical mechanics kinetic energy definition [16], (as well as consistency with the required kinetic energy form consistent with Riemannian geometry [17]), when seen in its variational context? First, consider the $v \ll c$ behavior:

$$E = \sqrt{(mvc)^2 + (mc^2)^2} \cong mc^2\left(1 + \frac{1}{2}\left(\frac{v}{c}\right)^2\right),$$

And dropping constant terms from the energy when used in a variational context, we have energy given as:

$$E \cong \frac{1}{2}mv^2,$$

precisely as required.

***Compton Effect (1924) [54]***

In 1924 Compton considers the scattering of radiation (light) off of electrons (residing in a metallic foil). When expressing energy conservation, use will be made of the four-vector energy momentum described above. If the light scatters as a classical wave, with re-radiation of light off of electrons that have been forced to oscillate, there results a re-radiation intensity variation with angle that should go as:

$$(1 + \cos^2\theta)$$

and without any dependency on the wavelength of the incident radiation. The observed scattering involves 're-radiation' at the same frequency as well as radiation scattering at a second frequency that is dependent on the scattering angle. Compton is able to explain this using simple rules of particulate matter scattering, where the light is treated as a 'photon' corpuscle of energy according to Einstein's relation for the photoelectric effect and momentum according to DeBroglie's relation. Let's denote the photon momentum with lowercase p and the electron momentum (initially zero) by uppercase P. Conservation of momentum gives: $\vec{p} = \vec{p}' + \vec{P},\ \ |p| = h\nu/c$. Conservation of energy gives: $h\nu + mc^2 = h\nu' + \sqrt{(mc)^2 + (Pc)^2}$. With some algebra this becomes:

$$h\nu\nu'(1 - \cos\theta) = mc^2(\nu - \nu')$$

or

$$\lambda' - \lambda = \frac{h}{mc}(1 - \cos\theta),$$

where the "Compton wavelength" for the electron is

$$\frac{h}{mc} = 2.4 \times 10^{-10} cm$$

***Bohr (1913)[55]***

Bohr eliminated the classical radiative collapse problem by saying that the energy levels of electrons in atoms are discrete, and that electrons can move between levels with emission or absorption of an associated quantum of photon energy. The nature of the discrete levels was such that the electrons were described as stationary waves that could exist at particular (circular) orbits [55]. This is equivalent to a quantization on angular momentum in the model. The model would recover the Balmer-

Rydberg formulae and other observations as well as make many accurate predictions.

### 2.5.2 Bohr Model

Recall that for a classical orbit in a central potential we have:

$$E = \frac{1}{2}\mu v^2 - \frac{e^2}{r} \quad \textit{and} \quad \frac{\mu v^2}{r} = \frac{e^2}{r^2}$$

for the classical circular orbit on a hydrogenic atom. If we then add a quantization condition on the angular momentum in the orbit:

$$\mu v r = n\hbar$$

we arrive at the stable orbit descriptions with:

$$v_n = \left(\frac{e^2}{\hbar}\right)\frac{1}{n}$$

$$r_n = a_o n^2 \quad \textit{where} \;\; a_o = \left(\frac{\hbar^2}{\mu e^2}\right) = 0.52\ \text{Å}$$

$$E_n = E_o \frac{1}{n^2} \quad \textit{where} \;\; E_o = \frac{\mu}{2}\left(\frac{e^2}{\hbar}\right)^2 = 13.6\ eV.$$

Some important relations:

$$\alpha = \frac{e^2}{\hbar c} = \frac{I}{137}$$

$$m_e c^2 = 0.51 MeV$$

$$\frac{\hbar}{m_e c} = 3.9\text{x}10^{-11}\text{cm}$$

$$\frac{\hbar}{m_e c^2} = 1.3\text{x}10^{-21}\text{sec}$$

***Sommerfeld-Wilson (1915,1916)***

Sommerfeld [56] and Wilson [57] would retain the quantization on angular momentum of the original Bohr model, but now have electron motion on elliptical orbits satisfying an additional radial quantization condition:

$$\int_0^T p dq = nh,$$

where $p$ is the radial momentum canonically conjugate to the radial coordinate $q$, and $T$ is the time for one period of the orbit. Interestingly, the fundamental quantization on angular momentum is revealed in even the simplest models. We will see why angular momentum should be so readily quantized, as a direct consequence of canonical quantization, in Section 3.2. Note that, as discussed in Section 3.5, the mere existence of a wave description of matter, by Benedick's theorem, will give rise to an uncertainty relation, from which canonical quantization can be shown (Lennard [58]), and thus angular moment quantization (shown in Section 3.2).

***Stern-Gerlach experiment – intrinsic spin (1922) [59]***
The Stern Gerlach Experiment was conducted in 1922 and it revealed that particles possess an intrinsic angular momentum that is quantized. Electrons, in particular, have an intrinsic spin of $\pm 1/2$. In Book 5 quantum field theory [15] we will see that electron spin properties are merely due to matter being spinorial, which is a representation of the Lorentz group along with the more familiar vector fields that underlie gauge forces.

## 2.6 Examples

### 2.6.1 Calculating the surface temperature of the Sun

Let $u(\lambda, \mathrm{T})$ be the energy density as a function of wavelength and temperature, and denote emissive power by $E(\lambda, \mathrm{T})$, then we know that:

$$u(\lambda, \mathrm{T}) = \frac{4E(\lambda, \mathrm{T})}{c}$$

and that total radiation energy per unit volume is:

$$U(\mathrm{T}) = \mathrm{aT}^4, \ \ where \ a = 7.56 \times 10^{-15} \, erg/cm^3.$$

So, we have:

$$E(\lambda, \mathrm{T}) = \frac{c}{4}(aT^4),$$

for the total rate of radiation per unit area of a BB.

If the Sun is overhead at Earth we have radiation emission from Sun as BB at temp T, where the emission energy is related to T by above to give:

$$E_{Sun}\{\pi R_{Sun}^2\}\left(\frac{Unit\ Area}{4\pi d_{Sun-Earth}^2}\right) = E_{Earth} \ \rightarrow \ E_{Sun} = \frac{R_{Sun}^2}{4d_{Sun-Earth}^2} E_{Earth} \cdot$$

Thus, an observation at Earth of energy falling per unit area of

$$E_{Earth} = 1.40 \times 10^6 \, erg.s/cm^2sec$$

Indicates an emission rate at the Sun's surface, per unit are, of:

$$E_{Sun} = 2.57 \times 10^{11} \, ergs/cm^2sec.$$

If we now use the relation $E(\lambda, \mathrm{T}) = \frac{c}{4}(aT^4)$, we obtain for temperature at the surface of the Sun:

$$T = 8{,}200^oK.$$

### 2.6.2 Wavelength of maximum energy density and relation to BB temperature

Recall the relation:

$$u(v,T) = \frac{8\pi h}{c^3}\frac{v^3}{e^{hv/kT}-1},$$

alternatively:

$$u(\lambda,T) = u(v,T)\left|\frac{dv}{d\lambda}\right| = \frac{8\pi hc}{\lambda^5}\frac{1}{e^{hc/\lambda kT}-1}.$$

Taking the derivative:

$$\frac{du}{d\lambda} = 0 \;\Rightarrow\; \frac{1}{\lambda}\frac{1}{e^{hc/\lambda kT}-1}\left(\frac{hc}{kT}\right) = 5e^{-hc/\lambda KT}$$

Which reduces to solving for

$$5e^{-x} = 5 - x,\;\; where\;\; x = \frac{hc}{\lambda kT}, for\; max\; \lambda\,.$$

Solving, we get:

$$\frac{hc}{\lambda_{max}kT} = 4.9651$$

From this we get the convenient relation for determining BB temperature once the wavelength of maximum energy density is determined:

$$T = \frac{0.29\; cm\; {}^oK}{\lambda_{max}}$$

For the Sun, we just computed the effective BB temperature at the surface to be $T = 8{,}200^oK$, thus the wavelength with the maximal energy density is:

$$\frac{0.29\; cm\; {}^oK}{8{,}200K} = 3.54\times 10^{-5} cm = 3{,}540\text{Å},$$

which is in the UV.

### 2.6.3 Electron beam diffraction off of crystal

Consider a beam of electrons of uniform energy 100 $eV$ falls in the z direction on a crystal with a cubic lattice of lattice spacing 5 $\dot{A}$ ($5\times 10^{-10}\; meter$) aligned with the xyz axes:

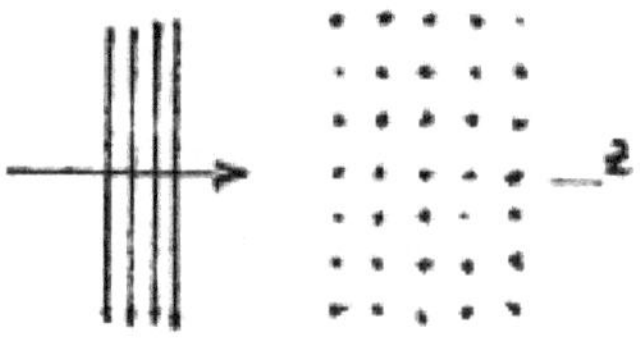

(a) Write a wave function in the form of a plane wave that can be used to represent the beam of electrons. Ignore normalization.
(b) Assuming that most of the beam reflected from the crystal is reflected by the plane of atoms on atoms on the surface, what will be the angle from the normal to the first maximum of the diffraction pattern of the reflected electrons?

Answer:
(a) $\psi_k(r,t) = Ae^{i(wt-kr)} = Ae^{i(wt-kz)}$. The deBroglie wavelength of an electron of electric energy $E(eV)$ is given by $\lambda = \frac{12.3\times10^{-8}cm}{E^{1/2}} = 1.23\times10^{-8}cm$ with $E = 100$ in units of $eV$. Thus $k = \frac{2\pi}{\lambda} = 5.11\times10^{8}\ cm^{-1}$ with $w = \frac{E}{\hbar}$.

(b) Phase difference $\Delta = 5\text{Å}\sin\theta = n\lambda$, so for $n = 1$ at the first maximum: $\theta = \sin^{-1}\left(\frac{\lambda}{5\text{Å}}\right)$. Thus, $\theta = \sin^{-1}\left(\frac{\lambda}{5\text{Å}}\right) = \sin^{-1}\left(\frac{1.23\times10^{-8}\ cm}{5\times10^{-8}\ cm}\right) = \sin^{-1}(.246) \rightarrow \theta = 14.2°$.

**2.6.4 Electron beam incident on a narrow slit**
Consider a uniform beam of electrons of momentum p falls at normal incidence on a screen with a narrow slit of which D. The electrons passing through the slit form a spot on a second screen a distance L from the first screen. Assume $L \gg D$.

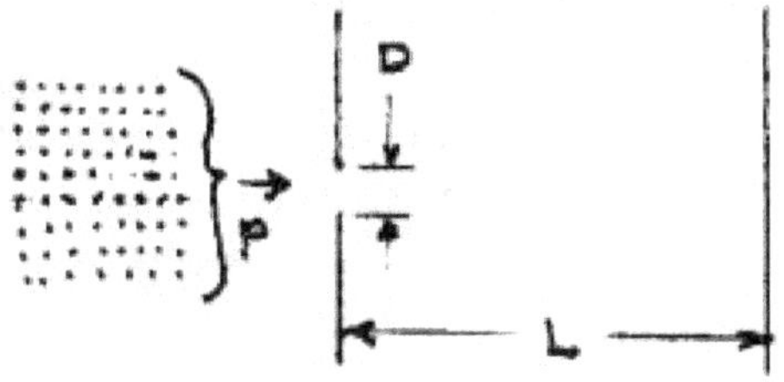


(a) What is the quantum mechanical wavelength $\lambda$ associated with the electrons?
(b) If $D \gg \lambda$, what is the approximately diameter $S_1$ of the spot formed on the second screen?
(c) If $\lambda \gg D$, what is the approximately diameter $S_2$ of the spot formed on the second screen?
(d) What width $D_m$ will give approximately the minimum diameter $S_{min}$ of the spot on the second screen?

Answer:

(a) $p = \hbar k = \frac{h}{2\pi} \cdot \frac{2\pi}{\lambda} = \frac{h}{\lambda}$ Thus, $\lambda = \frac{h}{p}$ $(De\ Broglie)$.

(b) $D \gg \lambda$ → have Geometric optics → Diameter image ≃ diameter hole: $S_1 \approx D$

(c) $\lambda \gg D$ → diffraction important, as $\frac{\lambda}{D} \to \infty$ → $S_D \to \infty$ .

(d) $D_0 = \sqrt{2L\lambda}$

**2.6.5 Electron beam incident on a diatomic molecule**

A uniform beam of electrons of momentum p moves in the x direction and scatters from diatomic molecules that are aligned in the x direction. The diatomic molecules can be approximated by a model consisting of two identical scattering centers paced a distance $d = 3\ \dot{A}$ along the x axis as shown above at left. One atom (1/2 of the molecule) alone produces a scattering intensity as a function of angle as shown above at center. The diatomic molecule gives a scattering intensity as shown at right.

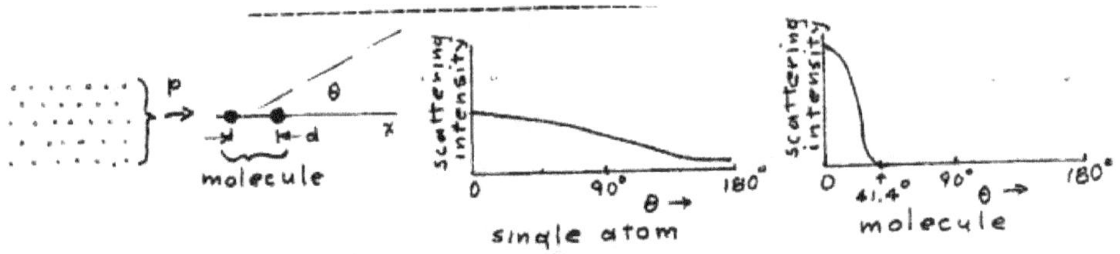


(a) Extend the scattering diagram for the diatomic molecule over the complete range of angle to $180°$ by an approximate (qualitative) sketch.

(b) If the energy of the electron beam is increased by a factor of 4, show qualitatively how the scattering curve will change.

(c) What is the energy E in electron volts of the electrons producing the scattering curve in (a)?

***Answer***

(a) Path length diff $= d - x_1 = d(1 - \cos\theta)$

$x_1 = d\cos\theta$

minima $\quad d(1 - \cos\theta_n) = {}^{\lambda n}/_{2} = {}^{n\pi}/_{k} \quad n = 1, 3, 5, \ldots$

maxima $\quad d(1 - \cos\theta_m) = \lambda m = {}^{2\pi m}/_{k} \quad m = 1, 2, 3, \ldots$

$n = 1 \Rightarrow \theta_n = 41.4^0 \Rightarrow \cos\theta_1 = {}^{3}/_{4} \quad \boxed{kd = 4\pi}$

Maxima $\theta_m = cos^{-1}(1 - {}^{m}/_{2})$ $\theta_1 = 60^\circ$ $\theta_2 = 90^\circ$ $\theta_3 = 120^\circ$ $\theta_4 = 180^\circ$

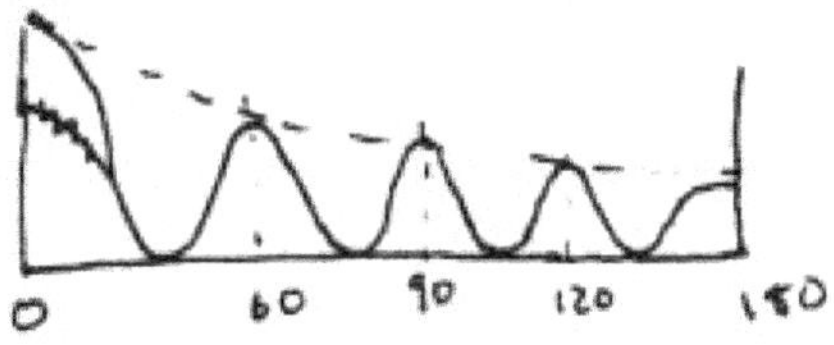


(b) If $E = 4E_0$ The $P = tk = 2p_0$ $k = 2k_0$ $\boxed{dk = 2d}$

The new condition for maxima is $\underline{\theta_m = cos^{-1}\left(1 - \frac{m}{4}\right)}$ so there are approximately twice as many maxima:

(c) $E = {}^{p^2}/_{2m} = \frac{\hbar^2}{2m}\left(\frac{4\pi}{d}\right)^2 =$

$$\frac{(1.055\times10^{-27})^2}{2(9.11\times10^{-28})}\frac{(erg.s)^2}{g}\frac{(4\pi)^2}{(3\times10^{-8})^2 cm^2}\frac{1}{(1.6\times10^{-12})erg/eV} \cong 68eV$$

**2.7 Questions**

(1) What percentage of Sun's energy is radiated in the visible (4000Å – 7000Å) at Earth?

(2) What is the BB radiation when T=2.73K?

(3) A 3,500Å wavelength light source falls on potassium producing a maximum energy photoelectron excitation of 1.6eV. What is the Work function for potassium? (assume Planck's constant is known).Reminder of photoelectric as photon hv – work function for given material, and how we can vary v while keeping work function the same, and thereby derive 'h' in a directly accessible experiment.

(4) There are two measurements: (i) with a 2,000Å wavelength light source on aluminum there is produced a maximum energy photoelectron excitation of 2.3eV. (ii) with a 3,130Å wavelength light source on aluminum there is produced a maximum energy photoelectron excitation of 0.9eV. What is the Work function for potassium? (assume Planck's constant is NOT known, and solve for that as well).

# Chapter 3. Modern Quantum Mechanics

## 3.1 Early History and Background

### 3.1.1 History

The early phase of discovery for quantum mechanics moved into the modern quantum mechanics formalism with the discovery of Heisenberg of the successful application of matrix mechanics and the resultant uncertainty principle (1925) [25]. In 1926, Schrodinger showed that the problem of finding a diagonal Hamiltonian matrix in the Heisenberg's mechanics is equivalent to finding wavefunction solutions to his wave equation [26]. An interpretation of the wavefunction was then clarified in 1927 by Born [28]. Dirac developed a manifestly relativistic formalism for the wavefunction and wave-equation for fermionic matter (1928) [29]. An axiomatic reformulation of quantum mechanics was then given by Dirac (1930) [5], laying the foundation for much of modern quantum notation and for critical issues such as self-adjointness. Dirac then described a formulation of a quantum propagation path, with quantum propagator having the familiar phase factor involving the action, in his paper "The Lagrangian in Quantum Mechanics" in 1933 [60]. In essence, Dirac had obtained a single path, in what would eventually be generalized by Feynman to all paths with the invention of the path integral formalism (1942 & 1948) [61,62]. The equivalence of a quantum mechanical formulation in terms of path integrals and the Schrodinger formalism was shown by Feynman in 1948 [62].

In a path integral description, the quantum mixture state, semiclassical physics, and classical trajectories are all given by the stationary phase dominated component. A stationary phase solution that is dominated by a single path is typical for a classical system. Thus, variational methods are fundamental to analysis of physical systems, whether it be in the form of Lagrangian and Hamiltonian analysis, or in various equivalent integral formulations.

Feynman's discovery of the path integral formalism wasn't solely based on the prior work of Dirac (1933) [60], although by appending that paper to his PhD thesis (1942) [61] its importance was clearly emphasized. Feynman also benefited from work going as far back as Laplace [23] for selection process based on highly oscillatory integral constructions that self-select for their stationary phase component. This branch of mathematics eventually became associated with Laplace's method of steepest descents, then to the work of Stokes and Lord Kelvin, then to the work of Erdelyi (1953) [63-65].

Feynman and others then invented quantum field theory for electromagnetism (QED) during 1946-1949 (more on this later). The extension to electroweak occurred in 1959, and to QCD in 1973, and to the "Standard Model" in 1973-1975. Thus, the impact of the path integral revolution in quantum physics was felt well into the 1970's, but this was only the beginning. At their inception path integrals were examined by Norbert Wiener, with the introduction of the Wiener Integral, for solving problems in statistical mechanics in diffusion and Brownian motion. In the 1970's this led to what is now known as "the grand synthesis" which unified quantum field theory and statistical field theory of a fluctuating field near a second-order phase transition, and where use of renormalization group methods enabled significant advances from quantum field theory to be carried over to statistical field theory.

The grand synthesis is one of many instances to come where we see analytic continuation of a constant or a parameter giving rise to familiar physics in the thermodynamic and statistical mechanics domains, showing a deeper connection (for further details, see App. D and Book 7 [21]). The Schrödinger equation, for example, can be seen to be a diffusion equation with an imaginary diffusion constant [66,67]. Likewise, the path integral can be seen to be an analytic continuation of the method for summing up all possible random walks [68].

### 3.1.2 Background

***Quantum Measurement Theory***

We have a Quantum Observable that interacts with a classical apparatus, to give a 'measurement'. We suppose that a classical apparatus is the limiting case of a Quantum process, and we require this limiting case to exist to have a complete formulation in terms of observable and apparatus. The existence of 'limiting cases', e.g., convergent sums, would occur in a theory that described a Martingale process. In Book 7 we see that the fundamental evolutionary process is Martingale, thus the quantum propagator formalism will have such limiting cases, thus the existence of the quantum measurement theory formulation is guaranteed from the Martingale Convergence properties.

**Canonical Quantization Shortcut**

Let's go back to the 1920's and suppose that the fundamental matter description is a wave-functional of some kind (as already indicated by early Quantum Mechanics, Bohr-Sommerfeld). The operative word being 'wave', and other than that we know that classical mechanics must somehow emerge. To be specific, lets consider the classical equation for the energy of a particle of mass $m$ and momentum $p$ moving in 1-D in a potential $V(x)$:

$$E = \frac{p^2}{2m} + V(x).$$

Let's now suppose that this equation is the characteristic equation that results from operation on a wavefunction with operators that yield the observed values:

$$\hat{E}\psi = \frac{\hat{p}^2}{2m}\psi + \widehat{V(x)}\psi \rightarrow \quad E = \frac{p^2}{2m} + V(x),$$

where $\hat{E}\psi = E\psi$, $\hat{p}\psi = p\psi$, and $\widehat{V(x)}\psi = V(x)\psi$. In the operator formalism, and the wavefunction that it acts on, we thus have that classical variable that commuted before, no longer do so now. Classically we have:

$$[x, p] = 0,$$

while x and p as operators $\hat{x}$ and $\hat{p}$ operating on a wavefunction satisfy:

$$[\,\hat{x}, \hat{p}] = i\hbar,$$

known as (Dirac) canonical quantization on the canonical pair $\{\, x, p \,\}$. Thus, if we now anachronistically jump to this single non-commutation feature, foundational in the later Dirac formulation, we have both he Schrödinger and Heisenberg formulations as well. For Schrödinger consider that we can satisfy $[\,\hat{x}, \hat{p}] = i\hbar$ if we choose an operator representation as follows:

$$\hat{x}\psi = x\psi \quad and \quad \hat{p}\psi = i\hbar\frac{\partial}{\partial x}\psi,$$

from which the Schrödinger equation results for our particle description above. From the canonical commutation relations it is also possible to prove an uncertainty principle on those canonical variables ([58]) known as the Heisenberg Uncertainty Principle. Note, however, that given the assumption of an underlying wave theory, we could have presumed an uncertainty principle would exist on that basis alone, as established in the mathematics of wave theory (Benedict's Theorem). From that required uncertainty relation on the 'operators' of the quantized theory, we could have guessed non-commutation on canonical variables and shown consistency on such solutions, thereby arriving as the canonical commutation relations for operators of the theory merely from it being a wave theory.

In what follows we will see the utility of the Schrödinger approach to quantization by examining bound-sate configurations that exist for particles in a 1-D potential well. Later this approach is generalized to the 3-D Hydrogen atom to get the prior spectral results from early Quantum Mechanics. Eventually, however, we will want to examine non-bound-state configurations, such as passing light through polarized screens in quantum state preparation experiments. In such descriptions we shift from the wavefunction formulation to a wavefunction state vector representation (Dirac) or an operator or matrix representation (Heisenberg). For example, consider the following experiment with light:

(unpolarized light)
→ [up filter] → (up polarized light)
→ [diagonal filter] → (diagonal polarized light)
→ [sideways filter] → (sideways polarized light)

In Quantum Mechanics, each filter action or 'measurement' is vector-space *projective* in nature (later we will introduce projection operators), while classically, each filter action is *selective* in nature. Thus, classically, the sideways polarized light at the end has zero intensity having been selectively filtered out at the first filter (passing only up polarization). Quantum mechanically, on the other hand, we project unpolarized through the up filter to get the up polarized light, we then project onto the diagonal filter to get diagonal polarized light. Inherent to this process is the underlying vector space mathematics where the up polarized light can be written as a linear superposition of the diagonal and anti-diagonal polarized light, and the diagonal polarization is passed by the diagonal filter. The third filter for the quantum process then allows for 1/8 amplitude of the original unpolarized light to pass (instead of zero classically).

In Schrödinger's bound-state solutions we will see obvious reasons for quantized solutions due to boundary condition constraints (for the Hydrogen-like atom solutions, this generates the periodic table). We will also see a perfect correspondence with the spectral results that drove the early Quantum Mechanics models (e.g., the Bohr-Sommerfeld Model). To understand the projective nature of quantum mechanics, however, it is best to work in Dirac's Hilbert space formulation that will follow the

Schrödinger examples. The Schrodinger, Heisenberg, and Dirac formulations are all equivalent in their areas of overlapping application, but for some problems some formulations are more suitable. For measurement theory problems, like above, we could capture the projective aspect of the theory in the Schrödinger formulation using overlap integrals on the wavefunctions, but this is far more complicated than simply representing the wavefunctions as Ket vectors, with projection simply accomplished by taking an inner product between Ket vectors.

### 3.2 Angular Momentum Quantization

Let's start with the usual canonical quantization relations (details on origins will be given later), where $[\hat{x},\hat{p}] = i\hbar$ for a pair of conjugate variables., and compute the commutators on angular momentum operators:

$$[L_x, L_y] = [YP_z - ZP_y, ZP_x - XP_z] = [YP_z, ZP_x] + [ZP_y, XP_z]$$
$$= YP_x[P_z, Z] + P_yX[Z, P_z] = (-YP_x + XP_y)(i\hbar)$$
$= i\hbar L_z$ then cyclic permutations to get:

$$[L_x, L_y] = i\hbar L_z$$
$$[L_y, L_z] = i\hbar L_x$$
$$[L_z, L_x] = i\hbar L_y$$

The commutation relations for the components of the angular momentum of a spinless particle (classical analog). The origin of commutation relations lies in the geometric properties of rotations in three-dimensional space. Adopting a more general point of view, define an angular momentum $\tilde{J}$ as any set of three observables $J_x, J_y, J_z$ which satisfies: $[J_x, J_y] = i\hbar J_z$ and cyclic permutations. Define

$$\tilde{J}^2 = J_x^2 + J_y^2 + J_z^2$$

then

$$[\tilde{J}^2, \tilde{J}] = 0$$

(not necessarily true for the general vector operator).

Angular momentum theory in quantum mechanics is founded entirely on the commutation relations. We seek the system of eigenvectors common to $\tilde{J}^2$ and $J_z$, and instead of using $J_x$ and $J_y$ it is more convenient to introduce.

$$J_+ = J_x + iJ_y$$
$$J_- = J_x - iJ_y$$

and $J_+^\dagger = J_-$, so not Hermitian.

$[J_z, J_+] = i\hbar J_y + i(-i\hbar)J_x = \hbar(J_x + iJ_0) = \hbar J_+$
$[J_z, J_+] = -\hbar J_-$
$[J_x, J_+] = 2\hbar J_z$
$[\tilde{J}^2, J_+] = [\tilde{J}^2, J_-] = [\tilde{J}^2, J_z] = 0$

$$J_+J_- = (J_x + iJ_y)(J_x - iJ_y) = J_x^2 + J_y^2 + iJ_yJ_x - iJ_xJ_y = \tilde{J}^2 - J_z^2 + \hbar J_z$$
$$J_-J_+ = J_x^2 + J_y^2 - \hbar J_z = J^2 - J_z^2 - \hbar J_z$$

Thus

$$\tilde{J}^2 = \frac{1}{2}(J_+J_- + J_-J_+) + J_z^2$$
$$< \Psi|\tilde{J}^2|\Psi> = < J_x^2 > + < J_y^2 > + < J_z^2 > \geq 0$$

Let the eigenvalue, $\lambda, of\ \tilde{J}^2$ be written $\lambda = j(j+1)\hbar^2$ (by convention) where $j \geq 0$. Then,

$$< \Psi|\tilde{J}^2|\Psi> = j(j+i)\hbar^2$$

( $\hbar$ has units of angular momentum, so, j is dimensionless).
Also, let $J_z|\Psi> = m\hbar|\Psi>$ where m is a dimensionless number (not necessarily positive). So,

$$\tilde{J}^2|k,j,m> = j(j+1)\hbar^2|k,j,m>$$
$$J_z|k,j,m> = m\hbar|k,j,m>$$
$$\left\|\tilde{J}|k,j,m>\right\|^2 = < k,j,m|J_{J+}|k,j,m> \geq 0$$

Thus,

$\left\|\tilde{J}|k,j,m>\right\|^2 = j(j+1)\hbar^2 - (m\hbar^2) - \hbar^2 m \geq 0$
$j(j+1) - m(m+1) \geq 0$
$(j-m)(j+m+1) \geq 0$
$\boxed{-(j+1) \leq m \leq j}$ for relation #1.

$\left\|\tilde{J}_-|k,j,m>\right\|^2 = < k,j,m|J_+J_-|k,j,m \geq 0$
$j(j+1) - m(m-1) \geq 0$
$(j-m+1)(j+m) \geq 0$
$\boxed{-j \leq m \leq j+1}$ for relation #2.

Combining the constraints on relations #1 and #2:

$$-j \leq m \leq +j.$$

Properties of $J_-|k,j,-j> = 0$

(i) If $m = -j,\ then\ J_-|k,j,-j> = 0$

(ii) If $m > -j, J_-|K,j,m>$ is a no-null eigenvector of $\vec{J}^2$ and $J_Z$ with the eigenvalues $j(j+1)\hbar^2$ and $(m-1)\hbar$ respectively.

Proof (i)

$\|J_-|K,j-j>\|^2 = (j(j+1) - m^2 + m)\hbar^2|_{m=-j} = j(j+1) - j^2 - j = 0$

$J_-|K,j,-j> = 0$

Proof (ii)

$[\vec{J}^2, J_-]|K,j,m> = 0$

So, $[\vec{J}^2, J_-]|K,j,m> = J_-\ j(j+1)\hbar^2|K,j,m>$, and $J_-|K,j,m>$ is in eigenvalue of $\vec{J}^2$ with value $j(j+1)\hbar^2$. Recall $[J_z, J_-] = -\hbar J_-$ .So,

$$J_Z J_-|K,j,m> = -\hbar J_-|K,j,m> + J_{J_Z}|K,j,m> = (-\hbar + m\hbar)J_-|K,j,m>$$
$$= (m-1)\hbar J_-|K,j,m>$$

Properties of $J_+|K,j,m>$

(i) If m=j then $J_+|K,j,m>= 0$

(ii) If m<j then $J_+|K,j,m>$ is a non-small eigenvector of $\vec{J}^2$ and $J_Z$ with eigenvalues $J(j+1)\hbar^2$ and $(m+1)\hbar$

Proof similar to above.

Determination of the spectrum of $J^2$ and $J_Z$:

Consider $|K,j,m>$ by notation chosen

$J^2|j,j,m>= j(j+1)\hbar^2|K,j,m>$

$J_z J_-|K,j,m>= m\hbar J_-|K,j,m>$

Properties of $J_\pm$ have revealed that the spectrum of $J_z$ is dependent on that of $J^2$: $-j \le m \le j$ .

Since $-j \le m \le j$ , we can find an integer, positive or zero, such that $-j \le m - p < -j + 1$. Consider the series of vectors:

$$|k.j,m>, J_-|K,j,m>,,,(J_-)^P|K,j,m> \quad for\ (n = 0, \dots P)$$

Each vector $(J_-)^{n-1}|K,j,m>$ has eigenvectors $j(j+1)\hbar^2$ and $(m-n)\hbar$. The proof is by iteration:

$$J_-(J_-)^{n-1}|K,j,m>$$

Where $(J_-)^{n-1}|K,j,m>$ has eigenvalues $j(j+1)\hbar^2$ and $(m-n)\hbar$, where

$$m - n + 1 > -j\ since\ m - n + 1 \ge m - p + 1 \ge -j + 1.$$

Thus, $(J_-)^{n-1}|K,j,m>$ is a non null eigenvector since

$$m - (n-1) > -j \quad for\ J_-\{(J_-)^{n-1}|K,j,m>\}$$

Now, let J_ act on $(J\_)^{P}|K,j,m>$. First consider the case where $m-p>-j$ in relation $-j\leq m-p-j+1$. Then, $J\_(J\_)^{n-1}|K,j,m>$ is non null and has eigenvalues:

$$j(j+1)\hbar^2 \ and \ (m-p-1)\hbar$$

But here we have a contradiction since

$m-p-1<(-j+1)-1$

$m-p-1<-j$ and we can't have any eigenvalues <-j.

Therefore, we must have $m-p=-j$, in which case:

$$(J\_)^{p}|K,j,m> \text{ has eigenvalue } (-j)\hbar.$$

Now, any further multiple of $J\_(J\_)^{p}|K,j,m>$ etc, is zero from the property of $J\_|K,j,-j>$. Thus, from the action of $J\_$ we have determined that there exists a positive or zero integer p such that $m-p=-j$. The analogous argument for $J_+$ reveals that $m+q=j$. Combining the two relations we find that $q+p=2j$. Thus, j is either integral or half integral.

So,

$J_z|K,j,m>=m\hbar|K,j,m>$

$J_+|K,j,m>=\hbar\sqrt{j(j+1)-m(m+1)}|K,j,m+1>$

$J_+|K,j,m>=\hbar\sqrt{j(j+1)-m(m-1)}|K,j,m-1>$

**3.2.1 Matrix representations**

Consider the $(J_n)^{(j)}$ matrices

Case (i) j=0 $(J_u)^{(0)}$ matrices are numbers, which the above relations show are zero.

Case (ii) $j=\frac{1}{2}$

Now,

$<K,j,m|J_z|K',j',m'>=m\hbar\delta_{kk'}\delta_{jj'}\delta_{mm'}$

$<K,j,m|J_\pm|K,j',m'>=\hbar\sqrt{j(j+1)-m'(m'\pm 1)}\delta_{kk'}\delta_{jj'}\delta_{mm'\pm 1}$

The subspaces $(k,j=1/2)$ are two dimensional $(m=1/2 \ or -1/2)$. Choose the basis vectors in descending order in m: $m={}^1/_2, m=-{}^1/_2$, we find:

$$(J_z)^{({}^1/_2)}=\frac{\hbar}{2}\begin{pmatrix}1 & 0\\ 0 & -1\end{pmatrix}$$

and

$$m\delta'_{mm} = \frac{1}{2}\delta_{1/2m'} + (-1/2)\delta_{-1/2m'} = \frac{1}{2}\begin{pmatrix}1 & 0\\ 0 & 0\end{pmatrix} - \frac{1}{2}\begin{pmatrix}0 & 0\\ 0 & 1\end{pmatrix}$$
$$= \frac{1}{2}\begin{pmatrix}1 & 0\\ 0 & -1\end{pmatrix}$$

Then

$$(J_+)^{(1/2)} = \hbar\begin{pmatrix}0 & 1\\ 0 & 0\end{pmatrix}$$

So,

$$< K, j, m|J_z|K', J', m' >= m\hbar\delta_{kk'}\delta_{jj'}\delta_{mm'}$$
$$< K, j, m|J_\pm|K', J', m' >= \hbar\sqrt{j(j+1) - m'(m' \pm 1)}\hbar\delta_{kk'}\delta_{jj'}\delta_{mm'\pm 1}$$

And

$$J_+ = J_x + iJ_y \quad and \quad J_- = J_x - iJ_y$$

are all we need to know in order to find the matrix associated with an arbitrary component $J_u$ in a standard basis. All we need to do is calculate $(J_u)^{(j)}$ what represent $J_u$ inside the subspaces $\varepsilon(k, j)$ for all possible $j$ ($j = 0, 1/2, 1, 3/2, \ldots$):

(i) $j = 0$, $(J_u)^{(0)}$ is a number (1-d matrix) which is zero by relations shown.

(ii) $j = 1/2$

$(J_z)^{(1/2)} = ?$

$notice\ (J_u)^{(1/2)}\begin{pmatrix}\{m' = 1/2\}\\ \{m' = -1/2\}\end{pmatrix} = \begin{pmatrix}\{m' = 1/2\}\\ \{m' = -1/2\}\end{pmatrix}$ in our state.

So, $(J_z)^{(1/2)} = m\hbar\delta_{mm'} = m\hbar\delta_{m(1/2)} + m\hbar\delta_{m(-1/2)}$

$$\boxed{(J_z)^{(1/2)} = \frac{\hbar}{2}\begin{pmatrix}1 & 0\\ 0 & -1\end{pmatrix}}$$

$$(J_+)^{(1/2)} = \sqrt{\frac{3}{4} - m'(m' + 1)\hbar\delta_{m'm} + 1}$$

$$\boxed{(J_+)^{(1/2)} = \hbar\begin{pmatrix}0 & 1\\ 0 & 0\end{pmatrix}} \quad \text{raising}$$

$$\boxed{(J_-)^{(1/2)} = \hbar\begin{pmatrix}0 & 0\\ 1 & 0\end{pmatrix}} \quad \text{lowering}$$

So,

$$(J_x)^{(1/2)} = \frac{1}{2}(J_+ + J_-) = \frac{\hbar}{2}\begin{pmatrix}0 & 1\\ 1 & 0\end{pmatrix}$$

$$(J_y)^{(1/2)} = \frac{1}{2}(iJ_+ - iJ_+) = \frac{\hbar}{2}\begin{pmatrix} 0 & -i \\ i & 0 \end{pmatrix}$$
$$\vec{J}^2 = J_x^2 + J_y^2 + J_z^2$$
$$\vec{J}^2 = \frac{3}{4}\hbar^2\begin{pmatrix} 1 & 0 \\ 0 & 1 \end{pmatrix}$$

(iii) $j = 1$

Denote $|K, j, m >$ *by* $|m >$ *and assume* $K' = K, j' = j$

$< m|J_z|m' >= m\hbar\delta_{mm'}$

$< m|J_\pm|m' >= \sqrt{j(j+1) - m'(m \pm 1)}\hbar\delta_{mm'\pm 1}$

For $j = 1$

$$\{\delta_{m,m'}\} = \begin{pmatrix} 1 & 0 & 0 \\ 0 & 1 & 0 \\ 0 & 0 & 1 \end{pmatrix}, \quad \delta_{mm'+1} = \begin{pmatrix} 0 & 1 & 0 \\ 0 & 0 & 1 \\ 0 & 0 & 0 \end{pmatrix}, \delta_{mm'-1} = \begin{pmatrix} 0 & 0 & 0 \\ 1 & 0 & 0 \\ 0 & 1 & 0 \end{pmatrix}$$

$$J_z^{(1)} = \hbar\begin{pmatrix} 1 & 0 & 0 \\ 0 & 0 & 0 \\ 0 & 0 & -1 \end{pmatrix}, J_+^{(1)} = \hbar\begin{pmatrix} 0 & \sqrt{2} & 0 \\ 0 & 0 & \sqrt{2} \\ 0 & 0 & 0 \end{pmatrix}, J_-^{(1)}$$
$$= \hbar\begin{pmatrix} 0 & 0 & 0 \\ \sqrt{2} & 0 & 0 \\ 0 & \sqrt{2} & 0 \end{pmatrix}$$

So,

$$J_z = \hbar\begin{pmatrix} 1 & 0 & 0 \\ 0 & 0 & 0 \\ 0 & 0 & -1 \end{pmatrix}, \quad J_x = \frac{\hbar}{\sqrt{2}}\begin{pmatrix} 0 & 1 & 0 \\ 1 & 0 & 1 \\ 0 & 1 & 0 \end{pmatrix},$$
$$J_Y = \frac{\hbar}{\sqrt{2}}\begin{pmatrix} 0 & -1 & 0 \\ i & 0 & -i \\ 0 & i & 0 \end{pmatrix}$$

and

$$(\vec{J}^2)^{(1)} = 2\hbar^2 I$$

An orthogonal law $\{|k, j, m >\}$ with

$\vec{J}^2|k, j, m > = j(j+1)\hbar^2|k, j, m >$

$J_z|k, j, m > = m\hbar|k, j, m >$

Is called a "standard basis" if

$J_\pm|k, j, m > = \sqrt{j(j+1) - m'(m' \pm 1)}\hbar\ |k, j, m' \pm 1 >$

### 3.2.2 Position Representation

In the $\{|r >\}$ representation we can write out the L's:

$$L_x = \frac{\hbar}{i}\left(y\frac{\partial}{\partial z} - z\frac{\partial}{\partial y}\right)$$
$$L_y = \frac{\hbar}{i}\left(z\frac{\partial}{\partial x} - x\frac{\partial}{\partial z}\right)$$
$$L_z = \frac{\hbar}{i}\left(x\frac{\partial}{\partial y} - y\frac{\partial}{\partial x}\right)$$

We are now in a cartesian coordinate system, it is usually beneficial to switch to spherical co-ordinates:
$x = rsin\theta \cos\theta$
$y = rsin\theta \sin\theta$
$z = r\cos\theta$

Now, $\frac{\partial}{\partial y} = ? \ = A(\theta,\varphi)\frac{\partial}{\partial r} + B\frac{1}{r}\frac{\partial}{\partial\theta} + C\frac{1}{rsin\,\theta}\frac{\partial}{\partial\varphi}$
$\frac{\partial y}{\partial y} = 1 = A\sin\theta\sin\theta + Bcos\theta\sin\theta + C\cos\theta$
$A = \sin\theta sin\varphi, B = \cos\theta, C = \cos\varphi$ works, so:

$$\frac{\partial}{\partial y} = \sin\theta\sin\varphi\frac{\partial}{\partial r} + \frac{\cos\theta\sin\varphi}{r}\frac{\partial}{\partial\theta} + \frac{\cos\varphi}{rsin\theta}\frac{\partial}{\partial\varphi}$$
$$\frac{\partial}{\partial x} = \sin\theta\cos\theta\frac{\partial}{\partial r} + \cos\theta\cos\varphi\frac{\partial}{\partial\theta} - \frac{\sin\varphi}{rsin\,\theta}\frac{\partial}{\partial\varphi}$$
$$\frac{\partial}{\partial z} = A\frac{\partial}{\partial r} + B\frac{1}{r}\frac{\partial}{\partial\theta} + c\frac{1}{rsin\,\theta}\frac{\partial}{\partial\varphi}$$
$$\frac{\partial z}{\partial z} = 1 = A\cos\theta + [-B\sin\theta] + C(0)$$
$$\frac{\partial x}{\partial z} = 0 = A\sin\theta\cos\varphi + B\cos\theta\cos\varphi + (-C\sin\varphi)$$
$$\frac{\partial y}{\partial z} = 0 = A\sin\theta\sin\varphi + Bcos\theta\cos\varphi + C\cos\varphi$$
Thus,
$$\frac{\partial}{\partial z} = \cos\theta\frac{\partial}{\partial r} - \frac{\sin\theta}{r}\frac{\partial}{\partial\theta}.$$
So,
$$L_x = \frac{\hbar}{i}\left((rsin\theta\sin\varphi)\left\{\cos\theta\frac{\partial}{\partial r} - \frac{\sin\theta}{r}\frac{\partial}{\partial\theta}\right\}\right.$$
$$\left. - rcos\theta\left\{\sin\theta\sin\varphi\frac{\partial}{\partial r} + \frac{\cos\theta\sin\varphi}{r}\frac{\partial}{\partial\theta} + \frac{\cos\varphi}{rsin\theta}\frac{\partial}{\partial\varphi}\right\}\right)$$
$$L_x = i\hbar\left(\sin\theta\frac{\partial}{\partial\theta} + \frac{\cos\varphi}{\tan\theta}\frac{\partial}{\partial\varphi}\right)$$

$$L_y = \frac{\hbar}{i}\left(r\cos\theta\left\{\sin\theta\cos\frac{\partial}{\partial t} + \cos\theta\cos\varphi\frac{\partial}{\partial\theta} - \frac{\sin\varphi}{r\sin\theta}\frac{\partial}{\partial\varphi}\right\}\right.$$
$$\left. - r\sin\theta\cos\varphi\left\{\cos\theta\frac{\partial}{\partial r} - \frac{\sin\theta}{r}\frac{\partial}{\partial\theta}\right\}\right)$$
$$L_y = i\hbar\left(-\cos\varphi\frac{\partial}{\partial\theta} + \frac{\sin\varphi}{\tan\varphi}\frac{\partial}{\partial\varphi}\right)$$

and

$$L_z = i\hbar\frac{\partial}{\partial\varphi}$$

$$L^2 = L_x^2 + L_y^2 + L_z^2 = -\hbar^2\left\{\frac{\partial^2}{\partial\theta^2} + \frac{1}{\tan\theta}\frac{\partial}{\partial\theta} + \frac{1}{tan^2\theta}\frac{\partial^2}{\partial\varphi^2} + \frac{\partial^2}{\partial\varphi^2}\right\} =$$
$$-\hbar^2\left\{\frac{\partial^2}{\partial\theta^2} + \frac{1}{\tan\theta}\frac{\partial}{\partial\theta} + \frac{1}{sin^2\theta}\frac{\partial^2}{\partial\varphi^2}\right\}$$

**3.2.3 Spherical Harmonics**

In spherical coordinates, $ds^2 = dr^2 + r^2(d\theta^2 + sin^2\theta d\varphi^2)$, we have:

$$\nabla^2 = \frac{1}{\sqrt{|g|}}\partial_\mu\left(\sqrt{|g|}g^{\mu\nu}\partial_\nu\right) \qquad g = r^4 sin^2\theta \qquad g_{\mu\nu} =$$
$$\begin{pmatrix} 1 & 0 & 0 \\ 0 & r^2 & 0 \\ 0 & 0 & r^2 sin^2\theta \end{pmatrix}$$
$$\nabla^2 = \frac{1}{r^2}\partial_r(r^2\partial_r) + \frac{1}{r^2 sin\theta}\partial_\theta(\sin\theta\,\partial_\theta) + \cdots$$
$$\nabla^2 = \frac{1}{r}\frac{\partial^2}{\partial r^2}r + \frac{1}{r^2}\{\dots\} = \frac{1}{r}\frac{d^2}{dr^2}r + \frac{1}{r^2}\left\{\frac{1}{\tan\theta}\frac{\partial}{\partial\theta} + \frac{\partial^2}{\partial\theta^2} + \frac{1}{sin^2\varphi}\frac{\partial^2}{\partial\varphi^2}\right\}$$
$$\nabla^2 = \frac{1}{r}\frac{d^2}{dr^2}r + \frac{\vec{L}^2}{r^2}$$

So,

$$\vec{L}^2 = -\hbar^2\left(\frac{\partial^2}{\partial\varphi^2} + \frac{1}{\tan\theta}\frac{\partial}{\partial\theta} + \frac{1}{sin^2\varphi}\frac{\partial^2}{\partial\varphi^2}\right)$$
$$L_+ = \hbar\left[(i sin\,\varphi + \cos\varphi)\frac{\partial}{\partial\theta} + \frac{(i cos\varphi - \sin\varphi)}{\tan\theta}\frac{\partial}{\partial\varphi}\right] = \hbar e^{i\varphi}\left(\frac{\partial}{\partial\theta} + i cot\theta\frac{\partial}{\partial\theta}\right)$$
$$L_- = \hbar e^{-i\varphi}\left(-\frac{\partial}{\partial\theta} + i cot\theta\frac{\partial}{\partial\theta}\right)$$

So,

$$\vec{L}^2\,\Psi(r,\theta\varphi) = l(l+1)\hbar^2\,\Psi(r,\theta,\varphi)$$

And

$$L_z\,\Psi(r,\theta,\varphi) = m\hbar\,\Psi(r,\theta,\varphi)$$

Gives us the equations:

$$-\left\{\frac{\partial^2}{\partial\theta^2} + \frac{1}{\tan\theta}\frac{\partial}{\partial\theta} + \frac{1}{sin^2\varphi}\frac{\partial^2}{\partial\varphi^2}\right\}\Psi(r,\theta,\varphi) = l(l+1)\,\Psi(r,\theta,\varphi)$$

and

$$-\frac{i\partial \Psi(r,\theta,\varphi)}{\partial\varphi} = m\,\Psi(r,\theta,\varphi)$$

Let's introduce spherical harmonics notation by denoting $Y_l^m(\theta,\varphi)$ the common eigenfunctions of $\vec{L}^2$ and $L_z$ are:

$$-\frac{i\partial Y_l^m}{\partial\varphi} = mY_l^m \rightarrow Y_l^m(\theta,\varphi) = F_l^m(\theta)e^{im\varphi}$$

Since $0 \leq \varphi \leq 2\pi$ , and since a wave function must be continuous at all points in space, we must have $Y_l^m(\varphi = 0) = Y_l^m(\varphi = 2\pi)$. Thus, $e^{2im\pi} = 1 \rightarrow m$ must be an integer. Thus, for orbital angular momentum, m (and thus l) must be an integer. We know

$$L_+Y_l^m(\theta,\varphi) = 0 \rightarrow \left\{\frac{d}{d\theta} - l\cot\theta\right\}F_l^l(\theta) = 0$$

Thus, $F_l^l(\theta) = C_l(\sin\theta)^l$ .
So, $Y_l^l|\theta,\varphi) = C_l(\sin\theta)^l\,e^{il\varphi}$ .
Through repeated action of $L_-$ we can construct the (2l+1) m values for the given l.

***Properties of spherical Harmonics***

From $L_\pm Y_l^m(\theta,\varphi) = \hbar\sqrt{l(l+1) - m(m\pm 1)}Y_l^{m\pm 1}(\theta,\varphi)$ we have the recurrence relations:

$$e^{i\varphi}\left(\frac{\partial}{\partial\theta} - m\cot\theta\right)Y_l^m(\theta,\varphi) = \sqrt{l(l+1) - m(m-1)}Y_l^{m+1}(\theta,\varphi)$$

And

$$e^{-i\varphi}\left(-\frac{\partial}{\partial\theta} - m\cot\theta\right)Y_l^m(\theta,\varphi) = \sqrt{l(l+1) - m(m-1)}Y_l^{m-1}(\theta,\varphi)$$

The eigenvalue equations for $Y_l^m$ determines them to within a constant factor. This factor is chosen so as to normalise the $Y_l^m$ where

$$\int_0^{2\pi} d\varphi \int_0^{\pi} \sin\theta d\theta\, Y_l^m(\theta,\varphi)^* Y_l^m(\theta,\varphi) = \delta_{l'l}\delta_{m'm}$$

We can expand $f(\theta,\varphi) = \sum_{l=0}^{\infty}\sum_{n=-l}^{l} C_{l,m}Y_l^m\,(\theta,\varphi)$.

For Closure

$$\sum_{l=0}^{\infty}\sum_{m=-l}^{l} Y_l^m\,(\theta,\varphi)Y_l^{m*}(\theta',\varphi') = \delta(\cos\theta - \cos\theta')\delta(\varphi - \varphi') = \frac{1}{\sin\theta}\delta(\varphi - \varphi')\delta(\varphi - \varphi')$$

For Parity with respect to origin and complex conjugation:

$Y_l^m(\pi - \theta, \pi + \varphi) = (-1)^l Y_l^m(\theta, \varphi)$
$(Y_l^m)^* = (-1)^m Y_l^{-m}(\theta, \varphi)$

### 3.2.4 Finite rotation operator follows from infinitesimal ones

We can 'build' a (finite) rotation operator from the infinitesimal rotation operation (described previously in regards to angular momentum). For rotation about the z-axis:

$$R_{\hat{e}_z}(\alpha + d\alpha) = R_{\hat{e}_z}(\alpha) R_{\hat{e}_z}(d\alpha) = R_{\hat{e}_z}(\alpha)\left[1 - \frac{i}{\hbar} d\alpha\, L_z\right].$$

Thus,

$$\frac{dR_{\hat{e}_z}}{d\alpha} \equiv \frac{R_{\hat{e}_z}(\alpha + d\alpha) - R_{\hat{e}_z}(\alpha)}{d\alpha} = -\frac{i}{\hbar} R_{\hat{e}_z}(\alpha) L_z\,,$$

and since $R_{\hat{e}_z}(\alpha)$ and $L_z$ commute the solution is formally an ordinary integration. Thus,

$$R_{\hat{e}_z}(\alpha) = e^{-\frac{i}{\hbar}\alpha L_z}\,,$$

and, in general:

$$R_{\vec{u}}(\alpha) = exp\left(-\frac{i}{\hbar}\alpha\, \vec{L} \cdot \vec{u}\right).$$

Note that $R_{\vec{u}}(2\pi) = I$ since $R_{\hat{e}_z}(2\pi) = \exp\left(-\frac{i(2\pi)}{\hbar} m\hbar\right) = 1$ since m is integer valued.

***Rotation of observables***

Consider:

$$|u_n' > = R|u_n >$$

Consider an observable $A$:

$$A|u_n > = a_n|u_n >$$

If we rotate system and measuring device we should have:

$$A'|u'_n > = a_n|u'_n >$$

Thus,

$$A'R|u_n > = a_n R|u_n >.$$

So,

$$R^\dagger A' R|u_n > = a_n|u_n >$$
$$A' = RAR^\dagger$$

For an infinitesimal rotation: $A' = \left(1 - \frac{i}{\hbar} d\alpha \vec{J} \cdot \vec{u}\right) A \left(1 + \frac{i}{\hbar} d\alpha \vec{J} \cdot \vec{u}\right)$

$$A' = A - \frac{i}{\hbar} d\alpha [\vec{J} \cdot \vec{u}, A]$$

Scalar observables have $A' = A$ thus, $[\vec{j} \cdot \vec{u}, A] = 0$. Scalar observables commute with the three components of the total angular momentum. The Hamiltonian is a scalar observable → conservation of angular momentums.

Vector observables, $\vec{V}$, have a set of observables $V_x, V_y, V_z$ which is transformed by rotations like a vector. $V_x$ is unchanged by rotation about the x-axis, so:

$$V_x' = V_x - \frac{i}{\hbar} d\alpha[J_x, V_x] \quad and \quad V_x' = V_x \ \ when\, \vec{j} \cdot \vec{u} = \vec{J}_x$$

So,

$$[J_x, V_x] = 0$$

Consider rotation about the y-axis:

$$V_x' = V_x - \frac{i}{\hbar} d\alpha[J_y, V_x]$$

We can write $\hat{e}'_x = \hat{e}_x + d\alpha(\hat{e}_y \times \hat{e}_x) = \hat{e}_x - d\alpha\hat{e}_z$, consequently, if $V_x$ is a vector observable:

$$V_x' = \vec{V} \cdot \hat{e}_x - d\alpha\vec{V} \cdot \hat{e}_z = V_x - d\alpha V_z.$$

Thus,

$$[V_x, J_y] = i\hbar V_z$$

and cyclic permutations (like Poincare group relation).

### 3.2.5 Rotation of Diatomic molecule

Consider a system of two nuclei, bound, rotating about their centre of mass, where, to a first approximation, we assume their separation to be constant. We, thus, have a rigid rotator.

***Classical study***

In the C.M frame we have $m_1 r_1 = m_2 r_2$ and:

$$I = m_1 {r_1}^2 + m_2 {r_2}^2, \qquad \mu = \frac{m_1 m_2}{m_1 + m_2}, \qquad \frac{r_1}{m_2} = \frac{r_2}{m_1} = \frac{r_e}{m_1 + m_2},$$

$$I = \mu re^2$$

The angular momentum is:

$$|L| = r_1(mr_1\omega_R) + r_2(mr_2\omega_R) = I\omega_R = \mu re^2\omega_R$$

And Hamiltonian:

$$H = \frac{1}{2} I\omega_R^2 = \frac{|L|^2}{2I}$$

This is equivalent to studying a fictitious particle of mass $\mu$ forced to remain at a distance $r_e$ from the origin, about which it rotates at $\omega_R$.

***Quantization of the rigid rotator***

Since $r_e$ is fixed the quantum mechanical state is described by a wave function $\Psi(\theta,\varphi)$ , normalization then gives.

$$\int_0^{2\pi} d\varphi \int_0^{\pi} d\theta \sin\theta \,|\,\Psi(\theta,\varphi)|^2$$

Thus, $\Psi(\theta,\varphi)|^2 \sin\theta d\theta\varphi$ represents the probability of finding the axis of the rotator pointing in the solid angle element $d\Omega = \sin\theta d\theta d\varphi$. Using Dirac rotation we associate with every $\Psi(\theta,\varphi)$ a $|\,\Psi>$ is $\varepsilon_\Omega$ . The quantum mechanical Hamiltonian follows easily from quantization rules.

$$H = \frac{|L|^2}{2I} \rightarrow \boldsymbol{H} = \frac{\vec{\boldsymbol{L}}^2}{2I}$$

So, the eigenvalue equation of interest is very simple:

$$H|\,\Psi> \leftrightarrow \; \frac{-\hbar^2}{2\mu r_e^2}\left[\frac{\partial^2}{\partial\theta^2} + \frac{1}{\tan\theta}\frac{\partial}{\partial\theta} + \frac{1}{sin^2\theta}\frac{\partial^2}{\partial\theta^2}\right]\Psi(\theta,\varphi)$$

Denote by $|l,m>$ the Ket of $\varepsilon_\Omega$ associated with $Y_l^m(\theta,\varphi)$. Thus,

$$H\,|l,m> \; = \frac{l(l+1)\hbar^2}{2\mu r_e^2}|l,m>$$

Let $B = \frac{\hbar}{4\pi I}$, then we have $E_l = Bhl(l+1)$ and:

$$E_l - E_{l+1} = 2Bhl.$$

The separation of two levels increases linearly in l (unlike the harmonic oscillation where the separation is constant in n). Recall that for z-position $Z|\,\Psi> \leftrightarrow r_e\cos\theta\,\Psi(\theta,\varphi)$. A Bohr frequency $\left(\frac{E_l - E_{l'}}{\hbar}\right)$ can appear in the function $<Z>(t)$ if $Z$ has a nonzero matrix element between a state $|l,m> \; and \; |l',m'>$ . This is shown next.

Consider

$$\cos\theta Y_e^m = \sqrt{\frac{l^2 - m^2}{4l^2 - 1}}Y_{l-1}^m + \sqrt{\frac{(l+1)^2 - m^2}{4(l+1)^2 - 1}}Y_{l+1}^m$$

Thus,

$$<l',m'|Z|l,m> = r_e\delta_{mm'}\left[\delta_{l',l-1}\sqrt{\frac{l^2 - m^2}{4l^2 - 1}} + \delta_{l',l+1}\sqrt{\frac{(l+1)^2 - m^2}{4(l+1)^2 - 1}}\right]$$

So, the selection rules for $Z$ are:

$$\Delta l = \pm 1, \Delta m = 0.$$

For X and Y we similarly have $\Delta l = \pm 1, \Delta m = \pm 1$. Since $E_l$ depends only on l, the Boht frequencies are the same for <X>,<Y>, and <Z>. Also, $\nu_{l(l-1)} = \frac{E_l - E_{l-1}}{\hbar} = 2Bl \rightarrow$ a series of equidistant frequencies.

To determine a classical state where <Z> behaves like z, one must superimpose a large number of states $|l, m>$ :

$$[\Psi(\dagger)\rangle = \sum_{\ell=0}^{\infty} \sum_{m=-l}^{+l} c_{l,m}(t)|\ell, m\rangle,$$

Where the probable value of state $|\ell, m\rangle$ is $|c_{l,m}(t)|^2$. Consider $|c_{l,m}(t)|^2$ for $l_M$ to be very large, also the spread $\Delta l\ of\ l$ may be very large in absolute value but small in relative value:

$$l_M, \Delta l \gg 1, \quad \frac{\Delta \ell}{\ell m} \ll 1.$$

Then $\langle \vec{L} \rangle^2 \simeq \langle \vec{L}^2 \rangle \simeq l_M(l_M + 1)\hbar^2 \simeq {l_M}^2 \hbar^2$, and

$$\nu_m \simeq \frac{2B|\langle \mathrm{L} \rangle|}{\hbar} = \frac{|\langle \vec{\mathrm{L}} \rangle|}{2\pi \mathrm{I}} \ (classical\ relation)$$

Because of the spread $\Delta l\ of\ l$ the wave packet becomes distorted over time, where the distortion becomes appreciable after time

$$\tau \cong \frac{I}{2B\Delta l}.$$

Since $\nu_m \tau \cong \frac{l_M}{\Delta \ell} \gg 1$ , the distortion is slow. Also, since the Bohr frequencies form a discrete series of equidistant frequencies, separated by interval 2B, the overall motion is periodic, with period $T=\frac{1}{2B}$. The overall periodicity is related to the fact that the wave packet evolves on the unit sphere – a bounded surface.

**Angular Momentum Summary**

$$[\mathrm{J_x}, \mathrm{J_y}] = i\hbar \mathrm{J_z}$$

This leads to:

$$[\vec{\mathrm{J}^2}, \vec{\mathrm{J}}] = 0.$$

Introducing shift operators:

$$\mathrm{J_+} = \mathrm{J_x} + i\mathrm{J_y} \quad and \quad \mathrm{J_-} = \mathrm{J_x} - i\mathrm{J_y}$$

We get:

$$\vec{\mathrm{J}^2} = -\hbar^2 \left\{ \frac{\partial^2}{\partial\theta} + \frac{1}{\tan\theta} \frac{\partial}{\partial\theta} + \frac{1}{\sin^2\theta} \frac{d^2}{d\varphi^2} \right\}$$

and

$$\overrightarrow{J^2}\,\varphi(x) = \ell(\ell+1)\hbar^2\varphi(x) \ \ and \ \ J_z\varphi = m\hbar\varphi$$

and

$$J_z|k,j,m> \ = m\hbar|k,j,m>$$
$$J_+|k,j,m> \ = \hbar\sqrt{j(j+1)-m(m+1)}|k,j,m+1>$$
$$J_-|k,j,m> \ = \hbar\sqrt{j(j+1)-m(m-1)}|k,j,m-1>$$

**Addition of angular momentum**

$$\vec{J} = \vec{J_1} + \vec{J_2}$$
$$\overrightarrow{J^2} = \vec{J}_1^2 + \vec{J}_2^2 + 2\vec{J_1}\cdot\vec{J_2} = \vec{J}_1^2 + \vec{J}_2^2 + 2J_{1z}J_{2z} + J_+J_- + J_-J_+$$

**Vector Operations**

$$[J_x, V_x] = 0$$
$$[J_x, V_y] = i\hbar V_z \ \ (plus\ permutations)$$
$$[J_x, V_z] = -i\hbar V_y \ \ (plus\ permutations)$$

**Projection Theorem**

$$\vec{V} = \frac{<\vec{J}\cdot\vec{V}>_{k,j}}{\langle\overrightarrow{J^2}\rangle_{k,j}}\,\vec{J}$$

**3.3 Superposition and the Pauli Exclusion Principle (1924) [69]**

**3.3.1 Superposition**

The superposition principle leads to the probabilistic interpretation of quantum mechanics.

The states of a quantum mechanical system form a linear space (linear space also called a vector space) mathematically a linear space satisfies certain properties:

$\underset{\smile}{u}, \underset{\smile}{V} \rightarrow \lambda, \underset{\smile}{U} + \lambda_2 \underset{\smile}{V}$ linear superposition.

$\underset{\smile}{u}, \underset{\smile}{V} = \underline{V} + \underline{U}$ commutative

$\underline{u} + \{\underline{V} + \underline{\omega}\} = (\underline{u} + \underline{v}) + \omega$ associative

$\underline{u} + \underline{O} = \underline{u}$ there exist a $\underline{O}$ element (identity)

$(a+b)\underline{u} = a\underline{u} + b\underline{u}$ distributive property on scalar multiplication

$a(b\underline{u}) = (ab)\underline{u}$

$a(\underline{u} + \underline{v}) = a\underline{u} + a\underline{v}$

$1\underline{u} = \underline{u}$ identity

$O\underline{u} = \underline{O}$

Scalar product : $(\underline{u}, \underline{v}) = \lambda$

Properties $(\underline{u}, \underline{v})^* = (\underline{v}, \underline{u})$

$(\underline{u}, \lambda, \underline{u}_1 + \lambda + \lambda_2 \underline{v}_2) = \lambda_1(\underline{u}, \underline{v}_1) + \lambda_2(\underline{u}, \underline{v}_2)$
$(\lambda_1, \underline{u}_1 + \lambda_2 \underline{v}_2) = \lambda_1^*(\underline{u}_1, \underline{v}) + \lambda_2^*(\underline{u}_2, \underline{v})$
$(\underline{u}, \underline{u}) \geq 0$ when $(\underline{u}, \underline{u}) = 0 \quad then\, \underline{u} = 0$
In analogy to the norm on a space $(\underline{u}, \underline{u}) = \|\underline{u}\|^2$.

### 3.3.2 Pauli Exclusion

The Pauli exclusion principle is that a quantum state occupancy in systems with identical particles of spin ½ (electrons) can be either 0 or 1 and no more, due to exclusion [69]. This rule, formulated to some extent in spectral analysis applications in 1924 and earlier, was able to succinctly explained the spectra observed. Later, in 1940, Pauli would extend the spin-exclusion principle to a more general spin-statistics theorem applicable to the fermion particles and boson particles in general [70]. Further discussion of spin-statistics in Section 7.3.

## 3.4 Schrodinger Wavefunction

### 3.4.1 Obtaining the Schrodinger Equation from Correspondence and deBroglie relation

We can obtain the Schrodinger equation from operator substitution (deBroglie rules) in the classical (non-relativistic) equation:

$$E = \frac{p^2}{2m} + V(r) \Rightarrow i\hbar\frac{\partial}{\partial t} = \frac{\left(\frac{\hbar}{i}\nabla\right)^2}{2m} + V(R)$$

Thus

$$i\hbar\frac{\partial}{\partial t}\varphi(\vec{r}, t) = \frac{-\hbar^2}{2m}\nabla^2\varphi(\vec{r}, t) + V(\vec{r}, t)\varphi(\vec{r}, t)$$

with Stationary solution:

$$\varphi(\vec{r}, t) = \phi(\vec{r})e^{-i\frac{E}{\hbar}t} \Rightarrow [\frac{-\hbar^2}{2m}\nabla^2 + V(\vec{r})]\phi(\vec{r}) = E\phi(\vec{r})$$

Note the similarity to the electromagnetic field in one dimension (optical analogy).

***Optical Analogy (1D)***

$$\left[\frac{d^2}{dx^2} + \frac{2m}{\hbar^2}(E - V)\right]\phi(x) = 0$$

becomes:

$$\left[\frac{d^2}{dx^2} - \frac{1}{(^c/_n)^2}\frac{d^2}{dt^2}\right]E(x,t) = 0 \;\Rightarrow\; \left[\frac{d^2}{dm^2} + \frac{n^2\Omega^2}{c^2}\right]E(x) = 0$$

if

$$\frac{2m}{\hbar^2}(E - V) = \frac{n^2\Omega^2}{c^2}$$

Note that duality is not complete: the evanescent electromagnetic wave oscillates as it dampens, while the quantum mechanical wave just has damping.

***Gaussian wave packet evolution***

Suppose we have a solution to Schrodinger's wave equation that consists of a (unit norm) gaussian superposition of wave solutions:

$$\varphi(x,t) = \frac{\sqrt{a}}{(2\pi)^{3/4}}\int_{-\infty}^{\infty} e^{-\frac{a^2}{4}(k-k_0)^2} e^{i[kx-w(k)t]}dk$$

which can be integrated:

$$\varphi(x,t) = \left(\frac{2a^2}{\pi}\right)^{1/4} \frac{e^{r\phi}}{\left(a^4 + \frac{4\hbar^2t^2}{m^2}\right)^{1/4}} e^{ik_0x} \exp\left\{\frac{-\left[x - \frac{\hbar k_0}{m}t\right]^2}{a^2 + \frac{2i\hbar t}{m^2}}\right\}$$

$$\text{where} \quad \phi = -\theta - \frac{\hbar k_0^2 t}{2m} \quad \textit{and} \quad \tan\theta = \frac{2\hbar t}{ma^2}.$$

From this we have:

$$\text{velocity: } v_0 = \frac{\hbar k_0}{m}$$

$$\text{width: } \Delta x(t) = \frac{a}{2}\sqrt{1 + \frac{4\hbar^2t^2}{m^2a^4}}$$

For stationary states in 1d potentials: $\varphi(x)$ is a solution to the Schrodinger equation in a given region and
$\varphi(x)$ and its at boundary between different regions must match. We will examine a number of problems of this type in Section 4.3.5 and 3.6.

***Canonical commutation relations***

$$[R_i, R_j] = 0$$
$$[P_i, P_j] = 0$$
$$[R_i, P_j] = i\hbar\delta_{ij}$$

### 3.4.2 A more complete Schrodinger Equation analysis

***The Time-dependent Schrodinger equation***

$$i\hbar\frac{\partial \Psi(\vec{r},t)}{\partial t} = \left[-\frac{\hbar^2}{2m}\nabla^2 + V(\vec{r},t)\right]\Psi(\vec{r},t)$$

For conservative systems: $V = V(\vec{r}),\ \ \Psi_E(\vec{r},t) = \Psi_E(\vec{r})\exp\left(-\frac{i}{\hbar}Et\right)$ and a general solution has the form:

$$\Psi(\vec{r},t) = \sum_E C_E\,\Psi_E(\vec{r})\exp\left(-\frac{i}{\hbar}Et\right)$$

This reduces to a Time-independent Schrodinger equation

***The Time-independent Schrodinger equation***

$$\nabla^2\Psi_E(\vec{r}) + \frac{2m}{\hbar^2}\left(E - V(\vec{r})\right)\Psi_E(\vec{r}) = 0$$

***Transmission of particles through potential Barriers***

Recall the probability current density is related to wavefunction by:

$$\vec{J}(\vec{r},t) = Re\left[\frac{\hbar}{im}\Psi^*(\vec{r},t)\nabla\Psi(\vec{r},t)\right]$$

Where $|\vec{J}_I| = |\vec{J}_T| + |\vec{J}_R|$ and the standard transmission and reflection coefficients are given by:

$$T = \frac{|\vec{J}_T|}{|\vec{J}_I|},\ R = \frac{|\vec{J}_R|}{|\vec{J}_I|},\qquad and \qquad T + R = 1.$$

***Motion in a central field***

$$H\Psi(r,\theta,\varphi) \equiv \left\{\frac{Pr^2}{2m} + \frac{L^2}{2mr^2} + V(r)\right\}\Psi(r,\theta,\varphi) = E\,\Psi(r,\theta,\varphi)$$

$$P_r = -ih\frac{1}{r}\frac{\partial}{\partial r}r \quad and \quad \vec{L}^2 = -\frac{\hbar^2}{sin^2\theta}\left[\sin\theta\frac{\partial}{\partial\theta}\left(\sin\theta\frac{\partial}{\partial\theta}\right) + \frac{\partial^2}{\partial\varphi^2}\right]$$

Thus,

$$\vec{L}^2 Y_l^m(\theta,\varphi) = l(l+1)\hbar^2 Y_l^m(\theta,\varphi)$$

$$l_z Y_l^m(\theta,\varphi) = m\hbar Y_l^m(\theta,\varphi)$$

$$\Psi_{E,l,m}(r,\theta,\varphi) = \frac{R_{E,l}(r)}{r} Y_l^m(\theta,\varphi)$$

$$\frac{d^2R_{E,l}}{dr^2} + \frac{2m}{\hbar^2}\left[E - \left(V(r) + \frac{l(l+1)\hbar^2}{2mr^2}\right)\right] R_{E,l} = 0 \ , \ \ R_{E,l}(0) = 0$$

***Separable potential***

If $V(\vec{r}) = V_1(x_1) + V_2(x_2) + V_3(x_3)$ we can decompose the Schrodinger equation as :

$\Psi(\vec{r}) = \Psi_1(x_1)\,\Psi_2(x_2)\,\Psi_3(x_3)$ and $E = E_1 + E_2 + E_3$

Time-independent Wave functions, i.e. solutions of the Schrodinger equation, correspond to bound or to unbound states according to whether they vanish or are merely bounded at infinity. In one dimensional problem, the energy spectrum of bound states is always non-degenerate.

***Oscillation theorem***

If the discrete eigenvalues of a one-dimensional Schrodinger equation are placed in order of increasing magnitude, $E_1 < E_2 < \cdots < E$ ...., then the corresponding Eigen functions will occur in increasing order of the number of zeroes, the n$^{th}$ Eigenfunction having n-1 zeros.

***Even-function potentials***

When the Hamiltonian, $H(x) = -\frac{\hbar^2}{2m}\frac{d^2}{dx^2} + V(x)$, has a potential that is an even function of $x, i.e. V(x) = V(-x)$ we have:

$$H\,\Psi(x) = E\,\Psi(x), H\,\Psi(-x)\,,$$

where $\Psi(x)\ and\ \Psi(-x)$ have same E. This results in two cases:

Case 1: E is non-degenerate: $\Psi(x) = C\Psi(-x) \rightarrow \Psi(x) = \pm\Psi(-x)$. Thus, the eigenfunctions are either even or odd. If all the energy levels are non-degenerate, then, if we write the energy eigenvalues in increasing order of magnitude, $E_1<E_2<E_3$..., the corresponding Eigenfunctions will occur in increasing order of the number of their zeros, the function corresponding to $E_n$ having n-1 zeros. Since the even (odd) functions have an even (odd) number of zeros, it follows that the Eigenfunctions will be alternatively even and odd, the ground state being always even.

Case 2: E is degenerate: The general solution to $H\varphi(x) = E\varphi(x)$ is then

$$\varphi(x) = C_1 \Psi(x) + C_2 \Psi(-x),$$

thus,

$$A[\Psi(x) + \Psi(-x)] + B[\Psi(x) + \Psi(-x)] = A\,\Psi_e(x) + B\,\Psi_o(x)$$

where $A + B = C_1$ *and* $A - B = C_2$, thus, the two Eigenfunctions having the same eigenvalue can be written in the form of a linear combination of two functions of well-defined parity, which are themselves Eigenfunctions with the same eigenvalue.

***Mean value of dynamical variable***

$$< A >= \frac{< \Psi, A\Psi >}{< \Psi, \Psi >}$$

***Uncertainty relations***

$$\Delta u \equiv \sqrt{< (u - < u >)^2 >} \quad = \quad \sqrt{< u^2 > - < u >^2}$$

Note that the relation:

$$\Delta U \Delta V \geq \frac{1}{2}\hbar \left| < -\frac{i}{\hbar}[U, V] > \right| \quad \rightarrow \qquad \Delta q_k \Delta p_k \geq \frac{1}{2}\hbar$$

(to show, start will $B = U + i\lambda V$ ... ).

The "Time-energy uncertainty relations": $\Delta t \Delta E \geq \frac{1}{2}\lambda$ has a different interpretation: $\Delta E$ is the difference between two values $E_1$ and $E_2$ at two different moments of time $t_1$ and $t_2$ $(\Delta t = t_2 - t_1)$ and is **not** the uncertainty in the energy at a given moment of time. In general:

$$i\hbar \frac{d < A >}{dt} =< [A, H] > + i\hbar \langle \frac{\partial A}{\partial t} \rangle$$

### 3.4.3 The probabilistic interpretation of the wave function

Let $\Psi(\vec{r}, t)$ denote the wavefunction, a probability (density) amplitude for a particle's presence at position $(\vec{r}, t)$, where $|\Psi(\vec{r}, t)|$ describes the probability density. The Schrödinger equation is then:

$$i\hbar \frac{\partial}{\partial t} \Psi(\vec{r}, t) = -\frac{\hbar^2}{2m} \Delta \Psi(\vec{r}, t) + V(\vec{r}, t)\,\Psi(\vec{r}, t).$$

In order to have the interpretation of $|\Psi|^2$ as probability density. We must have:

$$\int d^3r |\Psi(\vec{r}, t)|^2 = 1.$$

This necessarily leads us to the study square-integrable functions (a set called $L^2$). $L^2$ has the structure of a Hilbert space. From the physical point of view $L^2$ is too large since wave functions which are actually considered possess contain properties of regularly. To begin, we consider the Ψ's which are everywhere defined, continuous and infinitely differentiable, A function that is truly discontinuous at a given point in space, mathematically, would have no physical difference from a similar, not discontinuous, potential, since no experiment enables us to have access to real phenomena on a very small scale, say $10^{-30}$m. It is also convenient to confine ourselves to wave functions which have a bounded domain (the lab, etc.) However it is defined, let's denote by $\mathcal{F}$ the set of Ψ composed of sufficiently regular functions of $L^2$. Thus, $\mathcal{F}$ satisfies all the criteria of a vector space: We have a scalar product defined on $\mathcal{F}$:

$$(\varphi, \Psi) = \int d^3r\varphi^*(\vec{r})\,\Psi(\overrightarrow{r})$$

If $(\varphi, \Psi) = 0 \rightarrow \varphi \; and \; \Psi$ are said to be orthogonal.

$\sqrt{(\Psi, \Psi)}$ is called the norm of $\Psi(\vec{r}) \; and \; (\Psi, \Psi) = 0 \quad only\,if \; \Psi(\vec{r}) \equiv 0$.

Recall the Schwartz Inequality :

$$|(\Psi, \Psi)| \leq \sqrt{(\Psi_1, \Psi_1)}\sqrt{(\Psi_2, \Psi_2)}$$

with equality only if $\Psi_1 \propto \Psi_2$ .

A Linear Operator associates with every Ψ ∈ $\mathcal{F}$another Ψ', the correspondence being linear:

$$\Psi'(\vec{r}) = A\,\Psi(\vec{r})$$

where

$$A[\lambda_1\,\Psi_1(\vec{r}) + \lambda_2\,\Psi_2(\vec{r})] = \lambda_1 A\,\Psi_1(\vec{r}) + \lambda_2 A\,\Psi_2(\vec{r}).$$

***Additional Notes***

Partly operator $\pi$: $\pi\,\Psi(x, y, z) = \Psi(-x, -y, -z)$

Multiply-by-x op $X$: $X\,\Psi = x\,\Psi$ (acting on Ψ ∈ $\mathcal{F}$ can transform into Ψ ∉ $\mathcal{F}$)

Differential with respect to x op $D_x$: $D_x\,\Psi = \frac{\partial}{\partial x}\,\Psi$ (again, acting on Ψ ∈ $\mathcal{F}$ can transform into Ψ ∉ $\mathcal{F}$)

**3.4.4 Examples**

1. An arbitrary complex function $f(x)$ in the interval $-L/2 < x < L/2$ that repeats with the period L can be expressed as a sum in the form $f(x) = \sum a_n e^{i2\pi mx/L}$. Where $a_n$ are complex coefficients and the integer n ranges from $-\infty\ to + \infty$.

(a) Show that the set of functions $\psi_n = \exp(i2\pi nx/L)$ is an orthogonal set, that is $\int_{-L/2}^{L/2} \psi_m^* \psi_n dx = 0$ if $m \neq n$.

(b) Let $f(x) = A$ in the interval $-d/2 < x < d/2$ and zero in the rest of the period. A is real constant. Find the coefficients $a_n$.

(c) Let $k_n = i2\pi n/L$ and $\Delta k = 2\pi/L$, so that $k_n = n\Delta k = k$ and $f(x) = \sum_{-\infty}^{\infty}(a_n)(2\pi/L)\exp(ikx)\Delta k$. Consider the limit as A and L approach infinity and d and $\Delta k$ becomes small in such a way the $Ad = 1$. Write $f(x)$ for part (b) using an integral instead of sum. This is one representation of a $\delta$ – function.

**Answer**

(a) $\psi_n = \exp(i2\pi nx/L)$

$\psi_m = \exp(i2\pi mx/L)$ $\qquad\qquad \psi_m^* = \exp(i2\pi mx/L)$

$$\int_{-L/2}^{L/2} \psi_m^* \psi_n dx = \int_{-L/2}^{L/2} \exp\left(i2\pi \, {}^{x}/_{L} \, (n-m)\right) dx = \frac{L}{\pi} \frac{\sin \pi(n-m)}{(n-m)}$$

Which is zero except for when $n = m$, in which ease we can use L'Hopital's rule to get 'L', thus:

$$\int_{-L/2}^{L/2} \psi_n^* \psi_n dx = L$$

(b) $f(x) = \sum_{-\infty}^{\infty} a_n \, e^{i2\pi nx/L} = A$ in the interval $-d/2 < x < d/2$ and zero in the rest.

$$a_n = \frac{1}{v} \int_{-\infty}^{\infty} f(x) \, e^{i2\pi nx/L} dx = A \int_{-d/2}^{d/2} e^{-i2\pi nx/L} \, dx$$

$$= A\left(\frac{1}{-i2\pi nx/L}\right)\left(e^{-i\pi nd/L} - e^{i\pi nd/L}\right)$$

$$= \frac{AL}{Ln\pi} \cdot \sin(\pi nd/L)$$

$a_0 = Ad$ etc.

(c) $f(x) = \sum_{-\infty}^{\infty} a_n \frac{L}{2\pi} \exp(ikx)\, \Delta k = A$ , using part (b).

$$dA = \sum_{-\infty}^{\infty} d \left[\left(\frac{AL}{n\pi}\right) \cdot \sin(\pi nd/L)\right] \frac{L}{2\pi} \exp(ikx) \Delta K$$

$$1 = \int_{-\infty}^{\infty} \frac{L}{2\pi} \cdot \frac{\sin(\pi n d/L)}{\pi n/L} \exp(ikx) dK$$

$$1 = \int_{-\infty}^{\infty} \frac{L}{2\pi} \cdot d \cdot \exp(ikx) dk = \frac{A}{L} \frac{Ld}{2\pi} \int_{-\infty}^{\infty} e^{ikx}\, dk$$

2. The energy E of a relativistically moving particle is related to the momentum p and rest mass m by the relation $E^2 = \langle pc \rangle^2 + (mc^2)^2$ . The wave function $\psi(x)$ representing such a particle then satisfies the equation

$$-\hbar^2 \frac{\partial^2 \psi}{\partial t^2} = -\hbar^2 \frac{\partial^2 \psi}{\partial x^2} + (mc^2)^2 \psi.$$

(a) For waves of the form $\psi(x) = \exp[i(kx - wt)]$, what is the dispersion relation, that is, the relation between w and k?
(b) What is the group velocity $v_a$ of a wave packet formed from continuous waves in a narrow band of frequencies?
(c) Assume $p = \hbar k$, and find $v_a$ in terms of p, m, and c. find the nonrelativistic limit for $v_g$, when $pc \ll mc^2$.

**Answer**

(a) substituting in the wave equation we obtain

$$-\hbar^2 \frac{\partial^2 \psi}{\partial t^2} = -\hbar^2 c^2 + (mc^2)^2 \psi$$

*and with* $\psi(x) = \exp\big(i(kx - wt)\big) \rightarrow \hbar^2 w^2 \psi(x) = \hbar^2 c^2 k^2 \psi(x) + (mc^2)^2 \psi(x)$

So,

$$\hbar^2\, \omega^2 = \hbar c^2\, k^2 + m^2\, c^4 \quad \text{or} \quad \boxed{\omega^2 = c^2\, k^2 + \frac{m^2 c^4}{\hbar^2}} \qquad (*)$$

(b) by definition, $v_g = \frac{d\omega}{dk}$. Differentiating $(*)$ we get

$$\omega d\omega = c^2\, k dk \Rightarrow v_g = \frac{dw}{dk} = c^2\, \frac{k}{w} = c\, \frac{h}{\sqrt{k^2 + \frac{m^2 c^2}{\hbar^2}}}$$

(c) If $p = \hbar k$, we have $\boxed{v_g = c\, \frac{P}{\sqrt{p^2 + m^2\, c^2}}}$

When $pc \ll mc^2$, $\frac{P}{mc} \ll 1$, and so $v_g = \frac{P}{m} \frac{1}{\sqrt{\perp + \frac{p^2}{m^2\, c^2}}} = \frac{P}{m}\,(1 + \cdots)$

Therefore, in the non-relativistic limit $\boxed{v_g = \frac{P}{m}}$

3. (a) A free particle of mass m moves in the x direction. Write the Hamiltonian operator $\hat{H}$ in terms of
   x derivatives with respect to x.
   (b) The unnormalized wave function is $\varphi(x) = \exp(ikx)$. Where k is a real constant. What is the energy E?
   (c) Find $(\hat{H}\hat{x} - \hat{x}\hat{H})\,\psi(x)$ for the wave function in (b)
   (d) Express the operator $(\hat{H}\hat{x} - \hat{x}\hat{H})$in terms of $\hat{p}$ as the only operator.

**Answer**

a. $\hat{H} = \frac{\hat{p}^2}{2m} = \frac{1}{2m}\left(-i\hbar\frac{\partial}{\partial x}\right)^2 = \boxed{-\frac{\hbar^2}{2m}\frac{\partial^2}{\partial x^2}}$

b. $\hat{H}\psi(x) = \frac{\hbar^2}{2m}\frac{\partial^2}{\partial x^2}\,e^{ikx} = \frac{\hbar^2k^2}{2m}\,e^{ikx} \Rightarrow \boxed{E\frac{\hbar^2k^2}{2m}}$

(eigenvalue equation is linear, so normalization is irrelevant)

c. $\hat{H}\hat{x}\psi = -\frac{\hbar^2}{2m}\frac{\partial^2}{\partial x^2}\left(xe^{ikx}\right) = -\frac{\hbar^2}{2m}(2ik - xk^2)e^{i\,kx}$

$\hat{x}\hat{H}\psi = \frac{\hbar^2k^2}{2m}xe^{ikx} \Rightarrow (\hat{H}\hat{x} - \hat{x}\hat{H})\psi = \boxed{-i\frac{\hbar^2\,k}{m}\,\psi}$

d. $\hat{H}\hat{x}\psi = \frac{\hbar^2}{2m}\frac{\partial^2}{\partial x^2}(x\psi) = -\frac{\hbar^2}{m}\frac{\partial\psi}{\partial x} - \frac{\hbar^2}{2m}\times\frac{\partial^1\psi}{\partial x^2}$

$\hat{x}\hat{H}\psi = -\frac{\hbar^2}{2m}\times\frac{\partial^1\psi}{\partial x^2} \Rightarrow (\hat{H}\hat{x} - \hat{x}\hat{H})\psi = -\frac{\hbar^2}{m}\frac{\partial}{\partial x}\psi$

And since $\hat{p} = -i\hbar\frac{\partial}{\partial x} \Rightarrow \hat{H}\hat{x} - \hat{x}\hat{H} = \boxed{-i\frac{\hbar\hat{p}}{m}}$

### 3.4.5 "Step" Potentials Overview

Let's consider a potential step (or barrier), with the optical analogy we have:

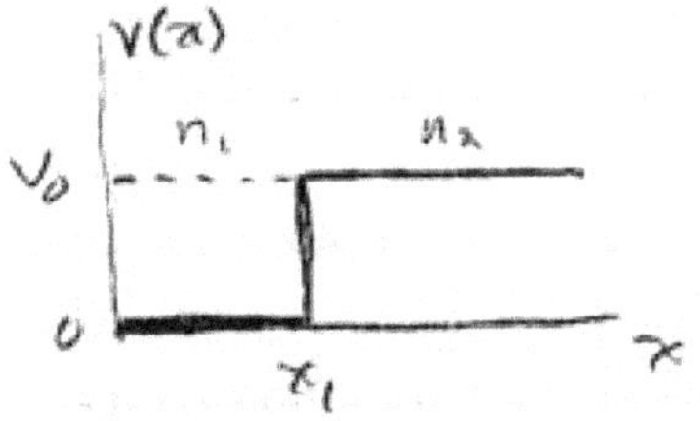


where:

$$n_1 = \frac{c}{\Omega\hbar}\sqrt{2mE}$$

$$n_2 = \frac{c}{\Omega\hbar}\sqrt{2m(E - V_0)}\,.$$

***If the energy of the incident wave is greater than the Potential Step:***
From the optical analogy, For an electromagnetic wave we have an incident wave coming from the left that splits into a reflected wave and transmitted wave. If we transpose to quantum mechanics (Quantum Mechanics) we have a probability amplitude wave that does the same, where the probability of being reflected is, say, P, and the probability of being transmitted is then (necessarily) 1-P.

***If the energy of the incident wave is less than the Potential Step:***
For the optical scenario there is total reflection, although there is penetration of electromagnetic field into the potential region with an exponential fall-off (the 'evanescent wave'). For the Quantum Mechanics scenario, the mere existence of a similar (non-oscillatory) evanescent wave means there is a non-sero *probability* of being transmitted into the region where classically the quantum of energy would never reside. This latter case is all the more apparent when we consider a potential barrier of finite extent (the step being a potential barrier of infinite extent), where the quantum transmittance is described as quantum 'tunneling'. Here is the potential barrier sketch:

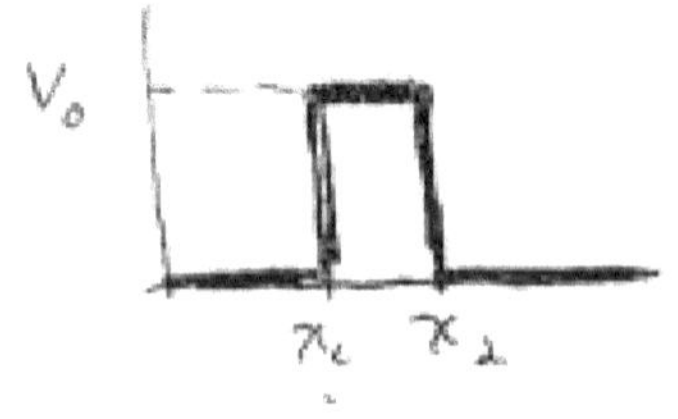


For the Potential Barrier we have the more interesting prospect of a tunneling effect, especially if the barrier is thin.

Another scenario is the potential well:

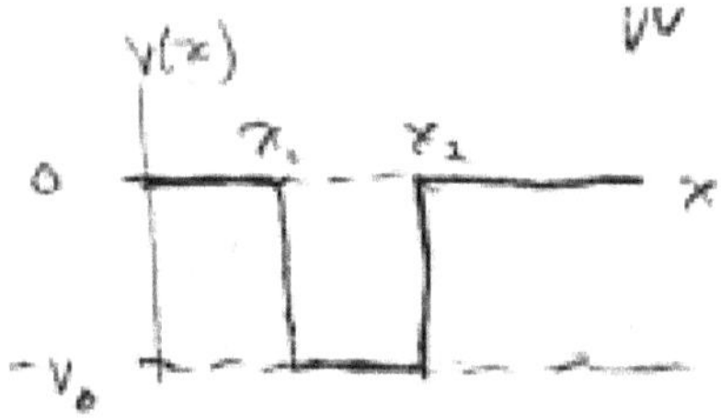

Classical mechanics predicts that a particle with energy $(-V_0 < E < 0)$, in such a potential, can only oscillate between $x_1$ and $x_2$. When the particle has positive energy and arrives from the left it undergoes an abrupt acceleration at $x_1$, then an equal deceleration at $x_2$, and then continues to the right.

Let's label the refractive indices for the regions before during and after the potential well. For $(-V_0 < E < 0)$ the $n_1$ and $n_3$ are imaginary.

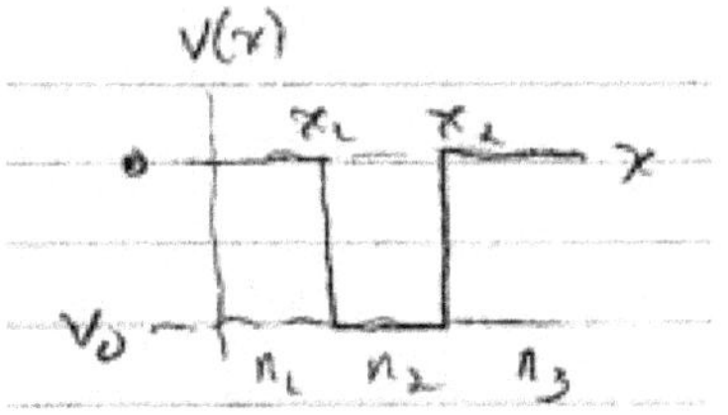


So, this situation is analogous to a layer of air between two reflecting media, with only "normal modes" with stable stationary waves allowed to form – classically the negative energies are thus quantized due to the boundary conditions (i.e., a system that isn't 'free'). If $(E < 0)$, then the indexes are real and the classical situation is analogous to optics for a layer of glass in air. In order to arrive at the reflected and transmitted waves it is necessary to consider and superpose an infinite number of waves that arise from successive reflections at the media-transitions (similar to the Fabry-Perot calculation mirror calculation). For certain incident frequencies the wave is entirely transmitted. From the quantum mechanical point of view there exist "resonant frequencies" for which the probability of transmission is 1.

Let's now specifically consider solutions to the 1D Schrodinger Equation for a variety of potentials to demonstrate the aforementioned high-level optical analogy overview with exact results. First, consider the square potential barrier, and make use of the time-independent Schrödinger Equation:

$$\left[\frac{-\hbar^2}{2m}\Delta + V(r)\right]\varphi(r) = E\varphi(r)$$

Which, in 1D, becomes:

$$\frac{d^2}{dx^2}\varphi(x) + \frac{2m}{\hbar^2}(E - V)\varphi(x) = 0.$$

Let's break this into the cases appropriate to energies above or below the barrier height.

Case (i) E>V: define $E - V = \frac{\hbar^2 k^2}{2m} \rightarrow \varphi(x) = Ae^{ikx} + A'e^{-ikx}$
Case (ii) E<V: define $V - E = \frac{\hbar^2 p^2}{2m} \rightarrow \varphi(x) = Be^{px} + B'e^{-px}$
Case (iii) E = V: $\rightarrow \varphi(x) = Cx + D$.

The full solution is found by matching the functions of the cases above by requiring $\varphi(x)$ and $d\varphi/dx$ to be continuous. Now to examine specific cases.

### 3.4.5.1 The simple step potential

Let's now consider the potential step in detail :

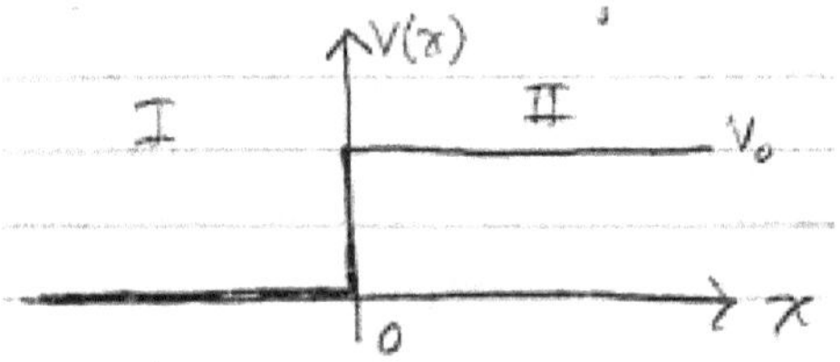


Case (i) $E > V$ :

$$\Psi(x,t) = \varphi(x)e^{-i\frac{E}{\hbar}t}$$

Potential step

$$E = \frac{P^2}{2m} + V \quad \rightarrow \quad i\hbar\frac{\partial \Psi}{\partial t} = \frac{-\hbar^2}{2m}\frac{\partial^2 \Psi}{\partial t} + V(x)\,\Psi(x)$$

We thus have the time-independent equation:

$$\frac{d^2}{dx^2}\varphi(x) + \frac{2m}{\hbar^2}(E - V)\varphi(x) = 0$$

Region I

$$\varphi(x) = Ae^{ikx} + A'e^{-ikx} \quad where \quad k = \sqrt{\frac{2mE}{\hbar^2}}$$

Region II

$$\varphi((x) = Be^{ik'x} + B'e^{-ik'x} \quad where \quad k' = \sqrt{\frac{2m(E - V_0)}{\hbar^2}}.$$

There are two degenerate solutions, consider only the wave coming from negative infinty, i.e., B'=0: $A + A' = B$ and $K(A - A') = K'B \quad \rightarrow$
$A - A' = \frac{K\prime}{k}B$ . Thus

$$2A = \left(1 + \frac{k'}{k}\right)B \;\rightarrow\; \frac{B}{A} = \frac{2k}{k + k'}$$

Thus

$$\frac{A'}{A} = \frac{B}{A} - 1 = \frac{K - K'}{K + K'}$$

And we have ($T \neq \left|\frac{B}{A}\right|^2$):

$$R = \left|\frac{A'}{A}\right|^2 \;\; ; \; T = \frac{K'}{K}\left|\frac{B}{A}\right|^2$$

Case (ii) $E < V_0$

Region I: $\varphi(x) = Ae^{ikx} + A'e^{-ikx}$

Region II: $\varphi(x) = Be^{px} + B'e^{-px}$

Where the $B'e^{-px}$ term describes the evanescent wave, and

$$p = \sqrt{\frac{2m(V_0 - E)}{\hbar^2}}$$

If we replace k' with ip and B with B' in the previous analysis:

$$\frac{B'}{A} = \frac{2k}{k + ip}$$

$$\frac{A'}{A} = \frac{K - ip}{K + ip} \;\rightarrow\; R = \left|\frac{A'}{A}\right|^2 = 1$$

Note: A'/A is complex. A phase shift appears upon reflection, which is due to the fact that the particle is delayed when it penetrates the x>0 region. Phase shift is analogous to that for light reflected from a metallic substance.

***Simple step with wave-packet***

Let's now reconsider the behavior of a **wave-packet** at a potential step (not a simple plane wave):

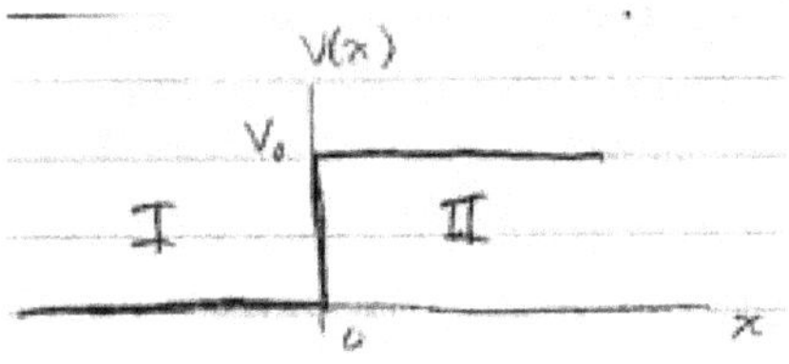


We have total reflection if $E < V_0$ , adopt solutions in the form:

$$\varphi_I(x) = A_1e^{ikx} + A_1'e^{-ikx}$$
$$\varphi_{II}(x) = A_2e^{ikx} + A_2'e^{-ikx}$$

where

$$K = \sqrt{\frac{2mE}{\hbar^2}} = K_1 \ previously \text{ and } K_0 = \sqrt{\frac{2mE}{\hbar^2}} = K_0 \ prev$$

$$\frac{A_1'}{A_1} = \frac{K_1 - K_2}{K_1 + K_2} \quad and \quad \frac{A_2}{A_1} = \frac{2k_1}{K_1 + K_2}$$

Construct a wave packet with $K < K_0$ $(for\ E < V_0\ case)$ :

Recall $E < V_0 \ \rightarrow \ p = \sqrt{\frac{2m(V_0-E)}{\hbar^2}}$ $and \ \varphi_{II}(x) = B_2 e^{p_2 x} + B_2' e^{-p_2 x}$,

where $B_2 = 0$ from boundary condition at positive infinity. Thus:

$$\frac{A_1'}{A_1} = \frac{K_1 - ip}{K_1 + ip} \quad and \quad \frac{B_2'}{A_1} = \frac{2k_1}{K_1 + ip}$$

And using the notation:

$$\frac{A_1'(k)}{A_1(k)} = e^{-2i\theta(k)} \rightarrow \tan\theta(k) = \frac{\sqrt{K_0^2 - K^2}}{K}.$$

We can now succinctly describe the wave solution in the region preceding the potential step (negative x region) as a combination of incident and reflected wave where their relative amplitudes are given by the above $\frac{A_1'(k)}{A_1(k)}$ relation. Consider a time evolved free wavepacket if there were no potential step:

$$\Psi(x,t) = \frac{1}{\sqrt{2\pi}} \int_{-\infty}^{\infty} g(k) e^{i[kx - \omega(k)t]}\, dk$$

At time t=0 we have:

$$\Psi(x,0) = \frac{1}{\sqrt{2\pi}} \int g(k) e^{i[kx]}\, dk$$

Let's now modify the solution for wavepacket incident from the left with the potential present such that there is reflection as indicated. The wave packet we are going to consider can, thus, be written, at time t=0, for negative x, as:

$$\Psi(x,0) = \frac{1}{\sqrt{2\pi}} \int dk g(k) \left[e^{ikx} + e^{-2i\theta(k)} e^{-ikx}\right] dk$$

Assume $|g(k)|$has a peak at $K = K_0$ of width $\Delta K$. Thus

$$\Psi(x,t) = \frac{1}{\sqrt{2\pi}} \int_0^{K_0} dk g(k) e^{i[kx-\omega(k)t]}$$

$$+ \frac{1}{\sqrt{2\pi}} \int_0^{K_0} dk g(k) e^{-i[kx-\omega(k)t+2\theta(k)]}$$

where $\omega(k) = \hbar k^2/2m$. By construction, the above is valid only for negative x.

**Stationary phase analysis**

Let's now consider the stationary phase analysis on $\Psi(x,t)$ with care:

$$X_{incident} = t\left[\frac{d\omega}{dk}\right]_{K=K_0} = \frac{\hbar k_0}{m} t \quad where\ V_\varphi = \frac{\omega}{k} \quad and \quad V_g = \frac{d\omega}{dk}$$

A trick is used to get $X_{reflected} = X_r$: differentiate $\tan\theta(k) = \frac{\sqrt{K_0^2 - K^2}}{K}$:

$$[1 + \tan^2\theta(k)]d\theta = \left[1 + \frac{K_0^2 - K^2}{K^2}\right] d\theta$$

$$= -\frac{dk}{K^2}\sqrt{K_0^2 - K^2} - \frac{dk}{\sqrt{K_0^2 - K^2}}$$

or

$$\frac{K_0^2}{K^2} d\theta = -\frac{K_0^2}{K^2} \frac{1}{\sqrt{K_0^2 - K^2}}$$

Thus,

$$X_r = -\left[\frac{d\omega}{dk} + 2\frac{d\theta}{dk}\right]_{K=k} = \frac{-\hbar K_0}{m} t + \frac{2}{\sqrt{K_0^2 - k_{\square}^2}}$$

And for negative time, there is no reflected wave packet.

Reflection introduces a delay $\tau$:

$$\tau = -2\left[\frac{d\theta/dk}{d\omega/dk}\right]_{K=k} = \frac{2m}{\hbar k_0 \sqrt{K_0^2 - k_{\square}^2}}$$

For an unbounded plane wave the "delay" is simply $\theta(k)$ which is just an overall phase shift upon reflection. For the delay of a wave packet this manifests in a more physical fashion, however, where the wave packet "spends" a time of order $\tau$ as an evanencent wave in the x>0 region.

Expand around $K=K_0$

$$\left\{-\left[\frac{d\theta}{dk}\right]_{K=K_0}-\left[\frac{d\omega}{dk}\right]_{K=K_0}t\right\}(K-K_0)=-\frac{\hbar K_0}{m}(K-K_0)\left(t-\frac{\tau}{2}\right)$$

$|\Psi|$ as a maximum for $t=\frac{\tau}{2}\rightarrow$ gives delay of $\tau$ as before. If $\Delta k$ is the width of $g(k)$, the waves go out of phase, and $|\Psi(x,t)|_{x>0}$ becomes negligible when;

$$\frac{\hbar K_0}{m}\Delta k\Delta t\geq 1$$

Thus, the wave packet as a whole remains in the x>0 region during an internal of time $\Delta t$ of order of

$$\Delta t=\frac{1/\Delta k}{\hbar k_0/m}$$

Which corresponds approximately to the time it takes, in the x<0 region, to travel a distance comparable to the width $1/\Delta k$.

Note: $\Delta t=\frac{1/\Delta k}{Kk_0/m}$ $and\ \tau=\frac{2m}{KK_0\sqrt{K_0^2-K_0^2}}$ .Since it is assumed $\Delta k\ll K_0$ then $\Delta t\gg\tau$ , thus, delay upon reflection involves a displacement which is much smaller than the wavepacket's width.

Let's now consider the partial reflection that results when encountering the potential step with $E>V_0$, where $g(k)$ has width $\Delta K\ at\ K=k_0>K_0, and\ g(k)$ is zero for $k<K_0$ :

$$\Psi(x,t)=\theta(-x)\frac{1}{\sqrt{2\pi}}\int_{K_0}^{\infty}dk g(k)e^{i[kx-\omega(k)t]}\qquad (A_1=1)$$

$$+\theta(-x)\frac{1}{\sqrt{2\pi}}\int_{K_0}^{\infty}dk g(k)A_1^1(k)e^{i[kx-\omega(k)t]}$$

$$+\theta(x)\frac{1}{\sqrt{2\pi}}\int_{K_0}^{\infty}dk g(k)A_2(k)e^{i\left[\sqrt{K^2-K_0^2}x-\omega(k)t\right]}$$

where

$$x_i=\frac{\hbar k_0}{m}t,\qquad x_r=\frac{-\hbar k_0}{m}t\ ,\ x_t=\hbar\frac{\sqrt{k_0^2-K_0^2}}{m}t.$$

For negative t, only $x_i$ exists. There is no delay, either upon reflection or transmission.

Assume $\Delta K$ to be small such that we can reflect the variation of $A_1'(k)$. Then, the reflected wave packet has the same form as the incident wave packet. Similarly, replace $A_2$ by $A_2(k_0)$ and expand $\sqrt{k^2 - K_0^2}$ :

$$\sqrt{k^2 - K_0^2} = \frac{\sqrt{k_0^2 - K_0^2}}{q_0} + (k - K_0)\frac{k_0}{q_0}$$

So,

$$\Psi_t(x,t) \cong A_2(k_0)e^{iq_0 x}\frac{1}{\sqrt{2\pi}}\int_{K_0}^{\infty} dkg(k)e^{i\left[(k-K_0)\frac{k_0}{q_0}x-\omega(k)t\right]}$$

Compare with $\Psi_i(x,t)$

$$|\Psi_t(x,t)| \cong A_2(K_0)\left|\Psi_i\left(\frac{K_0}{q_0}x,t\right)\right| \quad and \quad A_2(K_0) > 1$$

$(\Delta x)_t = \frac{q_0}{K_0}\Delta x \quad \leftarrow width\ less$

$T = \frac{q_0}{K_0}|A_0(K_0)|^2 \ \leq 1\ overall$

We can find the velocity:

$$V_t = \frac{\Delta x}{\Delta t}\frac{(\Delta x)_t}{\Delta x} = \frac{\hbar k_0}{m}\frac{q_0}{k_0} = \frac{\hbar q_0}{m}$$

***Simple step in two-dimensions (aka, oblique reflection, examined again in Section 4.1.2.5)***

Suppose steady-state two-dimensional waves represent the wave function of a particle of mass m and energy E scattered by a discontinuous potential step: $V(x) = V_0$ for $x < 0$ and $V(x) = 0$ for $x > 0$. Assume $E > V_0$ and $V_0 > 0$.

(a) Given the direction $\theta_1$ of the incident wave in the region $V(x) = V_0$, find the directions $\theta_1$ and $\theta_2$ of the reflected and transmitted waves.
(b) Find the probability that the particles will be transmitted across the potential step.
(c) In the corresponding classical mechanics problem, an incident mass sliding without friction over the same potential step $V_0$ changes direction by amount that can be calculated using conservation of momentum and energy. Find $\theta_2$ and $\theta_1$ and compare with the quantum mechanical result in (a).

**Answer**

(a) Regard the figure:

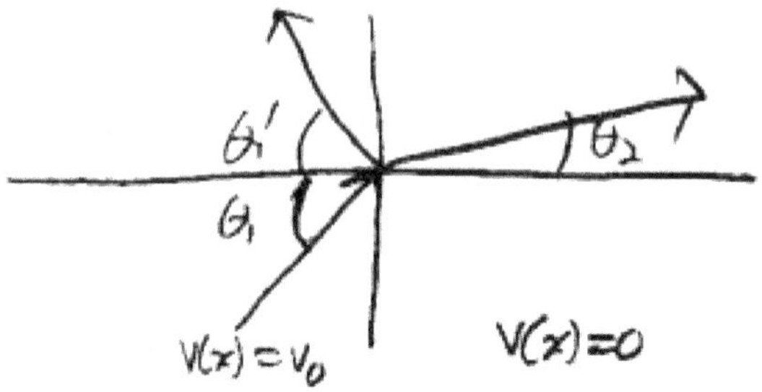


$\theta_1' = \theta_1$ and $\frac{\hbar^2 k_1^2}{2m} = E - V_0$ , $\frac{\hbar^2 k_2^2}{2m} = E \rightarrow k_1 = \sqrt{\frac{2m(E-V_0)}{\hbar^2}}$ *and* $k_2 = \sqrt{\frac{2m\,E}{\hbar^2}}$. We also have:

$$k_2 \sin\theta_1 = k_1 \sin\theta_1$$

Thus

$$\theta_2 = \sin^{-1}\left(\frac{k_1}{k_2}\sin\theta_1\right) = \sin^{-1}\left(\sqrt{1-\frac{V_0}{E}}\cdot\sin\theta_1\right)$$

(b) $A + B = C$

$A - B = \frac{k_2}{k_1}\, C$

$\frac{C}{A} = \frac{2}{1+{k_2}/{k_1}}$

$T = \left|\frac{C}{A}\right|^2 \frac{k_2}{k_1} = \frac{2}{1+{k_2}/{k_1}}\left(\frac{k_2}{k_1}\right) = \frac{2}{\sqrt{1+\frac{v_0}{E}}+1}$

Probability or transmission, $p$, is $p = \frac{2}{\sqrt{1+\frac{v_0}{E}}+1}$

(c) $mv_1 \sin\theta_1 = mv_2 \sin\theta_2$ and $\frac{(mv_1)^2}{2m} + v_0 = \frac{(mv_2)^2}{2m} \rightarrow v_1 = \sqrt{v_2^2 - 2\,v_0/m}$. Thus

$$\theta_2 = \sin^{-1}\left(\frac{v_1}{v_2}\sin\theta_1\right)$$

$$\theta_2 = \sin^{-1}\left(\frac{\sqrt{v_2^2-\frac{2v_0}{m}}}{v_2^2}\sin\theta_1\right)$$

$$\theta_2 = \sin^{-1}\left(\sqrt{1-\frac{v_0}{E}}\cdot\sin\theta_1\right)$$

The classical mechanics and quantum mechanics derivations give the same result.

### 3.4.5.2 The Barrier Potential

Let's now consider the square potential barrier in detail.

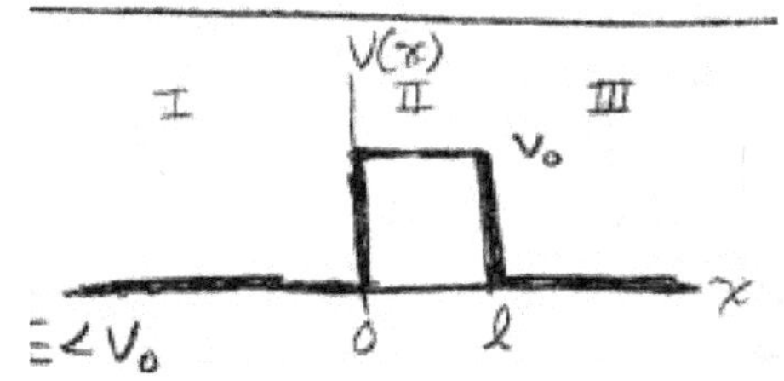


Case (i) $E < V_0$

I: $\varphi(x) = Ae^{ikx} + A'e^{-ikx}$

II: $\varphi(x) = Be^{ikx} + B'e^{-ikx}$

III: $\varphi(x) = Ce^{ikx} + C'e^{-ikx}$

$K = \sqrt{\frac{2mE}{\hbar^2}}$ and $K' = \sqrt{\frac{2m(E-V_0)}{\hbar^2}}$

Consider only the incoming wave: C'=0:

$$A + A' = B + B' \qquad Be^{ik\prime l} + B'e^{-ik\prime l} = Ce^{ikl}$$

$$K(A - A') = K'(B - B') \qquad K'\left(Be^{ik'l} - B'e^{-ik'l}\right) = KCe^{ikl}$$

$$2A = \left(1 + \frac{K'}{K}\right)B + \left(1 - \frac{K'}{K}\right)B' \qquad \left(1 - \frac{K'}{K}\right)Be^{ik\prime l}$$

$$+\left(1 + \frac{K'}{K}\right)B'^{e^{-ik'l}} = 0$$

$$B = \frac{-\left(1 + \frac{K'}{K}\right)}{\left(1 - \frac{K'}{K}\right)}e^{-2ik\prime l}B'$$

$$\frac{A}{C} = \frac{1}{2}\left\{\left(1 + \frac{K'}{K}\right)\frac{B}{C} + \left(1 - \frac{K'}{K}\right)\frac{B'}{C}\right\} = \frac{1}{2}\left\{\left(1 + \frac{K'}{K}\right)\frac{B'}{C}\frac{B}{B'} + \left(1 - \frac{K'}{K}\right)\frac{B'}{C}\right\}$$

$$\frac{A}{C} = \left\{\cos k'l - i\frac{(K^2 + K'^2)}{2KK'}\sin k'l\right\}e^{ikl}$$

Let's now find $\frac{A'}{C}$:

$$2A' = \left(1 - \frac{K'}{K}\right)B + \left(1 + \frac{K'}{K}\right)B' \rightarrow \frac{A'}{C}$$

$$= \frac{\left(1 - \frac{K'}{K}\right) + \left(1 + \frac{K'}{K}\right)\left\{-\frac{\left(1 - \frac{K'}{K}\right)e^{i2k\prime l}}{\left(1 + \frac{K'}{K}\right)}\right\}}{2\left[e^{i(k'-k)l} - \frac{\left(1 - \frac{K'}{K}\right)}{\left(1 + \frac{K'}{K}\right)}e^{i(k'-k)l}\right]}$$

$$\frac{A'}{C} = \frac{i(K^2 - K'^2)}{2KK'}\sin K'e^{-ikl}$$

Thus, we can now group:

$$\left(\frac{A'}{C}\right)\left(\frac{C}{A}\right) = \frac{A'}{A} = \frac{(K^2 - K'^2)\sin k'l}{-i2KK'\cos K'l - (k^2 + k'^2)sink'l}$$

And since $R = \left|\frac{A\prime}{A}\right|^2$ we get:

$$R = \frac{(K^2 - K'^2)sin^2K'l}{4KK^2cos^2K'l + [(K^2 - k^2)^2 + 4K^2K'^2]\sin K'l}$$

Thus,

$$R = \frac{(K^2 - K'^2)sin^2K'l}{4K^2K'^2 + (K^2 - K'^2)sin^2K'l}$$

Also

$$T = \left|\frac{C}{A}\right|^2 = \frac{4K^2K^{,2}}{4K^2K^{,2} + (K^2 - K'^2)^2sin^2K'l}$$

Note that $R + T = 1$ as it should.

Using $K = \sqrt{\frac{2mE}{\hbar^2}}$ and $K' = \sqrt{\frac{2m(E-V_0)}{\hbar^2}}$ we can rewrite as:

$$T = \frac{4E(E - V_0)}{4E(E - V_0) + V_0^2 sin^2\left(\sqrt{2m(E - V_0)} \cdot l/\hbar\right)}$$

Let's examine the Transmission = 1 resonance that occurs when:

$$\sqrt{2m(E - V_0)}\, l/\hbar = n\pi \qquad n = 0,1,2 \ldots$$

Such resonance occurs for energies:

$$E = \frac{1}{2m}\left(\frac{n\pi\hbar}{l}\right)^2 + V_0$$

Note that at such resonance, the wave packets will spend long times in the barrier.

Case (ii) $E < V_0$; Tunnel effect $K' \rightarrow -ip \quad then\ p = \sqrt{\frac{2m(V_0-E)}{\hbar^2}}$

We no have:

$$T = \left|\frac{C}{A}\right|^2 = \frac{4E(V_0 - E)}{4E(V_0 - E) + V_0^2 sinh^2\left[\sqrt{2m(V_0 - E)} \cdot l/\hbar\right]}$$

For $pl \gg 1$ :

$$T \cong \frac{16E(V_0 - E)e^{-2pl}}{V_0^2}$$

For $l \lesssim {}^1/_p$ we have considerable probability of the "tunnel effect". Consider the electron:

$$\left(\frac{1}{p}\right)_{el} = \sqrt{\frac{\hbar^2}{2m_e}}\frac{1}{\sqrt{V_0 - E}} = \frac{\sqrt{4.2\ x\ 10^{-21}m^2}}{\sqrt{V_0 - E}} = \frac{.65\dot{A}}{\sqrt{V_0 - E}}$$

***Barrier inside an infinite well***

Consider a particle of mass m is in a potential $V(x)$ with $V(x) = 0$ in $0 < x < L/3$ and $2L/3 < x$; $V(x) = V_0$ in $L/3 < x < 2L/3$. $V(x)$ is infinite outside of these ranges.

(a) Write the differential equation for the wave function $\psi(x)$ in the regions I, II, and III shown.
(b) Write the general form (as specific as you can) of the solutions to the equations in (a) in the three regions.

**Answer**

(a) Region I: $0 < x < {}^L/_3$ $\quad V = 0 \quad \frac{-\hbar^2}{2m}\frac{\partial^2\varphi}{\partial x^2} = E\varphi(x)$

Region II: ${}^L/_3 \leq x \leq {}^{2L}/_3$ $\quad V = V_0 \quad \frac{-\hbar^2}{2m}\frac{\partial^2\varphi}{\partial x^2} =$ $(E - V)\varphi(x)$

Region III: ${}^{2L}/_3 < x < L$ $\quad V = 0 \quad \frac{-\hbar^2}{2m}\frac{\partial^2\varphi}{\partial x^2} = E\varphi(x)$

For $x < 0$ and $x > L$ : $V = \infty$. Thus

$\varphi_I = Ae^{ik_1x} + Be^{-ik_1x}, \quad \frac{\hbar^2k_1^2}{2m} = E$

$\varphi_{II} = Ce^{ik_2x} + De^{-ik_2x}, \quad \frac{\hbar^2k_2^2}{2m} = E - V_0$

$\varphi_{III} = Ee^{ik_3x} + Fe^{-ik_3x}, \quad \frac{\hbar^2 k_3^2}{2m} = E$

Continuity relations:
$A + B = 0$
$Ae^{ik_1 L/3} + Be^{-ik_1 L/3} = Ce^{ik_2 L/3} + De^{-ik_2 L/3}$
And
$Ee^{ik_3 2L/3} + Fe^{-ik_3 2L/3} = Ce^{-ik_2 2L/3} + De^{-ik_2 2L/3}$
$E + F = 0$

Also.
$k_1 = k_3 \quad k_2^2 = \frac{2m}{\hbar^2}(E - V_0) = k_1^2 - \frac{2mV_0}{\hbar}$
$k_2 = \sqrt{k_1^2 - \frac{2mV_0}{\hbar}}$

Similar for derivative continuity:
$A = -B, E = -F$
$e^{ik_2 L/3} Ai \sin k_1 {}^{L}/_{3} = Ce^{ik_2 L/3} + D$
$Ei \sin k_1 {}^{2L}/_{3} = Ce^{ik_2 2L/3} + De^{-ik_2 2L/3}$

(b)

| $Energy/region$ | $I$ | $II$ | $III$ |
|---|---|---|---|
| $E_1$ $(E < V_0)$ | $\sin(k_1 x)$ | $\sinh k'_1 x$ $\cosh k_1 x$ | $\sin k_1(x - L)$ |
| $E_2$ $(E < V_0)$ | $\sin k_2 x$ | $\sin k'_1 x$ $\cos k'_2 x$ | $\sin k_2(x - L)$ |

Linear combination of these solution form the general solution in each of the regions then gives:

$$\frac{\hbar^2 k_1^2}{2m} = E_1; \frac{\hbar^2 k_2^2}{2m} = E_2; \frac{\hbar^2 k_1'^2}{2m} = V_0 - E_1; \frac{\hbar^2 k_2'^2}{2m} = E_2 - V_0$$

***Triangle Barrier***

(a) A beam of particles of mass m and energy $E = U_o + \frac{\hbar^2}{2m}\frac{\pi^2}{a^2}$ is incident from the left on a rectangular potential barrier.

$$U = \begin{cases} U_0 & 0 \le x \le a \\ 0 & elsewhere \end{cases}$$

Find the probability that a particle is transmitted.

(b) A beam of particles of mass m and energy $E = \frac{1}{2} U_o$ incident on the triangular barrier

$$U(x) = \begin{cases} U_o\left(1 - \frac{|x|}{a}\right), & |x| < a \\ 0 & elsewhere \end{cases},$$

Find the probability that a particle is reflected, assuming that the change in the potential is small over one wavelength. Only the exponential factor need be given.

(a) $E = u_o + \frac{\beta^2}{2m}$

I and II have the same energies so don't need

$$\left[\frac{-\hbar^2\nabla^2}{2m} + V(x)\right]\varphi = E\varphi$$

Region (I)b$\frac{-\hbar^2}{2m}\frac{d^2}{dx^2}\varphi(x) = E\varphi x$ let $K = \sqrt{\frac{2mE}{\hbar^2}}$

$\varphi = Ae^{ikx} + Be^{-ikx}$

(I) $\varphi_2(x) = Ae^{ikx} + Be^{-ikx}$

(II) $let\ p = \sqrt{\frac{2m(E-V)}{\hbar^2}}$

$\varphi_{II} = A'e^{ipx} + B'e^{-ipx}$

(III) $\varphi_{III} = A"e^{ikx} + B"e^{-ikx}$ outgoing only, so B"=0

$\varphi_{III} = A"e^{ikx}$

Matching the wavefunctions

$B + A = A' + B'$

$A'e^{ipa} + B'e^{-ipa} = A"e^{ika}$

$-ikB + ikA = ipA' + ipB$ $A'ipe^{-ipa} - B'ipe^{-ipa} =$

$A"ike^{ika}$

$2A'^{e^{ipa}} = A"e^{ika}\left(1 + \frac{k}{p}\right)$

Want to know $\left|\frac{A"}{A}\right|$ =?

$2B'^{e^{ipa}} = A"e^{ika}\left(1 - \frac{k}{p}\right)$

$2A = \left(i + \frac{p}{k}\right)A' + \left(1 - \frac{p}{k}\right)B'$ $A' = \frac{1}{2}e^{i(k-p)a}\left(1 + \frac{k}{p}\right)$

$$2A = \left\{\left(1 + \frac{p}{k}\right)\frac{1}{2}e^{i(k-p)a}\left(1 + \frac{k}{p}\right) + \left(1 - \frac{p}{k}\right)\frac{1}{2}e^{i(k+p)a}\left(1 - \frac{k}{p}\right)\right\}A"$$

$$\frac{A}{A"} = \frac{1}{4}\left\{\left(1 + \frac{p}{k}\right)\left(1 + \frac{k}{p}\right)e^{i(k-p)a} + \left(1 - \frac{p}{k}\right)\left(1 - \frac{k}{p}\right)e^{i(k+p)a}\right\}$$

(a)

$$\left|\frac{A^a}{A^a}\right| = \left|\frac{1}{4}\left\{\left(1+\frac{p}{k}\right)\left(1+\frac{k}{p}\right)e^{-ipa} + \left(1-\frac{p}{k}\right)\left(1-\frac{k}{p}\right)e^{ipa}\right\}\right|$$
$$1+\frac{p}{k}+\frac{k}{p}+1$$
$$= \frac{1}{4}\left|\left(2+\frac{p}{k}+\frac{k}{p}\right)e^{-ipa} + \left(2-\frac{p}{k}-\frac{k}{p}\right)e^{ipa}\right|$$
$$= \frac{1}{4}\left|2\left(e^{+ipa}+e^{-ipa}\right) + \left(\frac{p}{k}+\frac{k}{p}\right)\left(e^{-ipa}-e^{ipa}\right)\right|$$
$$p = \sqrt{\frac{2m\left(\frac{\hbar^2}{2m}\right)}{\hbar^2}} = \frac{\pi}{a}$$
$$k^2 = \frac{2m}{\hbar^2}\left(u_o + \frac{\hbar^2}{2m}\frac{\pi^2}{a^2}\right)$$
$$k^2 = \frac{2mu_o}{\hbar^2} + \frac{\pi^2}{a^2}$$
$$= \sqrt{\cos^2(pa) + i\left(\frac{(p^2+k^2)}{4k^2p^2}\right)\sin^2(pa)}$$
$$T = \left|\frac{A''}{A}\right|^2 = \frac{1}{\left\{\cos^2(\pi) + \left(\frac{2\left(\frac{\pi}{a}\right)^2 + \frac{2mu_o}{\hbar^2}}{4\left(\frac{\pi}{a}\right)^2\left[\left(\frac{\pi}{a}\right)^2 + \frac{2mu_o}{\hbar^2}\right]}\right)\sin^2(\pi)\right\}}$$
T=1 (resonant transmission)

(b) R=?

$$T = \text{ from gamow factor } \left(\frac{\hbar^2}{2m}\frac{d^2}{dx^2} + V(x)\right) = E\varphi$$
$$\varphi(x) = Ae^{-px}$$
$$T = e^{-2\int pdx}$$
So, $\varphi \cong e^{-\int pdx}$
$$p = \sqrt{\frac{2m(V-E)}{\hbar^2}}$$
$$R = 1 - T - 1 - \exp\left\{-2\sqrt{\frac{2m}{\hbar^2}}\int_{\frac{a}{2}}^{\frac{a}{2}}\sqrt{V(x) - \frac{1}{2}u_o dx} = 1 - \exp\left(-2\sqrt{\frac{2m}{\hbar^2}}\,2\int_0^{\frac{a}{2}}\sqrt{u_o - u_o\frac{x}{a} - \frac{1}{2}u_o\,dx}\right)\right.$$
$$\int_0^{\frac{a}{2}}\sqrt{\frac{1}{2}u_o - \frac{x}{a}u_o}\,dx = \sqrt{\frac{1}{2}u_o}\int_0^{a/2}\sqrt{1-\frac{x}{(a/2)}}\,dx$$
$$= \sqrt{\frac{1}{2}u_o}\left(-\frac{2}{3(\varphi)}\right)\left(1-\frac{x}{(a/2)}\right)^{3/2}\Big|_0^{a/2}\frac{d}{x}\left(1-\frac{1}{2}x\right)^{1/2} = \frac{3}{2}(1-b_x)^{1/2}(-b)$$
$$= \sqrt{\frac{1}{2}u_o}\,\frac{2}{3\left(\frac{2}{a}\right)} = \sqrt{\frac{1}{2}u_o}\,\frac{a}{3}$$

$$R = 1 - \exp\left(-\frac{4a}{3}\sqrt{\frac{mU_o}{\hbar^2}}\right)$$

**Double Barrier with Resonance**

Consider a beam of massive particles incident from the left on the potential barriers shown. Obtain the transcendental equation for the incident particle energies at which no particles are reflected by the barriers. You do not need to solve this equation analytically and need only to consider the limiting case of $V_0 \to \infty$ and $d \to 0$ where $V_0 d$ is kept constant, i.e.

$$V(x) = \alpha[\delta(x) + \delta(x-a)]\ , \qquad \alpha = constant$$

You should indicate the solutions by a qualitative graphical plot for the special case where

$$\alpha = \frac{4\hbar^2}{2ma}.$$

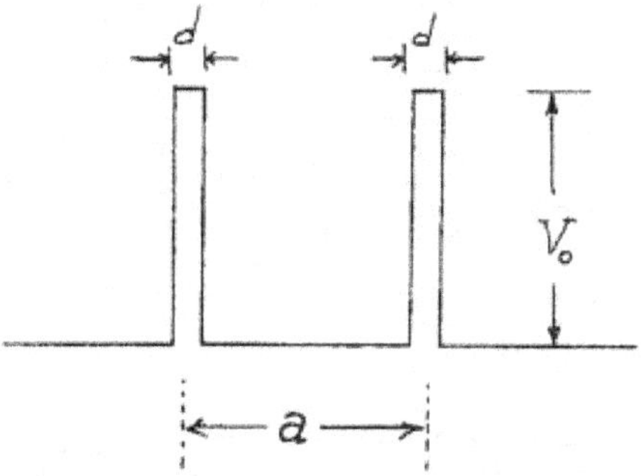


We have:

$V(x) = \alpha[\delta(x) + \delta(x-a)]$

$\left[\frac{-\hbar^2}{2m}\frac{d^2}{dx^2} + v(x)\right]\varphi(x) = E\varphi(x)$

$k^2 = \frac{2mE}{\hbar^2}$

$\gamma = \frac{2m\alpha}{\hbar^2}$

$\frac{d^2}{dx^2} + k^2 - \gamma(\delta(x) + \delta(x-a))\varphi(x) = 0$

Region (I) $\varphi_I(x) = Ae^{ikx} + Be^{-ikx}$

(II) $\varphi_{II}(x) = A'e^{ikx} + B'e^{-ikx}$

(III) $\varphi_{III}(x) = A''e^{ikx}$

$B'' = 0$ inclusive from left only

$A + B = A' + B'$ *continuity of* $\varphi$

$ik(A' - B') - ik(A - B) - \gamma(A' - B') = 0$

$$(A - B) = (A' - B') - \frac{\gamma}{ik}(A' + B')$$
$$A'^{e^{ika}} + B'^{e^{ika}} = A"e^{ika}$$
$$A"e^{ika} - ik\left(A'^{e^{ika}} - B'e^{-ika}\right) - \gamma A"e^{ika} = 0$$
$$\left(A'^{e^{ika}} - B'e^{-ika}\right) = A"e^{ika}\left(1 - \frac{\gamma}{ik}\right)$$

Thus,

$$2A = 2A' - \frac{\gamma}{ik}(A' + B') \qquad 2A'e^{ika} = A"e^{ika}\left(2 - \frac{\gamma}{ik}\right)$$
$$2B = 2B' - \frac{\gamma}{ik}(A' + B') \qquad 2B'e^{ika} = A"e^{ika}\left(\frac{\gamma}{ik}\right)$$
$$2B = 2B' + (2A' - 2A) = A\mathrm{e}^{2\mathrm{ika}}\left(\frac{\gamma}{\mathrm{ik}}\right) + \mathrm{A}\left(2 - \frac{\gamma}{ik}\right) - 2A$$

No reflection, $B = 0$

$$\cos(2ka) = -ka\sin\delta + 1$$
$$2A = A"\left(2 - \frac{\gamma}{ik}\right) - \frac{\gamma}{ik}\left(\frac{1}{2}\left(2 - \frac{\gamma}{ik}\right) + \frac{1}{2}e^{2ika}\left(\frac{\gamma}{ik}\right)\right)A"$$

$$0 = A"e^{2ika}\left(\frac{\gamma}{ik}\right) + A"\left(2 - \frac{\gamma}{ik}\right) - \left\{A"\left(2 - \frac{\gamma}{ik}\right) - A"\frac{\gamma}{2ik}\left[\left(2 - \frac{\gamma}{ik}\right) + e^{2ika}\left(\frac{\gamma}{ik}\right)\right]\right\}$$
$$0 = e^{2ika}\left(\frac{\gamma}{ik}\right) + \left(\frac{\gamma}{ik}\right)\frac{1}{2}\left[2 - \frac{\gamma}{ik} + e^{2ika}\left(\frac{\gamma}{ik}\right)\right]$$
$$0 = e^{2ika} + \frac{1}{2}\left[\left(2 - \frac{\gamma}{ik}\right) + e^{2ika}\left(\frac{\gamma}{ik}\right)\right]$$
$$\boxed{\alpha = \frac{4\hbar^2}{2ma}} \rightarrow \gamma = \frac{2m}{\hbar^2}\frac{2\hbar^2}{ma} = \frac{1}{a}$$
$$-1 = \left(1 - \frac{1}{2ika}\right)e^{2ika} + \frac{1}{2ika}$$
$$2ika = y$$
$$\left(\frac{1}{y} - 1\right)e^{-y} = \left(1 + \frac{1}{y}\right)$$
$$(1 - y)e^{-y} = (1 + y) \rightarrow \boxed{e^{-y} = \frac{1+y}{1-y}}$$
$$1 + e^{-y} = \frac{2}{1-y} \qquad 1 - e^{-y} = \frac{-2y}{1-y}$$
$$\frac{1+e^{-y}}{1+e^{-y}} = -y$$
$$\tan(ka) = -ka$$
$$k \cong \frac{n\pi}{2a}$$

### 3.4.5.3 The gap potential

Let's now consider the square well potential (finite) with particular attention on the bound states:

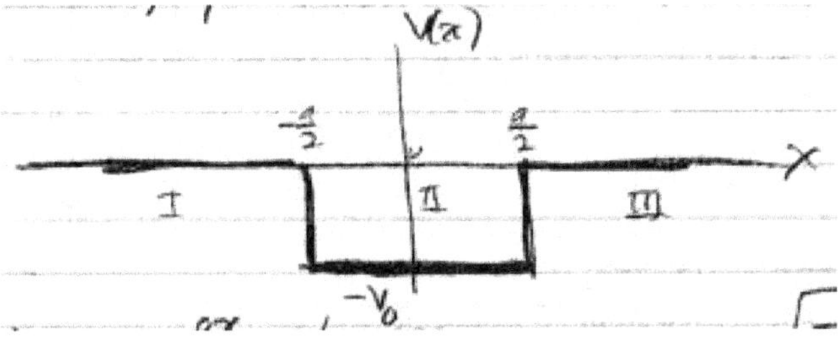


For bound states in the above square well potential (finite), where we have energy $-V_0 < E < 0$, the solutions in the three regions have the form:

$$\varphi_I(x) = B_1 e^{px} + B' e^{-px}$$
$$\varphi_{II}(x) = A_2 e^{ikx} + A_2' e^{-ikx}$$
$$\varphi_{III}(x) = B_3 e^{px} + B_3' e^{-px}$$

Where:

$$p = \sqrt{\frac{-2mE}{\hbar^2}} \quad and \quad K = \sqrt{\frac{2m(E+V_0)}{\hbar^2}}$$

At the $x = -\frac{a}{2}$ boundary:

$$B_1 e^{p\left(-\frac{a}{2}\right)} + B_1' e^{-p\left(-\frac{a}{2}\right)} = A_2 e^{-k\left(-\frac{a}{2}\right)} + A_2' e^{-ik\left(-\frac{a}{2}\right)}$$
$$pB_1 e^{p\left(-\frac{a}{2}\right)} - pB_1^1 e^{-p\left(-\frac{a}{2}\right)} = iKA_2 e^{ik\left(-\frac{a}{2}\right)} - ikA_2' e^{-ik\left(-\frac{a}{2}\right)}$$

Set $B_1' = 0$:

$$\frac{(p+iK)}{2iK} B_1 e^{-\frac{a}{2}p} = A_2 e^{-\frac{a}{2}ik} \quad \rightarrow \quad \boxed{A_2 = e^{(-p+ik)\frac{a}{2}} \frac{(p+ik)}{2ik} B_1}$$

$$\frac{(p-ik)}{-2ik} B_1 e^{-\frac{a}{2}p} = A_2' e^{ik\frac{a}{2}} \quad \rightarrow \quad \boxed{A_2' = -e^{-(p+ik)\frac{a}{2}} \frac{(p-ik)}{2ik} B_1}$$

At the $x = \frac{a}{2}$ boundary:

$$A_2 e^{ik\frac{a}{2}} + A_2' e^{-ik\frac{a}{2}} = B_3 e^{p\left(\frac{a}{2}\right)} + B_3' e^{-p\left(\frac{a}{2}\right)}$$
$$ikA_2 e^{ik\left(\frac{a}{2}\right)} - ikA_2' e^{-ik\left(\frac{a}{2}\right)} = pB_3 e^{p\left(\frac{a}{2}\right)} - pB_3' e^{-p\left(\frac{a}{2}\right)}$$

$$2pe^{p\left(\frac{a}{2}\right)}B_3 = (p+ik)e^{ik\left(\frac{a}{2}\right)}A_2 + (p-ik)e^{-ik\left(\frac{a}{2}\right)}A_2'$$
$$= \left[\frac{(p+ik)^2}{2ik}e^{(-p+2ik)\frac{a}{2}} - \frac{(p-ik)^2}{2ik}e^{(-p-2ik)}\right]B_1$$

Thus

$$\boxed{\frac{B_3}{B_1} = \frac{e^{-pa}}{4ikp}\left\{(p+ik)^2e^{ika} - (p-ik)^2e^{-ika}\right\}}$$

Alternatively:

$$\boxed{\frac{B_3'}{B_,} = \frac{(p^2+k^2)}{2kp}\sin ka}$$

Since bounded in III $B_3 = 0$ so

$$\boxed{\frac{(p+ik)^2}{(p+ik)^2} = e^{-i2ka}}$$

Since p and k depend on E, the above equation entails the quantization of energy. Two cases are possible:
Case (i):

$$\frac{(p-ik)}{(p+ik)} = -e^{ika} \quad \rightarrow \quad \frac{p}{k} = \tan\left(\frac{ka}{2}\right).$$

Let's define $K_0 = \sqrt{\frac{2mV_0}{\hbar^2}} = \sqrt{k^2+p^2}$ , then, $\frac{1}{cos^2\left(\frac{ka}{2}\right)} = 1 + tan^2\left(\frac{ka}{2}\right) = \frac{k^2+p^2}{k^2} = \left(\frac{k_0}{k}\right)^2$, and we have:

$$\left|cos\left(\frac{ka}{2}\right)\right| = \frac{k}{k_0}$$
$$\tan\left(\frac{ka}{2}\right) > 0$$

Case (ii):

$$\frac{(p-ik)}{(p+ik)} = e^{ika} \rightarrow \begin{Bmatrix} \left|sin\left(\frac{ka}{2}\right)\right| = \frac{k}{k_0} \\ \tan\left(\frac{ka}{2}\right) < 0 \end{Bmatrix}$$

If $k_0 \leq \frac{\pi}{a}$ $\left(i.e., \ V_0 < V_0 = \frac{\pi^2\hbar^2}{2ma^2}\right)$ there exists only one bound state of the particle, and this state has an even wavefunction.

Consider the wave packet from the other vantage point, i.e. in the x>0 region:

$$\Psi(x,r) = \frac{1}{\sqrt{2\pi}} \int_0^{k_0} dk g(k) B_2'(k) e^{-p(k)x} e^{-iw(k)t}$$

$$Let\ A_1 = 1 \rightarrow \frac{B_2'}{A_1} = B_2'$$

$$\frac{A_1'}{A_,} = \frac{k_1 - ip}{k_2 + ip} = \frac{(k_1 - ip)^2}{K_1^2 + P^2} \rightarrow arg\left(\frac{A_1'}{A_1}\right) = tan^{-1}\frac{-2ipk_1}{k_1^2 + p^2}$$

***Electron bound in a square well***

An electron is bound in a square potential well of width 1 $\hat{A}$ and depth 1 eV. That is, $V(x) = 1eV\ for - a < x < a,\ and\ V(x) = 0$ outside this region.

(a) Assume the even eigenstates corresponding to this potential have the following form: $\psi_2 = A\cos kx\ for\ - a < x < a,\ and\ \psi_\pi = B\exp(-kx)\ for\ a < x.$
Write the boundary conditions at $x = a$ needed to relate the unknown constants.

(b) From the results of (a), find a single equation in which the only unknown is the energy E of the even bound states.

(c) Sketch the form of the wave function for the lowest state, keeping the horizontal scale approximately correct. How many bound states are there?

**Answer**

(a) $\psi_I = A\cos kx$

$\psi_{II} = B\exp(-\bar{k}x)$

Continuity relationships:

$$\psi_I(a) = \psi_{II}(a)$$
$$\left.\frac{\partial\psi_I}{dx}\right|_a = \left.\frac{\partial\psi_{II}}{dx}\right|_a$$

(b) Using the continuity relationships:

$A\cos ka = B\exp(-\bar{k}a)$ $\qquad -Ak\sin ta = -\bar{k}\,B\exp(-\bar{k}a)$

$k\sin ka = \bar{k}\cos ka$

$k\tan k_a = \bar{k}$

Now, $\frac{\hbar^2\bar{k}^2}{2m} = |E|$ and $\frac{\hbar^2 k^2}{2m} = |V| - |E|$

$$\bar{k} = \sqrt{\frac{2m|E|}{\hbar^2}} \text{ and } k = \sqrt{\frac{2m|V|-|E|}{\hbar^2}}$$

$$\tan\left(\sqrt{\frac{2ma^2|V|-|E|}{\hbar^2}}\right) = \sqrt{\left|\frac{V}{E}\right| - 1}$$

(c) $p^2 = \frac{2ma^2|V|}{\hbar^2} = \frac{2(9.11\times10^{-28}g)(.5\dot{A})^2|-1ev|}{(1.055\times10^{-27}erg-s)^2} = .06556$

$p = .26$ since, $26 < \pi$ there is only <u>one bound state</u>

***First two bound states***

A particle of mass m is bound in a rectangular potential well of width 2a and depth $V_0$. That is, $V(x) = -V_0$ for $-a < x < a$ and zero otherwise.

(a) Find the minimum depth of the well $V_0$ such that there will be at least two bound states.

(b) The potential is changed for $x < 0$ so that $V(x)$ is infinite in this region. Find the minimum depth of the $V_0$ that will insure one bound state.

(c) The particle has a charge q. A function electric field $\varepsilon$ is applied in the region $x > 0$, which adds a potential $-q\varepsilon x$ in this region to the potential in part (b). Find the number of bound states for $q\xi > 0$ and for $q\xi < 0$.

**Answer**

(a)
$$\left.\begin{array}{ll} x < -a & \psi = -Be^{k_2 x} \\ -a < x < a & \psi = A\sin k_1 x \\ x > a & \psi = Be^{-kx} \end{array}\right\} \text{ use } \quad \begin{array}{l} k_1^2 = (V_0 - |E|)\frac{2m}{\hbar^2} \\ k_2^2 = \frac{|E|2m}{\hbar^2} \end{array}$$

Applying continuity of $\psi, \psi'$ at $x = a$, and using the relations

$$\lambda \equiv \frac{V_0 2ma^2}{\hbar^2}, \qquad y \equiv k_1 a \ , \ (k_2 a)^2 = \lambda - y^2$$

Then

$$\left.\begin{array}{r} A\sin k_1 a = Be^{-k_2 a} \\ Ak_1\cos k_1 a = -kBe^{-k_2 a} \end{array}\right\} \Rightarrow \tan k_1 a = -\frac{k_1 a}{k_2 a}$$

And

$$\tan y = -\frac{y}{\sqrt{\lambda - y^2}} \quad \text{or} \quad \cot y = -\sqrt{\frac{\lambda}{y^2} - 1}$$

The smallest $\lambda$(or $V_0$) for which two curves intersect corresponds to $y = \pi / 2$ and

$-\cot y = \sqrt{\frac{\lambda}{y^2} - 1} = 0$. Therefore

$$\lambda = \left({}^{\pi}/_{2}\right)^2 = \frac{V_0 2ma^2}{\hbar^2} \Rightarrow V_0 = \frac{\hbar^2 \pi^2}{8ma^2}$$

(b) Same as (a) since the second Energy level of the well in part a) matches the boundary condition for this well ($\psi = 0$ at $x = 0$) and it has no nodes.

c) $q\xi > 0$ $V < 0$ → No bound state
$q\xi < 0$ $V > 0$ → ∞ number of bound states.

### 3.4.5.4 The U- and W-potentials

***The Parabolic or U-shaped potential***

Consider a particle of mass m is in a one-dimensional potential well with $V(x) = Kx^2/2$ for $x > 0$, and $V(x) = \infty$ for $x < 0$.

(a) Write the Hamiltonian operator for this system.
(b) Give the boundary conditions on the wave function $\psi(x)$ at $x = 0$ and $x = \infty$. Sketch $\psi(x)$ for the lowest three eigenstates.
(c) Find the energy eigenvalues for the lowest three states
(d) Assume the particle is in the ground state with energy $E_{gnd}$.
Find the classical turning point $x = a$, that is, the limit beyond which the classical motion cannot extend for this energy. Then find approximately the probability the particle will be found between $x = a$ and $x = 1.01a$

**Answer**

(a) $H_x = \frac{p_x^2}{2m} + \frac{k_x^2}{2}$ $x > 0$
(b) $\psi(x) = 0$ at $x = 0$ and $x = \infty$ since $v(x) = \infty$ there.

(c) $E_n = \hbar w_0 \left(n + {}^1/_2\right)$ for the complete well, so, by symmetry: $E_n = \frac{1}{2}\hbar w_0 \left(n + {}^1/_2\right)$

$$E_0 = {}^{\hbar w_0}/_4$$
$$E_1 = {}^{3\hbar w_0}/_4$$
$$E_2 = {}^{5\hbar w_0}/_4$$

(d) Ground State $E_0 = \hbar w_0/4$
Turning point: $E = k\,x^2/2$

$$\hbar w_0/4 = kx_0^2/2$$
$$x_0^2 = \frac{\hbar w_0}{2k}$$
$$x_0 = \sqrt{\frac{\hbar w_0}{2k}}$$

$$\varphi_0(x) = \sqrt{2}\left(\frac{mw_0}{\hbar\pi}\right)^{1/4} \exp\left[-\left(\frac{mw_0}{\hbar\pi}x^2\right)/2\right]$$
$$P(x) = |\varphi_0(x)|^2 = 2\sqrt{\frac{mw_0}{\hbar k}} \exp\left[-\left(\frac{mw_0}{\hbar}x^2\right)\right]$$

Probability of finding the particle between $x = a$ and $x = 1.01a$ is

Probability $= 2\sqrt{\frac{mw_0}{2k}} \int_a^{1.01a} \exp\left(\frac{-mw_0}{\hbar\pi}x^2\right) \cdot xdx =$

$$2\sqrt{\frac{mw_0}{2k}}\left[\exp\left(\frac{-mw_0}{\hbar}x^2\right)/-2\,\frac{mw_0}{\hbar}\right]_a^{1.0.1a}$$
$$= \sqrt{\frac{\hbar}{\pi m w_0}}\left[\exp\left(\frac{-mw_0}{\hbar}a^2\right) - \exp\left(\frac{-mw_0}{\hbar}(1.02a^2)\right)\right]$$
$$= \sqrt{\frac{\hbar}{\pi m w_0}}\left[\exp\left(\frac{-mw_0}{\hbar}\cdot\frac{\hbar w_0}{2k}\right) - \exp\left(\frac{-mw_0}{\hbar}\cdot \frac{\hbar w_0}{2k}(1.02)\right)\right]$$
$$= \sqrt{\frac{\hbar}{\pi m w_0}}\left[\exp(-{}^1/_2) - \exp(-{}^{1.02}/_2)\right]$$
$$= (.006)\sqrt{\frac{\hbar}{\pi k}}$$

***The Quartic or W-shaped potential***

The potential energy for a particle of mass m in one-dimensional motion is given by $V(x) = Ax^a + Bx^4$, where A and B are real constants.

(a) Sketch $V(x)$ as a function of x, assuming $A < 0 \quad and\ B > 0$.
(b) Write the Hamiltonian function $H(p, x)$.
(c) Write Hamilton's two equations for the particle.
(d) Given that the total energy is $E_0$, what are the limits of the classical motion (turning points) for $E_0 > 0$? $and\ for\ E_0 < 0$?

**Answer**

(a)We have one-dimensional motion with potential energy $V(x) = Ax^2 + Bx^4$ , $A$ and $B$ real.
$V(x)$:

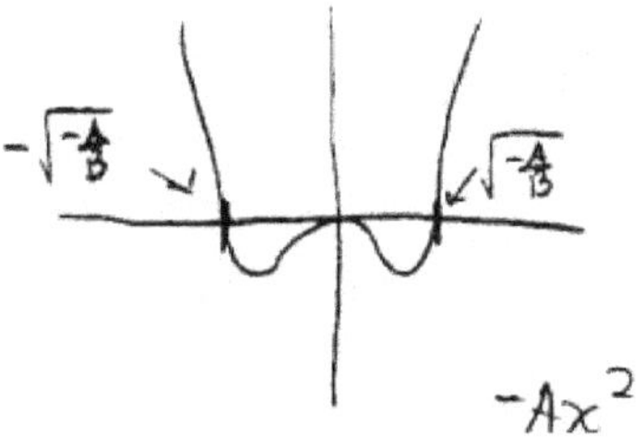

Where the zero-crossings are at $x = \pm\sqrt{\frac{-A}{B}}$,

(a) The Hamiltonian is: $H(p, x) = \frac{p^2}{2m} + V(x)$.
(b) We have $\frac{\partial H}{\partial x} = -\dot{p} = 2Ax + 4Bx^3$ and $\frac{\partial H}{\partial p} = \dot{x}$.
(c) $E = T + V$. For $E_0 > 0, T = E_0 - V = E_0 - Ax^2 - Bx^4$ have turning points at $T = 0$ with classical motion between A and D:

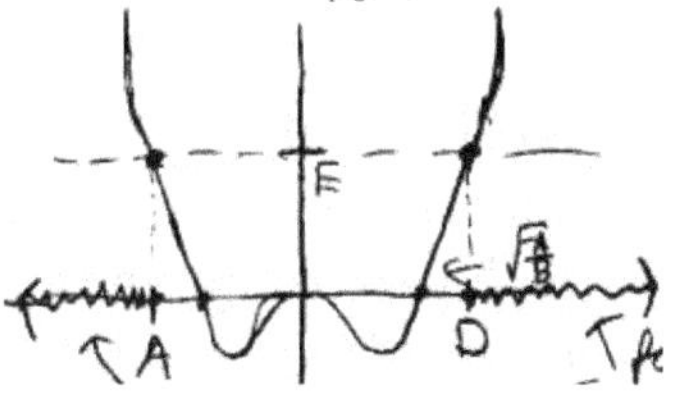

Limit (turning points) at $E_0 = Ax^2 + Bx^4$

$z = x^2 \qquad Bz^2 + Az - E_0 = 0$

$z = \frac{-A \pm \sqrt{A^2 + 4BE_0}}{2B}$

$x^2 = \frac{-A \pm \sqrt{A^2 + 4BE_0}}{2B}$

$x = \pm\sqrt{\frac{-A + \sqrt{A^2 + 4BE_0}}{2B}}$ turning points $E_0 > 0$

<u>**For $E_0 < 0$**</u>
We have:

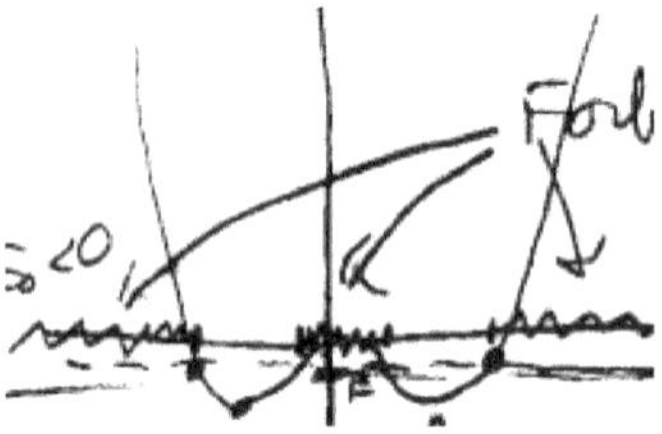

$x = \pm\sqrt{\frac{-A \pm \sqrt{A^2 + 4BE_0}}{2B}}$ turning points $E_0 < 0$,

$V(x) = Ax^2 + Bx^4$

$V' = 2Ax + 4Bx^3 = 0 \rightarrow$ minima at $x = \pm\sqrt{\frac{-A}{2B}}$

For $E_0 < V(x)$ *everywhere*, there is no classical motion (when $E_0 < \frac{-A^2}{4B}$).

### 3.4.5.5 Double Potential Well

Consider a particle of mass m that moves in one dimension. The potential energy is

$$V(x) = -\alpha\delta(x) - \alpha\delta(x - \ell)$$

where $\alpha$ and $\ell$ are constants. Calculate the bound states of the particle. You will need to give a graphic solution to find the bound states energies. (A rough qualitative sketch is sufficient).

**Answer**

Shift the coordinate origin so that the potential is countered:

$V(x) = -\alpha\delta\left(x + \frac{\ell}{2}\right) - \alpha\delta\left(x - \frac{\ell}{2}\right)$

$x \rightarrow x + \frac{\ell}{2}$

Now, since the potential is an even function x we commute the parity operator with the Hamiltonian. Thus, we can decompose the solutions into the common eigenstates of parity and the Hamiltonian, i.e. even and odd solutions

$$\left[-\frac{\hbar^2}{2m}\frac{d^2}{dx^2}+v(x)-E\right]\varphi=0$$

$$\varphi(x)=\begin{cases} e^{-kx} & x>\frac{\ell}{2} \\ Ae^{ikx+Be^{kx}} & -\frac{\ell}{2}<x<\frac{\ell}{2} \\ e^{kx} & x<-\frac{\ell}{2} \end{cases}$$

We can rewrite this knowing that $\varphi(x)$ has either even or odd parity:

$$\varphi(x)=\overbrace{\begin{cases} e^{-kx} & \frac{\ell}{2}<x \\ A\sin hkx & -\frac{\ell}{2}<x<\frac{\ell}{2} \\ e^{kx} & x<-\frac{\ell}{2} \end{cases}}^{ODD} \qquad \varphi(x)=\overbrace{\begin{cases} e^{-kx} & \frac{\ell}{2}<x \\ B\cosh\ kx & -\frac{\ell}{2}<x<\frac{\ell}{2} \\ e^{kx} & x<-\frac{\ell}{2} \end{cases}}^{EVEN}$$

For even case:

Continuity at $x=\frac{\ell}{2}$ $\quad gives\ e^{-kx}=B\cosh\left(k\frac{\ell}{2}\right)$

Discontinuity relation at $x=\frac{\ell}{2}$: $\quad \frac{d\varphi}{dx}\Big|_{x=\frac{\ell}{2}+\epsilon}-\frac{d\varphi}{dx}\Big|_{x=\frac{\ell}{2}-\epsilon}+\frac{2mx}{\hbar^2}\varphi(x)\Big|_{x=\frac{\ell}{2}}=0$

$$-ke^{-k\frac{\ell}{2}}-kB\sinh\left(k\frac{\ell}{2}\right)+\frac{2m\alpha}{\hbar^2}e^{-k\frac{\ell}{2}}=0$$

The eigenvalue condition is then

$$kB\sinh\left(k\frac{\ell}{2}\right)=-ke^{-k\frac{\ell}{2}}+\frac{2m\alpha}{\hbar^2}e^{-k\frac{\ell}{2}}$$

$$kB\cosh\left(k\frac{\ell}{2}\right)=ke^{-k\frac{\ell}{2}}$$

$$\boxed{\tanh\left(k\frac{\ell}{2}\right)=-1+\left(\frac{2m\alpha}{\hbar^2}\right)\frac{1}{k}}$$

$$\tanh y=\frac{\lambda}{y}-1$$

$$\boxed{\lambda = \frac{\ell m\alpha}{\hbar^2}}$$

$y < \lambda$

But since $\tanh y < 1$ $\quad \frac{\lambda}{y} < 1 \quad \rightarrow y > \frac{\lambda}{2}$

So, $\frac{\lambda}{2} < y < \lambda$ for a solution to exist

For odd case:

At $x = \ell/2$:

Continuity $e^{-\frac{\ell}{2}k} = A \sinh k\frac{\ell}{2}$

Discontinuity $\quad -ke^{-\frac{\ell}{2}k} - kA\cosh k\frac{\ell}{2} + \frac{2m\alpha}{\hbar^2}\varphi(x)|_{x=\frac{\ell}{2}} = 0$

$$-ke^{-\frac{\ell}{2}k} - KA\cosh k\frac{\ell}{2} + \frac{2m\alpha}{\hbar^2}e^{-k\frac{\ell}{2}} = 0$$

$$\boxed{\coth\left(k\frac{\ell}{2}\right) = -1 + \frac{2m\alpha}{\hbar^2}\frac{1}{k}} \rightarrow \coth y = \frac{\lambda}{y} - 1$$

$$\tanh y = \left(\frac{\lambda}{y} - 1\right)^{-1}$$

$$\frac{1}{\frac{\lambda}{y}-1} = \frac{y}{\lambda - y}$$

1 Solution in certain cases

A solution exists when the slope of tanh y is larger than that of $\left(\frac{\lambda}{y} - 1\right)^{-1}$ at the origin $\frac{d}{dy}\tanh y\,|_{y=0} = \left(1 - \frac{\sinh^2 y}{\cosh^2 y}\right)|_{y=0} = 1$ $\qquad \frac{d}{dy}\left(\frac{\lambda}{y} - 1\right)^{-1} = -\left(\frac{\lambda}{y} - 1\right)^{-3}\left(-\frac{\lambda}{y}\right) = \frac{\lambda}{(\lambda - y^2)^2}$

$1 > \frac{1}{\lambda} \rightarrow \lambda > 1$ for existence at $y = \lambda/2$ we reach $f(y) = 1$, so $y < {}^{\lambda}/_{2}$ exists.

### 3.4.5.6 Well with Step

What is the probability that a particle incident from the left or the barrier shown will be transmitted? You should provide an answer for the full range of allowed energies of the incident particle?

$T = ?$ For full range of allowed energies of excited particle

$$\left[-\frac{\hbar^2}{2m}\frac{d^2}{dx^2} + V(x) - E\right] \Psi(x) = 0$$

$$\left[\frac{d^2}{dx^2} + \frac{2m\big(E - V(x)\big)}{\hbar^2}\right] \Psi(x) = 0$$

Form: $\left[\frac{d^2}{dx^2} + K^2\right] \Psi(x) = 0$

$$k_1 = \sqrt{\frac{2mE}{\hbar^2}}$$

$$k_2 = \sqrt{\frac{2m(E + U_o)}{\hbar^2}}$$

$$k_1 = \sqrt{\frac{2m(E - V_o)}{\hbar^2}}$$

Region I: $\Psi_I = Ae^{ik_1x} + Be^{-ik_1x}$

II: $\Psi_{II} = A'e^{ik_2x} + B'e^{-ik_2x}$

III: $\Psi_{III} = A"e^{ik_3x} + B"e^{-ik_3x}$ $\qquad B" = 0,$

$inc. particles\ from\ left\ only$

$$J_I = \frac{\hbar}{2\mu i}\{\Psi_I^*\vec{V}\,\Psi_I - \Psi_I^*\vec{V}\,\Psi_I\}$$

$$\frac{\hbar}{2\mu i}\{(A^*e^{ik_1x} + B^*e^{ik_1x})(A_1K_1e^{ik_1x} - B_1K_1e^{ik_1x}) - (Ae^{-ik_1x} + Be^{-ik_1x})(-A^*ik_1e^{-ik_1x} + ik_1B^*e^{ik_1x})\}$$

$$\frac{\hbar}{2\mu i}\{|A|^2ik_1 - iA^*Bik_1e^{-2ik_1x} + B^*ik_1e^{2ik_1x} - |B|^2ik_1 - [-|A|^2ik_1 + AB^*ik_1e^{2ik_1x} - BA^*ik_1e^{-2ik_1x} + |B|^2ik_1]\}$$

$$= \frac{\hbar k_1}{\mu}\{|A|^2 - |B|^2\}$$

$$J_{II} = \frac{\hbar}{2\mu i}\{\Psi_{II}^*\nabla\Psi_{II} - \Psi_{II}\nabla\Psi_{II}^*\} = \frac{\hbar k_2}{\mu}\{|A'|^2 - |B'|^2\}$$

$$J_{III} = \frac{\hbar}{2\mu i}\{A"^*e^{ik_3x}(ik_3xA"e^{ik_3x}) - A"e^{ik_3x}(-ik_3xA"^*e^{-ik_3x})\}$$

$$= \frac{\hbar k_3}{\mu}|A"|^2$$

$$T = \frac{J_{III}}{J_I^{inc}} = \frac{\frac{\hbar k_3}{\mu}|A"|^2}{\frac{\hbar k_3}{\mu}|A|^2} = \frac{k_3}{k_1}\frac{|A"|^2}{|A|^2}$$

Matching at the origin:

$A + B = A' + B'$

$A - B = \left(\frac{k_3}{k_1}\right)(A' - B')$

$2A = \left(1 + \frac{k_2}{k_1}\right)A' + \left(1 - \frac{k_2}{k_1}\right)B'$

Matching at $x = a$:

$A'e^{ik_2x} + B'^{e^{-ik_2a}} = A"e^{ik_3a}$ $\qquad 2A'e^{ik_2a} = \left(1 + \frac{k_3}{k_1}\right)A'e^{ik_3a}$

$ik_2A'e^{ik_2a} - ik_2B'e^{-ik_2a} = ik_3A''e^{ik_2a}$

$\frac{k_2}{k_3}\{A'e^{-ik_2a} - B'e^{-ik_2a}\} = A"e^{-ik_3a}$ $\qquad 2B'^{e^{-ik_2a}} = \left(1 - \frac{k_3}{2_2}\right)e^{ik_3a}A"$

$2A = \left(1 + \frac{k_2}{k_1}\right)e^{-ik_2a}\left(1 + \frac{k_3}{k_2}\right)e^{ik_2a}A" + \left(1 - \frac{k_2}{k_1}\right)\frac{1}{2}e^{ik_2a}\left(1 - \frac{k_3}{k_2}\right)e^{ik_2a}A"$

$A = \left[\frac{1}{4}\left(1 + \frac{k_2}{k_1}\right)\left(1 + \frac{k_3}{k_2}\right)e^{i(k_3-k_2)a} + \frac{1}{4}\left(1 - \frac{k_2}{k_1}\right)\left(1 - \frac{k_3}{k_2}\right)e^{i(k_3+k_2)a}\right]$

$T = \frac{k_3}{k_1}\frac{1}{\left|\left\{\frac{1}{4}\left(1+\frac{k_2}{k_1}\right)\left(1+\frac{k_3}{k_2}\right)e^{i(k_3-k_2)a}+\frac{1}{4}\left(1-\frac{k_2}{k_1}\right)\left(1-\frac{k_3}{k_2}\right)e^{i(k_3+k_2)a}\right\}\right|^2}$

$= \left|\frac{1}{4}\left(1 + \frac{k_2}{k_1}\right)\left(\frac{k_3}{k_1} + \frac{k_3}{k_2}\right)e^{ik_2a} + \frac{1}{4}\left(1 + \frac{k_2}{k_1}\right) - \left(\frac{k_3}{k_1} - \frac{k_3}{k_2}\right)e^{ik_2a}\right|^{-2}$

$T = |T|^2 = \frac{k_3}{k_1}4\left|\left(1 + \frac{k_3}{k_1}\right)\left(\frac{e^{ik_2a}+e^{ik_2a}}{2}\right) + \left(\frac{k_2}{k_1} + \frac{k_3}{k_2}\right)\left(\frac{e^{-ik_2a}-e^{ik_2a}}{2}\right)\right|^{-2}$

$T = |T|^2 = \frac{4k_3}{k_1}\left|i\left(1 + \frac{k_3}{k_1}\right)\cos(k_2a) - i\left(\frac{k_2}{k_1} + \frac{k_3}{k_2}\right)\sin(k_2a)\right|^{-2}$

$= \frac{4k_3}{k_1}\frac{1}{\cos^2(k_2a)+i\left(\frac{k_2}{k_1}+\frac{k_3}{k_2}\right)^2\sin^2(k_2a)}$

When $E > V_0$ $\qquad k_3/k_1 = \sqrt{1 - V_0/E}$ , $k_3/k_2 = \sqrt{\frac{E-V_0}{E+V_0}}$ , ${}^{k_2}/_{k_1} = \sqrt{1 + {}^{u_o}/_E}$

$T|T|^2 = 4\sqrt{\frac{E-V_0}{E}}\left\{\left(1 + \sqrt{\frac{E-V_0}{E+V_0}}\right)^2\cos^2\left(a\sqrt{\frac{2m(E+u_o)}{\hbar^2}}\right) + \left(\sqrt{\frac{E+u_0}{E}} + \sqrt{\frac{E-V_0}{E+V_0}}\right)^2\sin^2\left(a\sqrt{\frac{2m(E+u_o)}{\hbar^2}}\right)\right\}$

When $E < V_o$ , $k_3$ is imaginary, let $k_3 \to ik_3$

$T|T|^2 = \frac{4ik_3}{k_1}\left|\left(1 + \frac{ik_3}{k_1}\right)\cos(k_2a) - 1\left(\frac{k_2}{k_1} + \frac{ik_3}{k_1}\right)\sin(k_2a)\right|^{-2}$

Thus the transmission coefficient is pure imaginary, which corresponds to pure reflection.

Probability of transmission = 0 for $E < V_0$.

**3.4.5.7 From Schrodinger to Canonical Quantization**

Consider the following:

(a) The operators $\hat{p}$ *and* $\hat{x}$ operate on $\psi(x)$, a continuous function of a. If $\psi(x) = \exp(ikx)$, find $\hat{x}\psi, \hat{p}\ \psi, \hat{p}^{-1}\ \psi$, and $\hat{p}^2\psi$ as functions of x.

(b) For $\psi(x) = \exp(ikx)$, find $(\hat{x}\hat{p} - \hat{p}\hat{x})\ \psi(x)$, and express the answer in terms of $\psi(x)$.

(c) Express the operator $(\hat{x}\hat{p} - \hat{p}\hat{x})$ as a complex number.

**Answer**

(a) $\psi(x) = \exp(ikx)$

$$\hat{x}\psi = x\exp(ikx)$$

$$\hat{p}\psi = -i\hbar(ik)\exp(ikx) = \hbar k\exp(ikx)$$

$$\hat{p}^{-1}\psi = \frac{1}{-i\hbar}\int^{x}\exp(ikx) \text{ to within an additive constant}$$

$$= \frac{1}{i\hbar}\exp(ikx)$$

$$\hat{p}^2\psi = -\hbar^2\frac{\partial^2}{\partial x^2}\exp(ikx) = (\hbar k)^2\exp(ikx)$$

(b) $(\hat{x}\hat{p} - \hat{p}\hat{x})\psi(x) = \hat{x}\hat{p}\ \psi(x) - \hat{p}\hat{x}\psi(x) = \hat{x}(\hbar k\exp(ikx)) - \hat{p}(x\exp(ikx))$

$$= x\hbar k\exp(ikx) + i\hbar\frac{\partial}{\partial x}(x\exp(ikx))$$

$$= x\hbar k\exp(ikx) + i\hbar\,(xik\exp(ikx) + \exp(ikx))$$

$$= i\hbar\exp(ikx)$$

$$= i\hbar\psi(x)$$

(c) $(\hat{x}\hat{p} - \hat{p}\hat{x}) = x\left(-i\hbar\,{}^{\partial}/_{\partial x}\right) + i\hbar\,{}^{\partial}/_{\partial x}(x)$

$$= i\hbar x\,{}^{\partial}/_{\partial x} + i\hbar x\,{}^{\partial}/_{\partial x} + i\hbar$$

$$= i\hbar$$

**3.5 Heisenberg Uncertainty (1925) and Matrix Mechanics (1927)**

Heisenberg uncertainty relation: $\Delta x \cdot \Delta p \geq \hbar$. Note that this uncertainty relation is a basic property of wave function solutions (Benedick's Theorem [24]).

**Root-mean –square deviations**

Let's show

$$\Delta A \cdot \Delta B \geq \frac{1}{2} < [A, B] >$$

Consider

$$|\varphi> = (A + i\lambda B)\,|\Psi>.$$

Then,

$$<\varphi|\varphi> = <A^2> + i\lambda <[A,B]> + \lambda^2 <B^2> \geq 0.$$

The discriminator of the quadratic is:

$$disc. = -<[A,B]>^2 - 4<B^2><A^2> \leq 0$$
$$|<[A,B]>|^2 \leq 4<A^2><B^2>$$

Consider, $\begin{cases} A' = A - <A> \\ B' = B - <B> \end{cases} \rightarrow [A',B'] = [A,B]$, thus,

$$(\Delta A)^2 = <A'^2>\ ,(\Delta B)^2 = <B'^2>.$$

Thus,

$$\Delta A \Delta B \geq \frac{1}{2}|<[A,B]>|,$$

the generalized thereby uncertainty relation.

For conjugate variables P,Q → [P,Q]=iħ, then:

$$\Delta P \Delta Q \geq \frac{\hbar}{2}$$

the Heisenberg uncertainty relation.

**Examples**

(1) A particle of mass m is confined in one dimension so that it is localized in a region of width d.

(a) Use the uncertainty relation $\Delta x\ \Delta p > \hbar$ to find an expression for the minimum possible kinetic energy $T_{min}$ of the particle in terms of h (Planck's constant) m, and d.
(b) Use (a) to estimate the lowest energy T $(\text{in } eV)$ of an electron confined in an atom with diameter d $10^{-10}$ meter.
(c) Repeat (b) for a proton confined in a nucleus where d is approximately $10^{-15}$ meter.

**Answer**

(a) $\Delta x = {}^{d}/_{2}$

$\Delta p \Delta x \geq \hbar$ and $T_{min} = \frac{p_{min}^2}{2m} \rightarrow \Delta p_{min} = \frac{\hbar}{\Delta x} = {}^{2\hbar}/_{d}$. The $p_{min}$ would be the closest to zero we can get with $\Delta p_{min} = {}^{2\hbar}/_{d}, thus\ p_{min} = {}^{2\hbar}/_{d}$, and

$$T_{min} = \frac{\left(2\hbar/d\right)^2}{2m} = \frac{4\hbar^2}{2md^2} = \frac{h^2}{2\pi^2 md^2}$$

(b) $d = 1 \times 10^{-8} cm$
$m = 9.11 \times 10^{-28} g$
$\hbar = 1.055 \times 10^{-27}\ erg \cdot s$

Thus: $T_{min} = \frac{2\left(1.055\times10^{-2}\ erg\ s\right)^2}{(9.11\times10^{-28} g)(1\times10^{-8} cm)^2} \cdot \left(\frac{1eV}{1.602\times10^{-12} erg}\right) = 15.24\ eV$

(c) $T_{min} = \frac{4\left(1.055\times10^{-2}\ erg\ s\right)^2}{(9.11\times10^{-24} g)(1\times10^{-13} cm)^2} \cdot \left(\frac{1eV}{1.602\times10^{-12} erg}\right) = 8.32 \times 10^7\ eV$

(2) A particle of mass m is in a uniform gravitational field $g$ in the x direction. At $t = 0$, the wave function is in the form of a wave packet localized around the origin in such a way that $\langle\psi: \hat{p}: \psi\rangle = p_0$.

(a) Find $d\langle\hat{x}\rangle/dt$ at $t = 0$
(b) Find $\langle\hat{x}\rangle$ as a function of term of time for $t > 0$

**Answer**

We have to start $\hat{H} = \frac{\hat{P}x^2}{2m} + mgx\ \ and\ \ \hat{P}x = -i\hbar\frac{\partial}{\partial x}$.

(a) $\frac{d\langle x\rangle}{dt} = \frac{i}{\hbar}\langle[\hat{H},\hat{x}]\rangle = \frac{i}{\hbar}\left\langle\left[\frac{\hat{P}x^2}{2m},\hat{x}\right]\right\rangle = \frac{i}{2m\hbar}\langle\hat{p}[\hat{p},\hat{x}] + [\hat{p},\hat{x}]\hat{p}\rangle =$
$\frac{i}{2m\hbar}\langle -2i\hbar p\rangle = \langle\frac{p}{m}\rangle = \frac{\langle p\rangle}{m}$.

Or, using $[\hat{p},\hat{x}] = -i\hbar, [\hat{p}^2,\hat{x}] = -2i\hat{p}$ we also get:

$$\frac{d\langle\hat{x}\rangle}{dt} = \frac{\langle\hat{p}\rangle}{m}.$$

(b) $\frac{d\langle\hat{p}\rangle}{dt} = \frac{i}{\hbar}\langle[\hat{H},\hat{P}]\rangle = mg \Rightarrow \langle\hat{p}\rangle = p_0 + mgt$ so $\langle\hat{x}\rangle = \frac{p_0}{m}t + \frac{1}{2}g\,t^2$

(3) A free particle of mass m is represented at a given time by a wave function such that $\psi^*\psi = A\exp[-(x/(2a))^2]$. The constant $a$ is a rough measure of the width of the packet and A normalizes the probability.

(a) Find the time T required for the wave packet to double in width.
(b) A large number of particles are contained in a one-dimensional box of width a. The average momentum is zero, but there is a spread of momenta between $p = -\hbar/(2a)$ and $p = +\hbar/(2a)$. At $t = 0$, the walls of the box vanish and the particles spread freely as classical

masses with different velocities. Find the time T for the size of the group of particles to double.

(c) Assume m = (mass electron) and a = 1Å. Find T in seconds in (a).

(d) Assume m = (mass person) = 70kg and a =. 001 cm. find T in seconds in (a).

**Answer**

Note that $a \rightarrow a(1+t^2/\tau^2)^{1/2}$, where the packet becomes wider in time by the factor $(1+t^2/\tau^2)^{1/2}$ where the time constant $\tau$ is measured by

$$\tau = \frac{k^2 a^2}{\omega} \quad and \quad \omega = \frac{\hbar k^2}{2m}$$

Thus

$$\tau = \frac{2mk^2 a^2}{\hbar k^2} = \frac{2ma^2}{\hbar}.$$

(a) $(1+t^2/\tau^2)^{1/2} = 2$

$t^2/\tau^2 = 3$

$t^2 = 3\tau^2$

$t = \sqrt{3}\tau = \frac{2\sqrt{3}\, ma^2}{\hbar}$

(b) $\langle p \rangle = 0$ assuming the spread of particles is Gaussian. Size doubles when the extremes $p = \hbar/2a = m\dot{x}$ travel the distance $a/2$ thus $\dot{x} = \frac{\hbar}{2am} \Rightarrow x = \frac{\hbar T}{2am} + \frac{a}{2} = a$, thus $\frac{\hbar T}{2am} = \frac{a}{2}$ or

$$T = \frac{ma^2}{\hbar}$$

(c) $T = 2\sqrt{3}(9.11 \times 10^{-28} g)\,(1 \times 10^{-8} cm)^2/(1.055 \times 10^{-27} erg.s) = 2.99 \times 10^{-16} sec$

(d) $T = 2\sqrt{3}\,(70{,}000g)\,(.001cm)^2/(1.055 \times 10^{-27}\ erg.s) = 2.30 \times 10^{26} sec \approx 2.30 \times 10^{19}\ yrs.$

(4) A wave packet representing the state of a particle of mass m and q moves in the x direction in a uniform electric field $\xi$. The classical force on the particle is $F = q\xi = -d\,(-q\xi x)/dx$.

(a) Find $d\,\langle p \rangle/dt$, the rate of change of the expectation value of the momentum p with respect to time.

(b) The uncertainty $\Delta p$ of the momentum of the particle in the electric field may or may not spread in time. Find which is true by calculating $d\,\langle \Delta p \rangle/dt$.

(c) Let the function $\varphi(p,t)$ represent the state of the particle in the momentum representation. Write the differential equation that determines the change in time of $\varphi(p,t)$.

**Answer**

(a) $\frac{d\langle P\rangle}{dt} = \frac{i}{\hbar}\langle[\hat{H},\hat{P}]\rangle = \frac{i}{\hbar}\langle\left[\frac{P^2}{2m} - q\xi\hat{x},\hat{P}\right]\rangle = \underline{\underline{q\xi}}$, since $[\hat{P}^2,\hat{P}] = 0, [\hat{x},\hat{P}] = i\hbar$ .

(b) $\Delta p^2 = \langle p^2\rangle - \langle p\rangle^2 \Rightarrow 2\Delta p \cdot \frac{d\Delta p}{dt} = \frac{d\langle p^2\rangle}{dt} - 2\langle p\rangle \cdot \frac{d\langle p\rangle}{dt}$

But $\frac{d\langle p^2\rangle}{dt} = \frac{i}{\hbar}\langle[\hat{H},\hat{P}^2]\rangle = -q\xi\frac{i}{\hbar}\langle[\hat{x},\hat{P}^2]\rangle = 2q\xi\langle p\rangle$, and using (a):

$$2\Delta p \cdot \frac{d\Delta p}{dt} = 2q\xi\langle p\rangle - 2\langle p\rangle \cdot q\xi = 0 \Longrightarrow \underline{\frac{d\Delta p}{dt} = 0.}$$

(c) In momentum representation: $\hat{p} = p, \hat{x} = i\hbar\frac{\partial}{\partial p}$, so

$$H\varphi = \left(\frac{\hat{p}^2}{2m} - q\xi\hat{x}\right)\varphi = i\hbar\frac{\partial\varphi}{\partial t} \Longrightarrow \frac{p^2}{2m}\varphi - i\hbar q\xi\frac{\partial\varphi}{\partial p} = i\hbar\frac{\partial\varphi}{\partial t}$$

**3.6 Particle in a Box Problems**

In this section are grouped "particle in a box" problems that are similar to gap potential problems, but these problems typically have added complications, such as change in box size.

***Example 1***

(a) A particle of mass m is in a one-dimensional "box". The potential $V(x) = 0$ in the interval $0 < x < L$ and $V(x)$ is infinite outside this range. The particle is in the lowest energy state. Find the normalized eigenfunction corresponding to this state.

(b) A $t = 0$ the box is suddenly expanded so that $V(x)$ is changed to $V(x) = 0$ in the interval $-L < x < L$, without changing the wavefunction. Calculate the probability for finding the particle in the 2nd energy state for $t > 0$.

**Answer**

(a) The eigenfunctions are $\psi_n = A_n \sin k_n x$ . We require $\psi_n(0) = \psi_n(L) = 0$, so $k_n L = n\pi$. Also,

$$1 = |<\psi_n|\psi_n>| = A_n^2\int_0^L \sin^2 k_n x dx = A_n^2\frac{L}{2}$$

Therefore, $A_n = \sqrt{\frac{2}{L}}$, and $\psi_n = \sqrt{\frac{2}{L}}\sin\frac{n\pi x}{L}$. The lowest state is

$$\psi_1 = \sqrt{\frac{2}{L}}\sin\frac{\pi x}{L}$$

(b) The energy eigenfunction are in general $\varphi_n(x) = A_n \sin k_n x + B_n \cos k_n x$. Requiring $\psi_n(-L) = \varphi_n(L) = 0$ and $|< \varphi_n|\varphi_n >| = 1$, we find that

$$\varphi_{2n} = \sqrt{\frac{1}{L}}\sin\frac{n\pi x}{L}\quad ,\varphi_{2n+1} = \sqrt{\frac{1}{L}}\cos\frac{\left(n+\frac{1}{2}\right)\pi x}{L}$$

The second level is $\varphi_2 = \sqrt{\frac{1}{L}}\sin\frac{\pi x}{L}$. At $t = 0$, the wavefunction is $\psi(x) = \begin{cases} 0, -L < x < 0 \\ \psi_1(x),\ 0 < x < L \end{cases}$. Therefore, the probability for finding the particle in the 2nd state is:

$$P_2 = |< \varphi_2|\psi >|^2$$

Now,

$$\langle\varphi_2|\psi\rangle = \int_{-L}^{L}\varphi_2(x)\psi(x)dx = \int_0^L \varphi_2(x)\psi_1(x)dx$$

$$= \int_0^L \sqrt{\frac{1}{L}}\sin\frac{\pi x}{L}\sqrt{\frac{2}{L}}\sin\frac{\pi x}{L}dx = \frac{\sqrt{2}}{L}\int_0^L \frac{1}{2}\left(1-\cos\frac{2\pi x}{L}\right)dx = \frac{\sqrt{2}}{2}$$

Therefore $P_2 = \left(\frac{\sqrt{2}}{2}\right)^2 = \frac{1}{2}$, where $\psi_1(x) = \sqrt{\frac{2}{L}}\sin\frac{\pi x}{L}$, $\psi_2(x) = \sqrt{\frac{2}{L}}\sin\frac{2\pi x}{L}$.

***Example 2***

(a) The particle in problem 1 (a) above has a wave function corresponding to the 1st eigenstate of that potential. Find the expectation value for the following measurable quantities: $x, p, x^2$, and $p^2$.

(b) Suppose the wave functions is $|\psi\rangle = (1/\sqrt{2})|\psi_1\rangle + ((1+i)/2)|\psi_2\rangle$ at a given time, where $|\psi_1\rangle$ and $|\psi_2\rangle$ are the normalized eigenfunctions for the 1st and second energy states. Find the expectation value $\langle p^2\rangle$, using Dirac notation where possible.

(c) For the wave function given in (b), what is the probability of finding the particle in the range $0 < x < L/2$?

**Answer**

a) Since the wavefunction is symmetric around $x = \frac{L}{2} \Rightarrow \langle x \rangle = \frac{L}{2}$.
Since the wavefunction is real $\Rightarrow \langle p \rangle = 0$. For the square terms:

$$\langle x^2 \rangle = \int_0^L x^2 \frac{2}{L} \sin^2 \frac{\pi x}{L}\, dx = \frac{2L^2}{n^3} \int_0^n y^2 \sin^2 y\, dy$$

$$= L^2 \left(\frac{1}{3} - \frac{1}{2\pi^2}\right)$$

$$\langle p^2 \rangle = \int_0^L \sqrt{\frac{2}{L}} \sin \frac{\pi x}{L} \left(-\hbar^2 \frac{d^2}{dx^2}\right) \sqrt{\frac{2}{L}} \sin \frac{\pi x}{L}\, dx$$

$$= \frac{2}{L} \hbar^2 \cdot \frac{\pi^2}{L^2} \int sin^2 \frac{\pi x}{L}\, dx = \left(\frac{\hbar \pi}{L}\right)^2$$

b) We know: $H|\psi_1\rangle = E_1|\psi_1\rangle = \frac{\hbar^2\pi^2}{2mL^2}|\psi_1\rangle$ and $H|\psi_2\rangle = E_2|\psi_2\rangle = 4\frac{\hbar^2\pi^2}{2mL^2}|\psi_2\rangle$

Also: $H = +\frac{\hat{p}^2}{2m} \Rightarrow \hat{P}^2|\psi_1\rangle = \frac{\hbar^2\pi^2}{L^2}|\psi_2\rangle,\ \hat{P}^2|\psi_2\rangle = 4\frac{\hbar^2\pi^2}{L^2}|\psi_2\rangle$, so:
$\langle \hat{P}^2 \rangle = \left(\frac{1}{\sqrt{2}}\langle\psi_1| + \frac{1-i}{2}\langle\psi_2|\right)\left(\frac{1}{\sqrt{2}}\frac{\hbar^2\pi^2}{L^2}|\psi_1\rangle + \frac{1+i}{2}\frac{4\hbar^2\pi^2}{L^2}|\psi_2\rangle\right) = \frac{1}{2}\frac{\hbar^2\pi^2}{L^2} + \frac{1}{2}\frac{4\hbar^2\pi^2}{L^2} = \frac{5\,\hbar^2\pi^2}{2L^2}$

where we used $\langle\psi_1|\psi_1\rangle = \langle\psi_2|\psi_2\rangle = 1,\ \langle\psi_1|\psi_2\rangle = \langle\psi_2|\psi_1\rangle = 0$.

c) $P\left(0 < x < \frac{L}{2}\right) = \int_0^{L/2} |\psi(x)|^2\, dx = \int_0^{L/2} \left|\frac{1}{\sqrt{2}}\psi_1 + \frac{1+i}{2}\psi_2\right|^2 dx$

$= \int_0^{L/2} \left(\frac{1}{2}\psi_1^2 + \frac{1}{2}\psi_2^2 + \frac{1}{\sqrt{2}}\psi_1\psi_2\right) dx$. The first two terms, by symmetry of $\psi_1^2, \psi_2^2$ around $\frac{1}{2}$, give $\frac{1}{2} \cdot \frac{1}{2}$ each. The last term is:

$$\frac{1}{\sqrt{2}} \cdot \frac{2}{L} \int_0^{L/2} \sin \frac{\pi x}{L} \sin \frac{2\pi x}{L} = \frac{1}{\sqrt{2}} \cdot \frac{L}{2} \int_0^{\frac{L}{2}} \frac{1}{2}\left(\cos \frac{\pi x}{L} - \cos \frac{3\pi x}{L}\right) dx$$

$$= \frac{4}{3\pi\sqrt{2}}$$

$$P\left(0 < x < \frac{L}{2}\right) = \frac{1}{2} + \frac{4}{3\pi\sqrt{2}}$$

*Example 3*

A particle of mass m is in a one-dimensional box of width L represented by an infinite square-well potential. A measurement of the energy gives the value $E = (4\pi\hbar/L)^2/(2m)$.

(a) What is the degeneracy of this energy eigenvalue? After the measurement what is the wave function? Give a different normalized wave function with the same energy eigenvalue. Is this second wave function linearly independent from the first?

(b) Find $\Delta p\ \Delta x$ for the state in (a)

(c) Find $\Delta E\ \Delta x$ for the state in (a). Explain how this answer fits in with the generalized uncertainty principle

**Answer**

(a) There is no higher order degeneracy in one-dimension. The degeneracy is 1. A wave function that fits the conditions is

$$\varphi_4 = \sqrt{\frac{2}{L}}\sin\left(\frac{4\pi x}{L}\right)$$

A different normalized wave function would simply be that of $\varphi_4$ differing by a phase factor, consider

$$\varphi_4' = -\sqrt{\frac{2}{L}}\sin\left(\frac{4\pi x}{L}\right)$$

$\varphi_4$ can be expressed by $A\varphi_4$ where $A = -1$, therefore the second wave function is linearly dependent on the first.

(b) $\quad \psi = \varphi_4 = \sqrt{\frac{2}{L}}\sin\left(\frac{4\pi x}{L}\right)$

$\langle x\rangle = \int_0^L x\ \frac{2}{L}\ \sin^2\left(\frac{4\pi x}{L}\right) dx = \frac{L}{2}$

$\langle x^2\rangle = L^2\left(\frac{1}{3} - \frac{1}{2(4\pi)^2}\right)$

$\langle p\rangle = 0$

$\langle p^2\rangle = \left(\frac{\hbar 4\pi}{L}\right)^2$

$\Delta x = \sqrt{\langle x^2\rangle - \langle x\rangle^2} = \sqrt{L^2\left(\frac{1}{3} - \frac{1}{2(4\pi)^2}\right) - {L^2}/{4}} = L\sqrt{\frac{1}{12} - \frac{1}{32\pi^2}}$

$\Delta p = \sqrt{\langle p^2\rangle - \langle p\rangle^2} = {\hbar 4\pi}/{L}$

$\Delta p \Delta x = \hbar 4\pi\sqrt{{1}/{12} - {1}/{32\pi^2}}$

(c) $\quad \Delta E = \sqrt{\langle E^2\rangle - \langle E\rangle^2}$

$\hat{E} = \frac{\hat{p}^2}{2m}$ so $\Delta E = \frac{1}{2m}\sqrt{\langle \hat{p}^4 \rangle - \langle \hat{p}^2 \rangle^2}$

Since

$\langle \hat{p}^2 \rangle = \left(\frac{\hbar 4\pi}{L}\right)^2$

$\langle \hat{p}^4 \rangle = \frac{2}{L}\int_0^L \sin\left(\frac{4\pi x}{L}\right)\left(-i\hbar\frac{\partial}{\partial x}\right)^4 \sin\left(\frac{4\pi x}{L}\right) dx = \frac{2}{L}\hbar^4\left(\frac{4\pi x}{L}\right)^4 \frac{L}{2} = \left(\frac{4\pi x}{L}\right)^4$

So,

$\Delta E = 0$

Using the generalized uncertainty principle: $[\hat{x}, \hat{H}] = i\hat{C}$ where

$$\frac{\hbar^2}{2m}\frac{\partial}{dx} = i\hat{C}$$

Thus,

$$\langle \hat{C} \rangle = \int_0^L \psi^* \hat{C}\psi = \frac{2}{L}\int_0^L \sin\frac{4\pi x}{L}\left(-\frac{i\hbar^2}{m}\cdot\frac{4\pi}{L}\cdot\cos\frac{4\pi x}{L}\right) dx$$
$$= -\frac{i\hbar^2}{m}\sin^2\frac{4\pi x}{L}\Big|_0^L = 0$$

***Example 4***

A particle of mass m slides on a circular ring of unit radius to form a one-dimensional quantum system with the angle $\theta$ as the coordinate of the system.

(a) Write the classical Hamiltonian, and find the normalized eigenstates and eigenvalues of the operator $\hat{H}$. (Note that the wavefunctions representing the eigenstates must be single-valued with a continuous derivative.)

(b) Find the eigenstates and eigenvalues of the operator $\hat{p}$, and give the degeneracy.

(c) Consider the following three unnormalized states: (1) $\exp(i2x)$, (2) $\exp(i2x) + \exp(i4x)/\sqrt{2}$, (3) $\cos 4x$, and (4) $\exp(i2x + b)$, where b is a real constant. Find the states that are linearly dependent, and demonstrate the linear dependency.

**Answer**

(a) Using cylindrical coordinates due to the symmetry, and $r = 1$:

$\hat{H} = \frac{\hat{p}_\theta^2}{2mr^2}$ due to cylindrical symmetry and $r = 1$ gives $\hat{H} = \frac{\hat{p}_\theta^2}{2m}$. Now, with $\hat{H}\psi = E_k\psi$ guess $\psi = Ae^{ik\theta}$ to get:

$$\frac{-\hbar^2}{2m}\left(-k^2 e^{ik\theta}\right) = E_k e^{ik\theta}, \quad 0 \le \theta < 2\pi$$

So, $E_k = \frac{\hbar^2 k^2}{2m}$ and $\int_0^{2\pi} \psi^* \psi d\theta = 1 \rightarrow \ A = \sqrt{\frac{1}{2\pi}}$, to give:

$$\psi_k = \sqrt{\frac{1}{2\pi}}\, e^{ik\theta}$$

$$\psi(\theta) = \psi(\theta + 2n\pi) \Rightarrow k = n,$$

and there is quantization from angular momentum alone (discrete solutions).

(b) $\quad \hat{p}\psi_k = p_k\psi \quad -i\hbar\frac{\partial}{\partial\theta}\left(\sqrt{\frac{1}{2\pi}}\, e^{ik\theta}\right) = p_k\sqrt{\frac{1}{2\pi}}\, e^{ik\theta} \rightarrow -\hbar(ik) =$

$p_k = \hbar k\ eigenvalues.$

The eigenstates are

$$\psi_k = \sqrt{\frac{1}{2\pi}}\, e^{ik\theta}, \quad 0 \le \theta < 2\pi.$$

The is degeneracy 1.

(c) $e^{i2x}$ is not linearly dependent on $\left(e^{i2x} + e^{i4x}\right)/\sqrt{2}$ due to the $e^{i4x}$ term. While $e^{i2x}$ is not linearly dependent on $\cos 4x = \left(e^{i4x} + e^{i4x}\right)/\sqrt{2}$ since we can't express the higher power term by a multiple of $e^{i2x}$. $\cos 4x = \left(e^{i4x} + e^{i4x}\right)/2$ is not linearly dependent on $\left(e^{i2x} + e^{i4x}\right)/2$ since $e^{i2k}$ can't be expressed as $e^{i4x}$ as before. The only linear dependency is between (1) $e^{i2x}$ and (4) $e^{i2x+b}$ since $e^{i2x}$ can be expressed as $e^{i2x+b}$ by a multiplicative constant: $Ae^{i2x} = e^{i2x+b} for\ A = e^b$.

***Example 5***

Two identical particles are in a rigid 1-dimensional box of width $\pi d$, with $d = 1$. The single-particle eigenstates are $\psi_n = \sqrt{2/\pi}\sin nx$.Suppose the unsymmetrized wave function for the two particles is $\psi_1(x_1)\psi_2(x_2)$.

(a) The two particles are identical fermions. Write a correctly symmetrized and normalized state corresponding to the wave function defined above.
(b) Repeat (a) assuming the two particles are identical bosons.

**Answer**
Normalization check:

$$\int_{-\infty}^{\infty}\int_{-\infty}^{\infty}\left|\left(\frac{2}{\pi}\right)^2 \sin^2 x_1 \sin^2 x_2\right| dx_1 dx_2 = \left(\frac{2}{\pi}\right)^2 \cdot \left(\frac{\pi}{2}\right)^2 = 1$$

(a) Fermions – antisymmetric: $\varphi_n(x_1, x_2) = -\varphi_n(x_2, x_1)$ . So

$$\varphi_A = \psi_{n_A}(x_1, x_2) = \frac{1}{\sqrt{2}}\left[\psi_{n_1}(x_1)\psi_{n_2}(x_2) - \psi_{n_1}(x_2)\psi_{n_2}(x_1)\right]$$

(b) Bosons – symmetric:

$$\varphi_S = \psi_{n_S}(x_1, x_2) = \frac{1}{\sqrt{2}}\left[\psi_{n_1}(x_1)\psi_{n_2}(x_2) + \psi_{n_1}(x_2)\psi_{n_2}(x_1)\right]$$

***Example 6***
Consider

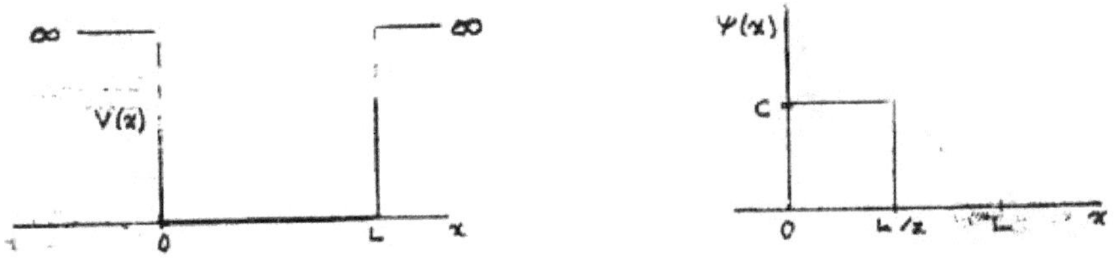


A one-dimensional potential $V(x)$ is zero for x in the range $0 < x < L$ and infinite outside this range as shown above at left. A particle of mass m is in this potential in a state described at $t = 0$ by the wave function at upper right, where $\psi(x) = C$ for x in the range $0 < x < L/2$ and zero outside this range. C is a real constant.

(a) Find the value of the real constant C to correctly normalize the wave function.
(b) Find the expectation value of the energy $\langle E \rangle$ and the position $\langle x \rangle$ for this state at $t = 0$.
(c) Find $\langle E \rangle$ and $\langle x \rangle$ at a later time $t = T$.

Note: $\sin mx \sin nx = (\cos(m - n)x - \cos(m + n)\,x)/2$

**Answer**

(a) We want $1 = \langle\psi|\psi\rangle = C^2 \int_0^{L/2} dx = C^2 \frac{L}{2} \Rightarrow \boxed{C = \sqrt{\frac{2}{L}}}$

(b) The eigenfunctions of the Hamiltonian are $\varphi_n = \sqrt{\frac{2}{L}} \sin\frac{n\pi x}{L}$ with eigenvalues $E_n = \frac{n^2\pi^2\hbar^2}{2mL^2}$

$\psi = \sum_{n=1}^{\infty} a_n \varphi_n$ , where $a_n = \langle\varphi_n|\psi\rangle = \int_0^{L/2} \frac{L}{2} \sin\frac{n\pi x}{L} = \frac{2}{n\pi}\left(1 - \cos\frac{n\pi}{2}\right)$.

Therefore $\langle E\rangle = \langle\psi|H|\psi\rangle = \sum_{n=1}^{\infty}|a_n|^2 E_n = \sum_{n=1}^{\infty}\frac{4}{n^2\pi^2}\left(1 - \cos\frac{n\pi}{2}\right)^2 \frac{n^2\pi^2\hbar^2}{2mL^2} = \frac{2\hbar}{mL^2}\sum_{n=1}^{\infty}\left(1 - \cos\frac{n\pi}{2}\right)^2$, which is infinite. Also, $\langle x\rangle = \langle\psi|x|\psi\rangle = \frac{2}{L}\int_0^{L/2} xdx = \frac{L}{4}$.

(c) At time $t > 0$, $\psi = \sum_{n=1}^{\infty} a_n\psi_n e^{-\frac{E_n t}{\hbar}} = \sum_{n=1}^{\infty} a_n\psi_n e^{-in^2 w_0 t}$, where $w_0 = \frac{\pi^2\hbar}{2mL^2}$.

So $\langle E(t)\rangle \sum_{n=1}^{\infty}\left|a_n e^{-in^2 w_0 t}\right|^2 E_n = \sum_{n=1}^{\infty}|a_n|^2 E_n = \langle E\rangle = \infty$

Also, $\langle x\rangle = \int_0^L dxx\left(\sum_{n=1}^{\infty} a_n e^{-in^2 w_0 t}\varphi_n\right)\left(\sum_{n=1}^{\infty} a_n e^{-im^2 w_0 t}\varphi_m\right)$

$$= \sum_{n=1}^{\infty}\sum_{n=1}^{\infty} e^{-i(n^2-m^2)w_0 t} A_{mn}\, a_m a_n$$

Where $A_{mn} = \int_0^L d\,x\,x\,\varphi_n\,\varphi_m = \frac{1}{L}\int_0^L d\,x\,x\left[\cos\frac{(n-m)\pi x}{L} - \cos\frac{(n+m)\pi x}{L}\right]$

Integrating by parts, we get $A_{mn} = -\frac{4mnL}{(m^2-n^2)^2\pi^2}\left(1 - (-1)^{m+n}\right)$

If $m \neq n$ and $A_{n\pi} = \frac{L}{2}$. Therefore,

$$\langle x\rangle = \frac{L}{2} + \sum_{n\neq m}\frac{16L}{\pi^4}\left(1 - \cos\frac{n\pi}{2}\right)\left(1 - \cos\frac{m\pi}{2}\right)(1 - (-1)^{m+n})\frac{e^{-i(n^2-m^2)w_0}}{(n^2-m^2)^2}$$

***Example 7***

A three-dimensional spherically-symmetric potential V(r) is defined by

$$V(r) = 0 \qquad r < R$$

$$V(r) = \infty \qquad r > R$$

a) Find the spherically symmetric wavefunctions, and the corresponding energy eigenvalues.

b) Initially, a particle is in the ground state of the potential V(r). The potential is then very slowly (adiabatically) changed, to new potential $\tilde{V}(r)$:

$$\tilde{V}(r) = 0 \qquad r < L$$
$$\tilde{V}(r) = \infty \qquad r > L$$

with L<R. What is the new wavefunction of the particle, and its energy?

c) Now the modified potential $\tilde{V}$ is suddenly (instantaneously) changed back to the original potential V. What is the probability that the particle is in the ground state?

d) What is the net change in the particle's mean energy during the whole cycle $(V \to \tilde{V} \to V)$?

**Answer**

$$V(r) = 0 \qquad r < R$$
$$V(r) = \infty \qquad r > R$$
$$\left\{\frac{-\hbar^2\nabla^2}{2m} + V(r)\right\}\varphi(r) = E\varphi(r)$$

$\nabla^2 = \frac{1}{r}\frac{d^2}{dr^2}r +$ angular terms

(a)

$$\frac{-\hbar^2}{2m}\frac{1}{r}\frac{d^2}{dr^2}r\varphi(r) = E\varphi(r) \qquad let\ \varphi(r) = \frac{u(r)}{r}$$

$$\frac{-\hbar^2}{2m}\frac{d^2}{dr^2}u(r) = Eu(r) \qquad \lim_{r\to 0} u(r) \to 0$$

$thus\ u(0) = 0$

Reduces 1-d potential problem:

$$kR = n\pi \qquad k = \frac{n\pi}{R}$$

$$k = \sqrt{\frac{2mE}{\hbar^2}}$$

$$u = A\sin kr$$

$$\int_0^R |\varphi|^2 r^2 dr = \int_0^R |u|^2 dr = A^2\frac{R}{2} \to A = \sqrt{\frac{2}{R}}$$

$$\varphi_n(r) = \frac{\sqrt{\frac{2}{R}}\sin\left(\frac{n\pi r}{R}\right)}{r} \qquad n = 1,2,3\ldots \quad E_n = \frac{-\hbar^2\left(\frac{n\pi r}{R}\right)^2}{2m}$$

(b) ground state: $n = 1$. adiabatic, the number of nodes don't change: n=1

$$\varphi_n(r)_{(L)} = \frac{\sqrt{\frac{2}{L}}\sin\left(\frac{n\pi r}{L}\right)}{r} \qquad (n = 1) \quad E_1 = \frac{-\hbar^2\left(\frac{\pi}{L}\right)^2}{2m}$$

(c) Want to project $\varphi_1(r)_{(L)} = \frac{\sqrt{\frac{2}{L}}\sin\left(\frac{n\pi r}{L}\right)}{r}$ into the former set of wavefunctions $\varphi_n(r) = \frac{\sqrt{\frac{2}{R}}\sin\left(\frac{n\pi r}{R}\right)}{r}$ in particular, what is the projection on the $n = 1$ state? (groundstate)

Denote probability in ground state P:

$$P = \int_0^L \frac{\sqrt{\frac{2}{R}}\sin\left(\frac{n\pi r}{R}\right)}{r}\frac{\sqrt{\frac{2}{L}}\sin\left(\frac{n\pi r}{L}\right)}{r} r^2 dr = \frac{2}{\sqrt{RL}}\int_0^L \sin\left(\frac{\pi r}{R}\right)\sin\left(\frac{\pi r}{L}\right) dr$$

$$\cos(A \pm B) = \cos A\cos B \mp \sin A \sin B$$

So, $\sin A \sin B = \frac{1}{2}(\cos(A - B) - \cos(A + B))$

$$\mathcal{P} = \frac{2}{\sqrt{RL}}\int_0^L \frac{1}{2}\left\{\cos\left(\pi\left(\frac{1}{R} - \frac{1}{L}\right)r - \cos\left(\pi\left(\frac{1}{R} - \frac{1}{L}\right)r\right)\right)\right\} dr$$

$$= \frac{2}{\pi}\left\{\frac{\sin\left(\pi\left(\frac{1}{R} - \frac{1}{L}\right)r\right)}{\pi\left(\frac{1}{R} - \frac{1}{L}\right)}\Big|_0^L - \frac{\sin\left(\pi\left(\frac{1}{R} + \frac{1}{L}\right)r\right)}{\pi\left(\frac{1}{R} + \frac{1}{L}\right)}\Big|_0^L\right\}$$

$$= \frac{2\sqrt{RL}}{\pi}\left\{\frac{\sin\left(\pi\frac{L}{R} - \pi\right)}{(L - R)} + \frac{\sin\left(\pi\frac{L}{R} - \pi\right)}{(L + R)}\right\}$$

$$= \frac{2\sqrt{RL}}{\pi}\left\{-\frac{\sin\left(\pi\frac{L}{R}\right)}{(L-R)} + \frac{\sin\left(\pi\frac{L}{R}\right)}{(L+R)}\right\} = \frac{2\sqrt{RL}\sin\left(\pi\frac{L}{R}\right)}{\pi}\left(\frac{2R}{R^2-L^2}\right)$$

$$P = \left(\frac{2}{\pi}\sin\left(\pi\frac{L}{R}\right)\frac{R^{3/2}\sqrt{L}}{R^2-L^2}\right)$$

(d) originally $E_1^{(R)} = \frac{\hbar^2\left(\frac{\pi}{R}\right)^2}{2m}$ then $E_1^{(L)} = \frac{\hbar^2\left(\frac{\pi}{L}\right)^2}{2m}$ Get : $\Delta_x$ *decreases*, $\Delta_p$ *increases*, *E increase*

Mean energy:

$$\langle \varphi^{(L)'} | H^{(R)} | \varphi^{(L)'} \rangle$$

$$= \int_0^L \left( \frac{\sqrt{\frac{2}{L}} \sin\left(\frac{\pi r}{L}\right)}{r} \right) \left\{ \frac{-\hbar^2}{2m} \frac{1}{r} \frac{d^2}{dr^2} r \right\} \left( \frac{\sqrt{\frac{2}{L}} \sin\left(\frac{\pi r}{L}\right)}{r} \right) r^2 dr$$

$$= \left(\frac{2}{L}\right) \frac{\hbar^2}{2m} = \int_0^L \sin\left(\frac{\pi r}{L}\right) \left(-\left[\frac{\pi}{L}\right]^2\right) \sin\left(\frac{\pi r}{L}\right) dr = \left(\frac{2}{L}\right) \left[\frac{\pi}{L}\right]^2 = \frac{\hbar^2 \left(\frac{\pi}{L}\right)^2}{2m}$$

Energy after sudden change remains the same

$$\left.\begin{matrix} V \\ E = \frac{\hbar^2 \left(\frac{\pi}{L}\right)^2}{2m} \end{matrix}\right\} \rightarrow \left.\begin{matrix} \tilde{V} \\ E = \frac{\hbar^2 \left(\frac{\pi}{L}\right)^2}{2m} \end{matrix}\right\} \rightarrow \begin{matrix} V \\ E = \frac{\hbar^2 \left(\frac{\pi}{L}\right)^2}{2m} \end{matrix}$$

Net change during cycle:

$$\boxed{\Delta E = \frac{\hbar^2 \pi^2}{2m} \left( \frac{1}{L^2} - \frac{1}{R^2} \right)}$$

## 3.7 Particle in a central potential – the hydrogen atom

***Stationary states of a particle in a central potential***

Classical Hamiltonian: $H = \frac{{P_r}^2}{2\mu} + \frac{L^2}{2\mu r^2} + V(r)$ where

$$V_{eff}(r) = V(r) + \frac{L^2}{2\mu r^2}$$

Quantum Mechanical Hamiltonian:

$$\left[ \frac{-\hbar^2}{2\mu} \Delta + V(r) \right] \varphi(\vec{r}) = \mathrm{E} \varphi(\vec{r})$$

$$\Delta = \frac{1}{r} \frac{\partial^2}{\partial r^2} r + \frac{1}{2} \left( \frac{\partial^2}{\partial \theta^2} + \frac{1}{tan\theta} \frac{\partial}{\partial \theta} + \frac{1}{sin^2\theta} \frac{\partial^2}{\partial \theta^2} \right) = \frac{1}{r} \frac{\partial^2}{\partial r^2} r + \frac{\vec{L}^2}{r^2} \left( \frac{-1}{\hbar^2} \right)$$

So,

$$H = \frac{-\hbar^2}{2\mu} \frac{1}{r} \frac{\partial^2}{\partial r^2} r + \frac{1}{2\mu r^2} \vec{L}^2 + V(r)$$

Now, to solve the eigenvalue equation

$$\left[ \frac{-\hbar^2}{2\mu} \frac{1}{r} \frac{\partial^2}{\partial r^2} r + \frac{\vec{L}^2}{2\mu r^2} + V(r) \right] \varphi(r, \theta, \varphi) = E \varphi(r, \theta, \varphi)$$

Since the components of $\vec{L}$ commute with $r$ $and$ $L^2$

$$\left[ H, \vec{L} \right] = 0$$

The components of $\vec{L}$ are constants of the motion. Although we now have four constants of the motion ($\vec{L}$ and $\vec{L}^2$) we can't use them all in the same eigenvalue equation because they don'tall commute. A basis is typically chosen with the commuting elements H, $\vec{L}^2, L_{2,}$ :

$$\left\{\begin{aligned} H\varphi(\vec{r}) &= E\varphi(\vec{r}) \\ \vec{L}^2\varphi(\vec{r}) &= \ell(\ell+\mathrm{I})\hbar^2\varphi(\vec{r}) \\ L_z\varphi(\vec{r}) &= m\hbar\varphi(\vec{r}) \end{aligned}\right\}$$

We already know the eigenfunctions common to $\vec{L}^2$ *and* $L_z$, namely the spherical harmonics $Y_l^m(\theta,\varphi)$. So, consider $\varphi(\vec{r}) = R(\vec{r})Y_l^m(\theta,\varphi)$ and let's solve the radial equation:

$$\left[\frac{-\hbar^2}{2m}\frac{1}{r}\frac{\partial^2}{\partial r^2}r + \frac{\ell(\ell+I)\hbar^2}{2\mu r^2} + V(r)\right]R(r) = ER(r)$$

The above equation depends on ℓ but not m, so let $R(r) = \frac{1}{r}U_{k,\ell}(r)$:

$$\left[\frac{-\hbar^2}{2\mu}\frac{d^2}{dr^2}r + \frac{\ell(\ell+\mathrm{I})\hbar^2}{2\mu r^2} + \mathrm{V(r)}\right]U_{k,\ell}(r) = E_{k,\ell}U_{k,\ell}(r)$$

Once again we have a one-directional problem with $V_{eff}(r) = \mathrm{V(r)}$ $+\ell(\ell+\mathrm{I})\hbar^2/2\mu r^2$. Also, $r \geq 0$. Note that in order for $U_{k,\ell}(r)$ to be a solution it must be sufficiently regular at the origin, and this is what eventually forces the quantization of possible states. Accordingly, let's consider the asymptotic behavior as we approach the origin. Assume V(r) $\to \infty$ less rapidily than $\frac{1}{\gamma}$ as r→ 0

(true for the case of a Coulomb potential). So, as r→ 0 *we may ignore* $V(r)$*to get*

$$\left[\frac{-\hbar^2}{2\mu}\frac{d^2}{dr^2}r + \frac{\ell(\ell+\mathrm{I})\hbar^2}{2\mu r^2}\right]U_{k,\ell}(r) \simeq E_{k,\ell}U_{k,\ell}(r)$$

Now, assume U behaves at the origin as $r^v$

$$R_{k,\ell}(r) \sim cr^s, \qquad then: \; -(S+\mathrm{I})S + \ell(\ell+\mathrm{I}) = 0 \quad (to\ a\ given\ order)$$

Consequently, either S=$\ell$ *or* $S = -(\ell+\mathrm{I})$. Acceptable solutions behave like

$$U_{k,\ell}(r) \sim Cr^{\ell+\mathrm{I}} for\ r \to 0.$$

Consequently we have the added condition $U_{k,\ell}(0) = 0$.

The range of r is 0 to $\infty$, *however,* thanks to the condition $U_{k,\ell}(0) = 0$, we can assume that we are actually dealing with a one-dimensional problem where the effective potential is infinite for all negative values of the variable. Thus, $U_{k,\ell}(0) = 0$ insures the continuity of the wave function at r=0.

Since V(r) is independent of $\theta$ *and* $\varphi$ it is possible to require the eigenfunctions of H to be simultaneous eigenfunctions of $\vec{L}^2$ *and* $L_z$:

$$\psi_{k,\ell,m}(\vec{r}) = R_{k,\ell} Y_\ell^m(\theta, \varphi)$$

We can replace the eigenvalue equation of H by a differential equation involving only r. This differential equation is 2nd order so has a priori two linearly independent solutions, but one of these solution in eliminated by $U_{k,\ell}(0) = 0$. Thus, H, $\vec{L}^2$, and $L_z$ contribute a complete set of commuting observables.

Essential degeneracies follow from H being independent of $L_z$. Accidental degeneracy results when $E_{k,\ell} = U_{k',\ell'}$ for a given potential.

***Motion of Center-of-Mass and relative motion for a system of two interacting particles***

Consider

$$\vec{R}_G = \frac{m_1 R_1 + m_2 R_2}{m_1 + m_2}$$

motion of C.M operator, where $\vec{R} = \vec{R}_1 - \vec{R}_2$ (the relative motion operator). Now,

$$[X_1, P_{1x}] = i\hbar, \qquad [X_2, P_{2x}] = i\hbar, \qquad etc.$$

Yields:

$$[X_G, P_G] = i\hbar \ and \ [X, P_X] = i\hbar$$

We can interpret $\overrightarrow{R_G}$ and $\overrightarrow{P_G}$ as being the position and momentum observables of two distinct fictitious particles.

$$\vec{J} = \overrightarrow{L_1} + \overrightarrow{L_2} = \overrightarrow{R_1} x \overrightarrow{P_1} + \overrightarrow{R_2}\, x \overrightarrow{P_2} = \overrightarrow{L_G} + \vec{L} = \overrightarrow{R_G} x \overrightarrow{P_G} + \vec{R} x \vec{P}$$

***Hydrogen atom***

$m_p = 1.7x10^{-27}$kg q=1.6x$10^{-19}$ *Coulomb*

$m_e = 0.911x10^{-30} kg$

$$V(r) = \frac{-q^2}{4\pi\epsilon_0}\frac{1}{r} = \frac{-e^2}{r}$$

$$H(\vec{r}, \vec{p}) = \frac{\vec{p}^2}{2\mu} - \frac{e^2}{r}$$

since $m_p \gg m_e$ $\mu = \frac{m_p m_e}{m_e + m_p} \simeq m_e\left(1 - \frac{m_e}{m_p}\right)$

***Bohr model***

(based on concept of a trajectory -- is incompatible with Q.M) The model is based on the hypothesis that the electron describes a circular orbit obeying the following equation:

$$\left.\begin{aligned} E &= \frac{1}{2}\mu v^2 - \frac{e^2}{r} \\ \frac{\mu v^2}{r} &= \frac{e^2}{r^2} \end{aligned}\right\} \mathit{Classical\ , circular\ orbit}$$

Bohr then requires

$$\mu v r = n\hbar; n \geq 0 \ and\ integer$$

a quantitation condition, introduced to explain the discrete energy levels. We have:

$$\mu v^2 = \frac{e^2}{r}, \qquad \frac{\mu v}{n\hbar} = \frac{\mathrm{I}}{r} \quad \rightarrow \quad \mu v^2 = \frac{e^2 \mu v}{n\hbar} \quad \rightarrow v_n = \frac{e^2}{n\hbar}$$

$$v_{(n)} = \left(\frac{e^2}{\hbar}\right)\frac{\mathrm{I}}{n}, \quad r_n = \frac{n\hbar}{\mu v} = n^2\left(\frac{\hbar^2}{\mu e^2}\right)$$

$$r_{(n)} = \left(\frac{\hbar^2}{\mu e^2}\right)n^2, \quad \left(\frac{\hbar^2}{\mu e^2}\right) = a_0 = 0.52\ \text{Å}$$

$$E_n = \frac{-1}{2}\mu\left(\frac{e^2}{\hbar}\right)^2 \frac{\mathrm{I}}{n^2}, \quad \frac{\mu}{2}\left(\frac{e^2}{\hbar}\right)^2 = 13.6eV$$

***Alternate Derivation***

Let's repeat the analysis of the Hydrogen atom with central potential by using dimensionless variables. as before, we have potential:

$$V = -\frac{e^2}{r},$$

thus,

$$\left[-\frac{\hbar^2}{2\mu}\Delta - \frac{e^2}{r}\right]\psi = E\psi$$

For a central potential we can separate:

$$\psi(r) = \frac{\mathrm{I}}{r}U_{k,l}(r)Y_l^m(\theta,\varphi)$$

To get:

$$\left[-\frac{\hbar^2}{2\mu}\left\{\frac{\mathrm{I}}{r}\frac{d^2}{dr^2}r + \frac{1}{r^2}\left(\frac{\vec{L}^2}{-\hbar^2}\right)\right\} - \frac{e^2}{r}\right]\frac{1}{r}U_{k,l}(r)Y_l^m(\theta,\varphi)$$
$$= \frac{E}{r}U_{k,l}(r)Y_l^m(\theta,\varphi)$$

which simplifies in the usual way to give:

$$\left[-\frac{\hbar^2}{2\mu}\frac{d^2}{dr^2}+\frac{l(l+1)\hbar^2}{2\mu r^2}-\frac{e^2}{r}\right]U_{k,l}(r)=U_{k,l}(r)\mathrm{E}_{k,l}$$

where $U_{k,l}(0)=0$. Analysis of the differential equation follows, but first, let's switch to dimensionless variables. Let:

$$\rho = r/a_0$$

$$\lambda_{k,l} = \sqrt{-E_{k,l}/E_I}$$

$$a_0 = \frac{\hbar^2}{\mu e^2}$$

$$E_I = \frac{\mu\, e^4}{2\,\hbar^2}$$

We now have:

$$\left[\frac{d^2}{d\rho^2}-\frac{\ell(\ell+l)}{\rho^2}+\frac{2}{\rho}-\lambda_{k,l}{}^2\right]U_{k,l}(\rho)=0$$

Let's examine the asymptotic (large $\rho$) behavior:

$$\left[\frac{d^2}{d\rho^2}-\lambda_{k,l}{}^2\right]U_{k,e}(\rho)\simeq 0 \rightarrow \quad U_{k,e}\cong e^{\pm\lambda_{k,l}\,\rho}$$

So, let $U_{k,e}(\rho)=e^{-\lambda_{k,l}\,\rho}y_{k,l}\,(\rho)$:

$$\frac{d^2}{d\rho^2}U=\frac{d}{d\rho}\left(e^{-\rho\lambda}\frac{dy}{d\rho}-\lambda e^{-\rho\lambda}y\right)$$

$$=-2\lambda e^{-\rho\lambda}\frac{dy}{d\rho}-e^{-\rho\lambda}\frac{d^2y}{d\rho^2}+\lambda^2e^{-\rho\lambda}y$$

Thus,

$$\left[\frac{d^2}{d\rho^2}-2\lambda\frac{d}{d\rho}+\left[\frac{2}{\rho}-\frac{l(l+1)}{\rho^2}\right]\right]y_{k,l}(\rho)=0$$

Now, try a power series solution of form:

$$y_{k,l}(\rho)=p^s\sum_{q=0}^{\infty}C_q\rho^q$$

By definition we take $C_0$ to be the first non-zero coefficient of the expansion, and get:

$$(q+s)(q+s-1)-\ l(l+1)=0$$

For $q=0$:

$$s(s-1)-\ l(l+1)=0\ \Rightarrow\ s=l+1\ or\ s=-l$$

where only $s=l+1$ is acceptable at the origin. The general characteristic equation has the form:

$$(q+s)(q-s-1)c_q-2\lambda(q-1+s)C_{q-1}+2C_{q-1}-\ l(l+1)C_q=0$$

If we let $s = l + 1$:

$$q(q + 2l + 1)C_q = 2[(q + l)\lambda_{k,l} - \text{I}]\, C_{q-1}$$

Fixing $C_0$, this gives us $C_q$ by recursion.

We want physically acceptable asymptotic behavior, such as boundedness, with the preceding recursion relation this boundedness will not occur unless the recursion terminates. In other words, we require a termination condition such as will occur with

$$(q + l)\lambda_{k,l} - \text{I} = 0 \text{ for some } k = q \rightarrow \quad \lambda_{k,\ell} = \frac{1}{k+l}$$

Thus,

$$E_{k,l} = \frac{-E_I}{(k + l)^2} \quad k = 1,2,3 \ldots$$

$$R_{k=1,l=0}(r) = 2(a_0)^{-3/2}\, e^{-r/d_0}$$

$$R_{k=1,l=0}(r) = 2(2a_0)^{-3/2}\left(1 - \frac{r}{2a_0}\right) e^{-r/2a_0}$$

$$R_{k=1,l=1}(\mathrm{r}) = (2a_0)^{-3/2} \frac{1}{\sqrt{3}} \frac{r}{a_0} e^{-r/2a_0}$$

***Stark effect***

Let's calculate the perturbation of the first two energy levels of a hydrogen atom in an electric field $\vec{E}$ :

$$H' = -\vec{d} \cdot \vec{E} = -e\vec{r} \cdot \vec{E} = -ez|E| \quad (\,choose\ \vec{E}\ along\ \vec{z})$$

The ground state |100> of hydrogen is non-degenerate. Because the corresponding wavefunction is even under the transformation $\text{Z} \rightarrow -\text{Z}$, we have $< 100|\text{Z}|100 >= 0$. For $n = 2$ we have:

$$n = 2: \quad k + l = 2 \quad Thus: \quad \begin{array}{c} k = 1, l = 1 \ \rightarrow m = -1,0,1 \\ k = 2, l = 0 \ \rightarrow m = 0 \end{array}$$

which is four fold degenerate. To study the perturbation of this level in first-order approximation the following linear combination must be formed:

$$\Psi^{(0)} = \sum_{k=1}^{4} a_k \Psi_K$$

Where $\Psi_1 = |200>$, $\Psi_2 = |210>$, $\Psi_3 = |211>$ $\Psi_4 = |21(-1)>$.

***The Hydrogen atom wavefunctions***

| | $n$ | $l$ | $m$ |
|---|---|---|---|
| | 1 | 0 | 0 |
| $\Psi_1$ | 2 | 0 | 0 |
| $\Psi_3$ | 2 | 1 | 1 |
| $\Psi_2$ | 2 | 1 | 0 |
| $\Psi_4$ | 2 | 1 | −1 |

$$\varphi_{100} = \frac{1}{\sqrt{\pi a^3}} \exp(-{}^{r}/_{a})$$

$$\varphi_{200} = \frac{1}{4\sqrt{4\pi a^3}} \left(2 - \frac{r}{a}\right) \exp(-{}^{r}/_{2a})$$

$$\varphi_{211} = \frac{1}{4\sqrt{4\pi a^3}} \left(\frac{r}{a}\right) \exp(-{}^{r}/_{2a}) \sin\theta e^{i\theta}$$

$$\varphi_{210} = \frac{1}{4\sqrt{4\pi a^3}} \left(\frac{r}{a}\right) \exp(-{}^{r}/_{2a}) \cos\theta$$

$$\varphi_{21-1} = \frac{1}{4\sqrt{4\pi a^3}} \left(\frac{r}{a}\right) \exp(-{}^{r}/_{2a}) \sin\theta e^{-i\theta}$$

The only non-vanishing matrix elements of the perturbation are:

$$H'_{12} = H^{I}_{21} = -eE < 200|\text{Z}|210 > = -3eaE$$

where $a = \hbar^2/me^2$ .

So,

$$\varepsilon = E_2 - E_2^{(0)} \quad and \quad \begin{vmatrix} -\varepsilon & -3eaE & 0 & 0 \\ -3eaE & -\varepsilon & 0 & 0 \\ 0 & 0 & -\varepsilon & 0 \\ 0 & 0 & 0 & -\varepsilon \end{vmatrix} = 0$$

Implies $(\varepsilon^2 - qe^2a^2E^2)\varepsilon^2 = 0$ with roots

$$\varepsilon_1 = 3eaE, \varepsilon_1 = -3eaE, \varepsilon_3 = 0, \varepsilon_4 = 0.$$

So,

$$\varepsilon = \varepsilon_1 \; and \; \begin{pmatrix} \varepsilon_1 & \varepsilon_1 & 0 & 0 \\ \varepsilon_1 & \varepsilon_1 & 0 & 0 \\ 0 & 0 & \varepsilon_1 & 0 \\ 0 & 0 & 0 & \varepsilon_1 \end{pmatrix} \begin{pmatrix} a_1 \\ a_2 \\ a_3 \\ a_4 \end{pmatrix} = 0$$

Thus,

$a_1 = -a_2$ and $a_3 = a_4 = 0$. Thus, $\Psi_1 = a(\varphi_1 - \varphi_2)$ (a determined by normalization). Similarly:

$$\varepsilon = \varepsilon_2 \rightarrow \Psi_2 = a(\varphi_1 + \varphi_2)$$

For $\varepsilon_3 = \varepsilon_4 = 0$ and $\left.\begin{aligned}\Psi_3 = a_3\varphi_3 + a_4\varphi_4\\ \Psi_3 = a'_3\varphi_3 + a'_4\varphi_4\end{aligned}\right\}$ which is still degenerate between the $m = \pm 1$ states.

Thus:

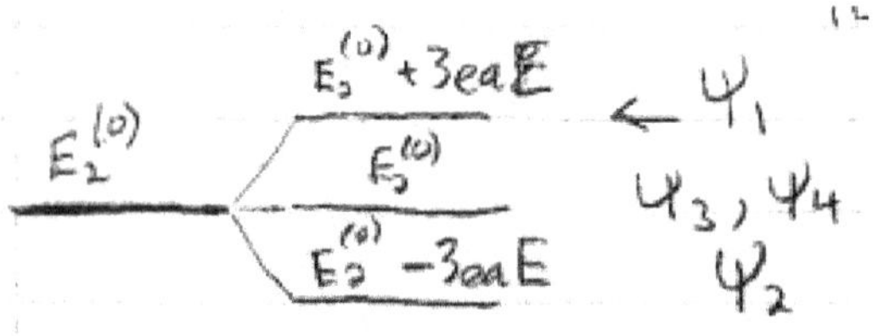


***Example 1***
Consider the motion of non-interacting particles in an axially symmetric harmonic oscillator potential:

$$V(x, y, z) = \frac{1}{2}mw_0^2[e^{\alpha}(x^2 + y^2) + e^{-2\alpha}z^2].$$

Add particles, one at a time, assume spin ½ so can have 2 particles per set of oscillator quantum numbers. For a given value of particle number under such a construction, the lowest energy state of the system can be obtained by considering the variation of energy with respect to the $\alpha$ parameter (related to the axial versus radial split optimization in the wavefunction). Now,
(1) Show that there are two shells and that $\alpha = 0$ is the optimum at each shell closure.
(2) Examine the variation of $\alpha$ with particle number for the two shells.

**Answer**
(1) The Hamiltonian is:

$$H = \frac{-\hbar^2\nabla^2}{2m} + V(x, y, z), where\ V(x, y, z) = \frac{1}{2}mw_0^2[e^{\alpha}(x^2 + y^2) + e^{-2\alpha}z^2].$$

The wavefunction will be separable, so let:

$$\psi(x, y, z) = f(x)g(y)h(z).$$

$$\frac{-\hbar^2}{2m}f(x)g(y)\frac{d^2h(z)}{dz^2} - \frac{\hbar^2}{2m}g(y)h(z)\frac{d^2f(z)}{dx^2} - \frac{\hbar^2}{2m}h(z)f(x)\frac{d^2g(y)}{dy^2} + V(x, y, z)\varphi = 0$$

Thus,

$$\frac{1}{h}\frac{d^2h}{dz^2}+\frac{1}{f}\frac{d^2f}{dx^2}+\frac{1}{g}\frac{d^2g}{dy^2}-\left[\left(\frac{mw_0}{\hbar}\right)^2\{e^{\alpha}x^2+e^{\alpha}y^2+e^{-2\alpha}z^2\}-\frac{2mE}{\hbar^2}\right]$$
$$=0$$

Let's start with the z-dependent part:

$$\frac{1}{h}\frac{d^2h}{dz^2}-\left(\frac{mw_0e^{-\alpha}}{\hbar}\right)^2 z^2+\frac{2mE}{\hbar^2}$$
$$=-\left\{\frac{1}{f}\frac{d^2f}{dx^2}+\frac{1}{g}\frac{d^2g}{dy^2}-\left(\frac{mw_0e^{\alpha/2}}{\hbar}\right)^2(x^2+y^2)\right\}=\frac{2m}{\hbar}C$$

Thus,

$$\frac{1}{h}\frac{d^2h}{dz^2}-\left(\frac{mw_0e^{-\alpha}}{\hbar}\right)^2 z^2=\frac{2m}{\hbar}(C-E).$$

Let $\mathfrak{z}=\alpha z$ and alpha bold (to match notation), $\boldsymbol{\alpha}=\sqrt{\frac{mw_0e^{-\alpha}}{\hbar}}$ to get:

$$\frac{d^2h}{d\mathfrak{z}^2}+\left[\frac{2m(E-C)}{\hbar^2\left(\frac{mw_0e^{-\alpha}}{\hbar}\right)}-\mathfrak{z}^2\right]h=0.$$

This is the Schrodinger equation for the harmonic oscillator, thus the solutions are quantized as follows:

$$\frac{2m(E-C)e^{\alpha}}{\hbar w_0}=(2n_z+1)\ ;n_z=1,2,\ldots$$

Let's now solve the x-dependent part similarly:

$$\frac{1}{f}\frac{d^2f}{dx^2}-\left(\frac{mw_0e^{\alpha/2}}{\hbar}\right)^2x^2+\frac{2mC}{\hbar^2}=\frac{2mC'}{\hbar^2}$$

Let $\mathcal{E}=\beta x$ and $\beta=\sqrt{\frac{mw_0e^{\alpha/2}}{\hbar}}$ and rewrite as:

$$\frac{d^2f}{d\mathcal{E}^2}+\left[\frac{2m(C-C')}{\hbar^2\left(\frac{mw_0e^{\alpha/2}}{\hbar}\right)}-\mathcal{E}^2\right]f=0$$

And similarly:

$$\frac{2m(C-C')}{\hbar w_0e^{\alpha/2}}=(2n_x+1)\ ;n_x=1,2,\ldots$$

And, finally:

$$\frac{1}{g}\frac{d^2g}{dy^2}-\left(\frac{mw_0e^{\propto/2}}{\hbar}\right)^2 y^2=\frac{-2mC'}{\hbar^2}$$

Let $\rho=\gamma y$ and $\gamma=\left(\frac{mw_0e^{\propto/2}}{\hbar}\right)^{1/2}$ to get:

$$\frac{d^2g}{d\rho^2}=\left[\frac{2m(C')}{\hbar^2\left(\frac{mw_0e^{\propto/2}}{\hbar}\right)}-\rho^2\right]g=0$$

Thus,

$$\frac{2mC'}{\hbar w_0e^{\alpha/2}}=(2n_y+1)\ ;n_y=1,2,\ldots$$

Combining the energy terms:

$$E=\frac{\hbar w_0}{2}\left[e^{-\alpha}(2n_z+1)+e^{\alpha/2}(2n_x+1)+e^{\alpha/2}(2n_y+1)\right]$$

Regrouping:

$$E=\hbar w_0\left[e^{\alpha/2}(n_x+n_y+1)+e^{-\alpha}\left(n_z+\frac{1}{2}\right)\right]$$

$$E_{n_xn_yn_z}=E$$

We now have:

$$E_{000}=\hbar w_0\left[e^{\alpha/2}+\frac{1}{2}e^{-\alpha}\right]$$

So,

$$\frac{dE_{000}}{d\alpha}=\frac{\hbar w_0}{2}\left[e^{\alpha/2}-e^{-\alpha}\right]=0\rightarrow\quad \alpha=0$$

So, for the first particle:

1st particle: $E=E_{000}=\frac{3}{2}\hbar w_0,\ \ \alpha=0.$

And the second particle has the same quantum numbers (different spin):

2nd particle: $E_{total}=2E_{000}=3\hbar w_0,\quad \alpha=0.$

When adding the third particle, there are two cases to consider:

3rd particle: first case

$$2E_{000}+E_{001}=2\hbar w_0\left[e^{\alpha/2}+\frac{1}{2}e^{-\alpha}\right]+\hbar w_0\left[e^{\alpha/2}+\frac{3}{2}e^{-\alpha}\right]$$

$$=\hbar w_0\left[3e^{\alpha/2}+\frac{5}{2}e^{-\alpha}\right]$$

Which yields: $\alpha=.341$ and $E_{total}=5.33\hbar w_0$.

3rd particle: second case

$$2E_{000} + E_{100} = \hbar w_0 \left[4e^{\alpha/2} + \frac{3}{2}e^{-\alpha}\right] \quad with \quad \alpha = -.192 \ and \ E_{total}$$
$$= 5.45$$

Since the first case has lower energy than 3rd particle will choose it over the second case. Thus, we have:

3rd particle: $E_{total} = 2E_{000} + E_{001} = 5.33\hbar w_0, \ and \ \alpha = .341.$

When adding a fourth particle, it is not clear if we should double the occupancy of the current level (where the 3rd particle resides) or ad into one of the other quantum states. Again, the resolution is by the lowest energy state (which corresponds to adding to the existing level to double-up):

4th particle :

First case

$2E_{000} + 2E_{001} = \hbar w_0\left[2e^{\alpha/2} + e^{-\alpha}\right] + \hbar w_0\left[2e^{\alpha/2} + 3e^{-\alpha}\right] =$
$\hbar w_0\left[4e^{\alpha/2} + 4e^{-\alpha}\right]$

$\alpha = .462 \qquad E_{total} = 7.56\ \hbar w_0$

Second case

$2E_{000} + E_{001} + E_{001} = \hbar w_0\left[2e^{\alpha/2} + e^{-\alpha}\right] + \hbar w_0\left[2e^{\alpha/2} + 3e^{-\alpha}\right] =$
$\hbar w_0\left[4e^{\alpha/2} + 4e^{-\alpha}\right]$

$\alpha = .123 \qquad E_{total} = 7.56\ \hbar w_0$

Third case

$2E_{000} + 2E_{001} = \hbar w_0\left[2e^{\alpha/2} + e^{-\alpha}\right] + \hbar w_0\left[2e^{\alpha/2} + 3e^{-\alpha}\right] =$
$\hbar w_0\left[4e^{\alpha/2} + 4e^{-\alpha}\right]$

$\alpha = 270 \qquad E_{total} = 7.56\hbar w_0$

4th particle uses first case: $E_{000} + 2E_{001} = 7.56\hbar w_0, \alpha = .462.$

5th particle (there are now four cases to consider):

First case

$$2E_{000} + 2E_{001} + E_{100} = \hbar w_0 \left[6e^{\alpha/2} + \frac{9}{2}e^{-\alpha}\right]$$
$$\alpha = .270 \quad and \quad E_T = 10.3\hbar w_0$$

Second case

$$2E_{000} + 2E_{001} + E_{002} = \hbar w_0 \left[5e^{\alpha/2} + \frac{13}{2}e^{-\alpha}\right]$$
$$\alpha = .637 \quad and \quad E_T = 10.31\hbar w_0$$

Third case

$$2E_{000} + 2E_{001} + E_{010} = \hbar w_0 \left[8e^{\alpha/2} + \frac{5}{2}e^{-\alpha}\right]$$
$$\alpha = -.313 \quad and \quad E_T = 10.26\hbar w_0$$

Fourth case

$$2E_{000} + 2E_{001} + E_{001} = \hbar w_0 \left[7e^{\alpha/2} + \frac{7}{2}e^{-\alpha}\right]$$
$$\alpha = 0 \quad and \quad E_T = 10.5\ \hbar w_0$$

5th particle $E_{total} = 2E_{000} + 2E_{001} + E_{010} = 10.26\ \hbar w_0,\ and\ \alpha = -.313$.

Similarly:
6th Particle $E_{total} = 2E_{000} + 2E_{001} + 2E_{010} = 12.65\ \hbar w_0,\ and\ \alpha = -.341$.
7th Particle $E_{total} = 2E_{000} + 2E_{001} + 2E_{010} + E_{001} = 15.4\ \hbar w_0,\ and\ \alpha = -.134$.
8th Particle $E_{total} = 2E_{000} + 2E_{001} + 2E_{010} + 2E_{001} = 18\ \hbar w_0,\ and\ \alpha = 0$.

The first shell is for $n_x = n_y = n_z = 0$. The first and second particles are placed in this shell and their energies are each $\frac{3}{2}\hbar w_0$. We have $\alpha = 0$ for the first and second particles, so it is shown that at the closure of the first shell, with the placement of the 2nd particle, $\alpha = 0$.

The second shell is for $n_x\ or\ n_y\ or\ n_z = 1$ the other two being zero. The closure of the second shell occurs with the placement of the 8th particle, where we again have $\alpha = 0$.

(2) is now done above aside from the tabulation:

| Particle number | $\alpha$ |
|---|---|
| 1 | 0 |
| 2 | 0 |
| 3 | .341 |
| 4 | .462 |
| 5 | .-313 |
| 6 | -.341 |
| 7 | -.134 |
| 8 | 0 |
| | |

***Example 2***

Consider a hydrogen atom in a very strong, constant magnetic field $B_0$ along the z-axis (corresponding to a hydrogen atom near the surface of a neutron star). To eliminate center-of-mass effects, let the proton be so heavy that only the electron motion matters (and neglect the electron spin). The Hamiltonian for the electron is:

$$H = \frac{1}{2m}\left\{\left(P + \frac{eA}{c}\right)^2 - \frac{e^2}{r}\right\}$$

Where $P$ is the momentum operator and $A = -\frac{1}{2}r \times B$ is a vector potential that produces the magnetic field.

(1) Show that the angular momentum of the electron is not conserved (e.g., that $L_z$and $L^2$ do not both commute with the Hamiltonian).

(2) For $B_0$ very large, show that the electron can't get very far from the origin in the x-y plane.

(3) The result from (2) permits replacing $\frac{e^2}{r}$ with $\frac{e^2}{z}$, which then allows separable solution in either cylindrical or cartesian coordinates. Working with this approximate form, find the lowest energy state of the system (find eigen energy, wavefunction, and quantum numbers.

(4) Repeat (3) by obtaining the general solution to the approximate form in cylindrical coordinates.

**Answer**

The momentum operator is: $P = -i\hbar\nabla$.

$$A = -\frac{1}{2}r \times B = -\frac{1}{2}\left(yB_z - zB_y\,, zB_x - xB_z, xB_y - yB_{x,}\right)$$

Since $B_z = B_0\ and\ B_x = B_y = 0$:

$$A = -\frac{1}{2}(yB_0\,, -xB_0, 0)$$

(Part 1)

$$H = \frac{1}{2m}\left\{P^2 + \frac{e}{c}PA + \frac{e}{c}AP + (\frac{e}{c})^2A^2 - \frac{e^2}{r}\right\}$$

$$PA = (-i\hbar\nabla)\left(-\frac{1}{2}(yB_0\,, -xB_0, 0)\right) = \left[-\frac{1}{2}(yB_0\,, -xB_0, 0)\right](-i\hbar\nabla)$$

$$= AP = \frac{B_0}{2}L_z$$

$$A^2 = \frac{1}{4}[y^2B_0^2 + x^2B_0^2] = \frac{B_0^2}{4}(x^2 + y^2)$$

Thus,

$$H = \frac{1}{2m}\left\{-\hbar^2\nabla^2 + \frac{2e}{c}\frac{B_0}{2}L_z + \left(\frac{e}{c}\right)^2\frac{B_0^2}{4}(x^2+y^2) - \frac{e^2}{\sqrt{x^2+y^2+z^2}}\right\}$$

To calculate $[L_z, H]$ we will first see how $L_z$ commutes with the individual components of the Hamiltonian. We have the usual

$$[L_z, P^2] = 0\,, \quad \left[L_z, \frac{2e}{c}\frac{B_0}{2}L_z\right] = 0, \quad \left[L_z, \left(\frac{e}{c}\right)^2\frac{B_0^2}{4}(x^2+y^2)\right] = 0.$$

So consider:

$$\left[L_z, \frac{e^2}{\sqrt{x^2+y^2+z^2}}\right] = -ihe^2\left[\left(x\frac{\partial}{\partial y} - y\frac{\partial}{\partial x}\right), (x^2+y^2+z^2)^{-\frac{1}{2}}\right] = 0.$$

Thus, $[L_{z,}H] = 0$.

Now , to calculate $[L^2, H]$, starting with the $[L^2, P^2]$ term. Recall that $P^2 = -\hbar^2\nabla^2$ and that the Laplacian in spherical coordination is

$$\nabla^2 = \frac{1}{r^2}\frac{\partial}{\partial r}\left(r^2\frac{\partial}{\partial r}\right) + \frac{1}{r^2 sin\theta}\frac{\partial}{\partial\theta}\left(sin\theta\,\frac{\partial}{\partial\theta}\right) + \frac{1}{r^2 sin\theta}\frac{\partial^2}{\partial\theta^2}$$

$$= \frac{1}{r^2}\frac{\partial}{\partial r}\left(r^2\frac{\partial}{\partial r}\right) + \frac{1}{r^2}\left(\frac{-1}{\hbar^2}\right)L^2$$

since

$$\frac{1}{-\hbar^2}L^2 = \frac{i}{sin\theta}\frac{\partial}{\partial\theta}\left(sin\theta\,\frac{\partial}{\partial\theta}\right) + \frac{1}{sin\theta}\frac{\partial^2}{\partial\theta^2}$$

Thus,

$$[L^2, \nabla^2] = \left[L^2, \frac{1}{r}\frac{\partial}{\partial r}\left(r^2\frac{\partial}{\partial r}\right)\right] = 0$$

And so,

$$[L^2, P^2] = 0.$$

We also know that $[L^2, L_{z,}] = 0$.

Consider

$$[L^2, x^2+y^2] = [L_x^2 + L_y^2 + L_z^2, x^2+y^2]$$

$$[L^2, x^2+y^2] = (2\hbar^2 + i4\hbar z P_z)(x^2+y^2) - 4\hbar z^2(\hbar - i(xp_z + yp_z))$$

We know $\left[L_z, \frac{1}{r}\right] = 0$, thus $\left[L_z^2, \frac{1}{r}\right] = 0$. Similarly for $\left[L_x^2, \frac{1}{r}\right] = 0$ and $\left[L_y^2, \frac{1}{r}\right] = 0$, so we have $\left[L^2, \frac{1}{r}\right] = 0$. Combing the results shows that $[L^2, H] \neq 0$.

(Part 2) If $B_0$ is very large compared to the Coulomb potential, then the Coulomb potential may be ignored and we have:

$$H = \frac{1}{2m}\left\{-\hbar^2\nabla^2 + \frac{eB_o}{c}\,L_z + \left(\frac{eB_o}{2c}\right)^2 (x^2+y^2)\right\}$$

The presence of the $(x^2 + y^2)$ term suggest cylindrical coordinates, with Laplacian:

$$\nabla^2 = \frac{\partial^2}{\partial z^2} + \frac{\partial^2}{\partial \rho^2} + \frac{1}{\rho}\frac{\partial}{\partial \rho} + \frac{1}{\rho^2}\frac{\partial^2}{\partial \varphi^2}$$

For $\varphi(\vec{r}) = u_n(\rho)e^{in\theta}e^{ikz}$ and H$\varphi$=E$\varphi$ which gives a different equation satisfied by $u_n(\rho)$:

$$\frac{d^2u}{d\rho^2} + \frac{1}{\rho}\frac{du}{d\rho} - \frac{n^2}{\rho^2}u - \frac{e^2B^2}{4\hbar^2c^2}\rho^2u + \left(\frac{2mE}{\hbar^2} - \frac{eB\hbar n}{\hbar^2 c}k^2\right)u = 0$$

Let $r = \sqrt{\frac{eB}{2\hbar C}}\,\rho$, and managing behaviour at infinity and at the origin, we have

$$u(r) = r^{|n|}e^{-\frac{r^2}{2}}G(r)$$

And

$$\frac{d^2G}{dr^2} + \left(\frac{2|n|+1}{r} - 2r\right)\frac{dG}{dr} + (\lambda - 2 - 2|n|)G = 0$$

where

$$\frac{1}{4}\lambda - \frac{1+|n|}{2} = n_r \; as\ the\ eigenvalue\ condition\colon n_r = 0,1,2,3,,,$$

Thus,

$$E - \frac{\hbar^2k^2}{2m} = \frac{eB\hbar}{2mc}\,(2n_r + 1 + |n| + n)$$

and

G(ʒ)=$L_{n_r}$(ʒ)

Thus, u(r) =

$$u(r) = r^{|n|}e^{-\frac{r^2}{2}}L_{n_r}^{|n|}(r^2)$$

Now to the radius of an orbit as determined by the peaking of the radial probability distribution.

In the ground state n=I, $n = 1, n_r = 0 \Rightarrow L_{n_r}^{|n|}(r^2) = constant$. Thus,

$$P(r) = r^{2|n|}e^{-r^2}$$

$$\frac{dP}{dr} = \left(2|n|r^{2|n|-1} - 2r\cdot r^{2|n|}\right)e^{-r^2} = 0 \rightarrow r = \sqrt{|n|}\,.$$

Thus,

$$\rho = \sqrt{\frac{2\hbar cn}{eB_0}}$$

For ground state (n=1) we then have with a typical neutron star with $B_0 \sim 10^{12}\ gauss$. Thus,

$$\rho \sim \sqrt{\frac{2\ x\ 1.054\ x\ 10^{-27} dyne\ \ cm/sec\ \ x\ 3\ x\ 10^{10} cm\ sec^{-1}}{4.803\ x\ \ 10^{-10} esu\ x\ 10^{12} dyne\ ese^{-1}}} \sim 3.63x10^{-10} cm$$

The Bohr radius is $a_o = 5.29\ x\ 10^{-9} cm$. Thus, the electron doesn't even get away as far as the lowest Bohr orbit, in the ground state. The election cannot get very far from the in the ground state. Since $x^2 + y^2$ is so small $(x^2 + y^2 + z^2) \ \simeq z^2, thus, r \simeq z.$

(Part 3) Now, $H = \frac{1}{2m}\left\{-\hbar^2 \nabla^2 + \frac{eB_o}{c} L_z + \left(\frac{eB_0}{c}\right)^2 (x^2 + y^2) - \frac{e^2}{z}\right\}$

The wave function with the combination of an Associated Laguerre polynomial, due to r=$x^2 + y^2$, and other factors. The Laguerre polynomial of lowest order is $L_0^0 = 1$:

$$u(r) = r^{|n|} e^{-r^2/2} G(r),\ G(r) = L_{n_r}^{|n|}(r),\quad \rho = \sqrt{\frac{2\hbar c}{eB_0}}\ \mathrm{r}$$

The lowest energy in the radial direction gives $n = n_r = 0$: u(r) = $e^{-r^2/2}$ and E=$\frac{e0\hbar}{2mc}$+$E_z$

The equation remaining is $\frac{-f''}{f} - \frac{e^2}{k^2}\frac{1}{z} = -\frac{2mE_z}{k^2}$, thus

$$E_z = \frac{e^4}{8m\hbar^2}$$

Since $L_z$ commutes with the Hamiltonian and must be normalized to one: $g(\varphi) = e^{in\varphi}$. Thus $\psi(\vec{r}) = u(\rho)g(\varphi)f(z)$ and the quantum numbers are n=0 and nr=0. Energy:

$$\boxed{E = \frac{eB_0\hbar}{2mc} + \frac{e^4}{8m\hbar^2}}$$

And wave equation:

$$\psi(\vec{r})_{100} = exp\left(-\frac{eB_0\rho^2}{4\hbar c}\right) exp\left(-\frac{e^2}{\hbar^2}\frac{z}{2}\right)\left(z\frac{e^2}{\hbar^2}\right)$$

(Part 4) H=$\frac{1}{2m}\left\{-\hbar^2\nabla^2 + \frac{eB_o}{C} L_z + \left(\frac{eB_o}{2C}\right)^2 (x^2 + y^2) - \frac{e^2}{z}\right\}$

In cylindrical coordination :

X= p$\cos\theta$ $\qquad L_z = -i\hbar\frac{O}{\partial o}$

Y=p $\sin\theta$ $\qquad \nabla^2 = \frac{\partial}{\partial x^2} + \frac{\partial^2}{dp^2} \sim \frac{I}{p}\frac{O}{\partial p} + \frac{I}{P^2}\frac{\partial^2}{d\varphi^2}$

$$-\hbar\left(\frac{f''}{f} + u''/u + \frac{i}{p}\frac{u''}{u} - \frac{n^2}{p^2}\right) + \frac{eB_o}{C} + \left(\frac{eB_o}{2C}\right)^2 \rho^2 - \frac{e^2}{\mathfrak{z}} - 2mE = 0$$

$$\frac{f''}{f} + \frac{u''}{u}\frac{i}{p}\frac{u'}{u} - \frac{n^2}{p^2} - \frac{eB_o n}{\hbar C} - \left(\frac{eB_o}{2\hbar C}\right)^2 p^2 + \frac{e^2}{\hbar^2}\frac{1}{\mathfrak{z}} + \frac{2mE}{\hbar^2} = 0$$

This is separable

$$\frac{u''}{u} + \frac{i}{p}\frac{u'}{u} - \frac{n^2}{p^2} - \frac{eB_o n}{\hbar C} - \left(\frac{eB_o}{2\hbar C}\right)^2 p^2 + \frac{2mE}{\hbar^2} = \frac{-f''}{f} - \frac{e^2}{\hbar^2}\frac{1}{2}$$

$$\frac{d^2u}{dp^2} + \frac{i}{p}\frac{du}{dp} - \frac{n^2u}{p^2} - \frac{eB^2{}_o}{4\hbar^2C^2}p^2u + \frac{2mE}{\hbar^2} = \frac{-f''}{f} - \frac{e^2}{\hbar^2}\frac{1}{2}$$

Let $r = \sqrt{\frac{eB_\circ}{2\hbar c}}p$

$$\frac{d^2u}{dr^2} + \frac{i}{r}\frac{du}{dr} - \frac{n^2u}{r^2} - \frac{eB^2{}_o}{4\hbar^2C^2}p^2u + \frac{2mE}{\hbar^2} = \frac{-f''}{f} - \frac{e^2}{\hbar^2}\frac{1}{2}$$

Let $u(r) = r^{|n|}e^{-r^2/2}G(r)$ $\qquad\qquad \lambda = \frac{4mc}{eB_o}(E + C') - 2n$

$$\frac{d^2G}{dr^2} + \left(\frac{2|n| + I}{r} - 2r\right)\frac{dG}{dr} + (\lambda - 2 - 2|n|)G = 0$$

Let $\mathfrak{z} = r^2$

$$\frac{d^2G}{d\mathfrak{z}^2} + \left(\frac{|n|+I}{g} - I\right)\frac{dG}{d\mathfrak{z}} + \frac{\lambda-2-2|n|}{4\mathfrak{z}}\ G=0$$

E-C'$= \frac{eB_\circ\hbar}{2mc}(2(n_r + I + |n| + n$

$G(\mathfrak{z}) = L_{n_r}^{|n|}(\mathfrak{z})$ → Associated Legendre Polynomial

$$\frac{-f''}{f} - \frac{e^2}{\hbar^2}\frac{1}{\mathfrak{z}} = +\frac{2mC'}{\hbar^2}$$

$$F'' + \frac{e^2}{\hbar^2}\frac{f}{\mathfrak{z}} - \frac{2mC'}{\hbar^2}f = 0$$

Let $\mathcal{E} = A\mathfrak{z}$ $\qquad \frac{d\mathcal{E}}{A} = d\mathfrak{z}$

$$A^2\frac{d^2f}{d\mathcal{E}^2} + \frac{e^2}{\hbar^2}\frac{fA}{\mathcal{E}} + \frac{2mC'}{\hbar^2}f = 0$$

$$\frac{d^2f}{d\mathcal{E}^2} + \frac{e^2}{\hbar^2}\frac{f}{\mathcal{E}A} + \frac{2mC'}{\hbar^2A^2}f = 0$$

$$\frac{d^2f}{d\mathcal{E}^2} + \frac{f}{\mathcal{E}} + \frac{2m\hbar^2C'}{e^4}f = 0$$

$$\frac{d^2f}{d\mathcal{E}^2} + \left(\frac{1}{\mathcal{E}} - k\right)f = 0$$

For large $\mathcal{E}$

$$\frac{d^2f}{d\mathcal{E}^2} - kf = 0$$

Let

$$F = e^{-\sqrt{k\mathcal{E}}} G(\mathcal{E})$$

$$f' = e^{-\sqrt{k\mathcal{E}}} G' - \sqrt{k} e^{-\sqrt{k\mathcal{E}}}$$

$$f'' = e^{-\sqrt{k\mathcal{E}}} G'' - 2\sqrt{k} e^{-\sqrt{k\mathcal{E}}} G' + k e^{-\sqrt{k\mathcal{E}}} G$$

$$f'' + \frac{f}{q} - kf = e^{-\sqrt{k\mathcal{E}}} G'' - 2\sqrt{k} e^{-\sqrt{k\mathcal{E}}} G + \sqrt{k} e^{-\sqrt{k\mathcal{E}}} G + e^{-\sqrt{k\mathcal{E}}} \frac{G(\mathcal{E})}{\mathcal{E}} -$$

$$kG(\mathcal{E}) e^{-\sqrt{k\mathcal{E}}} = 0$$

$$G'' - 2\sqrt{k} G' + \frac{G}{\mathcal{E}} = 0$$

$$\frac{2m\hbar^2 C\prime}{e^4} = \frac{1}{4}$$

$$C' = \frac{e^4}{8m\hbar^2}$$

Thus

$$E = \frac{eB_0 \hbar}{2mc}(2n_r + 1 + |n| + n) + \frac{e^4}{8m\hbar^2}$$

$$\psi(\vec{r}) = u(p) e^{in\varphi} f(z)$$

$$\boxed{\psi(\vec{r})_{nnr} = \left(\sqrt{\frac{eB_0\hbar}{2\hbar c}} p\right)^{|n|} \exp\left(-\frac{eB_\circ}{4\hbar c} p^2\right)^{\square} L_{n_r}^{|n|}\left(\frac{eB_\circ\hbar}{4\hbar c} p^2\right)^{\square}}$$

$$\boxed{e^{in\varphi}\left(\frac{e^2}{\hbar^2} 2\right) \exp\left(\frac{e^2}{\hbar^2}\frac{2}{2}\right)}$$

$$\psi(\vec{r})_{0,0} = \frac{e^2}{\hbar^2} 2 \exp\left(-\frac{e^2 2}{\hbar^2 2}\right) \exp\left(-\frac{eB_\circ}{4\hbar c} p^2\right) \quad E = \frac{eB_\circ\hbar}{2mc} + \frac{e^4}{8m\hbar^2}$$

$$\psi(\vec{r})_{0,1} = \left(I - \frac{eB_\circ}{2\hbar c} p^2\right)\psi(\vec{r})_{0,0} \quad E = \frac{3eB_\circ\hbar}{2mc} + \frac{e^4}{8m\hbar^2}$$

## Example 3

Consider the following model of a H atom in the lowest atomic state, is, where the spin of the electron $\vec{S}$ interacts both with the nuclear spin, $\vec{I}$, (i.e. the spin of the proton) and with a weak external magnetic field. The proton, like the electron, is a particle of spin ½ . take as a model $H = \alpha S_z + \beta \vec{S} \cdot \vec{I}$, where $\alpha = 2\mu_B B_0$. (The interaction of the proton's magnetic moment with the external magnetic field has been neglected.) Solve for the eigenstates and eigenvalues of this spin system for the following cases:

1) $\alpha = 0$
2) $\beta = 0$

3) $\alpha \ll \beta$
4) $\beta \ll \alpha$
5) $\alpha \approx \beta$

For each case tell what quantum numbers you are using (beyond $s = \frac{1}{2} = i$ in $S^2 = s(s+1)$ and $I^2 = i(i+1)$ and the degree of degeneracy for each eigenvalue you find.

**Answer**

$$H' = \alpha S_z + 3\vec{S} \cdot \vec{I}$$

$[S_z, \vec{S} \cdot \vec{I}] \neq 0$ already looks bad

Two choices for a basis: (neither will commute entirely with $H'$!)

$J^2, J_z, S^2, I^2$ $or$ $S^2, S_z, I^2, I_z$

Now, $\boxed{J^2, S_z} \neq 0$

$[J_2, S_z] = 0$

$[S^2, S_z] = 0$

$[I^2, S_z] = 0$

$\vec{J} = \vec{S} + \vec{I}$

$\vec{S} \cdot \vec{I} = \frac{1}{2}\{\vec{J}^2, \vec{S}^2\} = 0$

$[\vec{J}^2, \vec{S} \cdot \vec{I}] = 0$ $\qquad$ $[S^2, \vec{S} \cdot \vec{I}] = 0$

$[\vec{J}_z, \vec{S} \cdot \vec{I}] = 0$ $\qquad$ $[I^2, \vec{S} \cdot \vec{I}] \neq 0$

$[\vec{S}^2, \vec{S} \cdot \vec{I}] = 0$ $\qquad$ $[S_z, \vec{S} \cdot \vec{I}] \neq 0$

Either way we have to use degenerate perturbation theory

$\left.\begin{matrix}[J^2, S_z]70 \\ [J_z, S_z] = 0\end{matrix}\right\}$ $perturbation\ mixes\ the\ j = 0, m = 0\ and\ j = 1, m = 0\ states\ need\ a\ 2x2\ matrix$

$\left.\begin{matrix}[S_z, \vec{S} \cdot \vec{I}] \neq 0 \\ [I_z, \vec{S} \cdot \vec{I}] \neq 0\end{matrix}\right\}$ $perturbation\ mixes\ the\ m_s = \frac{1}{3}, \frac{1}{3}\left(S = \frac{1}{2}\right)\ states\ and\ the\ m_I = \frac{1}{2}, -\frac{1}{2}\ \left(I = \frac{1}{2}\right)\ states\ amongst\ themselves\ need\ 2\ \ 2x2\ \ matrices.$

$[S^2, \ldots] = 0$

$[I^2, \ldots] = 0$

So, choose the $J^2, J_z, S^2, I^2$ basis!

$J^2, J_z, S^2, I^2$ basis: $|j, m_j, S, i> = |j, m_j>$

For $j = 1, m = 1$ we have $(E_{jm})$: $\quad m = 1\ so\ m_z = \frac{1}{2}, m_I = \frac{1}{2}$

$E'_{11} =< 1,1|H'|1,1 >= \alpha m\hbar + \beta\frac{1}{2}\{j(j+1) - s(s+1) = i(i+1)\}\hbar^2$

$= \frac{1}{2}\alpha\hbar + \frac{1}{2}\beta\left\{2 - \frac{3}{2}\right\}\hbar^2 \qquad |j = 1, m = 1 >= |m_2 = \frac{1}{2}, m_I = -\frac{1}{2} >$

$E'_{1\,-1} =< 1,-1|H'|1,-1 >= -\frac{1}{2}\alpha\hbar + \frac{1}{4}\beta\hbar^2$

$|j = 1, m = -1 >= |m_s = -\frac{1}{2}, m_I = -\frac{1}{2} >$

For j=1 , m=0 and j=0, m=0 , we have, as perennially maintained a mixing of the corresponding degenerate states:

$< j, m|S_Z|j, m >=?$

We need to express $|j, m >$ *in the* $|m_s, m_I >$ basis

$|1,1 >= |\frac{1}{2}, \frac{1}{2} >$

$J_-|1,1 >= J_-|\frac{1}{2}, \frac{1}{2} >= (S_- + I_-)|\frac{1}{2}, \frac{1}{2}$

$J_-|1,1 >= \hbar\sqrt{j(j+1) - m(m-1)}|1,1 >= \hbar\sqrt{2}|1,0 >$

$(S_- + I_-)\left|\frac{1}{2}, \frac{1}{2} >= \hbar\sqrt{s(s+1) - m_s(m_s - 1)}\right| - \frac{1}{2}, \frac{1}{2} >$

$+\hbar\sqrt{i(i+1) - m_i(m_i - 1)}|\frac{1}{2}, -\frac{1}{2}$

$= \hbar\left|-\frac{1}{2}, \frac{1}{2} > +\hbar\right|\frac{1}{2}, -\frac{1}{2}$

So, $\sqrt{2}|1,0 >= | -\frac{1}{2}, \frac{1}{2} > +|\frac{1}{2}, -\frac{1}{2}$

$|1,0 >= \frac{1}{\sqrt{2}}\left\{| -\frac{1}{2}, \frac{1}{2} > +|\frac{1}{2}, -\frac{1}{2} >\right\}$

Since $|0,0 >$ must be orthogonal

$|0,0 >= \frac{1}{\sqrt{2}}\left\{| -\frac{1}{2}, \frac{1}{2} > -|\frac{1}{2}, -\frac{1}{2}\right\}$

$< 1,0|S_Z|1,0 >= \frac{1}{2}\left\{< -\frac{1}{2}, \frac{1}{2}| + < \frac{1}{2}, \frac{1}{2}|\right\} S_Z\left\{| -\frac{1}{2}, \frac{1}{2} > +|\frac{1}{2}, -\frac{1}{2}\right\}$

$= \frac{\hbar}{2}\left\{-\frac{1}{2} + \frac{1}{2}\right\} = 0$

$< 0,0|S_Z|0 >= \frac{\hbar}{2}\left\{-\frac{1}{2} + \frac{1}{2}\right\} = 0$

$< 1,0|S_Z|0,0 >= \frac{1}{2}\left\{< -\frac{1}{2}, \frac{1}{2}| + < \frac{1}{2}, -\frac{1}{2}\right\} S_Z\left\{1 - \frac{1}{2}, \frac{1}{2} > -|\frac{1}{2}, -\frac{1}{2} >\right\}$

$= \frac{1}{2}\left\{-\frac{\hbar}{2} - \frac{\hbar}{2}\right\} = -\frac{\hbar}{2}$

$< 0,0|S_Z|1,0 >= \frac{1}{2}\left\{-\frac{1}{2}, \frac{1}{2}| - < \frac{1}{2}, -\frac{1}{2}|\right\} S_Z\left\{| -\frac{1}{2}, \frac{1}{2} > +|\frac{1}{2}, -\frac{1}{2}\right\}$

$= \frac{1}{2}\left\{-\frac{\hbar}{2} - \frac{\hbar}{2}\right\} = -\frac{\hbar}{2}$

So, the submatrix: $< 1,0|\vec{S}\cdot\vec{I}|1,0 >= \frac{\hbar^2}{2}\left(2 - \frac{3}{2}\right) = \frac{\hbar^2}{4}$

$< 0,0|\vec{S}\cdot\vec{I}|0,0 >= \frac{\hbar^2}{2}\left(-\frac{3}{2}\right) = -\frac{3}{2}\hbar^2$

$$P(m=0)H'^{(m=0)} = \begin{pmatrix} <0,0|H'|0,0><1,0|H'|0,0> \\ <1,0|H'|0,0><1,0|H'|1,0> \end{pmatrix}$$

$$(H')_{restricted} = \begin{pmatrix} \left(-\frac{3}{4}\hbar^2\beta\right) & \left(-\alpha\frac{\hbar}{2}\right) \\ \left(-\alpha\frac{\hbar}{2}\right) & \left(\frac{\hbar^2}{4}\beta\right) \end{pmatrix}$$

Eigenvalues:

$$\left[\left(-\frac{3}{4}\hbar^2\beta\right)-E\right]\left[\frac{\hbar^2}{4}\beta-E\right]-\left(\alpha\frac{\hbar}{2}\right)^2=0$$

$$E^2+E\left(\frac{3}{4}-\frac{1}{4}\right)\hbar^2\beta-\frac{3}{16}\hbar^4\beta^2-\left(\alpha\frac{\hbar}{2}\right)^2=0$$

$$E=-\frac{1}{2}\hbar^2\beta\pm\sqrt{\left(\frac{1}{2}\hbar^2\beta\right)^2-4\left[-\frac{3}{10}\hbar^4\beta^2-\left(\alpha\frac{\hbar}{2}\right)^2\right]}$$

$$E=-\left(\frac{\hbar}{2}\right)^2\beta\pm\sqrt{\left(\frac{1}{4}\hbar^2\beta\right)^2+\frac{3}{16}\hbar^4\beta^2+\left(\alpha\frac{\hbar}{2}\right)^2}$$

$$E_I=-\left(\frac{\hbar}{2}\right)^2\beta\pm\sqrt{\left(\frac{1}{4}\right)\hbar^4\beta^2+\frac{1}{4}\hbar^2\alpha^2}=-\left(\frac{\hbar}{2}\right)^2\beta\pm\frac{\hbar}{2}\sqrt{\left(\frac{1}{2}\beta\right)^2+\alpha^2}$$

For the eigenvectors:

For $E_t$: $\left(-\frac{1}{2}\hbar^2\beta-\frac{\hbar}{2}\sqrt{\frac{\hbar\beta}{\lambda}+\alpha^2} \quad \left(-\alpha\frac{\hbar}{2}\right)\right)\binom{a}{b}=0$

$$-\left(\frac{1}{2}\hbar^2\beta+\frac{1}{2}\hbar\sqrt{\left(\frac{1}{2}3\right)^2+\alpha^2}\right)a-\left(\alpha\frac{\hbar}{2}\right)b=0$$

$$b=-\frac{\left(\frac{1}{2}\hbar^2\beta+\frac{1}{2}\hbar\sqrt{(\hbar\beta)^2+\alpha^2}\right)}{\left(\frac{1}{2}\hbar\alpha\right)}a=-\left(\hbar\left[\frac{\beta}{\alpha}\right]+\sqrt{\left(\frac{\hbar}{2}\frac{\beta}{\alpha}\right)^2+1}\right)$$

$|+>=|0,0>-\left(\left(\hbar\frac{\beta}{2}\right)+\sqrt{\left(\frac{\hbar}{2}\frac{\beta}{\alpha}\right)^2+1}\right)|1,0>$ un-normalized.

For $E\_$: $|->=|0,0>-\left(\left(\hbar\frac{\beta}{2}\right)-\sqrt{\left(\frac{\hbar}{2}\frac{\beta}{\alpha}\right)^2+i}\right)|1,0>$ un-normalised

| The cases: | eigenvalue | | eigenstate |
|---|---|---|---|
| | $E'_{1,1}=\frac{1}{4}\beta\hbar^2$ | $\rightarrow$ | $\|1,1>$ |
| (1) $\alpha=0$ | $E'_{1\,-1}=\frac{1}{4}\beta\hbar^2$ | $degeneracy$ | $\|1,-1>$ |
| | $E'_{1,0}=\frac{1}{4}\beta\hbar^2$ | $\rightarrow$ | $\|1,0>$ |
| | $E'_{0,0}=-\frac{3}{4}\beta\hbar^2$ | $degeneracy\ =1$ | $\|0,0>$ |

(2) $\beta =$

$$E'_{1,1} = \frac{1}{2}\alpha\hbar \qquad degeneracy = 1 \qquad |1,1>$$

$$E'_{1\,-1} = \frac{1}{2}\alpha\hbar \qquad degeneracy = 2 \qquad |1,-1>$$

0

$$E'_{+} = \frac{1}{2}\alpha\hbar \qquad degeneracy = 2 \qquad \frac{1}{\sqrt{2}}\{|1,0> -|0,0>\}$$

$$E'_{+} = \frac{1}{2}\alpha\hbar \qquad degeneracy = 2 \qquad \frac{1}{\sqrt{2}}\{|1,0> +|0,0>\}$$

(3) $\alpha \ll \beta$ $\quad E'_{11} = \frac{1}{4}\beta\hbar^2 + \frac{1}{2}\alpha\hbar \qquad |1,1>$

$$E'_{1-1} = \frac{1}{4}\beta\hbar^2 + \frac{1}{2}\alpha\hbar \qquad |1,-1>$$

$$E^1_{+} = -\left(\frac{\hbar}{2}\right)^2\beta + \frac{\hbar}{2}\left(\frac{\hbar}{2}\beta\right)\left[1 + \frac{\alpha^2}{\left(\frac{\hbar}{2}\beta\right)^2}\right]^{1/2}$$

$$= \frac{1}{2}\frac{\alpha^2}{\beta} + \frac{1}{4}\beta\hbar^2 \qquad |+>\cong \left|0,0> - \frac{3}{2}\hbar\frac{\beta}{2}\right| 1,0>$$

$$E'_{-} = -\frac{3}{4}\hbar^2\beta - \frac{1}{2}\frac{\alpha^2}{\beta} \qquad |->\cong |0,0> + \frac{1}{\left(\hbar^{\beta}/_{\alpha}\right)} \; |1,0>$$

(4)$B \ll \alpha$ $\qquad E'_{11}\,, E^1_{1,-1}$ same

$$E^1_{+} = \frac{\hbar}{2}\alpha \qquad |+>= \left|0,0> - \left(1 + \frac{3}{2}\hbar\frac{\beta}{\alpha}\right)\right| 1,0>$$

$$E^1_{-} = -\frac{\hbar}{2}\alpha \qquad |->= \left|0,0> + \left(1 - \frac{3}{2}\hbar\frac{\beta}{\alpha}\right)\right| 1,0>$$

(5) $\alpha = \beta$ already derived.

***Example 4***

Consider the 2p3s configuration of the $_6$C atom.

a) Draw a schematic energy-level diagram for this configuration. Use the same (exaggerative) scale for the fine structure splitting for all the levels within multiplets. Label each level with the spectroscopic notation.

b) On the previous energy-level diagram draw to the same scale (highly exaggerated) the Zeeman effect splitting for each level assuming that the atom is now under the influence of a weak magnetic field. Derive any formula you use to calculate this splitting.

**Answer**

$H = H_0 + H_1$ in the presence of magnate field

$$\mathfrak{M} = \gamma[\vec{L} - 2\vec{S}] = \gamma(\vec{J} + \vec{S})$$

$$H_1 = \omega_1(L_z + 2S_z) \qquad \omega_1 = -\frac{qB}{2m}$$

Consist $H_1$ inside the subsurface $\varepsilon(E_0, L, S, J)$ associated with the multipost under study.
Inside $\varepsilon$:

$\vec{L} = \frac{\langle L\cdot J\rangle \vec{J}}{\langle J\cdot J\rangle} = \frac{\langle L\cdot J\rangle}{J(J+1)\hbar^2}$ projection theorem (vector model)

$\vec{S} = \frac{\langle S\cdot J\rangle}{J(J+1)\hbar^2}$

So, $\vec{L}\cdot\vec{J} = \vec{L}\cdot(\vec{L}+\vec{S}) = \vec{L}^2 + \vec{L}\cdot\vec{S} = \vec{L}^2 + \frac{1}{2}(\vec{J}^2 - \vec{L}^2 - \vec{S}^2)$

$\vec{S}\cdot\vec{J} = \vec{S}^2 - \frac{1}{2}(\vec{J}^2 - \vec{L}^2 - \vec{S}^2)$

So,

$$\langle L\cdot J\rangle = L(L+1)\hbar^2 + \frac{\hbar^2}{2}[J(J+1) - L(L+1) - S(S+1)]$$

$$\langle S\cdot J\rangle = S(S+1)\hbar^2 - \frac{\hbar^2}{2}[J(J+1) - L(L+1) - S(S+1)]$$

Thus,

$$H_1 = W_1\left(\frac{\langle L\cdot J\rangle J_z}{J(J+1)\hbar^2} + \frac{2\langle S\cdot J\rangle J_z}{J(J+1)\hbar^2}\right) = w_1 J_z \underbrace{\left\{\frac{3}{2} + \frac{S(S+1)-L(L+1)}{2J(J+1)}\right\}}_{g_J}$$

$$E_1 = g_J M\hbar\omega_L$$

***Example 5***

Let $\alpha$ be the $m_s = +\frac{1}{2}$ spin wave function, $\beta$ the $m_s = -\frac{1}{2}$ spin wave function, $Y_{\ell,m}(\Omega)$ the normal spherical harmonics, and $R_{n,\ell}(r)$ the normal hydrogen radial wave functions. (For ease of writing we suppress the r and $\Omega$ from here on. Then $\psi_{n\ell jm_j} = R_{2,1}\left[\sqrt{\frac{2}{3}}\alpha Y_{1,0} + \frac{1}{\sqrt{3}}\beta Y_{1,1}\right]$ is a hydrogen atom wave function. What are the probabilities of finding the electron (in this state) with:

a) $m_\ell = 1$
b) $m_\ell = 0$
c) $m_\ell = -1$
d) $m_s = \frac{1}{2}$
e) $m_\ell = 1,\ m_s = \frac{1}{2}$
f) $m_\ell = 1,\ m_s = -\frac{1}{2}$
g) What is $\bar{L}_z$?
h) If the electron is found near the poles ($\theta = 0, \pi$) is its spin more likely to be up or down?
i) What is $m_j$ for this state?
j) What is j for this state?

k) For this value of $\ell$ what other value of j is possible? Call if j'.
l) What is $\Psi_{n,\ell,j\prime,m_j}$ ?
m) What is $\Psi_{n,\ell,j,m_j-1}$?

**Answer**

$\alpha \leftrightarrow m_s = {}^1/_2 \ \ wave\ function$

$\beta \leftrightarrow m_2 = -\ {}^1/_2$

$Y_{l,m}: spherical\ harmonics$

$R_{n.e}(r): Hydrogen\ radial\ wave\ function$

$$\Psi_{nljm_j} = R_{2,1}\left[\sqrt{\frac{2}{3}}\alpha Y_{1,0} + \frac{1}{\sqrt{3}}\beta Y_{1,1}\right]$$

Probabilities of given state:

(a) $m_\ell = 1\ ,{}^1/_3$
(b) $m_\ell = 0\quad ,{}^2/_3$
(c) $m_\ell = -1\ \ ,0$
(d) $m_s = {}^1/_2\ \ ,{}^2/_3$
(e) $m_\ell = 1\ \ and\ m_s = {}^1/_2\ \ ,0$
(f) $m_\ell = 1\ \ and\ m_s = -\,{}^1/_2\ \ ,\ {}^1/_3$
(g) $\bar{L}_Z = (\wp_{m_\ell})m_\ell\hbar = {}^1/_3\,\hbar$
(h) $Y_{1,0} \sim \cos\theta\ , Y_{1,\Sigma 1} \sim \sin\theta$
 if the friction is near the poles it is more likely to be in $Y_{1,0}$
 Thus, spin is $\alpha$ is up, more likely
(i) $m_j = m_s + m_\ell = \frac{1}{2}$
(j) $j = ?\quad either\ \frac{3}{2}\ or\ \frac{1}{2}$
 If $j = {}^3/_2\ \ we\ know\ |\frac{3}{2},\frac{3}{2} >= |1,\frac{1}{2};1,t>$
 Thus lower this to get $j = {}^3/_2\ ,l = {}^1/_2$

$$J_-|J,M >= \sqrt{j(j+1) - m_j \pm 1}|J,M-1>$$

$$J_- = L_- + S_-$$

$$J_-\left|\frac{3}{2},\frac{3}{2}\right> = \sqrt{\frac{3}{2},\left(\frac{5}{2}\right) - \frac{3}{2}\left(\frac{1}{2}\right)}\left|\frac{3}{2},\frac{1}{2}\right> \ = \sqrt{3}|\frac{3}{2},\frac{1}{2}>$$

$$\sqrt{3}\left|\frac{3}{2},\frac{1}{2}\right> = J_-\left|\frac{3}{2},\frac{3}{2}\right> = (L_- + S_-)\left|\frac{3}{2},\frac{3}{2}\right> = (L_- + S_-)\left|1,\frac{1}{2};1,\frac{1}{2}\right>$$

$$= \sqrt{\ell(\ell+1) - m_\ell(m_\ell - 1)}\left|1,\frac{1}{2};0,\frac{1}{2}\right> + \sqrt{s(s+1)\ldots}\left|1,\frac{1}{2};1,-\frac{1}{2}\right>$$

$$= \sqrt{2}\left|1,\frac{1}{2};0,\frac{1}{2}\right> + \sqrt{\frac{3}{4} - \left(-\frac{1}{4}\right)}\left|1,\frac{1}{2};1,-\frac{1}{2}\right>$$

$\left|\frac{3}{2},\frac{1}{2}>= \sqrt{\frac{2}{3}}\right| 1,\frac{1}{2};0,\frac{1}{2}> +\frac{1}{\sqrt{3}}|1,\frac{1}{2};1-\frac{1}{2}>$ if $1\frac{1}{2},\frac{1}{2}>$ the + sign would have opposite sign, etc.

So, (j) $j = {}^{3}/_{2}$!

(k) $j = {}^{1}/_{2}$

(l) $\Psi_{n,\ell,j'={}^{1}/_{2},m_j}$ =? Preceding comment of use now

$\left|\frac{1}{2},\frac{1}{2}>= \alpha\right| 1,\frac{1}{2};0,\frac{1}{2}> +\beta|1,\frac{1}{2};1,-\frac{1}{2}>$

$since\ <\frac{1}{2},\frac{1}{2}|\frac{1}{2},\frac{1}{2}>= 1$

$|\alpha|^2+|\beta|^2=1$

Also orthogonality:

$\alpha\sqrt{\frac{2}{3}}+\beta\frac{1}{\sqrt{3}}=0$

$\alpha=-\frac{1}{\sqrt{2}}\beta$ $\qquad\qquad \frac{1}{2}|\beta|^2+|\beta|^2=1$

$|\beta|=\left|\sqrt{\frac{2}{3}}\right|$

Choose $\alpha$ real and positive

$\alpha=\frac{1}{\sqrt{3}},\beta=-\sqrt{\frac{2}{3}}$

Thus, $\Psi_{n\ell,j',m_j}=\sqrt{j(j+1)-m_j(m_j-1)}\,\Psi_{n,\ell,j,m_j-1}$ $\qquad j=\frac{3}{2},m_j=\frac{1}{2}$

$=\sqrt{\frac{5}{4}+\frac{1}{7}}\,\Psi_{n\ell,j,m_j-1}=2\,\Psi_{n\ell,j,m_j-1}$

$J_-\Psi_{n\ell,j,m_j}=(L_-+S_-)\Psi_{n\ell,j,m_j}=(L_-+S_-)\left\{\sqrt{\frac{2}{3}}\left|1,\frac{1}{2};0,\frac{1}{2}>\right.\right.$
$\left.+\frac{1}{\sqrt{3}}\left|1,\frac{1}{2};1,-\frac{1}{2}>\right.\right\}$

$=\sqrt{\frac{3}{2}}\sqrt{2}\left|1,\frac{1}{2}-1,\frac{1}{2}>+\frac{1}{\sqrt{3}}\right|1,\frac{1}{2};0,-\frac{1}{2}>+\sqrt{\frac{2}{3}}\left|1,\frac{1}{2};0,\frac{1}{2}>\right.$
$+\frac{1}{\sqrt{3}}\sqrt{\frac{3}{2}}\left|1,\frac{1}{2};1\right.$

$2\,\Psi_{n\ell,j,m_j-1}=\frac{2}{\sqrt{3}}\left|1,\frac{1}{2};-1,\frac{1}{2}>+2\sqrt{\frac{2}{3}}\right|1,\frac{1}{2};0,-\frac{1}{2}>$

$$\Psi_{n\ell,j,m_j-1} = R_{2,1}\left[\frac{1}{\sqrt{3}}\alpha Y_{1,-1} + \sqrt{\frac{2}{3}}\beta Y_{1,0}\right]$$

## 3.8 Exercises

*Exercise 3.1*

(a) Write the time independent Schrodinger equation for the non-relativistic hydrogen atom.

(b) In the case when the stationary state wave function $\psi$ depends on r alone (spherically symmetric s state), write the above equation in spherical coordinates. Note: In spherical coordinates: $\nabla^2 = \frac{1}{r^2}\frac{\partial}{\partial r}\left(r^2\frac{\partial}{\partial r}\right)$ +(angular terms).

(c) For a large r find the function A(r) such that $\psi(r) \sim r^p A(r)$ for large r (you need not determine the power p). A(r) gives the asymptotic form of $\psi(r)$ for large r.

(d) Write $\psi(r) = f(r)A(r)$, and find the equation satisfied by f(r).

(e) The requirement that the norm of the wave function be finite implies that the power series expansion of f(r) terminates. Write

$$f(r) = n_0 + a_1 r + \cdots + a_n r^n$$

And find the condition relating n to the energy E. From that condition, write the expression for the S-state energy levels of hydrogen. Are these all the non-relativistic energy levels of hydrogen? Why?

*Exercise 3.2*

The angular momentum operator $\vec{L}$ is the generator of rotations in quantum mechanics. For a rotation through and angle $\delta\varphi$ about an axis $\hat{n}$, we have the operator:

$$e^{-\frac{i}{\hbar}\delta\varphi\hat{n}\cdot\vec{L}}$$

Show that a composition of rotations $e^{\frac{i}{\hbar}\frac{\pi}{2}L_x}e^{-\frac{i}{\hbar}\beta L_z}e^{-\frac{i}{\hbar}\frac{\pi}{2}L_x}$ is equal to a single rotation $e^{-\frac{i}{\hbar}\beta L_y}$ (hence rotations combine just as they do in classical mechanics).

*Exercise 3.3*

The incident flux $\vec{j}$ of a beam of particles has components only in the x-y plane. The beam is incident at an angle $\theta$ (with respect to the x-axis) on a semi-infinite step potential)

$$V(x) = \begin{cases} 0 & x < 0 \\ V_0 & x > 0 \end{cases}$$

Take the incident particle flux to have unit magnitude and the kinetic energy of the particles in the beam to be $K.E. = 4V_0$. Compute the transmitted and reflected particle fluxes (magnitude and direction) as a function of the incident angle for all $\theta$ such that $0 < \theta \leq \pi/2$.

*Exercise 3.4*

A) Consider a diatomic molecule to be a quantized rigid rotator, starting from the condition for quantized angular momentum, $I\omega = \sqrt{J(J+1)}\hbar$, derive an expression for the quantized rotational energy levels.

B) Write an expression for the population of the J$^{th}$ energy level relative to the J=0 level, i.e. get an expression for $N_J/N_0$ considering the Boltzmann temperature distribution and the degeneracy of the rotation levels.

C) Write an expression for the intensity of the Second Stokes line relative to the First anti-Stokes line in the Raman spectrum of this molecule, insofar as these intensities depend on population. (Selection for Stokes lines is $\Delta U = +2$; for anti-stokes lines, $\Delta U = -2$. The 1$^{st}$ stokes and anti-stokes lines are those closest to the unshifted frequency, the 2$^{nd}$ ones come next, etc.)

D) Derive an expression that will give the value of J for the most highly populated energy level at some temperature T.

*Exercise 3.5*

a) Sketch the distribution of number vs energy for electrons emitted from a surface bombarded by a beam of electrons of energy $E_p$.

b) Discuss the features seen in the graph indicating mechanisms, relative amounts, etc.

# Chapter 4. Advanced Quantum Mechanics

## 4.1 Dirac Axiomatic Approach (1930) and Born Probability (1927)

In Dirac's axiomatic approach [5] he starts by introducing a new way to represent the dynamical state of the system (the familiar wavefunction in the position representation) with use of "ket vectors", for which he will introduce the "ket" notation. In the process of doing this, Dirac carefully reviews the observed state superposition properties to arrive at what is truly a vector space (infinite dimensional) [5]. Dirac then describes an inner product between state vectors, and thereby establishes the basis for what is known as a Hilbert space. In Section 4.1.1 this mathematical formulation will be derived in detail, with many quantum mathematical properties placed in the Appendix that will be used throughout.

### 4.1.1 A more sophisticated mathematics and notation

***State Space – Dirac Notation***

Notice that the various bases, and components of $\Psi(\vec{r})$ that can be chosen in that (infinite) basis, are exactly analogous to choosing a coordinate system (finite basis) and specifying the coordinates for a point (where $\Psi(\vec{r})$ corresponds to a set of coordinates in the continuous $\varepsilon_{\vec{r}_0}(\vec{r})$ basis). This motivates the state space and Dirac notation.

Each quantum state of a particle will be characterized by a state vector, belonging to an abstract space, $\varepsilon_{\vec{r}_{\square}}$, called the state space of a particle. Since $\mathcal{F}$ is a subspace of $L^2$, $\varepsilon_{\vec{r}_{\square}}$ is a subspace of a Hilbert space. Such a State Space generalization to the theory will make it complete, at the nonrelativistic level at least. This is because the (non-relativistic) Schrödinger Equation with wavefunction description is incomplete in the sense that not all physical systems or properties have a wavefunction description. This is the case when spin degrees of freedom are taken into account (even for a single particle).

Let's develop the vector calculus for the $\varepsilon_{\vec{r}_{\square}}$ spanning a Hilbert subspace in what's kn
own as the (Dirac) 'Bra' and 'Ket' notation.

### 'Ket' vectors

Any element (vector) of the $\varepsilon_{\vec{r}_{\square}}$ space will be called a Ket vector and denoted: $|\Psi>$. Generalizing from the wavefunction analysis we have:

$$\Psi(\vec{r}) \in \mathcal{F} \iff |\Psi> \in \varepsilon_{\vec{r}_{\square}}$$

In a particular basis $\Psi(\vec{r})$ is $|\Psi>$ where $\vec{r}$ plays the role of a continuous index.

### Scalar Product

For a pair of Kets, $|\Psi>$ and $|\varphi>$, we associate a complex number to their scalar product:

$$(|\varphi>,|\Psi>) \equiv <\varphi|\Psi>$$

### Dual Space – 'Bra' vectors

As with tensor theory, where the vector space has a dual (covariant and contravariant), the existence of a linear functional here as well, will also lead to a dual vector space, referred to as the 'Bra' space with notation $<\varphi|$ such that the scalar product is written in 'Bra'-'Ket' operation as $<\varphi|\Psi>$.

In general, the dual space $\varepsilon_{\vec{r}}^{*}$ of $\varepsilon_{\vec{r}}$ is not isomorphic to $\varepsilon_{\vec{r}}$ (except if $\varepsilon_{\vec{r}}$ is finite-dimensional). In other words, for each Ket $|\varphi>$ of $\varepsilon_{\vec{r}}$ there corresponds a Bra vector in $\varepsilon_{\vec{r}}^{*}$, but the converse is not true. Even though we started with all of $L^2$ to begin with, a Hilbert space (infinite dimensional) that is isomorphic to it's dual, we have taken a subspace of $\mathcal{F}$ of $L^2$ to represent physical outcomes (see prior discussion). Since $\mathcal{F}$ (the Ket Vectors) is a subset of $L^2$, it isn't surprising that the dual space $\mathcal{F}^*$ (the Bra Vectors) is "larger" than $\mathcal{F}$. Thus, not every Bra Vector has a corresponding Ket vector. For calculational convenience it is convenient to have every Bra Vector correspond to Ket Vector or a generalized Ket Vector. This introduces generalized Ket Vectors without finite norm (infinite plane waves, etc.) that are convenient in intermediate calculations, but are not physical so are not exhibited in the final result, which will only be comprised of (regular) Ket vectors. In the intermediate calculations, even though the scalar product of a generalized Ket with itself (the norm) might be infinite, the scalar product of the generalized Ket with any regular Ket is finite.

Note that Vector space duality is induced by existence of a linear vector space and the existence of a finite norm. For tensor analysis in 4D spacetime we have vector space duality on a finite vector space [17], thus the dual vectors spaces (covariant and contravariant) are isomorphic. In quantum mechanics we are now shifting to a formulation in terms of the aforementioned Ket Space that is a subspace of a Hilbert Space (infinite dimensional, but exhibiting linearity and finite norm).

**Hermitian Conjugation**

Let's now consider linear operators acting on out Bra and Ket vectors. Consider a linear operator '$A$' acting on a Ket vector $<\varphi|$ : $<\varphi|A$. Since $<\varphi|A$ is linear vector operator in its own right, we must be able to associate with every Ket $|\Psi>$ a number that depends linearly on $|\Psi>$. This means that $<\varphi|$ with operation by $A$ must define a new linear functional on Kets. Thus, we have: $(<\varphi|A)\,|\Psi> = <\varphi|A\,(|\Psi>) \equiv <\varphi|A|\Psi>$.

**The Adjoint Operator of a given Operator**

Suppose we have an operator acting on the Ket $|\Psi>$ to give $|\Psi'> = A|\Psi>$ , let's define the adjoint operator by defining a dual relation on Bra vectors: $<\Psi'| = <\Psi|A^\dagger$. Since a scalar product satisfies $<\Psi'|\varphi> = <\varphi|\Psi'>^*$, we must have:

$$<\Psi|A^\dagger|\varphi> = <\varphi|A|\Psi>^*.$$

Consider what this means for evaluation of $(AB)^\dagger$ :

$$|\varphi> = AB|\Psi> = A(B|\Psi>)$$

With dual:

$$<\varphi| = <\Psi|(AB)^\dagger = (B|\Psi>)^\dagger A^\dagger = <\Psi|B^\dagger A^\dagger.$$

Thus,

$$(AB)^\dagger = B^\dagger A^\dagger.$$

From this we can now define the Hermitian conjugate of any expression by exchanging Bras and Kets, operators with their adjoints, and constants with their complex conjugates, and reverse the order of the (operator) factors.

**Hermitian Operators: $A = A^t$**

Let's now consider Hermitian operators (a.k.a., self-adjoint operators). Thus, for a Hermitian operator: $<\Psi|A|\varphi> = <\varphi|A|\Psi>^*$ . alternatively:

$$<A\varphi|\Psi> = <\varphi|A\Psi>.$$

Consider the Projection operator:

$$P_\Psi^\dagger = |\Psi><\Psi| = P_\Psi^{\square},$$

Thus it is Hermitian.

Note that the product of two Hermitian operators A and B is Hermitian only if $[A, B] = 0$ :

$$A = A^{\dagger}, B = B^{\dagger}, \ \ (AB)^{\dagger} = B^{\dagger}A^{\dagger} = BA$$

$$[A, B] = AB - BA \ \ if\ (AB)^{\dagger} \neq AB \rightarrow AB \neq BA \rightarrow [A, B] \neq 0$$

**Orthonormalization relation**

$< u_i|u_j >= \delta_{ij}$ discrete indexing.

$< w_{\alpha}|w_{\alpha'} >= \delta(\alpha - \alpha')$ continuous indexing.

**Closure relation**

Suppose:

$|\Psi\geq \sum_i c_i|u_i >$, if discrete,

or

$|\Psi>= \int d\alpha\, c(\alpha)|w_{\alpha} >$, if continuous.

For the projection operator this becomes:

$$P_{\{u_i\}} = \sum_i |u_i >< u_i | = 1 \ ,$$

or

$$P_{\{w_{\alpha}\}} = \int d\alpha\ |w_{\alpha} >< w_{\alpha}| = 1$$

**Representation of Kets and Bras**

In the $\{|u_i >\}$ basis, the Ket $|\Psi>$ is represented by the set of its components, $c_i =< u_i|\Psi>$. These numbers can be arranged vertically to form a one column matrix (with, in general, a countable infinity of rows):

$$\begin{pmatrix} < u_1|\Psi> \\ < u_2|\Psi> \\ \vdots \\ < u_i|\Psi> \\ \vdots \end{pmatrix} \text{ for a discrete basis}$$

And

$$\begin{pmatrix} \vdots \\ < w_{\alpha}|\Psi> \\ \vdots \end{pmatrix} \text{ for a continuous basis.}$$

For a Bra representation we have a one-row matrix:

$$< \varphi| =< \varphi|1 =< \varphi|P_{\{u_i\}} = \sum_i < \varphi|u_i >< u_i |$$

**Representation of operators**

Representation of operator A is by a "square matrix" given by:

$$A_{ij} =< u_i|A|u_j >$$

Or

$$A(\alpha, \alpha') =< w_\alpha|A|w_\alpha >$$

The representation of $|\Psi' >= A|\Psi' >$, knowing the rep's of A and $|\Psi' >$, follows by direct matrix calculus. Notice that $|\Psi' >< \Psi'|$ becomes

$$\begin{pmatrix} c_1 \\ c_2 \\ \vdots \\ c_i \\ \vdots \end{pmatrix} \left(c_1^* \, c_2^* \dots c_j^* \dots\right) = \begin{pmatrix} c_1c_1^* & c_1c_2^* & \dots & c_1c_j^* \dots \\ c_2c_1^* & c_2c_2^* & \dots & c_2c_j^* \dots \\ & & \vdots & \\ c_ic_1^* & c_ic_2^* & \dots & c_ic_j^* \dots \\ & & \vdots & \end{pmatrix}$$ *which is the rep.*

*of an operator*

Matrix rep. of a adjoint, $A^\dagger$, of $A$:

$$(A^\dagger)_{ij} = < u_i|A^\dagger|u_j > = < u_j|A|u_i >^* = A_{ji}^* .$$

A Hamilton operator is then represented by a Hamilton matrix:

$$A_{ij} = A_{ji}^*,$$

thus, diagonals always real.

**Change of representation**

Consider changing from one discrete orthogonal basis $\{|u_i >\}$ to another $\{|t_k >\}$, the change is defined by specifying the components $< u_i|t_k >$. Denote $S_{ik} =< u_i|t_k >$, thus also have:

$$(S^\dagger)_{ki} = (S_{ik})^* =< t_k|u_i >.$$

Calculations can be performed easily, by using:

$$P\{u_i\} = \sum_i |u_i >< u_i| = 1 \quad , \qquad < u_i|u_j > = \delta_{ij}$$

$$P\{t_i\} = \sum_k |t_k >< t_k| = 1 \quad , \qquad < t_k|t_e > = \delta_{kl}$$

Thus, can express any $< t_k|\Psi>$ knowing $S_{ik}$:

$$< t_k|\Psi> =< t_k| \left( \sum_i |u_i >< u_i| \right) |\Psi> = \sum_i S^\dagger < u_i|\Psi> .$$

Thus,

$$< t_k|A|t_l >= \sum_{i,j} < t_k|u_i >< u_i|A|u_j >< u_j|t_l >$$

$$A_{kl} = \sum_{i,j} S^\dagger_{kj} A_{ij} S^i_{jl}$$

**Eigenvalue equations – Observables**

We define $|\Psi>$ to be an eigenvalue of operator $A$ if

$$A|\Psi> = \lambda|\Psi>$$

where $\lambda$ is a constant, called the eigenvalue of the operator. In general, this equation only possesses solutions when the eigenvalue takes on certain values. The set of eigenvalues is called the spectrum of the operator.

Note that there is an ambiguity in eigenvalues according to an overall constant multiplier 'a', since

$$A(a|\Psi>) = a(A|\Psi>) = a\lambda\,|\Psi> = \lambda(a|\Psi>).$$

This is partly eliminated if we are concerning ourselves with eigenvectors that have unit norm:

$$< \Psi|\Psi> = 1,$$

which is often the case if dealing with wavefunctions. Even with unit normalization there is still a phase factor of ambiguity left. Dealing with this final ambiguity will arise in various ways, right up to quantum field theory regularization via renormalization that is described in Book5 [15]).

If there are at least two Ket vectors (linearly independent) with the same eigenvalue, then the eigenvalue is said to be degenerate. It's degree (or order) of degeneracy is the number of linearly independent eigenvectors that can be associated with it (the order can be infinite).

Knowing $A|\Psi>$, we a priori know nothing of $< \Psi|A$ . If we are told $A$ is Hermitian (self-adjoint), which will be the case for any physical observable, then we have:

$$A|\Psi> = \lambda|\Psi> \;\; \rightarrow \;\; < \Psi|A^\dagger = \\ < \Psi|\lambda^* \;\; (from\; duality\; transform)$$

and

$$A^\dagger = A \;\; (from\; hermitian).$$

Thus,

$$< \Psi|A = < \Psi|\lambda^*.$$

**Finding the eigenvalues and eigenvectors of an operator**
We start by choosing a eigenvector representation that is complete: $\{|u_i >\}$. We then project onto the eigenvalue equation:

$$< u_i|A|\Psi> = \lambda < u_i|\Psi>,$$

and using the closure relation we get:

$$\sum_j [A_{ij} - \lambda\delta_{kl}]c_j = 0.$$

A non-trivial solution (must have $c_j \neq 0\ for\ all\ j$) occurs IFF $Det[A - \lambda I] = 0$, which is called he characteristic equation or secular equation from the theory of differential equations. Thus, the eigenvalues of an operator are the roots of its characteristic equation.

**Observables**
If a vector space is finite dimensional, it is always possible to form a basis with the eigenvectors of a Hermitian operator. If a vector space is infinite dimensional it is not always possible to form a basis from a Hermitian operator's eigenvectors. When such a Hermitian operator permits such a basis to be formed, that operator will be considered to be an 'observable' of the theory. If we consider the projection operator 'P' on a basis indexed by 'n', we can then rewrite any Hermitian operator in terms of its (projection) eigenvalues:

$$A = \sum_n a_n P_n$$

**Commuting Observables**
Theorem 1: If two operators $A$ and $B$ commute, and if $|\Psi>$ is an eigenvector of $A$, then $B|\Psi>$ is also an eigenvector of $A$.

Theorem 2: If two operators $A$ and $B$ commute, then every eigensubspace of $A$ is globally invariant under the action of $B$.

Theorem 3: If two observables $A$ and $B$ commute, and $|\Psi_1>$ and $|\Psi_2>$ are two eigenvalues of $A$, then $< \Psi_1|B|\Psi_2> = 0$.

Theorem 4: If two observables $A$ and $B$ commute, one can construct an orthonormal basis of the state space with eigenvectors common to $A$ and $B$.

**Complete Set of Commuting Observables**
A set of observables is a complete set of commuting observables if there exists a unique orthonormal basis of common eigenvalues.

The position and momentum representations have been mentioned previously, each of which can be directly related to Fourier Transform theory, where completeness is well established for the representations indicated. Recall the position representation with basis: $\{\varepsilon_{r_0}(\vec{r})\}$ and momentum basis $\{v_{p_0}(\vec{r})\}$, where $\varepsilon_{r_0}(\vec{r}) = \delta(r - r_0)$ and $v_{p_0}(\vec{r}) = (2\pi\hbar)^{-3/2}\, e^{\frac{i}{\hbar}\overrightarrow{P_0}\cdot\overrightarrow{r_0}}$ (with generalized Ket's as mentioned). The mathematical formulation with the generalized Ket's is equivalent to that examined in Fourier Transform theory, will well established mathematical procedures to obtain results (where the generalized Ket's only appear in intermediate calculations, and never as a final answer). The scalar product between wave functions now become more strongly established, theoretically, in the formation:

$$< \varphi|\Psi> = \int d^3r < \varphi|r >< r|\Psi> = \int d^3r\, \varphi^*(r)\,\Psi(r)$$

$$= \int d^3p < \varphi|p >< p|\Psi> = \int d^3r\, \bar{\varphi}^*(r)\,\bar{\Psi}(r)$$

Changing from the $\{|r>\}$ rep. to the $\{|p>\}$ rep:

$$< r_0|p_0 > = < p_0|r_0 >^* = \left[\int d^3r\, v_{p_0}(\vec{r})\varepsilon_{r_0}(\vec{r})\right]^* = (2\pi\hbar)^{-3/2} e^{\frac{i}{\hbar}\overrightarrow{P_0}\cdot\overrightarrow{r_0}}$$

Since,

$$< r|\Psi> = \int d^3p < r|p >< p|\Psi>$$

We have

$$\Psi(\vec{r}) = \int d^3p\, (2\pi\hbar)^{-3/2} e^{\frac{i}{\hbar}\overrightarrow{P_0}\cdot\overrightarrow{r_0}}\, \bar{\Psi}(\vec{p})$$

Thus, $\Psi(r)$ *and* $\bar{\Psi}(\vec{p})$ are related by Fourier transform as mentioned.

$< \vec{r}'|A|\vec{p} > = A(\vec{r}', \vec{r})$ (defined, but a matrix with uncountable indices)

$$< \vec{p}'^{|A|\vec{p}} > = A(\vec{p}', \vec{p}) = \int d^3r \int d^3r' < \vec{p}'|\vec{r}' >< \vec{r}'^{|A|\vec{r}} >< \vec{r}|\vec{p} >$$

$$= \int d^3r \int d^3r' \frac{1}{(2\pi\hbar)^3} e^{\hbar(-\vec{p}\cdot\vec{r}' + \vec{p}\cdot\vec{r}\square)} A(\vec{r}', \vec{r})$$

**R and P operators**

In the $\{|p>\}$ basis the $\vec{R}$ operator is simply $\vec{R}|\vec{r} = \vec{r}|\vec{r}>$ , while in the $\{|p>\}$ basis the $\vec{P}$ operator is $\vec{P}|\vec{p}> = \vec{P}|\vec{p}$. What about $\vec{P}$ in the $\{|\vec{r}>\}$ representation? Consider an arbitrary Ket $|\Psi>$ , then

$$<\vec{r}|P_x|\Psi> = \int d^3p <r|p><p|P_x|\Psi>$$

$$= (2\pi\hbar)^{-3/2} \int d^3p e^{\frac{i}{\hbar}(\vec{p}\cdot\vec{r}^{\square})} p_x \vec{\Psi}(\vec{p}) = \frac{\hbar}{i}\frac{\partial}{\partial x}\Psi(\vec{r})$$

Therefore

$$<\vec{r}|\vec{p}|\Psi> = \frac{\hbar}{i}\nabla <\vec{r}|\Psi>.$$

**Tensor Product of State Spaces**

Just as with tensor product spaces in Geometry/General Relativity (Book 2 [17] & Book 3 [18]) much is the same with the Ket spaces of interest, which are subsets of Hilbert space as mentioned, possibly infinite dimensional. This will naturally arise when we consider a physical particle which will have intrinsic spin, thus we have the product of a position space representation, say ($\{\varepsilon_{r_0}(\vec{r})\}$ in the above), and the spin state space representation $\{\varepsilon_{spin}\}$. The larger vector space that results is simply the product space:

$$\varepsilon = \varepsilon_{r_0} \otimes \varepsilon_{spin},$$

Where the linearity and distributive properties are retained.

**Derivation of the Schwartz inequality**

We want to show

$$|<\varphi_1|\varphi_2>|^2 \leq <\varphi_1|\varphi_1><\varphi_2|\varphi_2>,$$

which is the Schwartz inequality in the Dirac notation adopted.

We know that $<\Psi|\Psi> \geq 0$ , let's consider rewriting the Ket vector as a linear combination of two other Ket's:

$$|\Psi> = |\varphi_1> + \lambda|\varphi_2>$$

Then,

$$<\Psi|\Psi> = <\varphi_1|\varphi_2> + \lambda <\varphi_1|\varphi_2> + \lambda^* <\varphi_1|\varphi_2> + \lambda\lambda^* <\varphi_1|\varphi_2> \geq 0.$$

If we now choose

$$\lambda = -\frac{<\varphi_2|\varphi_1>}{<\varphi_2|\varphi_2>}$$

We get:

$$< \Psi | \Psi > = < \varphi_1 | \varphi_1 > - \frac{< \varphi_1 | \varphi_2 >< \varphi_1 | \varphi_2 >^*}{< \varphi_2 | \varphi_2 >} \geq 0$$

Thus,

$$|< \varphi_1 | \varphi_1 > |^2 \leq < \varphi_1 | \varphi_1 >< \varphi_2 | \varphi_2 >,$$

As desired.

**Unitary operators**

We now come to unitary operators. These are especially important as the fundamental quantum theory can be expressed in terms of unitary propagation, as is explicit in the PI formulation. (In seeking a deeper theory in Book7 [21] we relax the unit norm unitary propagation of Quantum Mechanics and quantum field theory to the unit norm 'emanation' of emanator theory.)

An operator '$u$' is unitary if:

$$u^\dagger u = uu^\dagger = 1.$$

Let's explore why this condition should be so special. If $|\widehat{\Psi}_1 > = u|\Psi_1 >$ and $|\widehat{\Psi}_2 > = u|\Psi_2 >$, then $< \widehat{\Psi}_1|\widehat{\Psi}_2 > = < \Psi_1|u^\dagger u|\Psi_2 > =< \Psi_1|\Psi_2 >$, e.g., the scalar is product conserved under a unitary operator.

Note that if $A$ *is Hermitian* then $T = e^{iA}$ is unitary since:

$$T^\dagger = e^{-iA} = T^{-1} \rightarrow T^t T = TT^t = 1.$$

Note that the product of two unitary operators as also unitary:

$$(uV)^\dagger (uV) = V^\dagger u^\dagger uV = V^\dagger V = 1.$$

In ordinary 3D space of real vectors operators which preserve the scalar product (rotations, symmetry op.) are said to be orthogonal. Unitary operators constitute the generalization of orthogonal operators to complex spaces.

A necessary condition for an operator $u$ to be unitary is that the vectors of an orthogonal basis, transformed by $u$, constitute another orthogonal basis. Consider $u_{ij} =< V_i|u|V_j >$ what are it properties?

$$< V_i|u^\dagger u|V_j > = < V_i|V_j > = \delta_{ij}$$

So,

$$\sum_k < V_i|u^\dagger|u_k >< u_k|u|V_j > = \sum_k < u_k|u|V_i >* < u_k|u|V_j >$$

$$= \sum_k u_{ki}^* u_{kj} = \delta_{ij}$$

When a matrix is unitary, the sum of the products of the elements of one column and the complex conjugates of the elements of another column is:

(i) Zero if the two columns are different

(ii) One if they are the same.

**The infinitesimal unitary operator**

We saw previously that if $A$ *is Hermitian* then $T = e^{iA}$ is unitary. Let's now consider any unitary operator in the neighborhood of the unit identity operator, where we will see that the first order deviance from unity is proportional to a Hermitian operator (e.g., it takes the form of $T = e^{iA}$ at first order).

Consider $u(\epsilon)$ a unitary operator which depends on an infinitesimally small real quantity:

$$U(\epsilon) \rightarrow 1 \; when \; \epsilon \rightarrow 0.$$

Let's write this in terms of a power expansion on $\epsilon$ and only follow at first order:

$u(\epsilon) = 1 + \epsilon G + \cdots$

$u^\dagger(\epsilon) = 1 + \epsilon G^\dagger + \cdots$

Now $u(\epsilon)u^\dagger(\epsilon) = 1 + \epsilon(G + G^\dagger) + \cdots = 1$, which implies that $G + G^\dagger = 0,\; so \; G$ is anti- Hermitian. Let $F = iG \rightarrow iF + iF^\dagger = 0 \; or \; F = F^\dagger$, so F is Hermitian. So,

$$u(\epsilon) = 1 - i\epsilon F,$$

where $F$ is Hermitian and $\epsilon$ is infinitesimal.

Note, if $\tilde{A} = UAU^\dagger$, we have:

$$\tilde{A} = UAU^\dagger = (1 - i\epsilon F)A(1 - i\epsilon F) = A + i\epsilon[A, F],$$

thus

$$\tilde{A} - A = -i\epsilon[F, A].$$

**The Schrodinger Equation in the $\{|r >\}$ representation**

Let's revisit the Schrodinger equation, switching to Dirac notation, for the (standard) $\{|r >\}$ representation. For the time evolution we have:

$$i\hbar \frac{d}{dt} | \Psi(t) > H | \Psi(t) >.$$

While for a spinless particle in a scalar potential we have:

$$H = \frac{1}{2m}P^2 + V(R).$$

Projecting into the position representation via scalar product with Bra $<\vec{r}|$, we have:

$$i\hbar\frac{\partial}{\partial t} < \vec{r}\,|\,\Psi(t) > = \frac{1}{2m} < \vec{r}|P^2\,|\,\Psi(t) > + < \vec{r}|V(R)|\,\Psi(t) >$$

To relate back to Schrodinger wavefunction notation we can clearly see:

$$< \vec{r}|\,\Psi(t) > = \Psi(\vec{r}, t)$$

$$< \vec{r}|V(R)|\,\Psi(t) > = V(\vec{r})\,\Psi(\vec{r}, t)$$

$$< \vec{r}|P^2|\,\Psi(t) > = < \vec{r}|P_x^2 + P_y^2 + P_z^2|\,\Psi(t) >$$

$$= \hbar^2\left(\frac{\partial^2}{\partial x^2} + \frac{\partial^2}{\partial y^2} + \frac{\partial^2}{\partial z^2}\right)\Psi(x, y, z, t) = -\hbar^2 V^2\,\Psi(\vec{r}, t)$$

So, in $|r>$ representation:

$$i\hbar\frac{\partial}{\partial t}\Psi(\vec{r}, t) = \left[-\frac{\hbar^2}{2m}\nabla^2 + V(\vec{r})\right]\Psi(\vec{r}, t),$$

which is the standard form of the Schrodinger wave equation.

**The Schrodinger Equation in the $\{|p>\}$ representation**

The Schrodinger Equation in the $\{|\vec{P}>\}$ Rep. -- again, using $i\hbar\frac{d}{dt}|\,\Psi> = H|\,\Psi>$ we have:

$$i\hbar\frac{\partial}{\partial t} < \vec{p}|\,\Psi> = < \vec{p}|\frac{P^2}{2m} + V(R)|\,\Psi>$$

and have:

$$< \vec{p}|\,\Psi> = \overline{\Psi}(\vec{p}, t)$$

$$< \vec{p}|V(R)|\,\Psi> = \int d^3p' < \vec{p}|v(R)|\vec{p}' > < \vec{p}'|\,\Psi>$$

$$= \int d^3p'\left\{\int d^3r < \vec{p}|v(R)|\vec{r} > < \vec{r}|\vec{p}' >\right\}\overline{\Psi}(\vec{p}')$$

$$= \int d^3p'\left\{\int d^3r\, v(r)\frac{e^{\frac{1}{\hbar}(\vec{p}'\cdot r - \vec{p}\cdot r)}}{(2\pi\hbar)^3}\right\}\overline{\Psi}(\vec{p}')$$

Let

$$\bar{V}(p) = (2\pi\hbar)^{-3/2}\int d^3r V(r) e^{-\frac{1}{\hbar}\vec{p}\cdot\vec{r}}$$

Then

$$< \vec{P}|v(R)|\Psi>$$
$$= \int d^3p' \left\{ \frac{1}{(2\pi\hbar)^{-3/2}} \left[ \frac{1}{(2\pi\hbar)^{-3/2}} \int d^3r V(r) e^{-\frac{i}{\hbar}(\vec{p}-\vec{p'})\cdot\vec{r}} \right] \right\} \vec{\Psi}(\vec{p}')$$

Thus,

$$< \vec{P}|v(R)|\Psi> = \int d^3p' \frac{1}{(2\pi\hbar)^{-3/2}} \bar{V}(\vec{p}-\vec{p}')\vec{\Psi}(\vec{p}',t)$$

Thus, in the $|\overrightarrow{p>}$ rep.:

$$i\hbar\frac{\partial}{\partial t}\vec{\Psi}(\vec{p},t) = \frac{\vec{P}^2}{2m}\vec{\Psi}(\vec{p},t) + \frac{1}{(2\pi\hbar)^{-3/2}} \int d^3p' \bar{V}(\vec{p}-\vec{p}')\vec{\Psi}(\vec{p}',t)$$

**General properties of canonical observables**

Let's consider the general properties of two observables, Q and P, whose commutator $[Q,P] = i\hbar$ (e.g., canonical quantization). Define $S(\lambda) = e^{-i\lambda p/\hbar}$, where λ real, $S$ is unitary since P is Hermitian (an observable). We thus have:

$$[Q,S(\lambda)] = [Q,P]\left(-\frac{i\lambda}{\hbar}\right)S(\lambda) = \lambda S(\lambda)$$

Thus,

$$QS = S[Q+\lambda] \; and \; note \;\; S(\lambda)S(\mu) = S(\lambda+\mu)$$

***Spectrum of Q***

Assume for some $|q>$, $that \;\; Q|q> = q|q>$:

$$QS|q> = S(Q+\lambda)|q> = S(q+\lambda)> = (q+\lambda)S|q>$$

i.e. $QS|q> = (Q+\lambda)S|q>$ thus, $S|q>$ is another non-zero eigenvector with eigenvalue $(q+\lambda)$, since λ is arbitrary Q has a continuous spectrum.

***Degree of degeneracy***

All eigenvalues of Q must have the wave degree of degeneracy.

**What is $\vec{\mathbf{P}}$ in the position $\{<|\mathbf{q}>\}$ representation?**

In the $\{|q>\}$ representation we have:

$$\Psi(q) =< q|\Psi>$$

So,

$$< q|Q|\Psi> = q\,\Psi(q)$$

and

$$< q|S(\lambda)|\Psi> =< q-\lambda|\Psi> = \Psi(q-\lambda) \rightarrow S$$

which is the translation operator. The infinitesimal action of P in $\{<|q>\}$ representation is:

$$S(-\epsilon)=e^{i\lambda P/\hbar}=1+i\frac{\epsilon}{\hbar}\vec{P}+O(\epsilon^2).$$

So

$$<q|S(-\epsilon)|\Psi>=\Psi(q)+i\frac{\epsilon}{\hbar}<q|P|\Psi>+O(\epsilon^2)=<q+\epsilon|\Psi>$$
$$=\Psi(q+\epsilon)$$

So,

$$\Psi(q+\epsilon)=\Psi(q)+i\frac{\epsilon}{\hbar}<q|P|\Psi>+O(\epsilon^2)$$

Thus,

$$<q|P|\Psi>=\frac{\hbar}{i}\lim_{\epsilon\to 0}\frac{\Psi(q+\epsilon)-\Psi(q)}{\epsilon}=\frac{\hbar}{i}\frac{d}{dq}\Psi(q)$$

So,

$$\vec{P}\leftrightarrow\frac{\hbar}{i}\frac{d}{dq}$$

in the $\{<|q>\}$ representation.

**Review of the symmetrical nature of canonically conjugate observables**

In the $\{<|q>\}$ representation we just showed $<q|P|\Psi>=\frac{\hbar}{t}\frac{d}{dq}\Psi(q)$ using only $[Q,P]=i\hbar$. The derivation was made in thee steps:

(i) defining $S(\lambda)=e^{-i\lambda p/\hbar}$

(ii) establishing that $S(\lambda)$ is the translation operator

$QS=S[Q+r]\quad\to\quad QS|q>=(q+\lambda)S|q>\quad\to\quad S|q>=|q+\lambda>$

(iii) performing an infinitesimal analysis: $S(-\epsilon)=1+i\frac{\epsilon}{\hbar}P\rightarrow$

$$<q|S(-\epsilon)|\Psi>=\Psi(q+\epsilon)=\Psi(q)+i\frac{\epsilon}{\hbar}<q|P|\Psi>$$

(iv) correspondence with derivative definition:

$$<q|\vec{P}|\Psi>=\frac{\hbar}{i}\lim_{\epsilon\to 0}\frac{\Psi(q+\epsilon)-\Psi(q)}{\epsilon}=\frac{\hbar}{i}\frac{d}{dq}\Psi(q)$$

Which gives the interpretation $\vec{P}=\frac{\hbar}{i}\frac{d}{dq}$.

Now, using $<q|\vec{P}|\Psi>=\frac{\hbar}{i}\frac{d}{dq}\Psi(q)$, let $\Psi=p$ : $<q|\vec{P}|p>=p<q|p>$ $=\frac{\hbar}{i}\frac{d}{dq}\Psi(q)$. Thus,

$$\frac{\hbar}{i}pq=ln<q|p>+const\rightarrow<q|p>=Ne^{\frac{i}{\hbar}pq},$$

Which can't be normalized, but N is chosen, by conversion, to match that of a Fourier transform:

$$< q|p >= \frac{1}{(2\pi\hbar)^{1/2}} e^{\frac{i}{\hbar}pq}$$

Thus,

$$|p >= (2\pi\hbar)^{-1/2} \int_{-\infty}^{\infty} dq e^{\frac{i}{\hbar}pq} |q >$$

$$\bar{\Psi}(p) = (2\pi\hbar)^{-1/2} \int_{-\infty}^{\infty} dq e^{\frac{i}{\hbar}pq} \Psi(q)$$

the standard Fourier transform.

**States and Spectra**

Definition: $\underline{u} = \underline{v} \leftrightarrow \underline{u} - \underline{v} = 0 \leftrightarrow \|\underline{u} - \underline{v}\| = 0$

Consider a sequence of vectors $\underline{u_n}$

$\lim_{n\to\infty} \underline{u_n} = \underline{u} \leftrightarrow \lim_{n\to\infty} \|\underline{u_n} - \underline{u}\| = 0$

Proof, Can't have two limits to a sequence

$\lim_{n\to\infty} \underline{u_n} = \underline{v}$

$\|\underline{u} - \underline{v}\| = \|\underline{u} - u_n + u_n - \underline{v}\| \leq \|\underline{u} - u_n\| + \|\underline{u_n} - \underline{v_n}\|$

$\lim_{n\to\infty} \|\underline{u} - u_n\| \to 0$

$\|\underline{u} - \underline{v}\| \leq \lim_{n\to\infty} \|\underline{u_n} - \underline{v_n}\| \to \underline{u} = \underline{v}$

$\|\underline{u_n} - \underline{u_m}\| = \|\underline{u_n} - \underline{u} + \underline{u} - \underline{u_m}\| \leq \|\underline{u_n} - \underline{u}\| + \|\underline{u_n} - \underline{u_m}\|$

$\lim_{n-m\to\infty} \|\underline{u_n} - u_m\| = 0$ (Cauchy sequence)

Not reversible, Cauchy sequence doesn't require a limit point to be in its set.

When all Cauchy sequences converge to a vector in the space we have a complete space. Key in the issue of continuity. A vector space with a scalar product which is complete is called a Hilbert Space. Quantum Mechanics is not just based on a Hilbert Space, it's based on a special kind of Hilbert Space.

Countable basis $\{\underline{u_n}; n = 1, \dots, N\}$ linearly independent. $\underline{\Psi} = \sum_{n=1}^{N} C_n \underline{u_n}$ is a separable space. In Quantum Mechanics $N \to \infty$ and $\|\underline{\Psi} - \sum C_n u_n\| = 0$ . The space of Quantum Mechanics states is a separable Hilbert space. One definition of scalar product, which satisfies the complete space requirement is

$(\underline{u},\underline{v}) = \int_{-\infty}^{\infty} dx u^*(x) v(x)$
$\|\underline{u}\|^2 = \int_{-\infty}^{\infty} dx |u(x)|^2$
Harmonic oscillator with above scalar product
$u_n(x) = H_n(x) e^{-x^2/2}$
$\Psi(x) = \sum_n C_n u_n(x) \leftarrow stricter\ restriction\ on\ \Psi(x)$
$\underline{u} = \underline{v} \leftrightarrow \underline{u} - \underline{v} = 0 \leftrightarrow \|\underline{u} - \underline{v}\| = 0$
$\int_{-\infty}^{\infty} dx \left| \Psi(x) - \sum_{n=0}^{\infty} C_n H_n(x) e^{-x^2/2} \right|^2 = 0$
A looser restriction then $\Psi(x) = \sum_n C_n u_n(x)$, then:
$\int dx |\ldots|^2 = 0$ even if the $\Psi(x)$ and $\sum_{n=0}^{\infty}$ differ at an infinite set of discrete points.

**States are vectors in a separable Hilbert space**
$(\underline{u},\underline{v}) = \langle u|v \rangle$ ; $\underline{\Psi} = \sum_{n=1}^{N} C_n U_n$, $(\underline{u_n}, \underline{u_m}) = \delta_{n,m}$, $C_n \equiv (\underline{u_n}, \Psi)$, thus:

$$\underline{\Psi} = \sum_{n=1}^{N} (\underline{u_n}, \Psi) \underline{u_n} \ldots$$

Any Cauchy sequence converges.

Dual of a vector space = $\{space\ of\ linear\ functionals: F(\underline{u})\}$
$F(u) \in C$ takes a$(\underline{u})$ to a constant C.
$(F_1 + F_2)(\underline{u}) = F_1(\underline{u}) + F_2(\underline{u})$
and
$(aF)(\underline{u}) \equiv aF(\underline{u})$

If the vector space is a Hilbert space we already have a relation between $\underline{u}$ and C via the scalar product:

For any fixed $\underline{\Psi}$ : $F_{\underline{\Psi}} \equiv (\underline{\Psi}, \underline{u})$.

$F_{\underline{\Psi}_1 + \underline{\Psi}_2} = (\underline{\Psi}_1 + \underline{\Psi}_2, \underline{u}) = (\underline{\Psi}_1, \underline{u}) + (\underline{\Psi}_2, \underline{u}) = F_{\underline{\Psi}_1} + F_{\underline{\Psi}_2}$
$aF_{\underline{\Psi}} = F(a^* \underline{\Psi})$
$(aF_{\underline{\Psi}})(a) = F_{a^* \underline{\Psi}}(\underline{u}) = (a^* \underline{\Psi}, \underline{u}) = a(\underline{\Psi}, \underline{u}) = aF_{\Psi}(u)$

Consider:
$F(\underline{u}) = F_{\underline{\Psi}}$

Then $\underline{F} = \sum_{n=1}^{N}[F(\underline{u}_n)]^* \underline{u}_n$
$\underline{u} = \sum_n(\underline{u}_n, \underline{u})$ $\qquad F(\underline{u}) = F(\sum_n(\underline{u}_n, \underline{u})\underline{u}_n)$
Linearity $F(\underline{u}) = \sum_n(\underline{u}_n, \underline{u})F(\underline{u}_n) = (\underline{\Psi}, \underline{u})$
Is $\|\underline{\Psi}\| < \infty$? $as\ N \to \infty$ an issue if F is continuous i.e. if $F(\underline{u}_n) \to F(\underline{u})\ \ as\ \underline{u}_n \to \underline{u}$

So, the bra's ( <...| ) are elements of the dual space. These is a one correspondence between a bra and a ket. Need to show it is linear or antilinear:
$< \underline{u}| \leftrightarrow \underline{u}$
$\lambda < u| \leftrightarrow \lambda^* \underline{u} = \lambda^*|u >$

Correspondence between v space and its dual is special to Hilbert space.

Now let's consider Operators $\underline{u} \to \underline{v}$ such a mapping is associated with an operator $A\underline{u} = \underline{v}$

Linear operators are of interest in quantum mechanics:

$$\left.\begin{aligned} A(\underline{u} + \underline{v}) &= A\underline{u} + A\underline{v} \\ A(\lambda \underline{u}) &= \lambda A\underline{u} \end{aligned}\right\}$$

note that the operators themselves define a Hilbert space.

Antilinear operator (time reversal)
$A(\lambda \underline{u}) = \lambda^* A\underline{u}$
Well defined when operator acts on infinite element Hilbert space where $\|A\underline{u}\| < \infty$ with u in the domain of $A$ , $D(A)$ .
Recall $\int_{-\infty}^{\infty} dx\, |\Psi(x)|^2 < \infty$ is a square integrable functions
$x|\Psi(x)|_n^2 \xrightarrow[|x|\to\infty]{} 0\ \ thus\ \ \|\Psi(x)\|^2 \sim \frac{1}{x^{1+\epsilon}}$

Consider $Q\,\Psi(x) = x\,\Psi(x) \equiv \Psi'(x)$
Now $\|\Psi^1(x)\|^2 \sim \frac{1}{x^{1+\epsilon}}$
Thus, $x^3\|\Psi(x)\|^2 \to 0$ , so Q is not defined on all of $\Psi(x)$ only that portion that has $x^3\|\Psi(x)\|^2 \to 0, thus, \|\Psi(x)\|^2 \sim \frac{1}{x^{3+\epsilon}}$ , thus:

$$D(Q) = \left\{\Psi(x)|x^{3/2}\,\Psi(x) \xrightarrow[|x|\to\infty]{} 0\right\}$$

So operators may not be defined for every element of the available Hilbert space but only some sub-domain. There is a class of operators that are defined for every element of the Hilbert space:

Bounded operators $\|A\underline{u}\| \leq C\|\underline{u}\|$ where C is independent Of $\underline{u}$ . The smallest C which satisfies this inequality is called the norm of $A$, $\|A\|$.

Consider an N is dimensional space (N finite)
$\underline{u}_n \quad n = 1, \dots N$
$A\underline{u}_n = \sum_{m=1}^{N} A_{mn}\underline{u}_m$
Suppose
$\underline{\Psi} = \sum_n C_n\underline{u}_n \quad then\ A\underline{\Psi} = \sum_n C_n A_{mn}\underline{u}_m$
$\|A\underline{\Psi}\|^2 \leq \sum_{nm'} \sum_{n'm'} C_n^m C_{n'} A_{mn}^* A_{mn'}$ using $(\underline{u}_{m'}, \underline{u}_{m'}) = \delta_{nm}$
$b = \max\{|\sum_m A_{mn}^* A_{n'm}| for\ all\ n, n'\}$
$\|A\underline{\Psi}\|^2 \leq \sum_{n,n'}|C_n||C_{n'}|b = b(\sum_n |C_n|)^2$
Use $(\sum_{n=1}^{N} x_n)^2 \leq N\sum_{n=1}^{N}(x_n)^2$ :

$\|A\underline{\Psi}\|^2 \leq Nb\sum_{n=1}^{N}|C_n|^2 = Nb\|\underline{\Psi}\|^2 \rightarrow so\ C = Nb$ finite.

Proof clearly doesn't go over to infinite dimension because $N \rightarrow \infty$ means $C \rightarrow \infty$. In a finite dimensional space all operators are bounded.

If we have:
$\underline{u} \overset{A}{\rightarrow} \underline{v}$
then
$\underline{v} \overset{A^{-1}}{\longrightarrow} \underline{u}\ , but\ A^{-1}\ can't\ always\ be\ defined$
$A^{-1}A = AA^{-1} = I$
Unitary operator $\mho$
$\|\mho\underline{\Psi}\| = \|\underline{\Psi}\|$ therefore U is a bounded operator with C=1.

Consider $\underline{\Psi}\underline{x}$ :
By definition of $U(\underline{\Psi} + \lambda\underline{x})$ :
$\left(U(\underline{\Psi} + \lambda\underline{x}) + U(\underline{\Psi} + \lambda\underline{x})\right) = \left((\underline{\Psi} + \lambda\underline{x}), (\underline{\Psi} + \lambda\underline{x})\right)$
$(U\underline{\Psi} + \lambda\underline{x}) + |\lambda|^2(\underline{Ux}) + \lambda(\underline{Ux}, U\underline{\Psi}) + \lambda^*(\underline{Ux}, U\underline{\Psi}) = (\underline{\Psi}, \underline{\Psi}) +$
$|\lambda|^2(\underline{x}, \underline{x}) + \lambda(\underline{\Psi}, \underline{x}) + \lambda^*(\underline{x}, \underline{\Psi})$

Consider

$\lambda = 1: (U\underline{\Psi}, U\underline{x}) + (U\underline{x}, U\underline{\Psi}) = (\underline{\Psi}, \underline{x}) + (\underline{x}, \underline{\Psi})$
$\lambda = i: (U\underline{\Psi}, U\underline{x}) - (U\underline{x}, U\underline{\Psi}) = (\underline{\Psi}, \underline{x}) - (\underline{x}, \underline{\Psi})$
add... $(u\underline{\Psi}, u\underline{x}) = (\underline{\Psi}, \underline{x})$.

Thus, scalar products are preserved as well as norms for unitary operators.

Unitary operators in Hilbert space synonymous with rotations in the regular space. Given any operator A we can define another operator called the adjourn of $A$, $A^{\dagger}$:
$(A\underline{u}, \underline{v}) = (\underline{u}, A^{\dagger}\underline{v})$ for every $\underline{u} \in D(A)$ . Does this define a unique association? $\underline{v} \rightarrow \underline{x} \equiv A^{\dagger}\underline{v}$ ? Thus, does $(A\underline{u}, \underline{v}) = (\underline{u}, \underline{x})$? A unique vector $\underline{x}$ needed? Or Does $(A\underline{u}, \underline{u}) = (\underline{u}, \underline{x}')$ $\underline{x}$ nonunique occur? If the domain of A is the whole Hilbert space, Then the above implies $(\underline{u}, \underline{x} - x') = 0$ for all (for uniqueness).

If $D(A) = H$ then choose $\underline{u} = \underline{x} - \underline{x}'$ to get $\|\underline{x} - \underline{x}'\| = 0 \rightarrow \underline{x} = \underline{x}'$ If $D(A)$ is a subspace of H then we cannot define an adjoint arbitrarily, $\underline{u} = \underline{x} - x'$ may not be in D(A) etc.

**Definition:** D(A) is dense in H if and only if for any $\underline{\Psi} \in H$ you can find a sequence of vectors $\underline{u_n} \in D(A)$ such that $\underline{u_n} \xrightarrow[n\rightarrow\infty]{} \underline{\Psi}$. So $\underline{\Psi}$ doesn't have to be in D(A) we just have to be able to get arbitrary "close" to it. If D(A) is dense then $(\underline{u_n} = \underline{x} - x') = 0$ can be found where $(\underline{u_n} = \underline{x} - x')$ and the proof above applies and we can define the adjoint.

Is the D(position operator, x) dense? Consider the oscillator $\Psi_N = \sum_{n=1}^{N} C_n H_n(x) e^{-x^2/2}$ all $\Psi_N \in D(Q)$ since $e^{-x^2/2}$ factor .

A basis in the Hilbert space so the position operator in an unbound operator with a dense domain i.e. an adjoint can be defined for the position operator.

**Definition of Symmetric**
'A' symmetric if $(Au, v) = (\underline{u}, A\underline{v})$ for all $\underline{u}, \underline{v}$ *in* $D(A)$. So, if $D(A) = D(A^{\dagger})$ *then* $A = A^{\dagger}$ and the operator is called self-adjoint or Hermitian.

Suppose $\Psi(x)$ square integrable functions defined in the domain $0 \leq x < \infty$ :

$$\int_{-\infty}^{\infty} dx \,|\, \Psi(x)|^2 < \infty$$

Define $A = i\frac{d}{dx}; D(A) - \{square\ int.\ \Psi^{\prime s}\ with\ \Psi(0) = 0\}$

If we pick any $\Psi, \varphi \in D(A)$

$(\Psi, A\varphi) = \int_0^\infty dx\ \Psi^* i\frac{d}{dx}\varphi = i|\Psi^*\varphi|_0^\infty - i\int_0^\infty A_x \frac{d\Psi^*}{dx}\varphi$

$= (A\Psi, \varphi) = (A^\dagger \Psi, \varphi)$ usual def. of adjoint define $A^\dagger = i\frac{d}{dx}$ but includes origin, so $D(A^\dagger) \supset D(A)$

**Eigenvalue problem for observables**

$A\underline{u}_n = \alpha_n \underline{u}_n$ if A is an observable, $\alpha_n$ will be real

In general $A\underline{u}_n^{(r)} = \alpha_n \underline{u}_n^{(r)}$ $\quad r = 1 \ldots R \quad \alpha_n$ is R – fold degenerate

Degenerate subspace

$P_n = \left\{\sum_{r=1}^{R} C_r \underline{u}_n^{(r)}\right\}$

$\sum_n P_n = I \qquad\qquad P_n P_{n'} = 0\ \ for\ n \neq n'$

$A = \sum_n \alpha_n P_n$

$E_x \equiv \sum_{\alpha_n \leq x} P_n = \sum_n \theta(x - x_n) P_n \leftarrow$ continuous democracy of prejection operators.

$E_x < E_y\ \ if\ x < y\ \ \rightarrow E_x E_y = E_y E_x = E_x$

$\lim_{x\to-\infty} E_x = 0 \qquad\qquad \lim_{x\to+\infty} E_x = I$

$E_{x+\epsilon} - E_x \rightarrow 0\ as\ \epsilon \rightarrow 0^+$

$E_x - E_{x-\epsilon} \rightarrow P_n\ \ if\ x = \alpha_n$

$\int_{-\infty}^{\infty} dE_x = I \qquad\qquad \int_{-\infty}^{\infty} dE_x = A$

$\left(\underline{\varphi}, \underline{\Psi}\right) = \int_{-\infty}^{\infty} d\left(\underline{\varphi}, E_x \underline{\Psi}\right) \qquad \left(\underline{\varphi}, A\underline{\Psi}\right) = \int_{-\infty}^{\infty} x d\left(\underline{\varphi}, E_x \underline{\Psi}\right)$

**Definition:** A family of projectors $E_x$ depending on a continuous parameter x is a spectral family if

(i) $E_x < E_y\ \ if\ x < y\ i.e.\ E_x E_y = E_y E_x = E_x$

(ii) $\lim_{x\to-\infty} E_x = 0\ \ ;\ \ \lim_{x\to+\infty} E_x = I$

(iii) $E_{x+\epsilon} - E_x \rightarrow 0 \qquad \epsilon \rightarrow 0^+$

Theorem: for each self-adjoint operator A there exists a unique spectra family $E_x$ such that

$\int_{-\infty}^{\infty} dE_x = I\ \ \ i.e. \left(\underline{\varphi}, \underline{\Psi}\right) = \int_{-\infty}^{\infty} d\left(\underline{\varphi}, E_x \underline{\Psi}\right)$

$\left(\underline{\varphi}, A\underline{\Psi}\right) = \int_{-\infty}^{\infty} x d\left(\underline{\varphi}, E_x\underline{\Psi}\right) \rightarrow A = \int_{-\infty}^{\infty} xE_x$ (Spectral decomposition of A)

**Position operator**

$\varphi\, \Psi(x) = x\, \Psi(x)$

$E_x\, \Psi(Z) = \begin{cases} \Psi(Z) & Z \leq x \\ 0 & Z > 0 \end{cases}$ definition of $E_x$ for position operator

$= \theta(x - Z)\, \Psi(Z)$

(i) $x < y \quad E_x E_y\, \Psi(Z) = E_x \theta(y - Z)\, \Psi(Z)$

$= \theta(x - Z)\theta(y - Z)\, \Psi(Z)$

$= \theta(x - Z)\, \Psi(Z) = E_x\, \Psi(Z)$

$\rightarrow E_x < E_y$

(ii) $[E_{x+\epsilon} - E_x]\, \Psi(Z) = \begin{cases} \Psi(Z) & x < Z < x+\epsilon \\ 0 & otherwise \end{cases}$

$\|(E_{x+\epsilon} - E_x)\, \Psi\|^2 = \int_x^{x+\epsilon} dZ\, |\, \Psi(Z)|^2 \xrightarrow[\epsilon \to 0]{} 0$ thus, (iii) satisfied too.

Note $\|(E_x - E_{x-\epsilon})\, \Psi\|^2 \rightarrow 0 \quad \epsilon \rightarrow 0$ also thus $E_x$ is a truly continuous function of x unlike other operators which are only "continuous one-way".

$\left(\underline{\varphi}, Q\underline{\Psi}\right) = \int_{-\infty}^{\infty} dZ Z \varphi^*(Z)\, \Psi(Z)$ wanted

But

$\left(\underline{\varphi}, Q\underline{\Psi}\right) = \int_{-\infty}^{\infty} x d\left(\underline{\varphi}, E_x\underline{\Psi}\right)$ from formalism

Let's show equivalence

$$\left(\underline{\varphi}, E_x\underline{\Psi}\right) = \int_{-\infty}^{\infty} dZ \varphi^*(Z)\, \Psi(Z)\, \theta(x - Z) = \int_{-\infty}^{x} dZ\, \varphi^*(Z)\, \Psi(Z)$$

$$= \int_{-\infty}^{x} x\, \varphi^*(x)\, \Psi(x) dx$$

**Let A be an observable and $E_x$ its spectral family**

$A = \int_{-\infty}^{\infty} x dE_x \quad E_x$ jumps in value at x=a iff a is an eigenvalue of A.

$E_a - E_{a-\epsilon} \rightarrow P_a(\epsilon \rightarrow 0^+) \quad \epsilon > \delta > 0$

$(E_a - E_{a-\epsilon})(E_a - E_{a-\delta}) = E_a - E_{a-\delta} + E_{a-\epsilon} = E_a - E_{a-\delta}$

$\left\|(E_a - E_{a-\epsilon})\underline{\Psi} - (E_a - E_{a-\delta})\underline{\Psi}\right\|^2 = \left((E_a - E_{a-\epsilon})\underline{\Psi} - (E_a - E_{a-\delta})\, \Psi\right), same$

$\left\|(E_a - E_{a-\epsilon})\underline{\Psi}\right\|^2 + \left\|(E_a - E_{a-\delta})\underline{\Psi}\right\|^2 - 2\left(\underline{\Psi}, (E_a - E_{a-\delta})\underline{\Psi}\right)$

$\underline{\Psi}(E_a - E_{a-\delta})\underline{\Psi} = (\underline{\Psi}, (E_a - E_{a-\delta})^2 \underline{\Psi}) = \left((E_a - E_{a-\delta})\underline{\Psi}, (E_a - E_{a-\delta})\underline{\Psi}\right) = \|(E_a - E_{a-\delta})\underline{\Psi}\|^2$

So, $\|(E_a - E_{a-\epsilon})\underline{\Psi}\|^2 - \|(E_a - E_{a-\delta})\underline{\Psi}\|^2 = \|(E_a - E_{a-\epsilon})\underline{\Psi}\|^2 - \|(E_a - E_{a-\delta})\underline{\Psi}\|^2$

Thus,

$\| \ldots \| = f(\epsilon) - f(\delta) > 0 \qquad (\epsilon > \delta)$

So,

(i) $\|(E_a - E_{a-\epsilon})\underline{\Psi}\|^2$ is a monotonically decreasing function of E
(ii) So, $(E_a - E_{a-\epsilon})\underline{\Psi}$ is a Cauchy sequence
(iii) Property of Hilbert space completeness property, the limit of the Cauchy sequence is also in the space. $\exists \underline{\Psi}_a$ s.t.

$(E_a - E_{a-\epsilon})\underline{\Psi} \rightarrow \underline{\Psi}_a \quad \epsilon \rightarrow 0$

(a) $\underline{\Psi}_a = 0$ (if for all choices of $\underline{\Psi}$ , $E_x$ is continuous)
(b) $\underline{\Psi}_a \neq 0 \ldots$

$x \geq a \rightarrow E_x(E_a - E_{a-\epsilon}) = E_a - E_{a-\epsilon}$

$x < a{-}\epsilon \qquad then\ E_x - E_x = 0$

$E_x\underline{\Psi}_a = \begin{cases} \underline{\Psi}_a & x \geq a \\ 0 & x < a \end{cases}$ ($\epsilon$ is arbitrary small)

Now $\left(\underline{\varphi}, A\underline{\Psi}_a\right) = \int_{-\infty}^{\infty} xd\left(\underline{\varphi}, E_x\underline{\Psi}_a\right)$

$\left(\underline{\varphi}, E_x\underline{\Psi}_a\right) = \int_{-\infty}^{x} dZ\varphi^*(Z)\underline{\Psi}_a(Z)$

Or $\left(\underline{\varphi}, E_x\underline{\Psi}_a\right) = \theta(x-a)\left(\underline{\varphi}, \underline{\Psi}_a\right)$

$\left(\underline{\varphi}, A\underline{\Psi}_a\right) = a\left(\underline{\varphi}, \underline{\Psi}_a\right)$

A derivable: $A = \int_{-\infty}^{\infty} xdE_x$

**Definition:** the spectrum of A is the set of all x's for which $E_x$ is not constant.

If $E_x$ is discontinuous at x=a, a is in the discrete spectrum.

If $E_x$ is continuous and increasing in the neighbourhood of x=a, a is in the continuous spectrum.

### 4.1.2 Clarity on Physical Concepts in Axiomatic Formulation

#### 4.1.2.1 Postulates of Quantum Mechanics

Now that we've familiarized ourselves with the concept of a Hilbert Space and the Dirac notation, let's move on to stating the postulates of Quantum Mechanics according to Dirac [5]:

$1^{st}$: At time $t$ we will specify the state of a physical description by a Ket vector $| \Psi(t) >$ belonging to the physical state space $\{\varepsilon_{r_0}(\vec{r})\}$ (that is a subspace of the $L^2$ Hilbert Space that covers distribution theory -- discussed previously).

$2^{nd}$: A measurable physical quantity $\mathcal{A}$ is the result of an operator $A$ acting on $\{\varepsilon_{r_0}(\vec{r})\}$. Known as an observable. To be a real-valued measurement requires that the associated operator is Hermitian (discussed previously).

$3^{rd}$: The possible results of a 'measurement' ($\mathcal{A}$ ) of a particular observable ($A$) will be one of the eigenvalues of the observable. If the spectrum of the observable $A$ is discrete, there will result explicit 'quantization' in measured values (e.g., a non-continuum of possibilities).

$4^{th}$: The probability of obtaining one of the aforementioned measurement results will be governed by the associated eigenfunction wavefunction squared (e.g., viewed as a probability amplitude [28]). There are two mathematical formulations according to whether the spectrum is discrete or continuous. Let's start with discrete.

*Discrete Spectrum:*
If we measure the quantity $a_n \in \mathcal{A}$ on a system in a normalized state $| \Psi >$, the associated probability is given by $\mathcal{P}(a_n)$, where

$$\mathcal{P}(a_n) = \sum_{i=1}^{g_n} | < u_n^i | \Psi > |^2$$

where $g_n$ is the degree of degeneracy of $a_n$ and $\{|u_n^i >\}$ on the subset $i = 1,2, \ldots, g_n$ describes an orthonormal set that forms a basis in the eigen subspace associated with the eigenvalue $a_n$ of $A$.

*Continuous spectrum, non-degenerate:*

$$d\mathcal{P}(\alpha) = | < v_\alpha | \Psi > |^2 \, d\alpha$$

$5^{th}$: So far the wave aspect is clearly revealed in the pre-measurement formulation, with possible quantization of measurement if spectrum is discrete. Post-measurement we have the distinctively quantum mechanical aspect of projection (or wave-collapse), where measurement of eigenvalue $a_n$ will result into projection (normalized) into the associated eigenfunction (or eigenspace) associated with $a_n$, which can be written in terms of the projection operator:

$$\frac{P_n|\Psi>}{\sqrt{<\Psi|P_n|\Psi>}}.$$

6th: The time evolution of the state vector $|\Psi(t)>$ is governed by the Time-Dependent Schrodinger equation:

$$i\hbar\frac{d}{dt}|\Psi(t)> \;=\; H(t)|\Psi(t)>,$$

where $H(t)$ is the observable associated with the total energy of the system. Notice that this indicates determinism in the time-evolution. Not at all what is brought to mind when thinking of quantum mechanics with its measurement oddities. The measurement event is where the probabilistic (and non-local) aspect comes into play.

### 4.1.2.2 Quantization Rules

The Schrodinger substitutions to arrive at the Schrodinger Equation are an explicit example of an application of the 'quantization rules' to arrive at a quantum formulation of the system. This works remarkably well in quantizing classical systems. Exception involve subtleties of gauge fields and such (Aharonov-Bohm effect [75]). There are quantum physical quantities which have no classical equivalent, the intrinsic spin found for all fundamental massive particles for example.

**The mean value of an observable**

The mean value for an observable $A$ in the state $|\Psi>$ is given by:

$$<A>_\Psi = \sum_n a_n\mathcal{P}(a_n) = <\Psi|A|\Psi>.$$

$<A>_\Psi$ gives a estimation of an observable when in state $|\Psi>$, as with any measurement, however, it is good to have an estimate of the dispersion as well, e.g., the root-mean-square deviation:

$$\Delta A = \sqrt{\langle(A-\langle A\rangle)^2\rangle} = \sqrt{\langle A^2\rangle - \langle A\rangle^2}$$

**Conservation of Probability**

Consider what happens to the probability of a state vector (full wavefunction) over time. We've indicated the necessity of normalizing to unity (unit probability) for the wavefunction interpretation in the aforementioned rules. Is this unit probability result maintained over time?

$$\frac{d}{dt} < \Psi(t)|\Psi(t) >$$
$$= \left[\frac{d}{dt} < \Psi(t)|\right] |\Psi(t) > \quad + \quad < \Psi(t)| \left[\frac{d}{dt} \Psi(t) >\right]$$

Thus,

$$\frac{d}{dt} < \Psi(t)|\Psi(t) > = -\frac{1}{i\hbar} < \Psi|H|\Psi> \; + \; \frac{1}{i\hbar} < \Psi|H|\Psi> = 0.$$

Thus, we have local conservation of probability → we have probability densities and probability currents (analogous to local charge conservation mathematics).

**Probability currents**

Consider

$$i\hbar\frac{d}{dt}|\Psi>= H(t)|\Psi> \text{ in the } \{|r\} \text{ rep. with } H = \frac{P^2}{2m} + V(R,t).$$

We then have:

$$i\hbar\frac{\partial}{\partial t}\Psi(\vec{r},t) = -\frac{\hbar^2}{2m}\nabla^2\Psi(\vec{r},t) + V(\vec{r},t)\Psi(\vec{r},t)$$

And taking the adjoint:

$$-i\hbar\frac{\partial}{\partial t}\Psi^*(\vec{r},t) = -\frac{\hbar^2}{2m}\nabla^2\Psi^*(\vec{r},t) + V(\vec{r},t)\Psi^*(\vec{r},t).$$

Write the probability density as, $p(\vec{r},t) = |\Psi(r,t)|^2$:

$$\frac{\partial}{\partial t}p(\vec{r},t) = \frac{\partial}{\partial t}\{\Psi(\vec{r},t)\Psi^*(\vec{r},t)\}$$
$$= \Psi\frac{1}{(-i\hbar)}\left\{\frac{-\hbar^2}{2m}\nabla^2\Psi^* + V\Psi^*\right\}$$
$$+ \Psi^*\frac{1}{(i\hbar)}\left\{\frac{-\hbar^2}{2m}\nabla^2\Psi + V\Psi\right\}$$
$$= \frac{-\hbar^2}{2m}\left(\Psi^*\nabla^2 + \Psi\nabla^2\Psi^*\right) = \frac{-\hbar^2}{2m}\nabla\cdot(\Psi^*\nabla\Psi - \Psi\nabla\Psi^*) = -\vec{\nabla}\cdot\vec{J}$$

This is the continuity form $\frac{\partial p}{\partial t} + \nabla\cdot\bar{J} = 0$ when J is defined by:

$$\bar{J} = \frac{\hbar}{2mi}(\Psi^*\nabla\Psi - \Psi\nabla\Psi^*) = \frac{1}{m}Re\left[\Psi^*\left(\frac{\hbar}{i}\nabla\Psi\right)\right]$$

Operator form:

$$\vec{K}(\vec{r}) = \frac{1}{2m}\{|r><r|\vec{P} + \vec{P}|r><r|\}$$

where $\frac{P}{m}$ is the velocity op. and mean value of $|r><r|$ is $p(\vec{r})$. $\vec{K}$ is the op. with the appropriate symmetrisation, from the product of the probability density and the velocity of the particle.

Note that a particle is an e-m field has:

$$\vec{J} = \frac{1}{m} Re\left[\Psi^* \left[\frac{\hbar}{i}\nabla - qA\right] \Psi\right].$$

**Evolution of the mean value of an observable**

Consider

$$\frac{d}{dt} < A > = \frac{1}{i\hbar} < [A, H(t)] > + < \frac{\partial A}{\partial t} >$$

Applying this relation to $\vec{R}$ *and* $\vec{P}$ gives Ehrenfest's theorem:

Let $H = \frac{\vec{P}^2}{2m} + V(\vec{R})$, then

$$\frac{d}{dt} < R > = \frac{1}{m} < P > \quad (correspondence\ with\ \frac{d}{dr}\vec{r} = \frac{1}{m}\vec{P})$$

and

$$\frac{d}{dt} < P > = -\frac{1}{m} < \nabla V(\vec{R}) > \quad (correspondence\ with\ \frac{d}{dr}\vec{P} = -\nabla V(\vec{r}))$$

where the classical correspondence is with Hamilton Jacobi equations [16], for which (re)combining yields:

$$\frac{d\vec{P}}{dt} = \frac{md^2\vec{r}}{dt^2} = \vec{F} = -\nabla v(\vec{r})$$

which is Newton's equation

Take $< \vec{R} >$ to be the trajectory of the "center" of the wave packet. If the scale of the wave packet is much less than any other scale of interest then there is no appreciable difference between the quantum and classical description of a particle.

Does the motion of the center of the wave packet obey the laws of classical mechanics in general? NO, as will be shown by Ehrenfest's theorem:

Quantum Mechanically: $m\frac{d^2}{dt^2} < R > = - < \nabla V(R) >$.

Classically: $m\frac{d^2}{dt^2} < R >= [-\nabla V(\vec{r})]|_{\vec{r}=<R>}$ , with interaction at a specific point, not involving the packet as a whole.

Consider $V(x) = \lambda x^n \rightarrow V(x) = \lambda X^n$. Then $< \nabla V(R) > = \lambda n < X^{n-1} >$ and

$$\left[\frac{d}{dx}(\lambda X^n)\right]_{<X>} = \lambda n < X >^{n-1}.$$

In general $< X^{n-1} > \neq < X >^{n-1}$ , however, for many familiar cases:
for n=0 (the free particle, the calc. is actually trivial constant case: $V(x) = \lambda$ ); while
for n=1 (uniform force field, again trivial); and
for n=2 (the harmonic oscillator, we have agreement), thus
for certain potentials the "center" of the wave packet does obey the classical laws of motion.

Also, if the wave-packet is sufficiently localized, to where $\nabla V(\vec{r})$ is practically constant, we again have:

$$\nabla V(\vec{r}) = \int d^3r\, \Psi^* \, \nabla V(\vec{r})\, \Psi \cong \nabla V(r_0) \int d^3r |\Psi|^2$$

$$\nabla V(R) \cong [\nabla V(r)]_{r=<R>}$$

In the macroscopic limit (where the de Broglie wavelengths are much smaller than the distances over which the potential varies), wave packets can be made sufficiently small to satisfy the adiabatic limit above while retaining a good degree of definition for the momentum. Thus, the equations of the center of mass follow from the Schrodinger equation in that certain limiting conditions satisfied, in particular, by most macroscopic systems.

**Conservative systems**

$$H|\varphi_{n,\tau} >= E_n|\varphi_{n,\tau} >$$

immediately yields

$$\Psi(t) > = \sum_n \sum_\tau c_{n,\tau}(t_0)e^{-iE_n(t-t_0)/\hbar}\varphi_{n,\tau} >$$

from Schrodinger.

Stationary states:

$$|\Psi(t_0) >$$

is itself an eigenstate of H.

Constants of the motion by definition satisfy (not necessarily conservative system, A an Op.):

$$\left\{\frac{\partial A}{\partial t} = 0 \quad and\ [A, H] = 0\right\}$$

For the evolution of the mean value of an operator that is a constant of the motion we therefore have: $\frac{d}{dt} < A >= 0$.

**Examples**

***Example 1***

The operator $\hat{A} = \hat{p} + ic\hat{x}$ is defined in terms of the usual operators $\hat{x}$ and $\hat{p}$, where c is a real constant.

(a) Find the Hermitian adjoint $\hat{A}^t$ in terms of $\hat{x}$ and $\hat{p}$.

(b) Find the operator $\hat{A}^t\hat{A} - \hat{A}\hat{A}^t$ represented as a number.

(c) Suppose $\psi_1$ is an eigenfunction of $\hat{A}^t\hat{A}$ with eigenvalue $b_1$ that is, $\hat{A}^t\hat{A}\,\psi_1 = b_1\psi_1$. Then show that $\hat{A}\psi_1$, is also an eigenstate of $\hat{A}^t\hat{A}$ and find the new eigenvalue $b_2$ such that $\hat{A}^t\hat{A}\,\left(\hat{A}\,\psi_1\right) = b_2\left(\hat{A}\,\psi_2\right)$. The value of $b_2$ is to be expressed in terms of $b_1\ ,\ c,\ and\ \hbar$.
(The operator $\hat{A}$ can be used repeatedly to generate a series of new eigenfunctions and eigenvalues.)

**Answer**

(a) $\hat{A} = \hat{p} + ic\hat{x} = -i\hbar\frac{\partial}{\partial x} + icx$

$\langle \hat{A}^t\psi_\ell|\psi_n\rangle = \left\langle\psi_\ell\middle|\hat{A}\psi_n\right\rangle = \langle\psi_\ell|(\hat{p} + ic\hat{x})\psi_n\rangle = \langle\psi_\ell|\hat{p}\psi_n\rangle + \langle\psi_\ell|ic\hat{x}\rangle$

Where:

$$\langle\psi_\ell|\hat{p}\psi_n\rangle = \int_{-\infty}^{\infty} \psi_\ell^*\left(-i\hbar\frac{\partial\psi_n}{\partial x}\right)dx$$

$$= -i\hbar[\psi_\ell^*\psi_n]_{-\infty}^{\infty} + i\hbar\int_{-\infty}^{\infty}\left(\frac{\partial}{\partial x}\psi_\ell^*\right)\psi_n dx$$

$$= \int_{-\infty}^{\infty}\left(-i\hbar\frac{\partial}{\partial x}\psi_\ell\right)^*\psi_n dx = \langle\hat{p}\psi_\ell|\psi_n\rangle$$

Similarly

$$\langle\psi_\ell|ic\hat{x}\psi_n\rangle = \int_{-\infty}^{\infty} \psi_\ell^* \,(icx)\psi_n dx = \int_{-\infty}^{\infty} icx\psi_\ell^*\,\psi_n dx$$

$$= \int_{-\infty}^{\infty} [(-icx^*)\,\psi_\ell]^*\psi_n dx = \langle -icx^*\psi_\ell|\psi_n\rangle$$

Thus,

$$\hat{A}^t = \hat{p} - ic\hat{x}^* = \hat{p} - ic\hat{x}$$

(b) $\hat{A}^t\hat{A} - \hat{A}\hat{A}^t = (\hat{p} - ic\hat{x}^*)(\hat{p} + ic\hat{x}) - (\hat{p} + ic\hat{x})(\hat{p} - ic\hat{x}^*)$
$= \hat{p}^2 - ic\hat{x}^*\hat{p} + ic\hat{x}\hat{p} - c^2\hat{x}^*\hat{x} - (\hat{p}^2 - ic\hat{p}\hat{x}^* + ic\hat{x}\hat{p} -$
$c^2\hat{x}\hat{x}^*)$
$= 2(-ic\hat{x}^*\hat{p}) - c^2\hat{x}^*\hat{x} + 2c\hat{p}\hat{x}^* + c^2\hat{x}^*\hat{x}$
$= ic(\hat{p}\hat{x}^* - \hat{x}^*\hat{p})$
$= ic\left(-i\hbar\frac{\partial}{\partial x}x^* - x^*(-i\hbar)\frac{\partial}{\partial x}\right)$
$= ic(-i\hbar)\left[x^*\frac{\partial}{\partial x} + \frac{\partial x^*}{\partial x} - x^*\frac{\partial}{\partial x}\right]$
$= c\hbar\frac{\partial x^*}{\partial x}$
$= 2c\hbar$

(c) We have $\hat{A}^t\hat{A}\psi_1 = b_1\,\psi_1$ , $\hat{A}^t\hat{A}(\hat{A}\psi_1) = b_1(\hat{A}\psi_1)$ , and express $b_2$ in terms of $b_1, c,$ and $\hbar$:
Making use of the identity in (3b): $(c\hbar + \hat{A}\hat{A}^t)(\hat{A}\psi_1) = b_2\,(\hat{A}\psi_1)$
Thus
$c\hbar(\hat{A}\psi_1) + \hat{A}\hat{A}^t\hat{A}\psi_1 = c\hbar(\hat{A}\psi_1) + \hat{A}[\hat{A}^t\hat{A}\psi_1] = b_2\,(\hat{A}\psi_1)$
$b_2\,(\hat{A}\psi_1) = c\hbar(\hat{A}\psi_1) + \hat{A}(b_1\psi_1) = c\hbar(\hat{A}\psi_1) + b_1(\hat{A}\psi_1)$

$$b_2\,(\hat{A}\psi_1) = (c\hbar + b_1)(\hat{A}\psi_1)$$

Thus, $b_2 = 2c\hbar + b_1$, and $\hat{A}\psi_1$ is also an eigenstate of $\hat{A}^t\hat{A}$.

***Example 2***

$\hat{A}$ , $\hat{B}$ *and* $\hat{C}$ are three operators that act on states in a Hilbert space and the commutator symbol is defined by $[\hat{A},\hat{B}] = \hat{A}\hat{B} - \hat{B}\hat{A}$.

(a) Find $[\hat{A},\hat{B} + \hat{C}]$ in terms of commutators of the operators $\hat{A}$ , $\hat{B}$ *and* $\hat{C}$.
(b) Find $[\hat{A},\hat{B}\hat{C}]$ in terms of $[\hat{A},\hat{B}]$ and $[\hat{A},\hat{C}]$.
(c) Given that $[\hat{A},\hat{B}] = 1$, show that $\exp(\hat{A})\quad(\hat{B}) = (\hat{B} + 1)\;\exp(\hat{A})$

**Answer**

(a) $[\hat{A}, \hat{B} + \hat{C}] = \hat{A}(\hat{B} + \hat{C}) + (\hat{B} + \hat{C})\hat{A} = \hat{A}\hat{B} + \hat{A}\hat{C} - \hat{B}\hat{A} - \hat{C}\hat{A}$

$$= [\hat{A}\hat{B} - \hat{B}\hat{A}] + [\hat{A}\hat{C} - \hat{C}\hat{A}] = [\hat{A}, \hat{B}] + [\hat{A}, \hat{C}]$$

(b) $[\hat{A}, \hat{B}\hat{C}] = \hat{A}(\hat{B}\hat{C}) - (\hat{B}\hat{C})\hat{A} = \hat{A}\hat{B}\hat{C} - \hat{B}\hat{C}\hat{A} = \hat{A}\hat{B}\hat{C} - \hat{B}\hat{C}\hat{A} - \hat{B}\hat{A}\hat{C} + \hat{B}\hat{A}\hat{C}$

$$= (\hat{A}\hat{B} - \hat{B}\hat{A})\hat{C} + \hat{B}(\hat{A}\hat{C} - \hat{C}\hat{A}) = [\hat{A}, \hat{B}]\hat{C} + \hat{B}[\hat{A}, \hat{C}]$$

(c) $(\exp(A))B = (\sum_{n=0}^{\infty} A^n)B$ by definition and $[\hat{A}, \hat{B}] = 1 \rightarrow AB = 1 + BA$:

$$(\exp(A))B = \left[1 + A + \frac{A^2}{2!} + \frac{A^3}{3!} + \frac{A^4}{4!} \cdots \right] B$$

$$= B + AB + \frac{A(AB)}{2!} + \frac{A^2(AB)}{3!} + \frac{A^3(AB)}{4!} \cdots$$

$$= B + (BA + 1) + \frac{A(BA + 1)}{2!} + \frac{A^2(BA + 1)}{3!} + \frac{A^3(BA + 1)}{4!} \cdots$$

$$= B + BA + 1 + \frac{BA^2 2A}{2!} + \frac{ABA^2 + A^2 + A^2}{3!} + \frac{A(AB)A^2 + A^3 + A^3}{4!} \cdots$$

$$= B + BA + 1 + \frac{BA^2 + 2A}{2!} + \frac{BA^3 + 3A^2}{3!} + \frac{BA^4 + 3A^3}{4!} \cdots$$

$$= B\left(1 + A + \frac{A^2}{2!} + \frac{A^3}{3!} + \frac{A^4}{4!} \cdots \right) + \left(1 + \frac{2A}{2!} + \frac{A^2}{3!} + \frac{A^3}{4!} \cdots \right)$$

$$= Be^A + e^A = (B + 1)e^A$$

Thus, $e^{\hat{A}}\hat{B} = (\hat{B} + 1)e^{\hat{A}}$.

***Example 3***

Consider the unnormalized quantum states of a free particle.

(a) Write $|x_0 >$ (a state of definite position $x_0$), $|p_0\rangle$ (a state of definite momentum $p_0$), and the operators $\hat{x}$ and $\hat{p}$ in terms of functions and derivatives of the variables x.

(b) Repeat (a) using functions and derivatives of p, that is, in the p representation.

(c) Write the eigenfunctions and corresponding eigenvalues for the operator $\hat{x}$ in the p representation.

**Answer**

(a) $|x_0\rangle = A\delta(x - x_0)$ , $|p_0\rangle = Be^{ip_0 x/\hbar}$ $(A, B$ constants), and $\hat{x} = x$ , $\hat{p} = -i\hbar \frac{\partial}{\partial x}$.

(b) $|x_0\rangle = Ae^{-ipx_0/\hbar}$, $|p_0\rangle = B\,\delta(p - p_0)$, and $\hat{p} = p$ and from $[\hat{x}, \hat{p}] = i\hbar \Rightarrow \hat{x} = i\hbar\frac{\partial}{\partial p}$

(c) $i\hbar\,\frac{\partial\psi}{\partial p} = x_0\psi \Rightarrow \psi = Ae^{-ix_0p/\hbar}$, with eigenvalue $x_0$.

***Example 4***

A particle of mass m is in a potential $V(x) = mw_0x^2/2$, which classically results in sinusoidal oscillations of frequency $w_0$. Quantum mechanically the state of the particle at $t = 0$ is $\psi(x,0) = \left(\psi_0(x) + \psi_1(x)\right)/\sqrt{2}$, where $\psi_0$ and $\psi_1$ are the 1st and 2nd energy eigenstates.

(a) Find the wave function $\psi(x,t)$ as a function of time.
(b) Find the probability density $P(x,t) = \psi^*(x,t)\psi(x,t)$ as a function of x and t.
(c) Find $d\langle x\rangle/dt$ as a function of time.
Note: $\hat{x} = \sqrt{\hbar/(2mw_0)}\ (\hat{a}^+ + \hat{a})$, and $\hat{p} = i\sqrt{mw_0\hbar/2}\,(\hat{a}^+ + \hat{a})$.
(d) A second particle is placed in the same harmonic oscillator potential with no interaction with the first. The two particles are identical fermions. A measurement of the total energy of the two particles gives $E = 2\ hw_0$. After this measurement, what is the wave function for the two-particle state?

**Answer**

(a) $\psi(x,t) = \frac{1}{\sqrt{2}}\left(\psi_0(x)e^{-i\frac{E_0}{\hbar}t} + \psi_1(x)e^{-i\frac{E_1}{\hbar}t}\right) = \frac{1}{\sqrt{2}}\left(\psi_0(x)e^{-\frac{1}{2}w_0\,t} + \psi_1(x)e^{-\frac{3}{2}w_0\,t}\right)$

(b) $P(x,t) = |\psi(x,t)|^2 = \frac{1}{2}\left(\psi_0^2 + \psi_1^2 + 2\psi_0\psi_1\cos\omega_0 t\right)$

(c) $\frac{d\langle x\rangle}{dt} = \frac{\langle p\rangle}{m} = i\sqrt{\frac{w_0\hbar}{2m}}\,\langle(a^\dagger - a)\rangle = i\sqrt{\frac{w_0\hbar}{2m}}\left(\langle a\rangle^* - \langle a\rangle\right)$, and

$$\langle a\rangle = \left\langle\psi(x,t)\middle|\frac{1}{\sqrt{2}}\psi_0(x)e^{-\frac{3}{2}w_0\,t}\right\rangle \quad (\text{since } a\psi_0 = 0, a\psi_1 = \psi_0)$$

So,

$$\langle a\rangle = \left\langle\frac{1}{\sqrt{2}}\psi_0(x)e^{-\frac{1}{2}w_0\,t}\middle|\frac{1}{\sqrt{2}}\psi_0(x)e^{-\frac{3}{2}w_0\,t}\right\rangle$$
$$= \frac{1}{2}\,e^{-iw_0\,t} \qquad (note\ \langle\psi_0|\psi_1\rangle = 0)$$

So

$$\frac{d\langle x\rangle}{dt} = i\sqrt{\frac{w_0\hbar}{2m}}\frac{1}{2}\left(e^{iw_0 t} - e^{-iw_0 t}\right) = -\sqrt{\frac{w_0\hbar}{2m}}\sin w_0 t$$

(d) For fermions, the wave function is antisymmetric. This means that they have to be in different states. Since $E = 2\hbar w_0 = E_0 + E_1$, they are in states $\psi_0$ & $\psi_1$. So the total wavefunction is

$$\psi(x_1, x_2) = \frac{1}{\sqrt{2}}\left(\psi_0(x_1)\psi_1(x_2) - \psi_0(x_2)\psi_1(x_1)\right).$$

**4.1.2.3 Bohr frequency of a system and selection rules**

For a conservative system consider

$$|\Psi(t)> = \sum_n \sum_\tau c_{n,\tau}(t_o)e^{-iE_n(t-t_0)/\hbar}|\varphi_{n,\tau}>$$

and

$$<\Psi(t)| = \sum_{n'} \sum_{\tau'} c^*_{n'\tau'}(t_o)e^{iE_{n'}(t-t_0)/\hbar} <\varphi_{n',\tau'}|$$

Consider an observable that does not depend on time, then we have:

$<B> = <\Psi(t)|B|\Psi(t)> = \Sigma_n\Sigma_\tau\Sigma_{n'}\Sigma_{\tau'}\ c_{n'\tau'}(t_0)c_{n\tau}(t) <\varphi_{n',\tau'}|B|\varphi_{n,\tau}> e^{i(E_{n'}-E_n)(t-t_0)/\hbar}$

Thus, the evolution is described by oscillating terms with freq. $\frac{1}{2\pi}\frac{|E_{n'}-E_n|}{\hbar}$

The oscillating terms are characteristics of the system under consideration but independent of B and of the initial state of the system. The frequencies are called the Bohr frequencies of the system. Thus, for an atom, the mean values of all the atomic quantities (electric and magnetic dipole moments, etc…) oscillate at the various Bohr frequencies of the atom. It is reasonable to imagine that only these frequencies can be radiated or absorbed by the atom. This allows an initiative understanding of the Bohr relation between the spectral frequencies emitted or absorbed and the difference in atomic energies …. To go deeper one must turn to field theory. Selection rules established by the matrix elements $<\varphi_{n,\tau'}|B|\varphi_{n,\tau'}>$, whichever ones are nonzero being selected.

**4.1.2.4 Time-energy uncertainty relation**

Consider $\Delta t$ : the time interval in which the system has evolved appreciably; $\Delta E$ : the energy uncertainty.

For a stationary state $\Delta E = 0$ , but since the system doesn't evolve $\Delta t = \infty$. Consider

$$|\Psi(t_0)>= c_1|\varphi_1> +c_2|\varphi_2>,$$

the superposition of the two eigenstates of H. Then,

$$|\Psi(t)> = c_1 e^{-iE_1(t-t_0)/\hbar}|\varphi_1> +c_2 e^{-iE_2(t-t_0)/\hbar}|\varphi_2>,$$

where energy varies between $E_1$ and $E_2$ , thus $\Delta E \cong |E_2 - E_1|$.

Consider an observable B with which to probe the system evolution, such a "probe" hopefully doesn't commute with H....

$$\begin{aligned} P(b_m, t) = & \; |< u_m|\Psi(t)>|^2 \\ & = |C_1|^2 \; |< u_m|\varphi_1>|^2 + |C_2|^2 \; |< u_m|\varphi_2>|^2 \\ & +2Re[c_2^* c_1 e^{-i(E_2-E_1)(t-t_0)/\hbar} < u_m|\varphi_2>^* < u_m|\varphi_1>] \end{aligned}$$

So, P oscillates between two extremes with Bohr frequency

$$\nu_{21} = \frac{|E_2 - E_1|}{\hbar} \rightarrow \Delta t \cong \frac{\hbar}{|E_2 - E_1|}$$

Thus, $\Delta t \Delta E \cong \hbar$.

**4.1.2.5 Probability currents and oblique reflection**

***Study of probability current notation for potential step***

Let's consider the step potential once again:

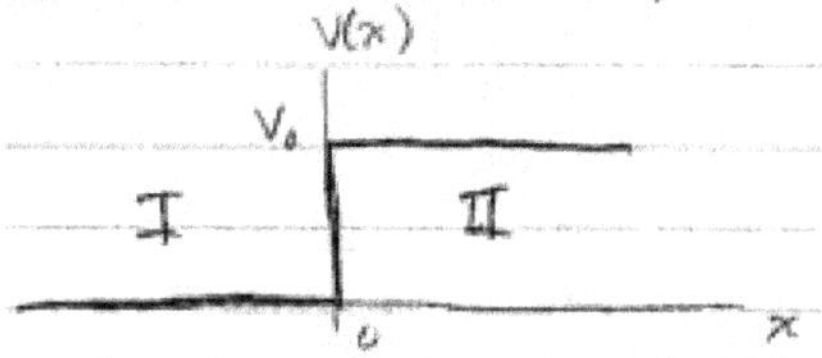


And let's examine the aforementioned probability current

$$\vec{J} = \frac{\hbar}{2mi}[\Psi^* \nabla \Psi - \Psi \nabla \Psi^*]$$

Case (i) $E > V_0$: $\Psi(x) = Ae^{ikx} + A'e^{-ikx}$, $E - V_0 = \frac{\hbar^2 k^2}{2m}$, thus,

$$\vec{J} = \frac{\hbar}{2mi}\Big[\big(A^* e^{-ikx} + A'^* e^{+ikx}\big)\big(Aike^{ikx} + A'(-ik)e^{-ikx}\big) - \Big(\big(A\, e^{ikx} + A'e^{-ikx}\big)\big(A^*(-ik)e^{ikx} + A'^*(ik)e^{ikx}\big)\Big)\Big]$$

Thus,

$$\vec{J} = \frac{\hbar}{2mi}[ik|A|^2 - ik|A'|^2 + same] = \frac{\hbar k}{m}[|A|^2 - |A'|^2]$$

Case (ii) $E' < V_0$: $\Psi(x) = Be^{px} + B'e^{-px}$, $V_0 - E = \frac{\hbar^2 k^2}{2m}$

$$\vec{J} = \frac{\hbar p}{m}[iB^* B' + c.c.]$$

***For incoming wave to potential***

Region I:

$$\Psi_1(x) = A_1 e^{ikx} + A_2' e^{-ikx} \qquad E = \frac{\hbar^2 k^2}{2m}$$

$$J_I = \frac{\hbar k_1}{m}[|A_1|^2 - |A_1'|^2]$$

Region II, case (i): $E > V_0$:

$$\Psi_2(x) = A_2 e^{ikx} + A_2' e^{-ikx},$$

where $A_2' = 0$ from boundary condition at infinity (and using $E - V_0 = \frac{\hbar^2 k^2}{2m}$), we get:

$$J_{II} = \frac{\hbar k_2}{m}[|A_2|^2].$$

$J_I$ is the sum of incident and reflected terms:

$$R = \left|\frac{A_1'}{A_1}\right|^2$$

$$T = \frac{J_{II}}{J_{I(inc)}} = \frac{\frac{\hbar k_2}{m}|A_2|^2}{\frac{\hbar k_1}{m}|A_1|^2} = \frac{k_2}{k_1}\left|\frac{A_2}{A_1}\right|^2$$

Region II, case (ii) : $E < V_0$:

$$\Psi_2(x) = Be^{px} + B^+ e^{-px},$$

where $B = 0$ to be bounded at infinity. So,

$$J_I = \frac{\hbar p}{m}[iB^* B' + c.c.] = 0$$

Thus, $T = 0$ and $R = 1$.

The probability current is zero but the probability of funding the particle in the region II is not. In the steady state, there are two probability currents in the region II: a positive current corresponding to the entrance into this region of part of the incident wave packet; and a negative current corresponding to the return towards region I of this part of the wave packet. These two currents are equal, so the overall result is zero.

In the case of a one dimensional problem, the structure of the probability current of the evanescent wave is therefore masked by the fact that the two opposite currents balance. This motivates considering the two-dimensional problem, for the case of oblique reflection, so as to obtain a nonzero result.

***Reflection from a two-dimensional potential step:***
Consider

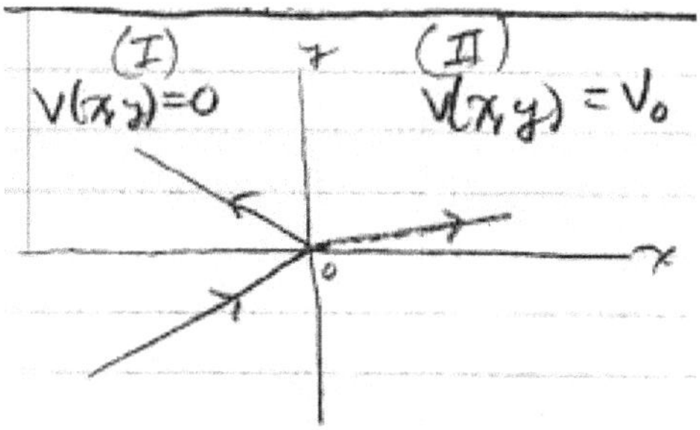


We have: $V(x, y) = V_1(x) + V_2(y)$, where

$$V_1(x) = \begin{Bmatrix} 0 \;\; for \;\; x < 0 \\ V_0 \;\; for \;\; x > 0 \end{Bmatrix} \qquad and \qquad V_2(y) = 0$$

Since

$$H = \frac{p^2}{2m} + V(r),$$

we have separability: $\varphi(x, y) = \varphi(x)\varphi(y) \;\; and \; E = E_1 + E_2$, with

$$\varphi_1(x) \to 1d \;\; step\,problem$$
$$\varphi_2(y) \to 1d \;\; free\,particle$$

Region I ($x < 0$) a free particle:

$$\varphi(x, y) = \alpha\left(e^{ik_x x} + \beta e^{-ik_x x}\right)e^{ik_y y} = Ae^{i(k_x x + k_y y)} + A'e^{i(-k_x x + k_y y)}$$

where $k_x = \sqrt{\frac{2mE_1}{\hbar^2}}$ and $k_y = \sqrt{\frac{2mE_2}{\hbar^2}}$.

Region II ($x > 0$): $(case\; V_0 > E)$ :

$$\varphi_{II}(x, y) = Be^{-P_x x}e^{ik_y y}$$

where $p_x = \sqrt{\frac{2m(V_0 - E_1)}{\hbar^2}}$ and letting $\tan\theta = \frac{P_x}{k_x} = \sqrt{\frac{V_0 - E_1}{E_1}}$.

Then,

$$\frac{A'}{A} = \frac{k_x - ip_x}{k_x + ip_x} = e^{-2i\theta}$$

Leads to:

$$J_I = \begin{cases} (J_I)_x = \frac{\hbar k_x}{m}[|A|^2 - |A'|^2] = 0 \\ (J_I)_y = \frac{\hbar k_y}{m}\left|A_2 e^{ik_x x} + A_2' e^{-ik_x x}\right|^2 = \frac{\hbar k_y}{m}|A|^2[2 + 2\cos(2k_x x + 2\theta)] \end{cases}$$

And:

$$\frac{B}{A} = \frac{2k_x}{k_x + ip_x} = 2\cos\theta e^{-i\theta}$$

Which leads to:

$$J_{II} = \begin{cases} (J_{II})_x = 0 \\ (J_{II})_y = \frac{\hbar k_y}{m}|B|^2 e^{-2p_x x} = \frac{\hbar k_y}{m} 4|A|^2 \cos^2\theta\, e^{-2p_x x} \end{cases}$$

The penetration of the particle into region II leads to a lateral shift upon reflection:

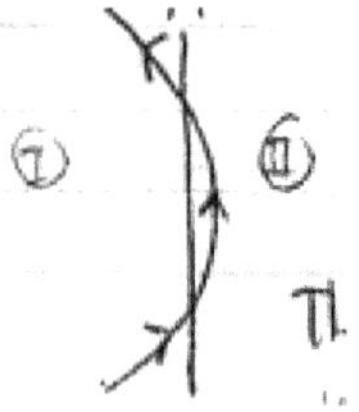

### 4.1.3 Applications

### 4.1.3.1 Recap

***Postulates of Quantum Mechanics (Dirac, axiomatic)***

(1) Physical state defined by a Ket in state space $\mathcal{E}$.

(2) Every measurable described by operation acting in $\mathcal{E}$

(3) Result of measurement is an eigenvalue of the operator.

(4) Probability of obtaining the eigenvalue $a_n$ is $P(a_n) = \sum_{i=1} \left|\langle u_n^i|\varphi\rangle\right|^2$

(5) State after measurement: $\frac{P_n|\varphi>}{\sqrt{<\varphi|P_n|\varphi>}}$

(6) $i\hbar\frac{d}{dt}|\varphi(t)\rangle = H(t)|\varphi(t)>$

The mean value of an observable: $< A >_\varphi = \sum a_n P(a_n) = \langle\varphi|A|\varphi\rangle$

The root mean square (rms) deviation: $\Delta A = \sqrt{\langle(A - < A >)^2\rangle}$.

There is an analogous formula for the continuity relation $\frac{dp}{dp} + \vec{\nabla}.\vec{j} = 0$.
Denote the probability density by

$$\rho = |\varphi(r,t)|^2$$

and the probability current by

$$\vec{j} = \frac{\hbar}{2mi}(\varphi^* \nabla\varphi - \varphi\nabla\varphi^*)$$

For a particle in an electromagnetic field, we have with covariant derivative:

$$\vec{j} = \frac{i}{m}\mathrm{Re}\left[\varphi^* \left\{\frac{\hbar}{i}\nabla - qA\right\}\varphi\right]$$

From Leibnitz' rule and the Schrodinger equation we can also show:

$$\frac{d}{dt} < A > = \frac{1}{i\hbar} < [A, H(t)] > + < \frac{\partial A}{\partial t} >$$

If applied to the canonical pair $\vec{R}$ and $\vec{P}$ we get Ehrenfest's theorem:

$$\frac{d < R >}{dt} = \frac{1}{m} < P >$$

and

$$\frac{d < P >}{dt} = - < \nabla V(\vec{R}) >$$

**Root – Mean – Square deviations**

Consider $|\phi> = (A + i\lambda\vec{\nabla})|\varphi>$ . Since

$$<\phi|\phi> \geq 0 \Rightarrow |<[A,B]>|^2 \leq 4 < A^2 >< B^2 >$$

And if we shift $A' = A - < A >$, and $[A', B'] = [A, B]$, then we get:

$$\Delta A\, \Delta B \geq \frac{1}{2}|<[A,B]>|$$

The generalized Heisenberg Uncertainty relation.

**Density formulation**

$$\rho = \sum_k p_k\, \rho_k \qquad < A > = Tr\{\rho A\}$$

$$i\hbar \frac{d}{dt}\rho(t) = [H(t), \rho(t)]$$

**Evolution operation**

$$|\varphi(t)> = U(t,t_0)|\varphi(t)>, \;\; where \;\; U(t,t_0) = e^{-iH(t-t_0)/\hbar}$$

And $H$ does not depend on time. In the Heisenberg representation, on the other hand, we have time-dependence in the operators:

$$i\hbar\frac{d}{dt}A_H(t) = [A_H(t), H_H(t)] + i\hbar\left(\frac{d}{dt}A_S(t)\right)_H$$

### 4.1.3.2 Two level system – preview

Consider a two-level spin system:

$$(S_u) = (S_x)sin\theta cos\varphi + (S_y)sin\theta sin\varphi + (S_z)cos\theta$$

$$= \frac{\hbar}{2}\begin{pmatrix} cos\theta & sin\theta e^{-i\varphi} \\ sin\theta e^{i\varphi} & -cos\theta \end{pmatrix}$$

$$|+>_{\vec{u}} = cos\frac{\theta}{2}\, e^{\left(-i\varphi/2\right)}\, |+> + sin\frac{\theta}{2}\, e^{\left(i\varphi/2\right)}\, |->$$

$$|->_{\vec{u}} = -sin\frac{\theta}{2}e^{\left(i\varphi/2\right)}|+> + cos\frac{\theta}{2}\, e^{\left(i\varphi/2\right)}\, |->$$

**Pauli Matrices**

Define the Pauli matrices to be:

$$\sigma_z = \begin{pmatrix} 1 & 0 \\ 0 & -1 \end{pmatrix}, \sigma_x = \begin{pmatrix} 0 & 1 \\ 1 & 0 \end{pmatrix}, \sigma_y = \begin{pmatrix} 0 & -i \\ i & 0 \end{pmatrix}$$

$$\sigma_j\sigma_k = i\sum_l \epsilon_{jk\ell}\sigma_\ell + \delta_{jk}I$$

$$(\vec{\sigma}\cdot\vec{A})(\vec{\sigma}\cdot\vec{B}) = \vec{A}\cdot\vec{B}I + i\vec{\sigma}\cdot(\vec{A}\times\vec{B})$$

**Evolution of spin $^1/_2$ particle in a uniform magnetic field**

$$\overrightarrow{B_0} = B_0 e_z$$

$$W = -\vec{\mathcal{M}}\cdot\overrightarrow{B_0} = -\mathcal{M}_z B_0 = -\gamma B_0\mathcal{L}_z = \omega_0\mathcal{L}_z,\ \ where\ \omega_0 = -\gamma B_0$$

Upon quantization we have $\mathcal{L}_z \rightarrow S_z$, the spin operator, and $W \rightarrow \mathcal{H}$, the Hamiltonian (Energy) operator. Then we have $\mathcal{H} = \omega_0 S_z$:

$$H|+> = \frac{+\hbar\omega_0}{2}|+>$$

$$H\Big|-> = \frac{-\hbar\omega_0}{2}\Big|->$$

Consider

$$|\psi(0)> = |+>_{\vec{u}} = cos\frac{\theta}{2}\, e^{\left(-i\varphi/2\right)}\, |+> + sin\frac{\theta}{2}\, e^{\left(i\varphi/2\right)}\, |->.$$

Then

$$|\psi(t)> = |+>_{\vec{u}}$$

$$= cos\frac{\theta}{2}\, e^{\left(-i\varphi/2\right)}e^{-\frac{iE_+t}{\hbar}}\, |+> + sin\frac{\theta}{2}\, e^{\left(i\varphi/2\right)}e^{-\frac{iE_-t}{\hbar}}\, |->,$$

where now there is Larmor precession, and the mean value of $S_z$, $S_x$, and $S_y$ follow from $|\psi(t)>$. The general two-level system is then treated as a spin ½ in a B-field formulation (above):

$$(\mathrm{H}) = \begin{pmatrix} E_1 + \omega_{11} & \omega_{12} \\ \omega_{21} & E_2 + \omega_{22} \end{pmatrix} = \begin{pmatrix} H_{11} & H_{12} \\ H_{21} & H_{22} \end{pmatrix}, \quad where\ H_{12} = H_{21}{}^*$$

Thus, can write:

$$(\mathrm{H}) = \frac{1}{2}(H_{11} + H_{22})\mathbf{1} + \frac{1}{2}(H_{11} - H_{22})\boldsymbol{K},$$

where

$$(\mathrm{K}) = \begin{pmatrix} 1 & \dfrac{2H_{12}}{H_{11} - H_{22}} \\ \dfrac{2H_{21}}{H_{11} - H_{22}} & -1 \end{pmatrix}$$

Let

$$\tan\theta = \frac{2|\mathrm{H}_{21}|}{\mathrm{H}_{11} - \mathrm{H}_{22}} \quad \text{and} \quad \mathrm{H}_{21} = |\mathrm{H}_{21}|\ e^{i\varphi}$$

then

$$(\mathrm{K}) = \begin{pmatrix} 1 & \tan\theta\ e^{-i\varphi} \\ \tan\theta\ e^{i\varphi} & -1 \end{pmatrix} = \frac{1}{\cos\theta}\begin{pmatrix} \cos\theta & \sin\theta\ e^{-i\varphi} \\ \sin\theta\ e^{i\varphi} & -\cos\theta \end{pmatrix}$$

$$= \frac{1}{\cos\theta}\frac{2}{\hbar}(S_u)$$

Thus

$$\mathrm{K}_+ = \frac{1}{\cos\theta} \quad \text{and} \quad \mathrm{K}_- = \frac{-1}{\cos\theta}$$

**4.1.3.3 One-dimensional harmonic oscillator – A preview**

From the classical theory, the Kinetic Energy in 1-D is given by the usual $\mathrm{P}^2/2m$, while the Potential Energy, with linear restoring force, is $m\omega^2\mathrm{X}^2/2$. The Hamiltonian is thus:

$$H = \frac{\mathrm{P}^2}{2m} + \frac{1}{2}m\omega^2\mathrm{X}^2$$

Let

$$\mathbf{X} = \sqrt{\frac{m\omega}{\hbar}}\mathrm{X}, \qquad \mathbf{P} = \frac{1}{\sqrt{\mathrm{m}\hbar\omega}}\mathrm{P}, \qquad \text{and} \quad \boldsymbol{H} = \frac{1}{\hbar\omega}H.$$

Then,

$$\boldsymbol{H} = \frac{1}{2}(\mathbf{X}^2 + \mathbf{P}^2).$$

Now that we have the standard, bare-bones, mathematical expression of a 1-D oscillator, let's make use of variables that express the energetics of the system in terms of modal excitation directly, via introduction of what are know as the 'creation' and 'annihilation' operators upon quantization:

Creation Operator function:

$$\mathrm{a} = \frac{1}{\sqrt{2}}(\mathbf{X} + \mathrm{i}\mathbf{P})$$

and Annihilation Operator function:

$$\mathrm{a}^{\dagger} = \frac{1}{\sqrt{2}}(\mathbf{X} - \mathrm{i}\mathbf{P})$$

To move to the quantum formulation for the system, we elevate the above functional relations to operator relations, where the $\{\mathbf{X}, \mathbf{P}\}$ canonical conjugate pair with its quantum commutator relations becomes for the creation and annihilation operators $\{\mathrm{a}, \mathrm{a}^{\dagger}\}$ the relation:

$$[\mathrm{a}, \mathrm{a}^{\dagger}] = 1$$

And we get

$$\boldsymbol{H} = N + \frac{1}{2}, \quad N = \mathrm{a}^{\dagger}\mathrm{a}$$

Where the 'annihilation' operator acts to reduce 'N', or when N=0, to act on the system wavefunction to give zero (as its name suggests). Using Dirac Ket notation for the wavefunction, his can be written:

$$\mathrm{a}|\varphi_0 > \ = 0$$

where the 'ground state' wavefunction $\varphi_o(x)$ can be solved:

$$\left[\sqrt{\frac{\mathrm{m}\omega}{\hbar}}\,\mathrm{x} + \sqrt{\frac{\hbar}{\mathrm{m}\omega}}\frac{\mathrm{d}}{\mathrm{dx}}\right]\varphi_0(x) = 0 \;\rightarrow\; \varphi_0(x) = Ae^{-\frac{1}{2}\frac{\mathrm{m}\omega}{\hbar}\mathrm{x}^2}$$

From this we can get all other states using the 'creation' operator:

$$|\varphi_n > \ = \frac{(\mathrm{a}^{\dagger})^{\mathrm{n}}}{\sqrt{\mathrm{n}!}}|\varphi_o >$$

And find that:

$$\mathrm{a}^{\dagger}|\varphi_n > \ = \sqrt{n+1}\,|\varphi_{n+1} >$$
$$\mathrm{a}|\varphi_n > \ = \sqrt{n}\,|\varphi_{n-1} >$$

and (with unit normalization)

$$\varphi_n(x) = \frac{1}{\sqrt{\mathrm{n}!}}\frac{1}{\sqrt{2^{\mathrm{n}}}}\left[\sqrt{\frac{\mathrm{m}\omega}{\hbar}}\,\mathrm{x} - \sqrt{\frac{\hbar}{\mathrm{m}\omega}}\frac{\mathrm{d}}{\mathrm{dx}}\right]^{\mathrm{n}}\left(\frac{\mathrm{m}\omega}{\pi\hbar}\right)^{1/4} \mathrm{e}^{-\frac{1}{2}\frac{\mathrm{m}\omega}{\hbar}\mathrm{x}^2}$$

Note that this is a special case of the Hermite differential equation:

$$\left[\frac{\mathrm{d}^2}{\mathrm{dx}^2} - 2\mathrm{x}\frac{\mathrm{d}}{\mathrm{dx}} + 2\mathrm{n}\right]\mathrm{h}(x) = 0.$$

#### 4.1.3.4 Evolution operator

Let's derive the form of the 'evolution operator', e.g., the operator that advances the state in 'time'. We've seen operations for translations in space, with concomitant conservation of momentum properties, here we will have energy (the Hamiltonian) and conservation of energy. Let's denote

$$|\Psi(t)> = u(t,t_0)|\Psi(t_0)>,$$

where $u(t_0,t_0) = 1$. From the Schrodinger quantization process we have that:

$$i\hbar\frac{\partial}{\partial t}u(t,t_0)|\Psi(t_0)> = H(t)u(t,t_0)|\Psi(t_0)>.$$

Thus,

$$i\hbar\frac{\partial}{\partial t}u(t,t_0) = H(t)u(t,t_0)$$

Written in integral form:

$$u(t,t_0) = 1 - \frac{i}{\hbar}\int_{t_0}^{t} H(t')u(t',t_0)dt'$$

The Chapman-Kolmogorov relation exists:

$$u(t,t'') = u(t,t')u(t',t'')$$

Where the existence of the inverse follows:

$$u(t,t')u(t',t'') = 1.$$

Using the definition of derivative:

$$d|\Psi(t) = |\Psi(t+dt)> - |\Psi(t)> = \left(-\frac{i}{\hbar}H(t)|\Psi(t)>\right)dt$$

Thus,

$$|\Psi(t+dt)> = \left(1-\frac{i}{\hbar}H(t)dt\right)|\Psi(t)>$$

Or,

$$u(t+dt,t) = 1 - \frac{i}{\hbar}H(t)dt.$$

For H that does not depend on time we simply have:

$$u(t,t_0) = e^{-H(t-t_0)/\hbar}.$$

***Schrodinger and Heisenberg pictures***

Schrodinger: time evolution entirely contained in state or wavefunction $|\Psi(t)>$, not in operator (position, momentum, and K.E. operators are independent of time).

Heisenberg: time evolution entirely contained in operators, not in ket's $|\Psi>$.

Denote state in Schrodinger representation by $|\Psi_S(t)>$ and in Heisenberg representation by $|\Psi_H>$, we then have:

$$|\Psi_S(t)> = u(t,t_0)|\Psi_S(t_0)>$$

and

$$|\Psi_H> = u^{\dagger}(t,t_0)|\Psi_S(t)> \quad and \quad A_H(t) = u^{\dagger}(t,t_0)A_S(t)u(t,t_0),$$

Note that when $[A_S, H_S] = 0 \rightarrow A_H = A_S$ .

For arbitrary $A_S$(t):

$$i\hbar\frac{d}{dt}A_H(t) = [A_H(t), H_H(t)] + i\hbar\left(\frac{d}{dt}A_s(t)\right)_H,$$

which is a more general than Ehrenfest's relation since it is a relation between operators.

When $H_S(t) = \frac{P_S^2}{2m} + V(X_S,t) \rightarrow H_H(t) = \frac{P_H^2}{2m} + V(X_H,t)$, and using in the modified Ehrenfest's relation:

$$\frac{d}{dt}X_H(t) = \frac{1}{m}P_H(t)\,; \quad \frac{d}{dt}P_H(t) = -\frac{\partial}{\partial X}V(X_H,t).$$

An advantage of the Heisenberg picture is that it leads to equations which are formally similar to those of classical mechanics.

#### 4.1.3.5 Gauge invariance

***Classically***

The dynamical variables describing a given motion may depend on the gauge chosen -- in the Hamiltonian formation this is often the case for the conjugate momenta. A "true" physical quantity is independent of gauge while a "non-physical" quantity is dependent on gauge.

***Quantum Mechanically***

As in classical mechanics, where the values of the dynamical variables generally change when the gauge is changed, a given physical system must be characterized by a mathematical state vector $|\Psi>$ which depends on the gauge. The transition between state vector in different gauges is affected by a unitary transformation. The form of the Schrodinger equation always remains the same (as do Hamilton's equations in classical mechanics). The simultaneous modification of the state vector and the observables is such that the physical content of the quantum mechanics does not depend on the gauge.

The representation of $\vec{R}$ *and* $\vec{P}$ *in the* $\{|\vec{r}>\}$ representation is the same in all gauges:

$$\vec{R}_{j\prime} = \vec{R}_j \;\; ; \;\; \vec{P}_{j\prime} = \vec{R}_j \;.$$

Under the gauge transformation $|\Psi'(t)> = T(t)|\Psi(t)>$, where $T(t)$ is a unitary operator. For a given unitary transformation an observable $K$ transforms as

$$\tilde{K} = T(t)\, K\, T^{\dagger}(t).$$

In Quantum Mechanics every true physical quantity has an operator $G_J(t)$ which satisfies:

$$\tilde{G}_J(t) = G_{J\prime}(t).$$

### 4.2 Two-Level Systems

Let's consider two-level systems, starting with Stern-Gerlach experiment, which demonstrates the quantization in the components of angular momentum of the system, neutral paramagnetic silver atoms originally, later electrons, etc., passing through an inhomogeneous magnetic field and observing the beam deflection(s) that occur.

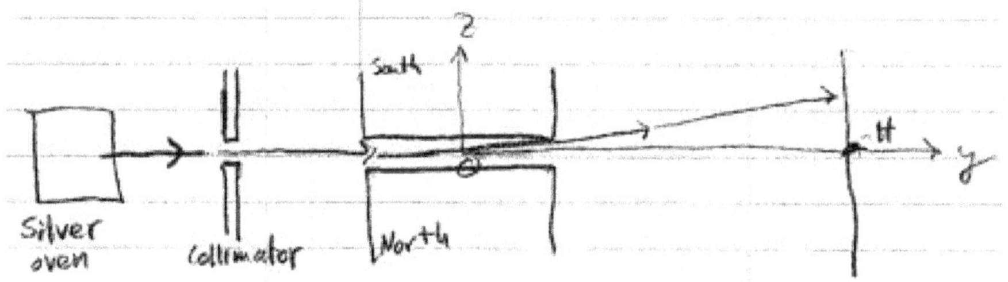


With inhomogeneity seen in cross-section:

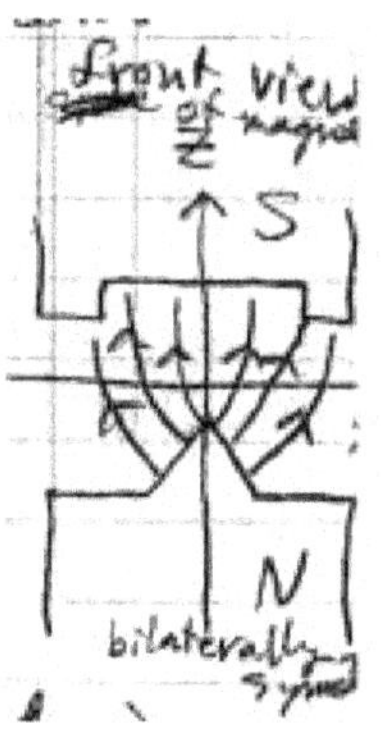

The beam spread is two-peaked, not the broad peak predicted by classical theory:

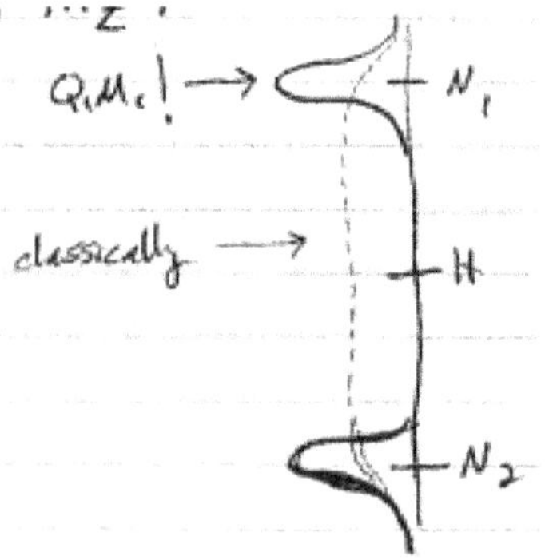

The internal angular momentum (spin) appears to be quantized. If we consider this as a quantum system with superposition of states and a measurement process, the results make sense. Let's define the up-down spin observable $S_z$ and the spin state space it resides in. From the classical theory we have:

$$\begin{Bmatrix} L_z \\ Physical\ quantity \end{Bmatrix} \rightarrow \begin{Bmatrix} S_z \\ Observable \end{Bmatrix} \rightarrow \{eigenvectors\ form\ a\ basis\}$$

According to experimental results, $S_z$ has 2 eigenvalues $\left\{\frac{\hbar}{2}, -\frac{\hbar}{2}\right\}$. Assuming the eigenvalues are not degenerate and are complete for the system, denote the eigenvectors by $|+>$ and $|->$, then:

$$S_z|+> = +\frac{\hbar}{2}|+> \ \ and\ \ S_z|-> = -\frac{\hbar}{2}|->$$

And

$$<+|+> = 1 = <-|->$$

And

$$<+|-> = 0 = <-|+>$$

Thus

$$|+><+| \ + \ |-><-| \ = \ 1.$$

A general state vector is thus:

$$|\Psi> = \alpha|+> + B|-> \quad where\ \ |\alpha|^2 + |B|^2 = 1$$

Using matrix notation in the $\{|+>, |->\}$ basis we have:

$$(S_z) = \frac{\hbar}{2}\begin{pmatrix} 1 & 0 \\ 0 & -1 \end{pmatrix}$$

$S_x$ and $S_y$ are represented in the $\{|+>, |->\}$ basis by 2 x 2 Hermitian matrices. The commutation relation and the rep chosen for $S_z$ then yield:

$$(S_z) = \frac{\hbar}{2}\begin{pmatrix} 0 & 1 \\ 1 & 0 \end{pmatrix} \quad and \quad (S_y) = \frac{\hbar}{2}\begin{pmatrix} 0 & -i \\ i & 0 \end{pmatrix}.$$

Let's now consider the general form of angular momentum with a given reference vector (other than unit direction on the z-axis):

$$L_{\vec{u}} = \vec{L} \cdot \vec{u} = L_x \sin\theta \cos\varphi + L_y \sin\theta sin\theta + L_z \cos\theta.$$

$$(S_u) = (S_x)\sin\theta\cos\theta + (S_y)\sin\theta\cos\theta + (S_y)cos\theta$$

$$(S_u) = \frac{\hbar}{2}\begin{pmatrix} \cos\theta & \sin\theta e^{-i\varphi} \\ \sin\theta e^{i\varphi} & -\cos\theta \end{pmatrix}$$

### 4.2.1 Background on Pauli Matrices

We can write $\vec{S} = \frac{\hbar}{2}\vec{\sigma}$ , where the $\vec{\sigma}$ "components" are the Pauli matrices, where $\vec{S}$ has $S_x, S_y, S_z$ in the $\{1+>, 1->\}$ basis of $S_z$:

$$\sigma_x = \begin{pmatrix} 0 & 1 \\ 1 & 0 \end{pmatrix}, \quad \sigma_y = \begin{pmatrix} 0 & -1 \\ i & 0 \end{pmatrix}, \quad \sigma_z = \begin{pmatrix} 1 & 0 \\ 0 & -1 \end{pmatrix}$$

All the $\sigma_s$ satisfy $\lambda^2 - 1 = 0$ the eigenvalues of $\sigma_x \sigma_y \; and \; \sigma_z$ are thus $\lambda = \pm 1$. Other properties include:

$$Det\,(\sigma_j) = -1$$

$$Tr\,(\sigma_j) = 0$$

$$\left\{ \begin{matrix} \lambda^2 = I \\ \sigma_x\sigma_y = -\sigma_y\sigma_x = i\sigma_z \end{matrix} \right\} \text{ commutation relations}$$

$$\boxed{\sigma_j\sigma_k = i\sum_l \varepsilon_{jkl}\sigma_l + \sigma_{jk}I}$$

$$\sigma_x\sigma_y\sigma_z = i\sigma_z\sigma_z = iI$$

$$\boxed{useful\,identity: \; (\vec{\sigma}\cdot\vec{A})(\vec{\sigma}\cdot\vec{B}) = \vec{A}\cdot\vec{B}I + i\vec{\sigma}\cdot(\vec{A}\times\vec{B})}$$

$$\vec{\sigma}\cdot\vec{A} = \sum_i \sigma_i A_i \,, \vec{\sigma}\cdot\vec{B} = \sum_j \sigma_j B_j$$

$$\textstyle\sum_i \sum_j \sigma_i A_i \sigma_j B_j = \sum_j A_i B_j \sigma_i \sigma_j \quad \text{(}A, B\text{ numbers in spin state space)}$$

$$\sum_{jk} A_i B_k \sigma_j \sigma_k = \sum_{jk} A_j B_k \left( i \sum_l \varepsilon_{jkl}\sigma_l + \sigma_{jk} I \right)$$

$$= i(A \times B)\cdot\vec{\sigma} + \vec{A}\cdot\vec{B}I$$

If A, B are operators in the full space, the order of A and B must be kept. It is possible to use the Pauli matrices and I as a basis of the 2 x 2 matrix space. Consider

$$M = \begin{pmatrix} m_{11} & M_{12} \\ m_{21} & m_{22} \end{pmatrix}$$

Then,

$$m = \frac{m_{11} + m_{22}}{2} I + \frac{m_{11} - m_{22}}{2} \sigma_z + \frac{m_{12} + m_{21}}{2} \sigma_x + i \frac{m_{12} - m_{21}}{2} \sigma_y$$
$$= a_0 I + \vec{a} \cdot \vec{\sigma}.$$

If we write $M = a_0 I + \vec{a} \cdot \vec{\sigma}$ , then M is Hermitian only if a and $\vec{a}$ are real. Also, to get a's conveniently.

$$a_0 = \frac{1}{2} Tr(M)$$

To get $\vec{a}$;

$$\vec{a} = \frac{1}{2} Tr(M\vec{\sigma})$$

### 4.2.2 Evolution of a spin 1/2 particle in a uniform magnetic field

Denote the Uniform field by $\vec{B}_0 = B_0 \tilde{e}_z$ , we can then write:

$$w = \vec{M} \cdot \vec{B}_0 = -\vec{M}_z \cdot B_0 = -\gamma B_0 L_z = \omega_0 L_z \quad where \quad \omega_0 = -\gamma B_0$$

Quantization:

$$L_z \rightarrow S_z \;\; , \;\; W \rightarrow H \;\; (here\ an\ internal\ degree\ of\ freedom)$$

where

$$H = \omega_0 S_z \quad (time\ indep.)$$

and

$$H|+>= +\frac{\hbar\omega_0}{2} |+> \qquad \rightarrow E_+ = {}^{+\hbar\omega_0}/_2$$
$$H|->= -\frac{\hbar\omega_0}{2} |-> \qquad \rightarrow E_- = {}^{-\hbar\omega_0}/_2$$

with "Bohr frequencies":

$$v_\pm = \frac{1}{h}(E_+ - E_-) = \frac{\omega_0}{2\pi}$$

Note, if

$$\left.\begin{matrix} \vec{B}_0 = \vec{B}_0 u \rightarrow w = \omega_0 L_{\vec{u}} \\ L_{\vec{u}} \rightarrow S_u \;, \qquad w \rightarrow H \end{matrix}\right\} \; then \;\; H = \omega_0 S_u$$

Also, silver $\gamma$ is negative, so it has $\omega_0$ positive.

***Example***

A neutral spin - ½ particle has magnetic moment $\vec{M} = \gamma\vec{S}$, where $\vec{S}$ is the spin operator and $\gamma$ is a constant. The only relevant degrees of freedom are the spin ones, and the particle is exposed to a uniform magnetic field $B_0$ in the x-direction. If initially the spin points in the +z-direction:

(i) Compute the state $|\Psi(t)>$ of the particle at a later time t.

(ii) Find the probability that the spin points in the +y-direction at time t.

**Answer**

(i) Spin initially in +Z-direction

Use as eigenstates $\vec{S}^2$ *and* $S_Z$ : $\{1+>, 1->\}$ *in* $L_Z$ basis

$|\Psi(0)>=|+>$

$H=-M\cdot B=-\gamma\vec{S}\cdot\vec{B}=-\gamma S_x B_o$

Now, to find the stationary states: $S_x$ commutes with the Hamiltonian

$S_{\pm}|S,m_s>=\hbar\sqrt{S(S+1)-m_s(m_s\pm 1)}|S,M_s\pm 1$

$S_+\left|S,m_s>=\hbar\sqrt{\frac{3}{4}-\left(-\frac{1}{2}\right)\left(\frac{1}{2}\right)}\right|S,M_s=\frac{1}{2}$

$\frac{S_+}{\hbar}=\begin{pmatrix}0 & 1\\ 0 & 0\end{pmatrix}$ $\qquad \frac{S_-}{\hbar}=\begin{pmatrix}0 & 0\\ 1 & 0\end{pmatrix}$

$S_+=S_x+iS_y \qquad S_{=S_x}-iS_y$

$2S_x=S_++S_{-1}$ $\boxed{\frac{\hbar}{2}\begin{pmatrix}0 & 1\\ 1 & 0\end{pmatrix}=S_x}$

What vector represents an eigenvector of $S_x$ in terms of the $S_Z$ basis.

$S_x\left|+>_x=\frac{\hbar}{2}\right|+>_x\rightarrow\begin{pmatrix}0 & 1\\ 1 & 0\end{pmatrix}\begin{pmatrix}\alpha\\ \beta\end{pmatrix}=\begin{pmatrix}\alpha\\ \beta\end{pmatrix}$

$1+>_x=\frac{1}{\sqrt{2}}\{|+>_Z-|->_Z\}$

$1->_x=\frac{1}{\sqrt{2}}\{|+>_Z-|->_Z\}$

So, $1+>_Z=\frac{\sqrt{2}}{2}\{1+>_x+1->_x\}=\frac{1}{\sqrt{2}}\{|+>_x+|->_x\}$

So, $|\Psi(0)>=\{1+>_x+1->_x\}$ $\qquad \frac{1}{2}\omega=-\frac{1}{2}\gamma B_o$

$\left|\Psi(t)>=e^{-i\frac{\mu}{\hbar}t}\right|\Psi(0)>$

$H|+>_x=-\gamma\frac{\hbar}{2}B_o=w\hbar/2$

$H|->_x=-w\hbar/2$

$$|\Psi(t)>=\frac{1}{\sqrt{2}}\left\{e^{-i\frac{\omega}{2}t}\left|+>_x+e^{+i\frac{\omega}{2}t}\right|->_x\right\}$$

$$=\frac{e^{-i\frac{\omega}{2}t}}{\sqrt{2}}\left\{1+>_x+e^{+i\frac{\omega}{2}t}|->_x\right\}$$

$$=\frac{1}{2}\left\{e^{-i\frac{\omega}{2}t}\frac{1}{\sqrt{2}}(1+>_z+|->_z)+e^{+i\frac{\omega}{2}t}\frac{1}{\sqrt{2}}(1+>_z-1->_z)\right\}$$

$$\left|\Psi(t)>=\cos\frac{\omega}{2}t\right|+>-i\sin\frac{\omega}{2}t\,1->$$

(ii) $2iS_y=S_+-S_-$

$S_y=\frac{\hbar}{2i}\begin{pmatrix}0&1\\-1&0\end{pmatrix}=\frac{\hbar}{2}\begin{pmatrix}0&-i\\i&0\end{pmatrix}$

$S_y\left|+>_y=\frac{\hbar}{2}\right|+>_y\rightarrow\begin{pmatrix}0&-i\\i&0\end{pmatrix}\begin{pmatrix}\alpha\\\beta\end{pmatrix}=\begin{pmatrix}\alpha\\\beta\end{pmatrix}$

$1+>_y=\frac{1}{\sqrt{2}}\{1+>_z+i1->_z\}$

${}_y<+|\Psi(t)>=\frac{1}{\sqrt{2}}\{{}_z<+1-i\,{}_z<-1\}\left\{\cos\frac{\omega}{2}t\,1+>_z-i\sin\frac{\omega}{2}t\,1->\right\}$

$=\frac{1}{2}\left(\cos\frac{\omega}{2}t+\sin\frac{\omega}{2}t\right)$

$\mathcal{P}=|<+|\Psi(t)>|^2=\frac{1}{2}\left(\cos^2\frac{\omega}{2}t+\sin^2\frac{\omega}{2}t+2\cos\frac{\omega}{2}t+\sin\frac{\omega}{2}t\right)=\frac{1+\sin(\omega t)}{2}$

### 4.2.3 Larmor Procession

Consider the state at t=0 to be $|+>_u$ , then:

$$|\Psi(0)>=\cos\frac{\theta}{2}e^{-i\varphi/2}|+>+\sin\frac{\theta}{2}e^{i\varphi/2}|->$$

$$|\Psi(t)>=\cos\frac{\theta}{2}e^{-i\varphi/2}e^{-iE_+\frac{1}{\hbar}t}|+>+\sin\frac{\theta}{2}e^{i\varphi/2}e^{-iE_-\frac{1}{\hbar}t}|->$$

$$=\cos\frac{\theta}{2}\exp(-i\,[\varphi+\omega_0 t]/2)\,|+>+\sin\frac{\theta}{2}exp(i\,[\varphi+\omega_0 t]/2)\,|->$$

So, $|+>_{\vec{u}}$ has:

$$\theta(t)=\theta$$

$$\varphi(t)=\varphi+\omega_0 t\qquad\leftarrow \textit{Larmor precession}$$

and

$$<\Psi(t)|S_z|\Psi(t)>=\frac{\hbar}{2}\left\{cos^2\frac{\theta}{2}-sin^2\frac{\theta}{2}\right\}=\frac{\hbar}{2}\cos\theta$$

$$< \Psi(t)|S_x|\Psi(t) > = \frac{\hbar}{2}\sin\theta\cos(\varphi + \omega_0 t)$$
$$< \Psi(t)|S_y|\Psi(t) > = \frac{\hbar}{2}\sin\theta\sin(\varphi + \omega_0 t)$$

Where the mean values behave like the components of a classical angular momentum undergoing Larmor precession.

**4.2.4 Two level systems as general model**

A system could be two-level to first approximation, such as two states with energies close together and very different from the energies of other states, and subject to small perturbation.

Every two-level system can be associated with a spin ½ (fictitious spin) system placed in a static field $\vec{B}$ and described by a Hamiltoman of identical form as before.

Let's denote the unperturbed Hamiltonian $H_0$ with two eigenstates $|\varphi_I >$ $and\ |\varphi_2 >$ :

$H_0|\varphi_I > \ = E_1|\varphi_I >$

$H_0|\varphi_2 > \ = E_2|\varphi_2 >$

The full Hamiltonian is (denoting eigenvalues as $|\,\Psi_\pm >\ and\ \pm E$ :

$H|\,\Psi_+ > = E_+|\,\Psi_+ >$

$H|\,\Psi_- > = E_-|\,\Psi_- >$

Now, consider $H = H_0 + W$:

$$(W) = \begin{pmatrix} W_{11} & W_{12} \\ W_{21} & W_{22} \end{pmatrix}, \qquad where\ \ W_{12} = W_{21}{}^*$$

(W is the perturbation or coupling).

Consequences of coupling:

(1) $E_1$ and $E_2$ are no longer the energies of the system, measurements now only yield $E_+$ or $E_-$.

(2) $|\varphi_I >\ and\ |\varphi_2 >$ are no longer stationary states. W therefore induces transitions between $|\varphi_I >\ and\ |\varphi_2 >$ .

***Static aspect: Effect of coupling on the stationary states***

Expressions for Eigen states and eigenvalues of H:

$$(H) = \begin{pmatrix} E_1 + W_{11} & W_{12} \\ W_{21} & E_2 + W_{22} \end{pmatrix} \qquad in\ the\{\,|\varphi_I >\ \ amd\ |\varphi_2 >\}basis$$

For eigenvalues, we diagonalize this 2 x 2. Hermitian matrix, there are some analytical techniques that are useful for this.

***Digression to discuss diagonalization of 2 x 2 Hermitian matrix***

Consider $(H) = \begin{pmatrix} H_{11} & H_{12} \\ H_{21} & H_{22} \end{pmatrix}$ $with \;\; H_{12} = {H_{21}}^{*}$, we then have:

$$(H) = \begin{pmatrix} \frac{1}{2}(H_{11}+H_{22}) & 0 \\ 0 & \frac{1}{2}(H_{11}+H_{22}) \end{pmatrix} + \begin{pmatrix} \frac{1}{2}(H_{11}-H_{22}) & H_{12} \\ H_{21} & \frac{1}{2}(H_{11}-H_{22}) \end{pmatrix}$$

$$= \frac{1}{2}(H_{11}+H_{22})1 + \frac{1}{2}(H_{11}-H_{22})K$$

$$(K) = \begin{pmatrix} 1 & \frac{2H_{12}}{H_{11}-H_{22}} \\ \frac{2H_{12}}{H_{11}-H_{22}} & -1 \end{pmatrix}$$

H and K have the same eigenvectors since the identity 1 is diagonal. Write:

$H|\Psi\pm> = E_{\pm}|\Psi\pm>$

$K|\Psi\pm> = K_{\pm}|\Psi\pm>$

From which

$E_{\pm} = \frac{1}{2}(H_{11}+H_{22}) + \frac{1}{2}(H_{11}-H_{22})K_{\pm}$

***Changing the eigenvalue origin***

We can eliminate the identity matrix factor by choosing for the eigenvalue origin $(H_{11}+H_{22})/2$, furthermore thus choice of origin will remain the same whatever the basis initially chosen since $(H_{11}+H_{22}) = Tr(H)$ is invalid under a change of orthogonal basis.

***Calculation of eigenvalues and eigenvectors***

Let $\tan\theta = \frac{2|H_{21}|}{H_{11}-H_{22}}$ and $H_{21} = |H_{21}|e^{i\varphi}$

Then,

$$(K) = \begin{pmatrix} 1 & \tan\theta\, e^{-i\varphi} \\ \tan\theta\, e^{i\varphi} & -1 \end{pmatrix} = \frac{1}{\cos\theta}\begin{pmatrix} \cos\theta & \sin\theta\, e^{-i\varphi} \\ \sin\theta\, e^{-i\varphi} & -\cos\theta \end{pmatrix}$$

$$= \frac{1}{\cos\theta}\left(\frac{2}{\hbar}\right)(S_u).$$

***The eigenvalues of K***

$$Det[(K) - KI] = -(1-K)(1+K) - \tan\theta = K^2 - 1 - tan^2\theta = 0$$

$$\rightarrow K_0 = \frac{1}{\cos\theta} \quad and \;\; K_- = -\frac{1}{\cos\theta}$$

where

$$\frac{1}{\cos\theta} = \frac{\sqrt{(H_{11}-H_{22})^2 + 4|H_{21}|^2}}{H_{11}-H_{22}}$$

Eigenvalues of H:

$$E_\pm = \frac{1}{2}(H_{11}+H_{22}) \pm \sqrt{(H_{11}-H_{22})^2 + |H_{21}|^2}$$

Comments:

(i) Eigenvalues can be obtained by direct analysis, no need to introduce $\theta$ *and* $\varphi$ angles, but the usefulness of the angles will become apparent in normalizing.

(ii) $E_\pm + E_- = H_{11} + H_{22} = Tr(H)$
$E_\pm E_- = H_{11}H_{22} - |H_{21}|^2 = Det(H)$

***Normalized Eigenvectors of H***

Denote $|\Psi_+ >= a|\varphi_1 > +b|\varphi_2 >$ . Then,

$$\begin{pmatrix} 1 & \tan\theta\, e^{-i\varphi} \\ \tan\theta\, e^{-i\varphi} & -1 \end{pmatrix}\begin{pmatrix} a \\ b \end{pmatrix} = \frac{1}{\cos\theta}\begin{pmatrix} a \\ b \end{pmatrix}$$

$$\rightarrow \left(1 - \frac{1}{\cos\theta}\right)a + \tan\theta\, e^{-i\varphi} b = 0$$

Thus

$$(\cos\theta - 1)a + \sin\theta\, e^{-i\varphi} b = 0$$

$$-2sin^2\frac{\theta}{2}a + 2sin\frac{\theta}{2}cos\frac{\theta}{2}e^{-i\varphi}b = 0$$

The normalized eigenvector $|\Psi_+ >$ is then:

$$|\Psi_+ > = \cos\frac{\theta}{2}e^{-i\varphi/2}|\varphi_1 > + \sin\frac{\theta}{2}e^{i\varphi/2}\,|\varphi_2 >$$

So, coupling gives us

$$(H) = \begin{pmatrix} E_1 + W_{11} & W_{12} \\ W_{21} & E_2 + W_{22} \end{pmatrix}$$

Eigenvalues:

$$E_\pm = \frac{1}{2}(E_1 + W_{11} + E_2 + W_{22})$$
$$\pm \frac{1}{2}\sqrt{(E_1 + W_{11} - E_2 - W_{22})^2 + 4|W_{12}|^2}$$

Eigenvectors:

$$|\Psi_+> = \cos\frac{\theta}{2} e^{-i\varphi} |\varphi_1> + \sin\frac{\theta}{2} e^{i\varphi/2} |\varphi_2>$$

$$|\Psi_-> = -\sin\frac{\theta}{2} e^{-i\varphi} |\varphi_1> + \cos\frac{\theta}{2} e^{i\varphi/2} |\varphi_2>$$

$$\tan\theta = \frac{2|W_{12}|^2}{E_1 + W_{11} - E_2 - W_{22}} \quad , \quad W_{21} = |W_{21}|^2 e^{i\varphi}$$

Let $E_1 + W_{11} \rightarrow E_1$ and $E_2 + W_{22} \rightarrow E_2$ , notationally, for convenience. Then define (for a graphical representation of the effect of coupling):

$$E_m = \frac{1}{2}(E_1 + E_2)$$

$$\Delta = \frac{1}{2}(E_1 - E_2)$$

$$E_{\pm} = E_m + \sqrt{\Delta^2 + |W_{12}|^2}$$

$$E_{\pm} = E_m - \sqrt{\Delta^2 + |W_{12}|^2}$$

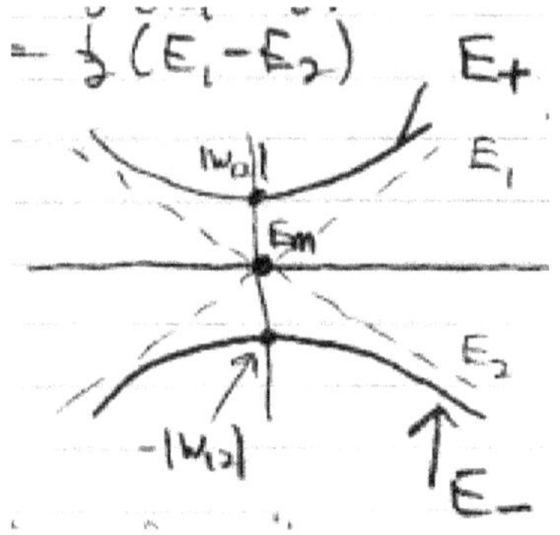


***Effect of coupling on the position of energy levels***

Graphically, the two levels "repel" each other, also, $|E_+ - E_-| > |E_1 - E_2|$ , thus, the coupling separates the normal frequencies further.

Near asymptotes ($|\Delta| \gg |W_{12}|$) :

$$E_+ \cong E_m + \Delta$$

$$E_- \cong E_m - \Delta$$

*At* $\Delta = 0$:

$$E_+ = E_m + |W_{12}|$$

$$E_- = E_m - |W_{12}|$$

The effect of coupling is most notable when $\Delta = 0$, as will be discussed next.

***Quantum Molecular Resonance***

Consider the situation where a molecule or physical system has ground state that is two-fold degenerate: $E_1 = E_2 = E_m$ and thus $\Delta = 0$, and where the ground state is sufficiently far from other levels that we may consider a two-state system in isolation, then, as previously, we see that any coupling between two corresponding states will be purely non-diagonal and will result in a lowering of the ground state of the system. (which becomes the most stable scenario). This new ground sate is known as the resonance stabilization state. Take for example the benzene molecule:

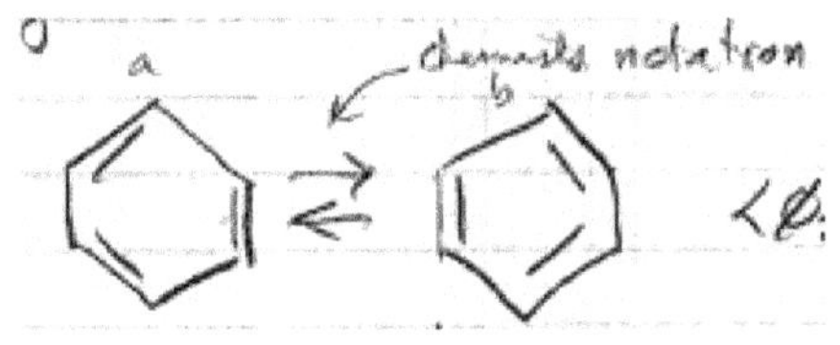

From the above we have for new ground state: $E_- = E_m - |W_{12}|$. This makes the benzene molecule more stable than expected from bond formation strengths considered separately.

***The nature of the chemical bond***

Consider an ionized hydrogen gas molecule (e.g., $H_2^+$), the lone remaining electron will be shared between the two hydrogen nuclei according to the two configurations shown:

Where on the left, with eigenstate $\varphi_1$, we have the electron residing near proton 1, and on the right we have eigenstate $\varphi_2$, with the electron residing near proton 2. As before, we have $E_1 = E_2 = E_m$ and thus $\Delta = 0$. Again, any coupling leads to a new ground state $E_- = E_m - |W_{12}|$. Here the new ground state corresponds to the electron being 'shared' between protons 1 and 2, e.g., the formation of the chemical bond. Thus, the 'chemical bond' is the lowing of potential energy that accompanies a delocalized electron, shared between the protons.

**4.2.5 Dynamical aspects of the two –state system and Rabi's formula**

Let's consider evolution of a state vector:

$$|\Psi(t) = a_1(t)|\varphi_1> +a_2(t)|\varphi_2>$$

where

$$i\hbar\frac{d}{dt}|\Psi(t)>=(H_0+W)|\Psi(t)>$$

and assume $W$ is non-diagonal. Project onto $|\varphi_1>$ and $|\varphi_2>$

$$i\hbar\dot{a}_1=E_1a_1+W_{12}a_2$$
$$i\hbar\dot{a}_2=W_{21}a_1+E_2a_2$$

The classical method of solution of the above is the solution of the eigenvalue equation:

$$|\Psi(0)>=\lambda|\Psi_+>+\mu|\Psi_->$$
$$|\Psi(t)>=\lambda e^{-iE_+/\hbar\, t}|\Psi_+>+\mu e^{-iE_-/\hbar\, t}|\Psi_->$$

then project onto $|\varphi_1>$ and $|\varphi_2>$. Consider $|\Psi(0)>=|\varphi_1>$ and calculate the probability $P_{12}(t)$ of finding $|\varphi_2>$ at the some time t:

$$|\Psi(0)>=|\varphi_1>=e^{i\varphi}\left[\cos\frac{\theta}{2}|\Psi_+>-\sin\frac{\theta}{2}|\Psi_->\right]$$

(from inverting previous equation in earlier section). Similarly:

$$|\Psi(t)>=e^{i\varphi}\left[\cos\frac{\theta}{2}e^{-iE_+/\hbar\, t}|\Psi_t>-\sin\frac{\theta}{2}e^{-iE_-/\hbar\, t}|\Psi_->\right]$$

$$P_{12}(t)=sin^2\theta sin^2\left(\frac{E_+-E_-}{2\hbar}t\right)$$

or

$$P_{12}(t)=\frac{4|W_{12}|^2}{4|W_{12}|^2+(E_1-E_2)^2}\,sin^2\left[\sqrt{4|W_{12}|^2+(E_1-E_2)^2}\frac{t}{2\hbar}\right]$$

which is known as Rabi's formula.

***Fictitious spin ½ associated with a two-level system***

Let's shift eigenvalues such that:

$$(H)=\begin{pmatrix}\frac{1}{2}(H_{11}-H_{22}) & H_{12}\\ H_{21} & -\frac{1}{2}(H_{11}-H_{22})\end{pmatrix}$$

If we can write the Hamilton as $H=H_0+w$ we will find that not only can be associate it with a magnetic field $\vec{B}$ acting on a spin ½ system but to each of $H_0$ and w, we can separately associate a $B_0$ and b such that $\vec{B}=\vec{B}_0+\vec{b}$. The Hamilton of the coupling between a spin ½ and magnetic field $\vec{B}$ can be written as.

$$\hat{H}=-\gamma B\cdot S=-\gamma\left(B_zS_z+B_yS_y+B_xS_x\right)$$

Using the matrices associated with $S_xS_yS_z$ (in $S_z$ *rep.*):

$$\hat{H}=-\frac{\gamma\hbar}{2}\begin{pmatrix}B_z & B_x-iB_y\\ B_x+iB_y & -B_z\end{pmatrix}$$

So, to make our matrices for (H) and $\hat{H}$ identical:

$$B_x = -\frac{2}{\gamma\hbar} ReH_{12}$$

$$B_y = -\frac{2}{\gamma\hbar} ImH_{12}$$

$$B_z = \frac{1}{\gamma\hbar}(H_{22} - H_{11})$$

where $B_{Perp.} = \frac{2}{\hbar}\left|\frac{H_{12}}{\gamma}\right|$ and where we get Rabi's formula from a purely geometrical interpretation.

If we write
$H = H_0 + w$ in the $\{|\varphi_1>, |\varphi_2>\}$ basis:

$$(H_0) = \begin{pmatrix} (E_1 - E_2)/2 & 0 \\ 0 & (E_1 - E_2)/2 \end{pmatrix} \; and \; (w) = \begin{pmatrix} 0 & w_{12} \\ w_{21} & 0 \end{pmatrix}$$

Thus,

$$B_{0z} = (E_2 - E_1)/\gamma\hbar \; ; \; b_z = 0$$

$$B_{0\,Perp.} = 0 \; and \; b_{Perp.} = \frac{2}{\hbar}\left|\frac{H_{12}}{\gamma}\right|$$

$$\vec{B} = \vec{B}_0 + \vec{b}$$

Strong coupling $|w_{12}| \gg |E_1 - E_2| \to |b| \gg |B_0|$
Weak coupling $|w_{12}| \ll |E_1 - E_2| \to |b| \ll |B_0|$

***Electron-Positron in uniform magnetic field***

Particle 1 (an electron) and particle 2 (a positron) interact in the presence of an external static uniform magnetic field $\vec{B}_0 = B_0\hat{z}$ through a Hamiltonian of the form

$$H = A\vec{S}_1 \cdot \vec{S}_2 + \frac{eB_0}{mc}(s_{1z} - s_{2z})$$

(a) Find the exact expressions for the eigenstates and eigenvalues of H.

(b) A second external sinusoidal magnetic field $\vec{B}_1 \sin\omega t$ is applied, with $(B_1 \ll B_0)$. We want this field to induce transitions between eigenstates with zero z-component of total spin. Should $\vec{B}_1$ lie in the $\hat{z}$ direction or the $\hat{x} - \hat{y}$ plane? What frequency $\omega$ results in the maximum transition rate?

**Answer**

$H = A\vec{S}_1 \cdot \vec{S}_2 + \frac{eB_0}{mc}(S_{1z} - S_{2z})$

$\vec{S}_1 \cdot \vec{S}_2 = S_{1z} - S_{2z} + S_{1x} - S_{2x} + S_{1y} - S_{2y}$

$S_{1-} = S_{1x} - iS_{1y} \qquad S_{1-}S_{2+} = S_{1x}S_{2x} + S_{1y}S_{2y}$

$S_{2-} = S_{2x} - iS_{2y} \qquad S_{2y}S_{1x} - iS_{1y}S_{2x}$

$S_{2+} = S_{2x} + iS_{2y} \qquad S_{1+}S_{2-} = S_{1x}S_{2x} + S_{1y}S_{2y}$

$iS_{1y}S_{2x} - is_{2x}s_{2y}$

$S_{1x}S_{2x} + S_{1y}S_{2y} = \frac{1}{2}(S_{1-}S_{2+} + S_{1+}S_{2-})$

$$S_{1z} = \begin{pmatrix} \frac{\hbar}{2} & 0 & \\ 0 & -\frac{\hbar}{2} & \varphi \\ \varphi & \hbar 1 & \varphi \end{pmatrix}$$

$$S_{2z} = \begin{pmatrix} \hbar 1 & \varphi & \\ \emptyset & \frac{\hbar}{2} & \frac{\hbar}{2} \end{pmatrix}$$

$$S_1 = 1\begin{pmatrix} 0 & 0 & \emptyset \\ k & 0 & \\ \emptyset & \hbar & \hbar 1 \end{pmatrix}$$

$$S_{1+} = 1\begin{pmatrix} 0 & \hbar & \emptyset \\ 0 & 0 & \\ \emptyset & \hbar 1 & \end{pmatrix}$$

Etc $\hat{S}_{1z} = \hat{S}_{1z} \otimes 1_2$

$$\frac{\vec{S}_1 \cdot \vec{S}_2}{\hbar^2} = \begin{pmatrix} \frac{1}{2} & & 0 \\ & -\frac{1}{2} & \frac{1}{2} \\ 0 & & -\frac{1}{2} \end{pmatrix} + \frac{1}{2}\left\{\begin{pmatrix} 0 & 0 & \\ 1 & 0 & \\ \emptyset & 0 & 1 \\ & 0 & 0 \end{pmatrix} + \begin{pmatrix} 0 & 1 & \\ 0 & 0 & \emptyset \\ \emptyset & 0 & 0 \\ & 1 & 0 \end{pmatrix}\right\}$$

$$\vec{S}_1 \cdot \vec{S}_2 = \frac{\hbar^2}{2}\begin{pmatrix} 1 & 1 & \emptyset \\ 1 & -1 & \\ & 1 & 1 \\ \emptyset & & \\ \emptyset & 1 & -1 \end{pmatrix} \quad (S_{1z} - S_{2z}) = \frac{\hbar}{2}\begin{pmatrix} 1 & & \emptyset \\ & -3 & \\ & -1 & \\ \emptyset & & \\ & & 3 \end{pmatrix}$$

Let $A\left(\frac{\hbar^2}{2}\right) = \alpha \quad and \quad \frac{eB_0\hbar}{2nc} = \beta$

$$(H) = \begin{pmatrix} (\alpha+\beta) & \alpha & \\ \alpha & -(\alpha+3\beta) & \\ & & (\alpha-\beta)\alpha \\ \emptyset & \alpha & -(\alpha-3\beta) \end{pmatrix}$$

$(\alpha+\beta)Det - \alpha Det$

$Det = -(\alpha+3\beta)Det$

$det = \alpha Det$

$(\alpha+\beta)(-[\alpha+3\beta])Det-\alpha(\alpha Det)=$
$\{-(\alpha+\beta)-[\alpha+3\beta]-\alpha^2\}Det=$ product of determinants!
$[-(\alpha+\beta)-(\alpha+3\beta)-\alpha^2][-(\alpha-\beta)-(\alpha-3\beta)-\alpha^2]=0$
Use J.M.

(b) 2nd field $\vec{B}=\vec{B}_1\sin\omega t$ applied $B_1\ll B_0$
$\vec{B}_1$ should lie in $\hat{x}-\hat{y}$ plane, for maximum transition rate $\omega=\frac{eB_0}{mc}$

***Probability to remain in state after perturbation***

A Hamiltonian $H_0$ has only two eigenstates, $|0>$ and $|1>$, with energies $\omega_0$ and $\omega_1$ respectively ($\hbar=1$). A particle is in the state $|0>$ until, at t=0, a constant perturbation V is applied suddenly. Show that the probability P that the particle will remain in the state $|0>$ at time t>0 is

$$P=1-\frac{|V_{10}|^2}{\sigma^2}\sin^2\sigma t$$

where $V_{jk}=<j|V|k>$ and

$$\sigma=\sqrt{|V_{10}|^2+\frac{1}{4}\gamma^2}$$
$$\gamma=V_{11}-V_{00}+(\omega_1-\omega_0)$$

**Answer**

$H_0=\begin{pmatrix}\omega_0 & 0\\ 0 & \omega_1\end{pmatrix}$ $in\ \{|0>,|1>\}\ basis\ \hbar=1$

$V=\begin{pmatrix}V_{00} & V_{01}\\ V_{10} & V_{11}\end{pmatrix}$ $H=H_0+U$

What are the stationary states? $\gamma=(V_{11}+\omega_1)-(V_{00}-\omega_0)$

$H=\begin{pmatrix}\omega_0+V_{00} & V_{01}\\ V_{10} & \omega_1+V_{11}\end{pmatrix}=\frac{1}{2}(\omega_0+\omega_1+V_{00}+V_{11})I+$
$\begin{pmatrix}-\frac{1}{2}\gamma & V_{01}\\ V_{10} & \frac{1}{2}\gamma\end{pmatrix}$

$H\begin{pmatrix}a\\b\end{pmatrix}=E\begin{pmatrix}a\\b\end{pmatrix}$ for a stationary states

$$H\begin{pmatrix}a\\b\end{pmatrix}=\frac{1}{2}(\omega_0+\omega_1+V_{00}+V_{11})\begin{pmatrix}a\\b\end{pmatrix}+\begin{pmatrix}-\frac{1}{2}\gamma & V_{01}\\ V_{10} & \frac{1}{2}\gamma\end{pmatrix}\begin{pmatrix}a\\b\end{pmatrix}=E\begin{pmatrix}a\\b\end{pmatrix}$$

$$\begin{pmatrix}-\frac{1}{2}\gamma & V_{01}\\ V_{10} & \frac{1}{2}\gamma\end{pmatrix}\begin{pmatrix}a\\b\end{pmatrix}=\left\{E-\frac{1}{2}(\omega_0+\omega_1+V_{00}+V_{11})\right\}\begin{pmatrix}a\\b\end{pmatrix}=\varepsilon\begin{pmatrix}a\\b\end{pmatrix}$$

So, $\begin{vmatrix} -\frac{1}{2}\gamma - \varepsilon & V_{01} \\ V_{10}^* & \frac{1}{2}\gamma - \varepsilon \end{vmatrix} = 0$ is the eigenvalue equation

$$-\left(\frac{1}{2}\gamma + \varepsilon\right)\left(\frac{1}{2}\gamma - \varepsilon\right) - |V_{01}|^2 = 0$$

$$\varepsilon^2 = \left(\frac{1}{2}\gamma\right)^2 + |V_{01}|^2$$

$$\varepsilon = \pm\sigma \qquad \sigma \ defined \ as \ \sigma = \sqrt{|V_{01}|^2 + \frac{1}{4}\gamma^2}$$

Eigenvectors? $\varepsilon = \sigma$ case

$$\begin{pmatrix} -\frac{1}{2}\gamma - \varepsilon & V_{01} \\ V_{10}^* & \frac{1}{2}\gamma - \varepsilon \end{pmatrix}\begin{pmatrix} a \\ b \end{pmatrix} = 0$$

$$\boxed{\frac{|\emptyset_1>}{N_1} = |0> + \left\{\frac{\left(\frac{1}{2}\gamma + \sigma\right)}{V_{01}}\right\} |1>}$$

$$-\left(\frac{1}{2}\gamma + \sigma\right) a + V_{01} b = 0$$

$$b = \frac{\left(\frac{1}{2}\gamma + \sigma\right)}{V_{01}} a$$

(3) $\varepsilon = -\sigma$ case : $\begin{pmatrix} -\frac{1}{2}\gamma - \varepsilon & V_{01} \\ V_{10}^* & \frac{1}{2}\gamma - \varepsilon \end{pmatrix}\begin{pmatrix} a \\ b \end{pmatrix} = 0$

Take 2nd equation:

Important choice, otherwise lots of work $V_{01}^* a + \left(\frac{1}{2}\gamma + \sigma\right) b = 0 \rightarrow a = -\frac{\left(\frac{1}{2}\gamma + \sigma\right)}{V_{01}^*} b$

$$\frac{|\emptyset_1>}{N_1} = -\frac{\left(\frac{1}{2}\gamma + \sigma\right)}{V_{01}^*} + |1>$$

$$\frac{1}{N_1^2} = 1 + \frac{\left(\frac{1}{2}\gamma + \sigma\right)^2}{|V_{01}|^2} \qquad \frac{1}{N_2^2} = 1 + \frac{\left(\frac{1}{2}\gamma + \sigma\right)^2 + 1}{|V_{01}|^2}$$

$$N_1^2 = N_2^2! \qquad N = \sqrt{N_1^2} = N_1 = N_2 \qquad N^2 = \frac{1}{1 + |a_1|^2}$$

$$|\emptyset_1 >= N|0> + Na_1|1>$$

$$|\emptyset_2 >= -Na_1^*|0> + N|1> \qquad -a_1|\emptyset_2 >= N|a_1|^2|0> - Na_1|1>$$

$$N'\{|\emptyset_1> - a_1|\emptyset_2>\} = N'\{N|0> + N|a_1|^2|0>\} = |0>$$

$$N'N(1 + |a_1|^2) = 1 \qquad N' = N \ !$$

$$|0 >= N\{|\emptyset_1> - a_1|\emptyset_2\}$$

$$|\Psi(t) >= e^{iHt}|0 >= N\{e^{-i(\delta+\sigma)t}|\emptyset_1> - a_1 e^{-i(\delta-\delta)t}|\emptyset_2>\}$$

$$= Ne^{i\delta t}\{e^{-0\sigma t}|\emptyset_1> - a_1 e^{i\sigma t}|\emptyset_2>\}$$

$< 0|\Psi(t) >= N^2 e^{i\sigma t}[< \emptyset_1| - a_1^* < \emptyset_2|][e^{-i\sigma t}|\emptyset_1 > -a_1 e^{i\sigma t}|\emptyset_2 >]$
$= N^2 e^{-i\delta t}\{e^{-i\sigma t} + |a_1|^2 e^{i\sigma t}\}(e^{i\sigma t} + |a_1|^2 e^{-i\sigma t})$
$\mathcal{P}_0(t) = |< 0|\Psi(t) > |^2 = N^4\{1 + |a_1|^4 + |a_1|^2(e^{2i\sigma t} + e^{-2i\sigma t})\}$

$\mathcal{P}_0(t) = \frac{1}{(1+|a_1|^2)^2}\left\{1 + |a_1|^4 + 2|a_1|^2 + |a_1|^2\left(e^{i\sigma t} - e^{-i\sigma t}\right)^2\right\} = 1 + \frac{|a_1|^2}{(1+|a_1|^2)^2}\left(e^{i\sigma t} - e^{-i\sigma t}\right)^2$

$a_1 = \frac{\left(\frac{1}{2}\gamma+\sigma\right)}{V_{01}} = 1 + \frac{\left(\frac{1}{2}\gamma+\sigma\right)^2 |V_{01}|^2}{\left(|V_{01}|^2+\left(\frac{1}{2}\gamma+\sigma\right)^2\right)^2}(\dots)1 + \frac{\left(\frac{1}{2}\gamma+\sigma\right)^2}{|V_{01}|^2}$

Recall $|V_{01}|^2 = \sigma^2 - \left(\frac{1}{2}\gamma\right)^2$

$\sigma^2 - \left(\frac{1}{2}\gamma\right)^2 + \left(\frac{1}{2}\gamma\right)^2 + \gamma\sigma + \sigma^2$

$\sigma^2(2\sigma + \gamma) = 2\sigma\left(\sigma + \frac{1}{2}\gamma\right) = 1 + \frac{\left(\frac{1}{2}\gamma+\sigma\right)^2 |V_{01}|^2}{(2\sigma)^2\left(\frac{1}{2}\gamma+\sigma\right)^2}(2i)^2 \sin^2 \sigma t$

$$\boxed{P = 1 - \frac{|V_{01}|^2}{\sigma^2}\sin^2 \sigma t}$$

**4.2.6 System of two spin ½ particles -- the irreducibility of tensor products**

$\varepsilon_S = \varepsilon_S(1) \otimes \varepsilon_S(2)$

$$\sum_{\varepsilon_1\varepsilon_2} |\varepsilon_1\varepsilon_2 >< \varepsilon_1\varepsilon_2|$$

$= |++><++| + |+-><+-| + |-+><-+| + |--><--|$

$$S_{1z}|\varepsilon_1\varepsilon_2 > = \frac{\hbar}{2}\varepsilon_1|\varepsilon_1\varepsilon_2 >$$
$$S_{2z}|\varepsilon_1\varepsilon_2 > = \frac{\hbar}{2}\varepsilon_2|\varepsilon_1\varepsilon_2 >$$

Taking a Ket of $\varepsilon_s(1)$ *and* $\varepsilon_s(2)$ we have:

$$|\varphi(1) > |x(2) > = \alpha_1\alpha_2| ++> + \alpha_1\beta_2|+-> + \alpha_2\beta_1 |-+> + \beta_1\beta_2|-->$$

But all kets of $\varepsilon_s$ are not tensor products:

General state:

$$|\Psi>= \alpha|++> +\beta|+-> +\gamma|-+> +\delta|-->$$
$$|\alpha|^2 + |\beta|^2 + |\gamma|^2 + |\delta|^2 = 1$$

Only gives a tensor product if $\frac{\alpha}{\beta} = \frac{\gamma}{\delta}$.

**Spin ½ density matrix**

***Pure case***

$$|\Psi>= |+>_u= \cos\frac{\theta}{2}e^{-i\varphi/2}|+> + \sin\frac{\theta}{2}e^{i\varphi/2}|->$$

In the $\{|+>, |->\}$ basis the density matrix for $|\Psi>$ is:

$$p(\theta,\varphi) = |\Psi><\Psi| = \begin{pmatrix} cos^2\frac{\theta}{2} & sin^2\frac{\theta}{2}cos^2\frac{\theta}{2}e^{-i\varphi/2} \\ sin^2\frac{\theta}{2}cos^2\frac{\theta}{2}e^{-i\varphi/2} & sin^2\frac{\theta}{2} \end{pmatrix}$$

***Statistical mixture: unpolarized spin***

$$p = \frac{1}{4\pi}\int d\Omega p(\theta,\varphi) = \frac{1}{4\pi}\int_0^{2\pi} d\varphi \int_0^{\pi} \sin\theta d\theta\, p\,(\theta,\varphi)$$

$$p = \begin{pmatrix} 1/2 & 0 \\ 0 & 1/2 \end{pmatrix}$$

$$< S_i >= Tr[pS_i] = \frac{1}{2}TrS_i = 0$$

***Spin ½ in thermodynamic equilibrium in static field***

Stationary states are $|+>$ *and* $|+>$ , with eigenvalues of $S_z$ when $\vec{B} = B_0\hat{e}_z$ where eigenvalues are $+\hbar\omega/2$ *and* $-\hbar\omega/2$.

Probability of state: $|+> = \frac{e^{-i\hbar\omega_0/2kT}}{Z}$ and $|-> = \frac{e^{+k\omega_0/2kT}}{Z}$ where

$$Z = e^{-i\hbar\omega/2kT} + e^{-i\hbar\omega_0/2kT}$$

So, $p = Z^{-1}\begin{pmatrix} e^{-i\hbar\omega_0/2kT} & 0 \\ 0 & e^{-i\hbar\omega_0/2kT} \end{pmatrix}$ and

$< S_x > = Tr(pS_x) = 0$

$< S_y >= 0$

$< S_z >= -\frac{\hbar}{2}\tanh\left(\frac{\hbar\omega_0}{2kT}\right)$

Thus, we can easily get Brillouin's formula:

$$\langle M_z \rangle = \gamma < S_z >= xB_0 \; = \gamma\left(-\frac{\hbar}{2}\tanh\left(\frac{\hbar\omega_0}{2kT}\right)\right)$$

And

$$x = \frac{\hbar\gamma}{2\sigma_0}\tanh\left(\frac{\hbar\gamma B_0}{2kT}\right)$$

where $(\omega_0 = -\gamma B_0)$ and where $x$ is paramagnetic susceptibility.

### 4.2.7 Spin ½ particle in a static magnetic field and a rotating field -- resonance

***Classical analysis – rotating reference frame***

Angular momentum: J

Magnetic moment: $\vec{m} = \gamma\vec{j}$ where $\gamma$ is the gyromagnetic ratio

Static magnetic field $\vec{B}_0$ exerts torque $\vec{\mu}x\vec{B}_0$ :

$$\frac{d\vec{m}(t)}{dt} = r\vec{m}(t)x\vec{B}_0$$

Scalar multiplication by $\vec{m}$ yields $\frac{d}{dt}[\vec{m}(t)]^2 = 0$, e.g. constant modulus by $\vec{B}_0$ yields $\frac{d}{dt}\left[\vec{m}(t)\cdot\vec{B}_0\right] = 0$ so constant angle $\theta$. Projecting the equation to the plane perpendicular to $\vec{B}_0$ we have

$$\frac{d\vec{m}_\perp}{dt} = \gamma m_\perp B_0$$

Larmor procession with $\omega_0 = -\gamma B_0$ (rotation counterclockwise of $\gamma$ is positive).

So, influence of a rotating field (still classical):

$\omega_0 = -\gamma B_0$

$\omega_1 = -\gamma B_1$ where $\vec{B}_1$ is perpendicular to $\vec{B}_0$ and rotates with angular velocity w.

$$\frac{d}{dt}\vec{m}(t) = \gamma\vec{m}(t) \times \left[\vec{B}_0 + \vec{B}_1(t)\right]$$

In rotating frame, we have

$$\left(\frac{d\vec{m}}{dt}\right)_{rel} = \frac{d\vec{m}}{dt} - \omega\hat{e}_z \times \vec{m}(t)$$

Let $\Delta\omega = \omega - \omega_0$

Then, $\left(\frac{d\vec{m}}{dt}\right)_{rel} = \vec{m}(t) \times [\Delta\omega\;\hat{e}_Z - \omega_1\hat{e}_x]$ and

$$\vec{B}_{eff} = \frac{1}{\gamma}[\Delta\omega\, \hat{e}_z - \omega_1 \hat{e}_x]$$

Resonance arises when $\Delta\omega = 0$. At resonance the magnetic moment can be completely flipped.

***Quantum Mechanics treatment***

$$|\Psi(t) >= a_T(t)|+> + a_-(t)|->$$

$$H(t) = -\vec{M}\cdot\vec{B}(t) = -\gamma\vec{S}\cdot\left[\vec{B}_0 + \vec{B}_1(t)\right]$$

$$= \omega_0 S_z + \omega_1\left[\cos\omega t\, S_x + \sin\omega t\, S_y\right]$$

$$H = \frac{\hbar}{2}\begin{pmatrix} \omega_0 & \omega_1 e^{-i\omega t} \\ \omega_1 e^{i\omega t} & -\omega_0 \end{pmatrix}$$

In the $\{|+>, |->\}$ basis. The Schrodinger equation then yields:

$$i\frac{d}{dt}a_+(t) = \frac{\omega_0}{2}a_+(t) + \frac{\omega_1}{2}e^{-i\omega t}a_-(t)$$

$$i\frac{d}{dt}a_-(t) = \frac{\omega_1}{2}e^{i\omega t}a_+(t) - \frac{\omega_0}{2}a_-(t)$$

Let, $b_+(t) = e^{-i\omega t/2}\ \ a_+(t),\ \ b_-(t) = e^{-i\omega t/2}\ \ a_-(t)$ (changing to rotating frame), then:

$$i\frac{d}{dt}b_+(t) = -\frac{\Delta\omega}{2}b_+(t) + \frac{\omega_1}{2}b_-(t)$$

$$i\frac{d}{dt}b_-(t) = \frac{\omega_1}{2}b_+(t) + \frac{\Delta\omega}{2}b_-(t)$$

Thus

$$i\hbar\frac{d}{dt}|\widetilde{\Psi}(t) > = \widetilde{H}|\widetilde{\Psi}(t) > \quad where\ \widetilde{\Psi}(t) = b_+(t)|+> + b_-(t)|->$$

and

$$\widetilde{H} = \frac{\hbar}{2}\begin{pmatrix} -\Delta\omega & \omega_1 \\ \omega_1 & \Delta\omega \end{pmatrix}$$

Now, consider $|\Psi(0) = |+>$ , the relations on the b yields $|\widetilde{\Psi}(0) = |+>$ . So, for $P_{+-}(t)$ we get:

$$P_{+-}(t) = |<-|\Psi(t)>|^2 = |a_-(t)|^2 = |b_-(t)|^2 = |<-|\widetilde{\Psi}(t)>|^2$$

Using the earlier Rabi formula and substituting:

$$|\varphi_1 > \to |+> \qquad E_1 \to -\frac{\hbar}{2}\Delta\omega \qquad W_{12} = \frac{\hbar}{2}\omega_1$$

$$|\varphi_2 > \to |+> \qquad E_2 \to -\frac{\hbar}{2}\Delta\omega$$

We get,

$$P_{+-}(t) = \frac{\omega^2}{\omega_1^2 + (\Delta\omega)^2}\ \ sin^2\left[\sqrt{\omega_1^2 + (\Delta\omega)^2}\,\frac{t}{2}\right]$$

When $|\Delta\omega| \gg |w_1|$ $\quad P_{\pm-}$ *remains almost zero*
When $|\Delta\omega| \approx 0$ $\quad P_{+-}$ *becomes large*
When $|\Delta\omega| = 0$ $\quad P_{+-} = 1$ *at times* $\frac{(2n+t)\pi}{\omega_1}$

**Effect of coupling between a stable and an unstable state**
Consider

$$H = \begin{pmatrix} \left(E_1 - i\frac{\hbar}{2}\gamma_1\right) & w_{12} \\ w_{21} & E_2 \end{pmatrix}$$

in the presence of coupling there is no longer a stable state.

***Weak coupling, Eigenvalues***

$$\varepsilon_1 \cong E_1 - i\frac{\hbar}{2}\gamma_1 + \frac{|w_{12}|^2}{E_1 - E_2 - i\frac{\hbar}{2}\gamma_1}$$

$$\varepsilon_2 \cong E_2 + \frac{|w_{12}|^2}{E_1 - E_2 + i\frac{\hbar}{2}\gamma_1}$$

Where the imaginary component on the latter gives instability.

Write $\varepsilon_2 = \Delta_2 - i\frac{\hbar}{2}\Gamma_2$

$$\Gamma_2 = \gamma_1 \frac{|w_{12}|^2}{(E_1 - E_2)^2 + \frac{\hbar^2}{4}\gamma_1{}^2} \qquad \textit{Bethe's formula}$$

When $E_1 = E_2$ above and $E_1$ is unstable, starting in state (2), the probability of finding the system in state (1):

$$P_{21}(t) = \frac{|w_{12}|^2}{|w_{12}|^2 - \frac{\hbar}{4}\gamma_1} e^{-\gamma_1 t/2} sin^2\left(\sqrt{|w_{12}|^2 - \left(\frac{\hbar}{4}\gamma_1\right)^2}\,\frac{t}{\hbar}\right),$$

Note the damping term.

**4.3 The one–dimensional harmonic oscillator**

For the classical harmonic oscillator we have $E = \frac{p^2}{2m} + \frac{1}{2}mw^2x^2$. For the oscillator in quantum mechanics, the classical quantities x and p are replaced by the operators x and p which satisfy:

$$[X, P] = i\hbar$$

And

$$H = \frac{P^2}{2m} + \frac{1}{2}mw^2X^2$$

Since it is time –independent the Q.M study of the harmonic oscillator reduces to the solution of the eigenvalue equation

$$H|\varphi >= E|\varphi >$$

In the $\{|+>\}$ representation:

$$\left[-\frac{\hbar^2}{2m}\frac{d^2}{dx^2} + \frac{1}{2}m\omega^2x^2\right]\varphi(x) = E\varphi(x)$$

Properties that can be deduced from the form of the potential:

(i) The eigenvalues are positive: $H = T + V \ so \ < H > = E =< T > + < V >$ but $< T > \geq 0$ and $< V > \geq V_m$ , so $E > V_m$ (uncertainty principle eliminates equality), and since we have chosen $V_m$ as the zero we then get $E > 0$.

(ii) Eigenfunctions of H have a definite parity since V(x) is even: $V(-x) = V(x)$

(iii) The energy spectrum is discrete.

### 4.3.1 Eigenvalues of the Hamiltonian

Define dimensionless operators for convenience.

$$\hat{X} = \sqrt{\frac{m\omega}{\hbar}}X \ \ and \ \ \hat{P} = \frac{1}{\sqrt{m\hbar\omega}}P$$

Now,

$$[\hat{X}, \hat{P}] = i$$

And $H = \hbar\omega\hat{H} \quad with\ \hat{H} = \frac{1}{2}(\hat{X}^2 + \hat{P}^2)$. Now we want solution to

$$\hat{H}|\varphi_v^i > = \varepsilon_V|\varphi_v^i >$$

where i labels degenerate vectors on v. $\hat{H}$ is dimensionless, as is $\varepsilon_V$. Since $\hat{X} \ and\ \hat{P}$ are operators we can't write $\hat{X}^2 + \hat{P}^2 \neq (\hat{X} - i\hat{P})(\hat{X} + i\hat{P})$. So. introducing the operators $\hat{a} \ \ and\ \hat{a}^+$ :

$$\left.\begin{aligned} a &= \frac{1}{\sqrt{2}}(\hat{X} + i\hat{P}) \\ a^\dagger &= \frac{1}{\sqrt{2}}(\hat{X} - i\hat{P}) \end{aligned}\right\} \rightarrow \qquad \begin{aligned} \hat{X} &= \frac{1}{\sqrt{2}}(a^\dagger + a) \\ \hat{P} &= \frac{1}{\sqrt{2}}(a^\dagger - a) \end{aligned}$$

$$[a, a^\dagger] = \frac{1}{2}[\hat{X} + i\hat{P}, \hat{X} - i\hat{P}] = \frac{1}{2}[\hat{X}, -i\hat{P}] + \frac{1}{2}[i\hat{P}, \hat{X},] = 1$$

$$[a, a^\dagger] = 1$$

Also,

$$a^\dagger a = \frac{1}{2}(\hat{X} - i\hat{P})(\hat{X} + i\hat{P}) = \frac{1}{2}\left(\hat{X}^2 + \hat{P}^2 - 1\right)$$

So,

$$\hat{H} = a, a^\dagger + \frac{1}{2}$$

Introduce operation N:

$$N = a, a^\dagger$$

Thus, eigenvectors of $\hat{H}$ are eigenvectors of N.

$$[N, a] = [a^\dagger a, a] = a^\dagger aa - aa^\dagger a = [a^\dagger, a]a = -a$$

$$[N, a^\dagger] = [a^\dagger a, a^\dagger] = a^\dagger aa^\dagger - a^\dagger a^\dagger a = a^\dagger[a, a^\dagger] = a^\dagger$$

Consider,

$$N|\varphi_j >= v|\varphi_v^i >$$

Then,

$$H|\varphi_j >= \left(v + \frac{1}{2}\right)\hbar\omega|\varphi_v^i >$$

### 4.3.2 Determination of the spectrum

I. The eigenvalues v of N are positive or zero

Proof

$\|a|\varphi_j >\| \geq 0$

$\|a|\varphi_j >\| =< \varphi_j|a^+ a|\varphi_j >=< \varphi_j|N|\varphi_j >= v \geq 0$

II. (i) if $v = 0$ $then\ a|\varphi_{v=0}^i >= 0$

(ii) $if\ v > 0\ then\ a|\varphi_v^i > \neq 0$ and is eigenvector of N with value v-1

Proof (i) $v = 0$ $then\ \|a|\varphi_0^i >\| = 0 \quad \rightarrow a|\varphi_0^i >= 0$

(ii) if $v > 0$ $then\ \|a|\varphi_v^i >\| > 0 \quad \rightarrow a|\varphi_v^i > \neq 0$

$N\left(a|\varphi_v^i >\right) = a^+ aa|\varphi_j >= (-1 + aa^+)a|\varphi_j{}^i >= (-a + aN)|\varphi_j{}^i$

$= (-1 + v)(a|\varphi_j >)$

So, $N\left(a|\varphi_v^i >\right) = (v - 1)(a|\varphi_v^i >)$

III. If $|\varphi_v^i >$ is a non zero eigenvector of N of eigenvalue v, then

(i) $a|\varphi_v^i >$ is always non-zero

(ii) $a^+|\varphi_v^i >$ is an eigenvector of N with eigenvalue v+1

Proof

(i) $\|a^+|\varphi_v^i >\| =< \varphi_v^i|aa^+|\varphi_v^i > = < \varphi_v^i|N + 1|\varphi_v^i >= (v + 1) \geq 1$

(ii) $N(a^{+}|\varphi_v^i>) = a^{+}aa^{+}|\varphi_v^i >= a^{+}(1+a^{+}a)|\varphi_v^i >=$
$(v+1)a^{+}|\varphi_v^i > \neq 0$

IV. The spectrum of N is composed of non-negative integers. If the eigenvalue of N on $|\varphi_v^i >$ is v and isn't integer valued, then we can write $n < v < n+1$ where n is an integer. Since (from II) $a^n|\varphi_v^i >$ is then an eigenvalue of N with strictly positive eigenvalue: v-n, we find that $a^{n+1}|\varphi_v^i >$ is eigenvalue. v-(n+1) which is strictly negative: thus, if V is non-integral, we can therefore construct a tech-zero eigenvector of N with a strictly negative eigenvalue but this contradicts (I), the hypothesis of non-integral V is thus negated.

So, let n<v<n+1 → n=V, what happens?
Rule II reveals that $a^n|\varphi_n^i >$ is non-zero with eigenvalue 0, thus, $a^{n+1}|\varphi_n^i >= a|\varphi_0 >= 0$.
Thus, V can only be a non-zero integer. Since $H|\varphi_v^i >= (v + {}^1/_2)\hbar\omega|\varphi_v^i >$, we get

$$E_n = \left(n + \frac{1}{2}\right)\hbar\omega$$

The energy of the harmonic oscillator is quantized, ground state non-zero: $a$ is the destruction or annihilation operator. $a^{\dagger}$ is the creation operator.

Degeneracy of eigenvalues? Now to show that the energy eigenvalues (levels) are not degenerate for the harmonic oscillator. Consider the eigenstates of $E_0 = \hbar\omega/2$ (eigenstates of N with n=0). Rule II reveals $a|\varphi_0{}^i >= 0$ , how many linearly indep, kets satisfy this relationship, that will reveal the degeneracy:

$$a|\varphi_0{}^i > = \frac{1}{\sqrt{2}}\left[\sqrt{\frac{mw}{\hbar}}X + \frac{i}{\sqrt{m\hbar\omega}}P\right]|\varphi_0{}^i >= 0$$

In the $\{|x\}$ representation:

$$\left(\frac{\hbar}{mw}x + \frac{d}{dx}\right)\varphi_0{}^i(x) = 0 \ where\ \varphi_0{}^i(0) =< x|\varphi_0{}^i >$$

The solution of which is a

$$\varphi_0{}^i(x) = Ce^{-\frac{1}{2}\frac{mw}{\hbar}x^2}$$

C is a const. integration resolved upon normalization. Consequently, there is only one ket $\varphi_0$ which satisfies $a|\varphi_0 >= 0$, and the ground state is not degenerate.

Using recurrence and the fact that the ground state is nondegenerate it is possible to slated that all of the states are non-degenerate: If we can show that $E_n = \left(n+\frac{1}{2}\right)\hbar\omega$ non-degenerate implies $E_{n+1} = (n+1+{}^1/_2)\hbar\omega$ is non-degenerate, then we are done. To begin, assume only one $|\varphi_n>$ satisfies $N|\varphi_n >= n|\varphi_n$ . Then, consider

$$N|\varphi_{n+1} > = (n+1)|\varphi_{n+1}^i >$$

From lemma II: $a|\varphi_{n+1}^i >= c^i|\varphi_n >$ . Now, $a^+a|\varphi_{n+1}^i >= c^i a^+|\varphi_n >$ and

$$|\varphi_{n+1}^i > = \frac{c^i}{n+1} a^+|\varphi_n >$$

thus all kets $|\varphi_{n+1}^i >$ are proportional to $a^+|\varphi_n^i >$ . They are therefore proportional to each other: the eigenvalue (n+1) is not degenerate. Thus, since n=0 is not degenerate, then the above recurrence relation shows that all of the states are nondegenerate.

### 4.3.3 Eigenstates of the Hamiltonian

N and H are observables, i.e. their Eigenvectors constitute a basis in the space $\varepsilon_x$. This could be proved by considering the wave functions associated with the eigenstates of N. Since none of the eigenvalues of N (or H) is degenerate, N alone constitutes a C.S.C.O in $\varepsilon_x$.

***Basis vectors in terms of*** $|\varphi_0>$

$a|\varphi_0 >= 0$ let $|\varphi_0 >$ be normalised (global $e^{i\theta}$ factor remains)
Now,
$|\varphi_1 >= c_1 a^+|\varphi_0 >$ choose $c_1$ by requiring $|\varphi_1 >$ to be normalised and such that the phase of $|\varphi_1 >$ relative to $|\varphi_0 >$ such that $c_1$ is real and positive. Then

$$< \varphi_1|\varphi_1 >= |c_1|^2 < \varphi_0|aa^+|\varphi_0 >= 1 \rightarrow c_1 = 1$$

$$|\varphi_1 > = a^+|\varphi_0 >$$

Using similar conventions for $|\varphi_2 >$:

$$|\varphi_2 >= c_2 a^+|\varphi_1 >$$

$$< \varphi_2|\varphi_2 >= |c_2|^2 < \varphi_1|aa^+|\varphi_1 >= 2|c_2|^2 = 1 \qquad c = \frac{1}{\sqrt{2}}$$

Thus, $|\varphi_2 > = \frac{a^+}{\sqrt{2}}|\varphi_1 >$.

Recursion gives

$$|\varphi_n>=\frac{a^+}{\sqrt{n}}\frac{a^+}{\sqrt{n-1}}\cdots\frac{a^+}{\sqrt{1}}|\varphi_0>=\frac{(a^+)^n}{\sqrt{n!}}|\varphi_0>$$

$$\boxed{|\varphi_n>=\frac{(a^+)^n}{\sqrt{n!}}|\varphi_0>}$$

**Orthonormalization and closure relations**

Since H is Hermitian $|\varphi_n>$ with different n are orthogonal, and since we have normalised we have $<\varphi_n|\varphi_n>=\delta_{nn'}$. Since H is also an observable, which by definition satisfies a closure relation, we have: $\sum_n|\varphi_n><\varphi_n|=1$.

Given the phase conventions chosen for the basis vectors the action of the $\hat{a}$ *and* $\hat{a}^+$ operators on the $\{|\varphi_n>\}$ basis is especially simple (in fact X and P, etc., should be calculated in terms of a and $a^+$ since it is much less work):

Since

$$\boxed{|\varphi_n>=\frac{1}{\sqrt{n}}a^+|\varphi_{n-1}>}$$

given our phase convention shift the index to get.

$$\boxed{a^+|\varphi_n>=\sqrt{n+1}|\varphi_{n+1}>}$$

Also

$$a|\varphi_n>=\frac{1}{\sqrt{n}}aa^+|\varphi_{n-1}>=\frac{n}{\sqrt{n}}|\varphi_{n-1}>=\sqrt{n}|\varphi_{n-1}>$$

$$\boxed{a|\varphi_n>=\sqrt{n}|\varphi_{n-1}>}$$

Now calculate X and P:

$$X|\varphi_n>=\sqrt{\frac{\hbar}{m\omega}}\frac{1}{\sqrt{2}}(a^++a)|\varphi_n>$$

$$=\sqrt{\frac{\hbar}{2m\omega}}\left[\sqrt{n+1}|\varphi_{n+1}>+\sqrt{n}|\varphi_{n-1}\right]$$

$$P|\varphi_n>=\sqrt{\hbar m\omega}\frac{1}{\sqrt{2}}(a^+-a)|\varphi_n>$$

$$=i\sqrt{\frac{m\hbar\omega}{2}}\left[\sqrt{n+1}|\varphi_{n+1}>-\sqrt{n}|\varphi_{n-1}\right]$$

Matrix elements in $\{|\varphi_n>\}$ representation

$< \varphi_{n'}|a|\varphi_n >= \sqrt{n}\delta_{n',n+1}$
$< \varphi_{n'}|a^+|\varphi_n >= \sqrt{n}\delta_{n',n+1}$
Wave functions associated with stationary states:

$$\varphi_n(x) = \ < x|\varphi_n > \ = \frac{1}{\sqrt{n!}} < x|(a^+)^n|\varphi_0 >$$

$$= \frac{1}{\sqrt{n!}}\frac{1}{\sqrt{2^n}}\left[\sqrt{\frac{m\omega}{\hbar}}x - \sqrt{\frac{\hbar}{m\omega}}\frac{d}{dx}\right]^n \varphi_0(x)$$

$$\varphi_n(x) = \left[\frac{1}{2^n n!}\left(\frac{\hbar}{m\omega}\right)^n\right]^{1/2}\left(\frac{m\omega}{\pi\hbar}\right)^{1/4}\left[\frac{m\omega}{\hbar}x - \frac{d}{dx}\right]^n e^{-\frac{1}{2}\frac{m\omega}{\hbar}x^2}$$

$\varphi_n(x)$ is the product of $e^{-\frac{1}{2}\frac{m\omega}{\hbar}x^2}$ and a polynomial of degree n and parity $(-1)^n$ called a Hermite polynomial.

Consider $\Delta x$ *and* $\Delta p$ for the harmonic oscillator:
$< \varphi_n|X|\varphi_n >= 0 \ , \ < \varphi_n|P|\varphi_n >= 0$
$(\Delta X)^2 = < \varphi_n|X^2|\varphi_n > - < \varphi_n|X|\varphi_n >^2 = < \varphi_n|X^2|\varphi_n >$
$(\Delta P)^2 = < \varphi_n|P^2|\varphi_n > - < \varphi_n|P|\varphi_n >^2 = < \varphi_n|P^2|\varphi_n >$
$X^2 = \frac{\hbar}{2m\omega}(a^+ + a)(a^+ + a) = \frac{\hbar}{2m\omega}\left((a^+) + a^+a + aa^+ + a^2\right)$
$P^2 = \frac{m\hbar\omega}{2}(a^+ + a)(a^+ + a) = -\frac{m\hbar\omega}{2}\left((a^+) + a^+a + aa^+ + a^2\right)$

$$\Delta X \Delta P = \hbar\left(n + \frac{1}{2}\right)$$

And $\Delta X \Delta P = \frac{1}{2}\hbar$ for the ground state (recall that the ground state is a Gaussian).

**Time evolution of mean values**
$|\Psi(0) > \ = \sum_{n=0}^{\infty} C_n(0)|\varphi_n >$
$|\Psi(t) > \ = \sum_{n=0}^{\infty} C_n(0)e^{-iE_n t/\hbar}|\varphi_n > \ = \sum_{n=0}^{\infty} C_n(0)e^{-iE_n t/\hbar}\,|\varphi_n >$
$< \Psi(t)|A|\Psi(t) > \ =$
$\sum_{n=0}^{\infty}\sum_{n=0}^{\infty} C_n{}^*(0)C_n(0)A_{mn}e^{-i(m-n)\omega t} \qquad A_{mn} =< \varphi_m|A|\varphi_n >$
Thus,

$$\frac{d}{dt} < X > = \frac{1}{i\hbar} < [X, H] > = \frac{< P >}{m}$$

$$\frac{d}{dt} < P > = \frac{1}{i\hbar} < [P, H] > = -m\omega^2 < X >$$

or

$$< X > (t) = < X > (0)\cos\omega t + \frac{1}{m\omega} < P > (0)\sin\ \omega t$$
$$< P > (t) = < P > (0)\cos\omega t - m\omega < X > (0)\sin\ \omega t$$

**Hermite Polynomials**

$$H_n(Z) = (-1)^n e^{Z^2}\frac{d^n}{dZ^n}e^{Z^2} \rightarrow H_n(Z) = \left(2Z - \frac{d}{dZ}\right)H_{n-1}(Z)$$

Generating function analysis on $F(Z+\lambda) = e^{-(Z+\lambda)^2}$ reveals an alternative definition:

$$H_n(Z) = \left\{\frac{\partial^n}{\partial\lambda^n}e^{-\lambda^2+2\lambda Z}\right\}_{\lambda=0} \quad where \quad \frac{d}{dZ}H_n(Z) = 2nH_{n-1}(Z)$$

Solving the eigenvalue equation of the Harmonic oscillation by the polynomial method:

$$H|\varphi> = E|\varphi> and\ H = \frac{P^2}{2m} + \frac{1}{2}m\omega^2X^2$$

In the $\{|x>\}$ rep:

$$\left[-\frac{\hbar^2}{2m}\frac{d^2}{dx^2} + \frac{1}{2}m\omega^2x^2\right]\varphi(x) = E\varphi(x)$$

Use dimensionless form $\hat{X} = \sqrt{\frac{m\omega}{\hbar}}X,\ \hat{P} = \frac{P}{\sqrt{m\omega\hbar}}$, $and \quad \varepsilon = \frac{E}{\hbar\omega}$ to get:

$$\frac{1}{2}\left[-\frac{d^2}{dx^2} - \hat{x}^2\right]\hat{\varphi}(\hat{x}) = \varepsilon\hat{\varphi}(\hat{x})$$
$$\left[\frac{d^2}{d\hat{x}^2} - (\hat{x}^2 - 2\varepsilon)\right]\hat{\varphi}(\hat{x}) = 0$$

Asymptotic analysis for large $\hat{x}$ :

$$\left[\frac{d^2}{d\hat{x}^2} - \hat{x}^2\right]\hat{\varphi}(\hat{x}) \cong 0 \rightarrow \hat{\varphi}(\hat{x}) \cong e^{\pm\hat{x}^2/2}$$

So, try $\hat{\varphi}(\hat{x}) = e^{-\hat{x}^2/2}h(\hat{x})$ substitute:

$$\frac{d^2}{d\hat{x}^2}h(\hat{x})\frac{d^2}{d\hat{x}^2}h(\hat{x}) + (2\varepsilon - 1)h(\hat{x}) = 0$$

Now do a series expansion. Thus, when $\hat{x} \rightarrow \infty$:

$$|h(\hat{x})| \underset{\hat{x}\rightarrow\infty}{\geq} \left|\frac{a_{2m}}{b_{2m}}\right|\hat{x}^p e^{\lambda\hat{x}^2}$$

Thus,

$$|\varphi(\hat{x})| \underset{\hat{x}\rightarrow\infty}{\gtrapprox} \left|\frac{a_{2m}}{b_{2m}}\right|\hat{x}^p e^{(\lambda-1/2)\hat{x}^2}$$

Since we can choose $0 < \lambda < 1$ and $\frac{1}{2} < \lambda < 1$ gives an unbounded $|\hat{\varphi}(\hat{x})|$ (which is physically unacceptable), there is only one possibility left for a solution and that is that the series terminate. Suppose the numerator of our recursion relation goes to zero when m=m₀ :

$a_{2m} \neq 0 \quad if \quad m \leq m_0$
$a_{2m} = 0 \quad if \quad m > m_0$
Thus,

$$4m_0 + 2p - 2\varepsilon + 1 = 0$$

$let\ 2m_0 + p =$
$n \quad an\ integer\ since\ m_0\ and\ p\ are\ an\ arbitrary\ positive$
Let $2m_0 + p = n$ be an integer. And

$$\varepsilon = \varepsilon_n = n + \frac{1}{2}$$

Where n is an arbitrary positive integer or zero. Then,

$$E_n = \left(n + \frac{1}{2}\right)\hbar\omega$$

as before. For arbitrary n:

$$\left[\frac{d^2}{d\hat{x}^2} - 2\hat{x}\frac{d}{dx} + 2n\right]h(\hat{x}) = 0$$

Which is the equation satisfied by the Hermite polynomial.

***For the isotropic oscillator we have***

$$E_n = (n + {}^3/_2)\hbar\omega$$

where $n = n_x + n_y + n_z$. H alone no longer forms a C.S.C.O. since the $E_n$ are degenerate.

Determining the degree of degeneracy.
Choose $n_x = 0 \ldots n$
$n_y + n_z = n - n_x \qquad (n - n_x + 1)\ \ possibilities$
$g_n = \sum_{n_x=0}^{n}(n - n_x + 1) = (n + 1)(n) - \sum_{n_x=0}^{n} n_x = (n + 1)h - \frac{1}{2}(n + 1)n$
$g_n = \frac{(n+1)(n+2)}{2}$

**4.3.4 Charged Harmonic Oscillator in a uniform Electric field**
Classical potential energy of a particle in a uniform field:

$$\omega(\varepsilon) = -q\varepsilon x \quad where \quad F = q\varepsilon \rightarrow w = -q\varepsilon x$$

(zero of potential energy placed at x=0).

So, $w(\varepsilon) = -q\varepsilon X$ going to quantum mechanics, thus

$$H' = \frac{P^2}{2m} + \frac{1}{2}m\omega^2 X - q\varepsilon X$$

Eigenvalue equation of $H'(\varepsilon)$ *in the* $\{|x>\}$ representation:

$$\left[-\frac{\hbar^2}{2m}\frac{d^2}{dx^2} + \frac{1}{2}m\omega^2 x^2 - q\varepsilon x\right]\varphi(x) = E\varphi(x)$$

Thus, only need to complete the square as follows to return to the form of a simple harmonic oscillator. This has a clear classical analogue in the shift of the oscillator point, but having the same frequency (energy shifts as well).

Thus,

$$\left[\frac{-\hbar^2}{2m}\frac{d^2}{dx^2} + \frac{1}{2}mw^2\left(x - \frac{q\varepsilon}{2m\omega^2}\right) - \frac{q^2\varepsilon^2}{2m\omega^2}\right]\varphi(x) = E\varphi(x)$$

$$\left[\frac{-\hbar^2}{2m}\frac{d^2}{du^2} + \frac{1}{2}m\omega^2 u^2\right]\varphi'(u) = E'\varphi(u)$$

where

$E' = E + \frac{q^2\varepsilon^2}{2m\omega^2}$

$E'_n = \left(n + \frac{1}{2}\right)\hbar\omega$

$E = \left(n + \frac{1}{2}\right)\hbar\omega - \frac{q^2\varepsilon^2}{2m\omega^2}$ (the entire spectrum is a shifted by the quantity $\frac{q^2\varepsilon^2}{2m\omega^2}$).

$\varphi'_n(x) = \varphi_n(x - \frac{q\varepsilon}{m\omega^2})$.

### 4.3.5 Coherent states of the harmonic oscillator

In constructing quantum mechanical states leading to physical predictions which are almost identical to the classical ones, we arrive at "quasi-classical states" known as coherent states. We are looking for a state vector where <X>, <P> and <H> are as close as possible to the corresponding classical values but, of course, keeping the deviations $\Delta X, \Delta P, and\ \Delta H$ negligible in the microscopic limit. A classical phase space analysis was the origin of the creation and annihilation operators for the harmonic oscillator:

Harmonic Oscillator: $\frac{d}{dt}x(t) = \frac{1}{m}p(t), \frac{d}{dt}p(t) = -m\omega^2 x(t)$

Dimensionless: $\frac{d}{dt}\hat{x} = \omega\hat{p}\ ,\ \frac{d}{dt}\hat{p} = -\omega\hat{x}$

Define: $\alpha(t) = 1/\sqrt{2}[\hat{x}(t) + i\hat{p}(t)]$

Then we have:

$$\frac{d}{dt}\alpha(t) = -i\omega\alpha(t) \quad \rightarrow \quad \alpha(t) = \alpha_0 e^{-i\omega t}$$

$$\hat{x} = \frac{1}{\sqrt{2}}\left[\alpha_0 e^{-i\omega t} + \alpha_0{}^* e^{i\omega t}\right], \quad \hat{p} = -\frac{i}{\sqrt{2}}\left[\alpha_0 e^{-i\omega t} - \alpha_0{}^* e^{i\omega t}\right]$$

$$H = \frac{1}{2m}p(0)^2 + \frac{1}{2}x(0)^2 = \frac{\hbar\omega}{2}\{[\hat{x}(0)]^2 + [\hat{p}(0)]^2\} = \hbar\omega|\alpha_0|^2$$

For a macroscopic oscillation $|\alpha_0|^2 \gg 1 \; and \; H \gg \hbar\omega.$

Consider $\hat{X} = \frac{1}{\sqrt{2}}(a + a^+)$ , $\hat{P} = -\frac{1}{\sqrt{2}}(a - a^+)$ , $H = \hbar\omega\left(a^+a + \frac{1}{2}\right)$.

Now,

$$i\hbar\frac{d}{dt} < a > (t) = \omega < a > (t) \rightarrow < a > (t) = < a > (0)e^{-i\omega t}$$

Similarly, (just take complex conjugate) $< a^+ > (t) = < a >^* (0)e^{i\omega t}$ .

Thus,

$< \hat{x} > (t) = \frac{1}{\sqrt{2}}\left[< a > (0)e^{-i\omega t} + < a >^* (0)e^{i\omega t}\right]$

$< \hat{P} > (t) = \frac{-i}{\sqrt{2}}\left[< a > (0)e^{-i\omega t} - < a >^* (0)e^{i\omega t}\right]$

In order to have $< \hat{x} > (t) = \hat{x}(H,) \; and < \hat{P} > (H) = \hat{P}(t) \rightarrow$

$\boxed{< a > (0) = \alpha_0}$

Thus,

$$< \Psi(0)|a|\Psi(0) > = \alpha_0$$

The condition from $< H > = H$ is: $< H > = \hbar\omega < a^+a > (0) + \frac{\hbar\omega}{2}$ (neglect last term since $|\alpha_0| \gg 1$ ). Thus, $< a^+a > (0) = |\alpha_0|^2$ and

$$< \Psi(0)|a^+a|\Psi(0) > = |\alpha_0|^2$$

Consider $b(\alpha_0) = a - \alpha_0$ :

$b^+(\alpha_0)b(\alpha_0) = a^+a - \alpha_0 a^* - \alpha_0^* a + \alpha_0^*\alpha_0$

$\rightarrow b(\alpha_0)|\Psi(0) >= 0$

$$\boxed{a|\Psi(0) > \; = \alpha_0|\Psi(0) >}$$

for coherent state, and

$$a|\alpha > \; = \alpha|\alpha >$$

Properties of $|\alpha >$ states: Expand on energy eigenstates: $|\alpha >= \sum_n C_n(\alpha)|\varphi_n >$

$a|\alpha >= \sum_n C_n(\alpha)\sqrt{n}|\varphi_{n-1} >$

$\alpha|\alpha >= \sum_n C_n(\alpha)\sqrt{n}|\varphi_{n-1} >$

$\rightarrow C_{n+1}(\alpha) = \frac{\alpha^n}{\sqrt{n+1}} C_n(\alpha)$

Form recurrence

$$C_n(\alpha) = \frac{\alpha^n}{\sqrt{n!}} C_0(\alpha)$$

thus, choose $C_0$ real and pos. and normalize, this determines $|\alpha >$ completely.

$$\sum_n |C_n(\alpha)|^2 = \sum_n \frac{|\alpha|^{2n}}{n!} C_0^2 = 1 = (c_0|\alpha|)^2 e^{|\alpha|^2}$$

So, $c_0(\alpha) = e^{-|\alpha|^2/2}$.

Energy in the $|\varphi >$ state.

$$E_n = (n + {}^1\!/_2)\hbar\omega \ \ and\ P_n(\alpha) = |C_n(\alpha)|^2 = \frac{|\alpha|^{2n}}{n!} e^{-|\alpha|^2}$$

The probability distribution is a Poisson distribution. Stationary analysis reveals that since

$$P_n(\alpha) = \frac{|\alpha|^2}{n} P_{n-1}(\alpha)$$

$P_n(\alpha)$ reaches its maximum value when $n = integral\ part\ of\ |\alpha|^2$ :

$< H >_\alpha = \sum_n P_n(\alpha) \left[n + \frac{1}{2}\right] \hbar\omega \ldots$

$< H > = \hbar\omega < \alpha| \left[a^+ a + \frac{1}{2}\right] |\alpha > = \hbar\omega \left[|\alpha|^2 + \frac{1}{2}\right] \cong \hbar\omega|\alpha|^2$

$< H^2 > = \hbar^2\omega^2 \left[3|\alpha|^2 + \frac{1}{4}\right]$

$\Delta H_\alpha = \sqrt{< H^2 > - < H >^2} = \hbar\omega|\alpha|^2$

$\frac{\Delta H_\alpha}{<\Delta H_\alpha>} \cong \frac{1}{|\alpha|} \ll 1$

$$\left.\begin{array}{r} < X >_\alpha = < \alpha|H|\alpha > = \sqrt{\frac{2\hbar}{m\omega}} Re(\alpha) \\ < P >_\alpha = < \alpha|P|\alpha > = \sqrt{2m\hbar\omega}\, Im(\alpha) \\ < X^2 >_\alpha = \frac{\hbar}{2m\omega}[(\alpha + \alpha^x)^2 + 1] \\ < P^2 >_\alpha = \frac{m\hbar\omega}{2}[1 - (\alpha - \alpha^x)^x] \end{array}\right\} \rightarrow \begin{array}{l} \Delta X_\alpha = \sqrt{\frac{\hbar}{2m\omega}} \\ \Delta P_\alpha = \sqrt{\frac{m\hbar\omega}{2}} \end{array}$$

Thus,

$$\boxed{\Delta X_\alpha \Delta P_\alpha = \hbar/2}$$

for any $|\alpha >$ state we obtain a Gaussian wave packet. The $|\alpha >$ states aren't orthogonal but do satisfy a closure relation.

### 4.4 Scattering, The Born Approximation (1926), and WKB (1926)

In 1926 Born solved a scattering problem by approximating, in a perturbation analysis, the field around the scatterer by the incident field alone [76]. This will be described in Section 4.4.1, followed by a more detailed analysis of perturbation theory (Section 4.4.2), and of the WKB (Wentzel–Kramers–Brillouin) [77] approximation method (Section 4.4.3) for finding approximate solutions to linear differential equations, with specific application to what's known as the semiclassical analysis. A general derivation of scattering theory is then given in Section 4.4.5. Before moving on, however, let's first review Picture and Representations as well as the dipole-dipole interaction in preparation for use with the perturbation example in Section 4.4.2.

***Pictures and Representations***

In Quantum Mechanics the state of a system at any time is described by a unit vector in a Hilbert space, in which sets of axes can be defined by the eigenvectors of complete sets of observables of the system. Any change with time in the state of the system can be investigated by keeping the axes fixed and allowing the state vector to rotate, or by keeping the state vector fixed and allowing the axes to rotate, or by permitting simultaneous rotation of the state vector and of the axes, using in each case the appropriate equations of motion of the vectors concerned. The three possibilities described above are called the Schrodinger, Heisenberg, and Interaction pictures, respectively.

***Equations of motion***

Schrodinger picture $\begin{cases} i\hbar\frac{\partial}{\partial t}|\Psi(t)> \; = H|\Psi(t)> \\ |\Psi(t)> \;\; = U(t,t_0)|\Psi(t_0)> \end{cases}$

Heisenberg picture $\begin{cases} \frac{\partial|\Psi_H>}{\partial t} = 0\,, i\hbar\frac{dA_H}{dt} = [A_H, H_H] + i\hbar U^+\frac{dA_H}{dt}U \\ |\Psi_H> = U^+|\Psi(t)>, \;\; A_H(t) = U^+(t,t_0)AU(t,t_0) \end{cases}$

Interaction picture

$$\begin{cases} i\hbar\frac{\partial}{\partial t}|\Psi_I(t)>= H_I'|\Psi_I(t) \\ i\hbar\frac{dA_I}{dt} = [A_I, H_{0I}] + i\hbar U^{(0)\dagger}\frac{dA_I}{dt}U^{(0)} \\ |\Psi_I(t)> \; = U^{(0)\dagger}(t,t_0)|\Psi(t)> \\ A_I(t) = U^{(0)\dagger}(t,t_0)AU^{(0)}(t,t_0) \\ i\hbar\frac{\partial U(t,t_0)}{\partial t} = HU(t,t_0), i\hbar\frac{\partial}{\lambda t}U^{(0)}(t,t_0) = H_0U^{(0)}(t,t_0) \end{cases}$$

$where\ for\ H_I, H_0\ time\ indep. U(t,t_0) = e^{-\frac{i}{\hbar}H(t,t_0)}$ and $U^{(0)}(t,t_0) = e^{-\frac{i}{\hbar}H_0(t,t_0)}$.

***Parity operator P:*** $\boldsymbol{PF(r,\theta,\varphi) = F(r,\pi-\theta,\varphi+\pi)}$

Consider that $[P,\vec{L}] = 0 \rightarrow$ spherical harmonic have well-defined party, which depends only on l.

We also have that

$$(\vec{\sigma}\cdot\vec{A})(\vec{\sigma}\cdot\vec{B}) = \vec{A}\cdot\vec{B} + i\sigma(\vec{A}x\vec{B})$$

The classical interaction between two dipoles can therefore be written:

$$V = V(r)\left[\frac{3(\vec{\sigma}_1\cdot\vec{r})(\vec{\sigma}_2\cdot\vec{r})}{r^2} - \vec{\sigma}_1\cdot\vec{\sigma}_2\right]$$

(the so called "tensor force"). The interaction energy between nucleons is similar in form:

$$S_{12} = \left[\frac{3(\vec{\sigma}_1\cdot\vec{r})(\vec{\sigma}_2\cdot\vec{r})}{r^2} - \vec{\sigma}_1\cdot\vec{\sigma}_2\right] = 2\left[\frac{3(\vec{S}\cdot\vec{r})}{r^2} - \vec{S}^2\right]$$

where $\vec{S} = \frac{1}{2}(\vec{\sigma}_1\cdot\vec{\sigma}_2)$.

Note:

$$(\vec{\sigma}_1\cdot\vec{\sigma}_2)^n A + B(\vec{\sigma}_1\cdot\vec{\sigma}_2) \leftarrow \begin{cases} (\vec{\sigma}_1\cdot\vec{\sigma}_2)X_{s=1} = X_{s=1} & triplet \\ (\vec{\sigma}_1\cdot\vec{\sigma}_2)X_{s=0} = -3X_{s=0} & singlet \end{cases}$$

Thus,

$$A = \frac{1}{4}[3 + (-3)^n], B = [1 - (-3)^n]$$

Schrodinger equation for an electron in a potential V and a magnetic field $\vec{H} = \nabla \times A\ is: i\hbar\frac{\partial\Psi}{\partial t} = H\Psi$, where:

$$H = \frac{1}{2m}\left(\vec{P} - \frac{e}{c}\vec{A}\right)^2 + V(r) - \left(\frac{e}{mc}\vec{S}\right)\cdot\vec{H}$$

for a homogenous field $\vec{H}$ we can write $\vec{A} = \frac{1}{2}(\vec{H}\times\vec{r})$.

### 4.4.1 Born Approximation

The Born approximation is a perturbation method for scattering problems. Recall the Schrodinger equation for a free particle:

$$-\frac{\hbar^2}{2m}\nabla^2\psi = E\psi.$$

Let's remain in the position representation but make use of the fact that the Energy will have the relation to wavenumber $k$ (momentum representation attribute) according to:

$$E = \frac{\hbar^2 k^2}{2m},$$

thus

$$\nabla^2 \psi = -k^2 \psi,$$

with the solutions denoted by $\psi_0$ for the free case. Now consider the particle not free, due to a potential $V(r)$, but it is a weak potential such that the scattering is weak, such that a perturbative solution is possible in term of a perturbation parameter $\lambda$:

$$\nabla^2 \psi + k^2 \psi = \lambda \frac{2m}{\hbar^2} V(r) \psi$$

$$\psi = \psi_0 + \lambda \psi_1 + \lambda^2 \psi_2 + \cdots$$

Satisfying the perturbative relation at each order then gives at first order:

$$(\nabla^2 + k^2)\psi_1 = \frac{2m}{\hbar^2} V(r) \psi_0$$

Green's function methods can be used to get the inhomogeneous solutions, starting with:

$$(\nabla^2 + k^2) G(r, r_0) = \delta(r - r_0)$$

and

$$\psi_1(r) = \frac{2m}{\hbar^2} \int V(r_0)\, \psi_0(r_0) G(r, r_0) d^3 r_0$$

Now, consider scattering of a plane wave $\psi_0 = \mathrm{e}^{\mathrm{ikz}}$ ($\lambda = 1$ then indicates that $\psi_1$ approximates the scattered, outgoing, wave). Solving $(\nabla^2 + k^2) G(r) = 0$: $G(r) = Ce^{\pm ikr}/4\pi r$, thus (choosing plus sign for outgoing):

$$G(r, r_0) = -\frac{e^{ik|r - r_0|}}{4\pi |r - r_0|}$$

The total wavefunction can then be written:

$$\psi \cong e^{ik' \cdot r} - \frac{m}{2\pi \hbar^2} e^{ikr} \int V(r_0)\, e^{i(k' - k) r_0} d^3 r_0$$

at first order. The scattering amplitude is thus:

$$f(\theta, \phi) \cong -\frac{m}{2\pi \hbar^2} \int V(r_0)\, e^{i(k' - k) r_0} d^3 r_0$$

The asymptotics for free spherical waves will be useful, so let's examine this using the following notation:

$$\varphi^{(0)}_{k,\ell,m}(\vec{r}) = \sqrt{\frac{2k^2}{\pi}}\, j_\ell(kr) Y_\ell^m(\theta,\phi)$$

with:

$$j_\ell(\rho) = (-1)^\ell \rho^\ell \left(\frac{1}{\rho}\frac{d}{d\rho}\right)^\ell \left(\frac{\sin\rho}{\rho}\right)$$

and

$$j_\ell(\rho) \sim \frac{\rho^\ell}{(2\ell+1)!!} \quad as\ \rho \to 0$$

Thus, $\rho^2 j_\ell^2(\rho)$ is small as long as $\rho < \sqrt{\ell(\ell+1)}$.

$$j_\ell(\rho) \sim \frac{1}{\rho}\sin\left(\rho - \ell\frac{\pi}{2}\right) \quad as\ \rho \to \infty$$

Or

$$u_{k,l} \sim \sin\left(kr - \ell\frac{\pi}{2}\right) \quad as\ \ r \to \infty.$$

***Partial waves in $V(r)$ with standard spherical wave expansion***

$$\varphi_{k,\ell,m}(\vec{r}) = R_{k\ell}(r) Y_\ell^m(\theta,\phi) \qquad R_{k,\ell}(r) = \frac{1}{r} u_{k,l}(r)$$

$$\left[-\frac{\hbar^2}{2\mu}\frac{d^2}{dr^2} + \frac{\ell(\ell+1)\hbar^2}{2\mu r^2} + V(r)\right] u_{k,l}(r) = \frac{\hbar^2 k^2}{2\mu} u_{k,l}(r)$$

With

$$u_{k,l} \sim \sin\left(kr - \ell\frac{\pi}{2} + \delta_\ell\right) \quad as\ \ r \to \infty.$$

***Finite Range Potentials***

$$v(r) = 0 \text{ for } r > r_0$$

Then

$$\sqrt{\ell_M(\ell_M+1)} \simeq k_0 r_0$$

And phase shifts $\delta_\ell$ are appreciable only for $\ell$ less than $\ell_M$ or $\sim\ell_M$. Thus:

$$f_k(\theta) = \frac{1}{k}\sum_{\ell=0}^{\infty} \sqrt{4\pi(2\ell+1)}\, e^{i\delta_\ell} sin\delta_\ell Y_\ell^0(\theta)$$

$$\sigma(\theta) = |f_k(\theta)|^2 = \frac{1}{k^2}\left|\sum_{\ell=0}^{\infty} \sqrt{4\pi(2\ell+1)}\, e^{i\delta_\ell} sin\delta_\ell Y_\ell^0(\theta)\right|^2$$

$$\sigma = \frac{4\pi}{k}\sum_{\ell=0}^{\infty}(2\ell+1)sin^2\delta_\ell$$

***Optical theorem***

$$\sigma_{total} = \frac{4\pi}{k} Im\, f_k(0)$$

***Scattering by a hard sphere***

Consider the scattering of a particle by a central potential such that

$$V(r) = 0 \quad ,r > r_0$$
$$V(r) = \infty \quad ,r < r_0$$

(i.e. scattering by a hard sphere).

Find an expression for the total scattering section in terms of $r_0$

(a) In the limit where the energy of the particle tends to zero.

(b) In the limit when the energy of the particle tends to infinity.

(You may need: $\int x\sin^2 nx\,dx = \frac{x^2}{4} - \frac{x\sin^2 ax}{4a} - \frac{\cos 2ax}{8a^2}$ and $\int \sin^2 nx \quad dx = \frac{x}{2} - \frac{\sin 2ax}{4a}$

**Answer**

(a) $E \to 0 \to kr_0 \ll 1$ $\qquad k = \frac{2mE}{\hbar^2}$

$$\left[-\frac{\hbar^2}{2m}\nabla^2 + V(r)\right]\Psi(r) = E\,\Psi(r)$$

Since we have a central potential consider partial wave analysis:

$$\left[-\frac{\hbar^2}{2m}\frac{1}{r}\frac{d^2}{dr^2}r - \frac{\hbar^2}{2m}\frac{\vec{L}^2}{(-\hbar^2 r^2)} + V(r)\right]\Psi(r) = E\,\Psi(r)$$

$$\Psi(r) = Y_\ell^m(\theta,\varphi)\varphi(r)$$

$$\left[\frac{d^2}{dr^2} - \frac{\ell(\ell+1)}{r^2} - \frac{2m(V-E)}{\hbar^2}\right]\varphi(r) = 0$$

$$\varphi(0) = 0$$

Asymptotic $\varphi(r) \cong \frac{1}{k}\sin\left(kr - \frac{\pi}{2}\ell - \delta_\ell(k)\right)$

$$\left[\frac{d^2}{dr^2} + k^2 - \frac{\ell(\ell+1)}{r^2}\right]\varphi(r) = 0$$

Largest at $r = r_0$

Impact parameter $b_\ell = \frac{|\ell|}{|P|} = \frac{\hbar\ell(\ell+1)}{k\hbar}$

So, $kr_0 \approx \sqrt{\ell_m(\ell_m+1)}$ *when* $kr_0 \ll 1 \to$ *only* $\ell = 0$

Thus, for $kr_0 \ll 1$ we just consider s-wave scattering, now to match at the boundary.

$$\left[\frac{d^2}{dr^2} + k^2\right]\varphi(r) = 0 \qquad \varphi(r) = 0$$

$r = r_0$
$\varphi(r) = A\sin(K[r - r_0]) \approx \sin(kr - kr_0)$
So, $\delta_0(k) = -kr_0$

$$\sigma(k) = \frac{4\pi}{k^2}\sum_{\ell=0}^{\infty}(2\ell + 1)\sin\delta_\ell(k)$$

$\sigma(k) = \frac{4\pi}{k^2}\sin(kr_0) \approx 4\pi r_0^2$

(b) Here $kr_0 \gg 1$ we know the free asymptotic form is $\varphi(r) \approx \frac{1}{k}\sin\left(kr - \frac{\pi}{2}\ell + \delta_\ell\right)$
Since $r_0$ is, thus, already in the asymptotic region we immediately have

$$kr_0 - \frac{\pi}{2}\ell + \delta_\ell = 0$$

So, $\delta_\ell = -kr_0 + \frac{\pi}{2}\ell$

$$\boxed{\delta_\ell(k) = -kr_0 + \frac{\pi}{2}\ell}$$

Thus, $\sigma(k) = \frac{4\pi}{k^2}\sum_{\ell=0}^{\infty}(2\ell + 1)\sin\frac{\pi}{2}\ell - kr_0$
Can't use the form directly!
$\sin(A + B) = \sin A\cos B + \sin B\cos A$

$$\sin\left(\frac{\pi}{2}\ell\right) = (-1)^{(\ell-1)/2}$$

$\delta_\ell = -kr_0 + \frac{\pi}{2}\ell$ where $kr_0 \gg \sqrt{\ell_m(\ell_m + 1)} \approx \ell_m$

$$\text{So, } \sin\delta_\ell = \begin{cases} \sin\left(\frac{\pi}{2}\ell\right) & \ell \gg kr_0 \gg 1 \\ \sin(-kr_0) & kr_0 \gg \ell \end{cases}$$

$$\sigma(k) \cong \frac{4\pi}{k^2}\int_0^{kr_0}(2\ell + 1)\sin^2\left(\frac{\pi}{2}\ell\right)dl$$

$$\sigma(k) \cong \frac{4\pi}{k^2}\left\{2\left[\frac{(kr_0)^2}{4} - \sigma(k)\right]\right\}$$

$\sigma(k) \cong 2\pi r_0^2$

**Coulomb Scattering**

The scattering amplitude for Coulomb scattering of distinguishable particles of charge e and momentum ђk in the center-of-mass system is, up to a constant factor:

$$f(\theta) = -\frac{\gamma}{2k\sin^2\frac{\theta}{2}}\exp\left(-i\gamma\ln\sin^2\frac{\theta}{2}\right)$$

where

$$\gamma = \frac{e_k^2\mu}{\hbar^2 k}$$

and $\mu$ is the reduced mass.

(i) Compute the differential cross section for the electron-electron scattering, both for the singlet and triplet spin states.

(ii) Show that the differential cross-section for unpolarized electron-electron scattering is given by:

$$\frac{d\sigma}{d\Omega} = \left(\frac{e^2}{4E}\right)^2\left[\frac{1}{\sin^4\frac{\theta}{2}} + \frac{1}{\cos^4\frac{\theta}{2}} - \frac{\cos\left(\gamma\ln\tan^2\frac{\theta}{2}\right)}{\sin^2\frac{\theta}{2}\cos^2\frac{\theta}{2}}\right]$$

**Answer**

$f(\theta) = \frac{-\gamma}{2k\sin^2\frac{\theta}{2}}\exp\left(-i\gamma\sin^2\frac{\theta}{2}\right)$ for distinguishable

Singlet state is of e-e scattering is undistinguishable and odd under uneven

Singlet $|00> = \frac{1}{\sqrt{2}}(1{+}{\rightarrow} -1 - +>)$

$$\text{Triplet}\begin{cases} |11> = |++> \\ |1-1> = |--> \\ |10> = \frac{1}{\sqrt{2}}\{|+-> + |-+>\} \end{cases}$$

(i)

$f_s = \frac{1}{2}[f(\theta) + f(\pi-\theta)]$ $\frac{d\sigma_0}{d\Omega} = |f_s|^2$

$f_T = \frac{1}{2}[f(\theta) - f(\pi-\theta)]$ $\frac{d\sigma_T}{d\Omega} = |f_\tau|^2$

(ii) unpolarised – statistically weighted sum.

$\frac{d\sigma}{d\Omega} = \frac{1}{4}|f_s|^2 + \frac{3}{4}|f_\tau|^2$

$\sin\left(\frac{\pi-\theta}{2}\right) = \cos\left(\frac{\theta}{2}\right)$

$= \frac{1}{16}\{f(\theta) - f(\pi-\theta)^2 + 3|f(\theta) + f(\pi-\theta)|^2\}$

$$= \frac{1}{16}\left(\frac{\gamma}{\partial k}\right)^2\left\{\left|\frac{-\exp\left(-i\gamma\ln\sin^2\frac{\theta}{2}\right)}{\sin^2\frac{\theta}{2}} + \frac{-\exp\left(-i\gamma\ln\cos^2\frac{\theta}{2}\right)}{\cos^2\frac{\theta}{2}}\right|^2 + \right.$$

$$\left. 3\left|\frac{-\exp\left(-i\gamma\ln\sin^2\frac{\theta}{2}\right)}{\sin^2\frac{\theta}{2}} + \frac{-\exp\left(-i\gamma\ln\cos^2\frac{\theta}{2}\right)}{\cos^2\frac{\theta}{2}}\right|^2\right\}$$

$$= \frac{1}{16}\left(\frac{\gamma}{\partial k}\right)^2 \left\{\frac{1}{\sin^4\frac{\theta}{2}} + \frac{1}{\cos^4\frac{\theta}{2}} + \frac{\exp\left(-i\gamma \ln \sin^2\frac{\theta}{2} + -i\gamma \ln \cos^2\frac{\theta}{2}\right)}{\sin^2\frac{\theta}{2}\cos^2\frac{\theta}{2}} + \right.$$

$$\left. \frac{\exp\left(-i\gamma \ln \sin^2\frac{\theta}{2} + -i\gamma \ln \cos^2\frac{\theta}{2}\right)}{\sin^2\frac{\theta}{2}\cos^2\frac{\theta}{2}} + 3|\ \ |^2\right\}$$

$$= \frac{1}{16}\left(\frac{\gamma}{\partial k}\right)^2 \left\{\frac{1}{\sin^2\frac{\theta}{2}} + \frac{1}{\cos^4\frac{\theta}{2}} + \frac{2\cos\left\{i\gamma \ln \sin^2\frac{\theta}{2} + -i\gamma \ln \cos^2\frac{\theta}{2}\right\}}{\sin^2\frac{\theta}{2}\cos^2\frac{\theta}{2}} + 3|\ldots|^2\right\}$$

$$\left|\frac{\exp\left(-i\gamma \ln \sin^2\frac{\theta}{2}\right)}{\sin^2\frac{\theta}{2}} - \frac{\left(-i\gamma \ln \cos^2\frac{\theta}{2}\right)}{\cos^2\frac{\theta}{2}}\right|^2$$

$$\left(\frac{1}{\sin^2\frac{\theta}{2}} + \frac{1}{\cos^4\frac{\theta}{2}} - \frac{2\cos\left\{i\gamma \ln \cos^2\frac{\theta}{2} - i\gamma \ln \sin^2\frac{\theta}{2}\right\}}{\sin^2\frac{\theta}{2}\cos^2\frac{\theta}{2}}\right)$$

(ii) $\frac{d\sigma}{d\Omega} = \frac{1}{16}\left(\frac{\gamma}{\partial k}\right)^2 \left\{\frac{1}{\sin^2\frac{\theta}{2}} + \frac{1}{\cos^4\frac{\theta}{2}} + \frac{+2-2(3)}{\sin^2\frac{\theta}{2}\cos^2\frac{\theta}{2}}\cos\left(\gamma \ln \tan^2\frac{\theta}{2}\right)\right\}$

$$= \frac{1}{4}\frac{\gamma^2}{k^2}\left\{\frac{1}{\sin^2\frac{\theta}{2}} + \frac{1}{\cos^4\frac{\theta}{2}} - \frac{\cos\left(\gamma \ln \tan^2\frac{\theta}{2}\right)}{\sin^2\frac{\theta}{2}\cos^2\frac{\theta}{2}}\right\}$$

$$\frac{\gamma^2}{k^2} = 4(e^2)^2\left(\frac{M}{k^2 k}\right)\frac{1}{k^2} = \frac{(e^2)^2}{E^2}$$

$$E^2 = \left(\frac{\hbar^2 k^2}{2\mu}\right)$$

$E_1$

$g_{3/2}\left(\frac{3}{2}\hbar\right)\omega_2 \qquad \omega_L = -\frac{qB}{2m}$

Dipole transition $\Delta\ell = \pm 1$
$\Delta m = 0, \pm 1$

$$g_{3/2} = 1 + \frac{\frac{3}{2}\left(\frac{5}{2}\right) + \frac{1}{2}\left(\frac{3}{2}\right) - 2}{2\left(\frac{3}{2}\left(\frac{3}{2}+1\right)\right)}$$

$$g_{3/2} = \frac{4}{3}$$

$$g_{1/2} = 1 + \frac{\frac{1}{2}\left(\frac{3}{2}\right) + \frac{1}{2}\left(\frac{3}{2}\right)}{2\left(\frac{1}{2}\left(\frac{3}{2}\right)\right)} = 2$$

$M = 3/2 \rightarrow \quad M = 1/2$

$$E = E_0 + \left(g_{3/2}\left(\frac{3}{2}\hbar\right)\omega_L - g_{1/2}\left(\frac{1}{2}\hbar\right)\omega\right)$$

$M_B = \frac{q\hbar}{2m_e} \qquad\qquad \omega_L = \frac{\mu_B B}{\hbar}$

$E = hV_0 + (\quad) = \frac{hc}{\lambda} + (\quad)$

$\lambda V = C \qquad V = \frac{c}{\lambda}$

$\lambda = \frac{c}{V}$

$= \frac{hc}{\lambda} + 2h\left(\frac{\mu_B B}{h}\right) - h\left(\frac{\mu_B B}{h}\right) = \frac{hc}{\lambda} + hcB\left(\frac{\mu_B}{hc}\right)$

$= \frac{hc}{5890x10^{-8}cm} + hc(.467cm^{-1}) = hc.cm^{-1}\left(\frac{.467+(10^{-8})}{.467(5890x10^{-8}cm)}\right)$

2p state with m=1 – circularly polarised in Z direction linearly polarised normal to field

M=0: linearly polarised

M=-1: reverse of phase of m=1

$m = {}^{3}/_{2} \ \ has\ m = 1 \rightarrow$ linearly polarised normal to field when viewed such

$m = {}^{1}/_{2} \ \ has\ m = 0\ \ or\ m = 1 \rightarrow$ linearly polarised at angle (real coeff. Between terms in expansion)

### 4.4.2 Stationary Perturbation Theory

***Stationary state perturbation theory***

We will have two cases to consider:

(1) The spectrum of $H_0$ is discrete and non-degenerate:

$E_n = E_n^{(0)} + H'_{nn} + \sum_{m\neq n} \frac{|H'_{mn}|^2}{E_n^{(0)} - E_m^{(0)}} \qquad |H'_{mn}| \ll |\,E_n^{(0)} - E_m^{(0)}|$

$$\Psi_n = \varphi_n + \sum_{m\neq n} \frac{|H'_{mn}|^2}{E_n^{(0)} - E_m^{(0)}} \varphi_m$$

$where\ \varphi_n\ \ and\ E_n^{(0)}$ are for the stationary state of the unperturbed Hamiltonian $H_0$ and , $H'_{mn} \equiv \ < m|H'|n > \ = \int \varphi_m^* H' \varphi_n dV$ .

(2) The spectrum of $H_0$ is discrete and degenerate:

Suppose $E_n^{(0)}$ is f-fold degenerate. The distinct energies which result from the initial level $E_n^{(0)}$ through the introduction of the perturbation are the solutions of the "secular eqn".

$$\begin{vmatrix} H_{11}^{(n)} - E & H_{12}^{(n)} & H_{13}^{(n)} \ldots \\ H_{21}^{(n)} & H_{21}^{(n)} - E & H_{23}^{(n)} \ldots \\ H_{31}^{(n)} & H_{32}^{(n)} & H_{33}^{(n)} - E \ldots \\ & \ldots\ldots\ldots. & \\ & \ldots\ldots\ldots. & \end{vmatrix} = 0$$

Where $H_{lk}^{(n)} = < nl|H'|nk > = \int \varphi_{nl}^* H' \varphi_{nk} dV$ . If E is any solution to the above matrix, then the corresponding wave function in zero-order approximation is

$$\Psi_n = \sum_{k=1}^{f} a_k \, \varphi_{nk}$$

Where the $a_k$ are determined by:

$$\sum_{k=1}^{f} \left( H_{lk}^{(n)} - E\delta_{lk} \right) a_k = 0.$$

***Electric dipole in electric field***

Consider a plane rigid rotator with an electric dipole moment $\vec{d}$ placed in a homogenous electric field $\vec{E}$. Determine the first non-vanishing correction to the energy levels of the rotator.:

Planar so:

$$E = \frac{1}{2} I\omega^2 = \frac{L^2}{2I} \quad and\ since\ planar = \frac{L_z^2}{2I}$$

$$So \quad H^{(0)} = \frac{L_z^2}{2I}$$

Since

$$H^{(0)} \Psi(\vec{r}) = E^{(0)} \Psi(\vec{r})$$

We have:

$$\frac{(m\hbar)^2}{2I} = E_n^{(0)} \qquad \psi_m^{(0)}(\varphi) = \frac{1}{\sqrt{2\pi}} e^{im\varphi} \qquad m = 0, \pm 1, \pm 2$$

where energies are doubly degenerate for $|m| > 0$.

Treating the electric field as a perturbation:

$$H' = H^{(0)} - \vec{E} \cdot \vec{d} = \frac{-\hbar^2}{2I} \frac{d^2}{d\varphi^2} - Edcos\,\theta$$

Since the potential is even we have a parity operator which commutes with both $H^{(0)}$ $and\ H'$, thus there will be no mixing of a state with –m with a +m state; so the perturbation theory for non-degenerate levels can be used:

$$H'_{mm'} = \frac{1}{2\pi} \int_0^{2\pi} e^{-im\varphi} (-Edcos\,\varphi) e^{im'\varphi} d\varphi$$

$$= \frac{-Ed}{2\pi}\int_0^{2\pi} e^{i(m'-m)\varphi}\cos\varphi d\varphi = \begin{Bmatrix} 0\,, & m' \neq m \pm 1 \\ -\frac{Ed}{2}\,, & m' = m \pm 1 \end{Bmatrix}$$

So, $E_m^{(1)} = H'_{mm} = 0$, and

$$E_m^{(2)} = \frac{\left|H'_{m,m-1}\right|^2}{E_m^{(0)} - E_{m-1}^{(0)}} + \frac{\left|H'_{m,m+1}\right|^2}{E_m^{(0)} - E_{m+1}^{(0)}} \left(\frac{Ed}{2}\right)^2 \frac{2I}{\hbar^2}\left(\frac{1}{m^2-(m-1)^2} + \frac{1}{m^2-(m+1)^2}\right)$$

$$= \left(\frac{Ed}{2\hbar}\right)^2 2I\left(\frac{1}{2m-1} + \frac{1}{-2m-1}\right) = \frac{IE^2d^2}{\hbar^2(4m^2-1)}$$

So, $E_m = E_m^{(0)} + E_m^{(1)} + E_m^{(2)} = \frac{\hbar^2 m^2}{2I} + \frac{IE^2d^2}{\hbar^2(4m^2-1)}$.

Now, lets solve the problem with the rotator no-longer restricted to the plane:

$$E = \frac{L^2}{2I}$$

$H^{(0)} = \frac{\vec{L}^2}{2I}$ with eigenstates $Y_l^m(\theta,\varphi)$ and eigenvalues $\frac{1}{2I}(\hbar^2 l(l+1)$ with degeneracy $(2l+1)$. Also have:

$$E_l^{(0)} = \frac{h^2}{2I} l(l+1) \qquad \vec{E}\ field\ along\ z\ axis$$

$$H' = -dE\cos\theta$$

The only non-zero matrix elements are:

$$< l,m|\cos\theta|\, l-1,m > = < l-1|\cos\theta|\, l,m >$$

$$= \left(\frac{l^2-m^2}{4l^2-1}\right)^2 \qquad \begin{matrix} m_1 = m_2 \\ l_1 = l_2 \pm 1 \end{matrix}$$

Since the free Hamiltonian has $H_0 = \frac{L^2}{2I}$, and the perturbed Hamiltonian $H=H_0+H'$, both commute with $L_z$. It follows that a perturbed state |l,m> which contain only unperturbed states |l',m> with the same quantum number m, and hence the problem can be solved by using only the perturbation theory for non-degenerate levels. Since $< l,m|H'|l,m >= 0$, the first-order energy corrections are zero, and at second order:

$$E_{lm}^{(2)} = \frac{2I}{\hbar^2}(Ed)^2 \sum_{l'=l} \frac{|< l,m|\cos\theta|l,m >|^2}{l(l+1) - l'(l'+1)}$$
$$= \frac{(Ed)^2}{E_l^{(0)}} \left\{ \frac{l(l+1) - 3m^2}{2(2l-1)(2l+3)} \right\}$$

Thus

$$E_{lm} \cong E_l^{(0)} \left[ 1 + \left( \frac{Ed}{E_l^{(0)}} \right)^2 \frac{l(l+1) - 3m^2}{2(2l-1)(2l+3)} \right]$$

Note that the initial degeneracy is only partially removed.

***Summary***

$$H(\lambda) = H_o + \lambda\omega$$
$$H(\lambda)|\psi(\lambda) > = E(\lambda)|\psi(\lambda) >$$
$$E(\lambda) = \varepsilon_o + \lambda\varepsilon_1 + \cdots + \lambda^q \varepsilon_q$$
$$|\psi(\lambda) > = |0\rangle + \lambda|1\rangle + \cdots + \lambda^q |q\rangle$$

***1st Order Corrections, nondegenerate***

$$E_n(\lambda) = E_n^0 + \langle\varphi_n|\omega|\varphi_n\rangle + \mathcal{O}(\lambda^2)$$

$$|\psi_n(\lambda)\rangle = |\varphi_n\rangle + \sum_{p\neq n}\sum_i \frac{\langle\varphi_p{}^i|\omega|\varphi_n\rangle}{E_n^0 - E_p^0} |\varphi_p{}^i\rangle + \mathcal{O}(\lambda^2)$$

***2nd Order Corrections in energy***

$$E_n(\lambda) = E_n^0 + \langle\varphi_n|\omega|\varphi_n\rangle + \sum_{p\neq n}\sum_i \frac{\left|\langle\varphi_p{}^i|\omega|\varphi_n\rangle\right|^2}{E_n^0 - E_p^0} + \mathcal{O}(\lambda^3)$$

***Perturbation of a degenerate state***

Project onto $|\varphi_n{}^i\rangle$ to get

$$\langle\varphi_n{}^i|\omega|0\rangle = \varepsilon_1\langle\varphi_n{}^i|0\rangle$$

***Perturbation Example 1***

A particle of mass m is confined to move in a circle (radius a) under the influence of a potential

$$V(\theta) = A\cos 3\theta.$$

(a) Find the Hamiltonian H and corresponding Schrodinger equation.
(b) Find a symmetry operator that commutes with H, and determine its eigenvalues.

Let $\alpha = \frac{ma^2 A}{\hbar^2}$ :

(c) Find the eigenfunctions and eigenvalues of H for $\alpha = 0$.
(d) Find the change to order $\alpha^2$ in the lowest energy state for $\alpha \ll 1$.
(e) Find the approximate eigenvalue and eigenfunction of the lowest energy state for $\alpha \gg 1$.

**Answer**

We have:

$V(\varphi) = A\cos 3\varphi \qquad 0 \le \varphi \le \frac{2\pi}{3}$

Spherical coordinates $r = a, \theta = \frac{\pi}{2}$

$ds^2 = dr^2 + r^2(d\theta^2 + \sin^2\theta d\varphi^2) \quad |g|^{1/2} = r^2\sin\theta$

$\nabla^2 = \frac{1}{|g|^{1/2}}\partial_\mu\left(|g|^{1/2} g^{\mu\nu}\partial_\nu\right) = \frac{1}{r^2}\frac{\partial}{\partial r}\left(r^2\frac{\partial}{\partial r}\right) + \frac{1}{r^2\sin\theta\partial\theta}\left(\sin\theta\frac{\partial}{\partial\theta}\right)$

$for \;\; r = a\,, \theta = \frac{\pi}{2} \qquad \nabla^2 = \frac{1}{a^2}\frac{\partial^2}{\partial\varphi^2}$

So, $H = \frac{P_\varphi^2}{2m} + V(\varphi) = \frac{-\hbar^2}{2m}\frac{1}{a^2}\frac{\partial^2}{\partial\varphi^2} + V(\varphi)$

(a)

$\left[\frac{-\hbar^2}{2ma^2}\frac{\partial^2}{\partial\varphi^2} + V(\varphi)\right]\Psi(\varphi) = E\,\Psi(\varphi)$

(b) since $V\left(\varphi + \left(\frac{2\pi}{3}\right)\right) = V(\varphi)$ we have a symmetry under a rotation by $\left(\frac{2\pi}{3}\right)$

$R_s\left(\frac{2\pi}{3}\right) = e^{-i\frac{L_z}{\hbar}\left(\frac{2\pi}{3}\right)} \qquad\qquad L_z = i\hbar\frac{\partial}{\partial\varphi}$

By construction $[R_z, V(\varphi)] = 0$

Also $[R_z, L_z^2] = \left[e^{-i\alpha L_z}, L_z^2\right] = 0 \quad since \quad [L_z, L_z^2]$

Thus, $R_z$ is our symmetry operator

$L_z = m\hbar \rightarrow R_z\left(\frac{2\pi}{3}\right) = w^{-im\left(\frac{2\pi}{3}\right)}$

Eigenvalues are $\left\{e^{-i\left(\frac{2\pi}{3}\right)}\,, e^{+i\left(\frac{2\pi}{3}\right)}\right\}$

R is unitary

(c) Let $\alpha = \frac{ma^2 A}{\hbar^2}$

$\left\{\frac{\partial^2}{\partial\varphi^2} - \frac{2ma^2 A}{\hbar^2}\cos 3\varphi + \frac{2ma^2 E}{\hbar^2}\right\}\Psi(\varphi) = 0$

$\left\{\frac{\partial^2}{\partial\varphi^2} - 2\alpha\cos 3\varphi + \frac{2ma^2 E}{\hbar^2}\right\}\Psi(\varphi) = 0$

$\beta^2 = \frac{2ma^2E}{\hbar^2}$

$\alpha = 0 \;:\; \left(\frac{\partial^2}{\partial\varphi^2} + \beta^2\right)\Psi(\varphi) = 0$

$\beta(2\pi) = n\pi \rightarrow \beta = \frac{n}{2}$

$\Psi(\varphi) = \left.\begin{matrix} B\sin\varphi \\ or\ B\cos\beta\varphi \end{matrix}\right\} \;;\; E = \frac{\hbar^2\beta^2}{2ma^2}$

$\beta(2\pi) = \left(n + \frac{1}{2}\right)\pi$

$\beta = \frac{1}{2}\left(n + \frac{1}{2}\right)$

$\int_0^{2\pi} |\Psi(\varphi)|^2 d\varphi = \frac{B^2}{\beta}\left(\frac{2\pi}{2}\right) = 1$

$B^2 = \frac{B}{\pi} \rightarrow B = \frac{B}{\pi}$

$\Psi(\varphi) = \sqrt{\frac{n}{2\pi}}\sin\left(\frac{n}{2}\varphi\right) \qquad E_n = \frac{\hbar^2 n^2}{8ma^2} \qquad n = 1,2,\ldots$

$= \sqrt{\frac{n+\frac{1}{3}}{2\pi}}\cos\left(\frac{n}{2}\varphi + \frac{1}{4}\varphi\right) \qquad E_n = \frac{\hbar^2\left(n+\frac{1}{2}\right)^2}{8ma^2} \qquad n = 0,1,\ldots$

(d) $\Psi_0(\varphi) = \frac{1}{\sqrt{4\pi}}\cos\left(\frac{1}{4}\varphi\right) \qquad E_0 = \frac{\hbar^2}{32ma^2}$

$H' = -2\alpha\cos 3\varphi$

$E' = \int_0^{2\pi}\frac{1}{4\pi}\cos^2\left(\frac{1}{4}\varphi\right)(-2\alpha\cos 3\varphi)d\varphi$

$= -\frac{\alpha}{2\pi}\int_0^{2\pi}\cos^2\left(\frac{1}{4\pi}\right)\cos(3\varphi)\,d\varphi$

$= 0$

(d) 2nd order

$$E' = \sum_{n\neq 0}\frac{\left|<\Psi_n^i|H'|\Psi_0>\right|^2}{E_0^{(0)} - E_n^{(0)}}$$

Since $V(\theta)$ has definite parity it will not mix the even and odd solutions:

$$E' = \sum_{\substack{n\neq 0\\ even}}\frac{\left|<\Psi_n^i|-2\cos 3\varphi|\Psi_0>\right|^2}{E_0 - E_n} =$$

$$\frac{\left(\left(\frac{2n+1}{4\pi}\right)\left(\frac{-2\alpha}{4\pi}\right)\right)\left|\int_0^{2\pi}\cos\left(\frac{n}{2}\varphi+\frac{1}{4}\varphi\right)\cos(3\varphi)\cos\left(\frac{1}{4}\varphi\right)d\varphi\right|}{\frac{\hbar^2}{ma^2}\left\{\frac{1}{32}-\frac{\left(n+\frac{1}{2}\right)^2}{8}\right\}}$$

For $n = 6 : \cos\left(3\varphi + \frac{1}{4}\varphi\right)\{\ldots\}$

Recall $\cos(A + B) = \cos A\cos B - \sin A\sin B$

$\cos(A - B) = \cos A\cos B + \sin A\sin B$

$$\cos A \cos B = \frac{1}{2}\{\cos(A+B) + \cos(A-B)\}$$

$$\cos 3\varphi \cos\frac{1}{4}\varphi = \frac{1}{2}\left\{\cos\left(3\varphi + \frac{1}{4}\varphi\right) + \cos\left(3\varphi - \frac{1}{4}\varphi\right)\right\}$$

For n=6 have a term (only one)

$$\int_0^{2\pi} \cos^2\left(\frac{1}{4}\varphi\right) d\varphi = 3\left(\frac{1}{2}\right) + \cos^2\left(\frac{1}{4}\varphi\right) d\varphi = \frac{3}{2} + \frac{1}{8} = \frac{7}{8}$$

$$E' = \left(\frac{3}{4\pi}\right)^2 \left(\frac{2\alpha}{4\pi}\right)^2 \frac{1}{4}\left(\frac{7}{8}\right)^2$$

$$E' = -\alpha^2 \frac{ma^2}{\hbar^2}\left\{\frac{(13)^2(2)^2(7)^2 8}{(4\pi)^4 + (8)^2(42)}\right\}$$

(e)$\alpha \gg 1$

$$\theta = \frac{2\pi}{3}$$

$$V\left(\frac{\pi}{3}\right) = -2\alpha\cos\left(3\left(\frac{\pi}{3}\right) + \varepsilon\right)$$

$$V(0) = -2\alpha$$

$$\approx +2\alpha\cos\varepsilon$$

$$-2\alpha\cos 3\varphi \to 2\alpha\cos 3\varepsilon$$

Lowest energy state is $-2\alpha\left(1 - \frac{3\varepsilon}{2}\right)^2 \approx 2\alpha\cos 3$

$$\left(\frac{\partial^2}{\partial\varepsilon^2} + 2\alpha\cos(3\varepsilon) + \beta^2\right)\Psi(\varepsilon) = 0$$

$$\cos\left(3\left[\frac{\pi}{3} + \varepsilon\right]\right)$$

$$-\cos 3\varepsilon$$

Assuming tunnelling is negligible on the time-scales of interest I'll look only at $\varphi = \frac{\pi}{3}$ minimum:

$\left[\frac{\partial^2}{\partial\varphi^2} - 2\alpha\cos 3\varphi + \beta^2\right]\Psi(\varphi) = 0$ small $\alpha$ negative given small $\varphi$ origin defined on $\varphi$ negative for bound.

$$\left[\frac{\partial^2}{\partial\varphi^2} - (2\alpha + \beta) + \alpha(3\varphi^2)\right]\Psi(\varphi) = 0$$

$$V = 2\alpha\left(1 - \frac{(3\varphi)^2}{2}\right)$$

$$\left(\frac{\partial^2}{\partial\varphi^2} + 9\alpha\varphi^2\right)\Psi(\varphi) = 0$$

$$\Psi(\varphi) = Ne^{-\frac{3\sqrt{\alpha}\varphi^2}{2}}$$

$E > V \qquad \beta = 2\alpha$

$$\Psi' = -\frac{3\sqrt{\alpha}}{2}2\varphi^2$$

$\Psi' = 9\alpha 4\varphi^2 \Psi$

***Example 2: anisotropic rotator***

An anisotropic rotator is described by the Hamiltonian $H = H_0 + H_1$ with

$$H_0 = AL^2 + BL_z^2 \quad ; \quad H_1 = CL_y^2$$

where A, B and C are constants.

a) Find the eigenvalues and eigenstates of $H_0$, ana determine the degeneracy of the various eigenvalues for all the states with $\ell \leq 2$.
b) If $H_1$ is treated as a perturbation, compute the first order shifts for $\ell = 1$ energy states.
   You may want to use the formula $L_\pm Y_{\ell,m} = \sqrt{\ell(\ell+1) - m(m+1)}\, Y_{\ell,m\pm1}$

(a) $H_0|\Psi\rangle = E_0|\Psi\rangle$

Let $|\Psi\rangle$ be the spherical harmonics with $\theta$ the angle from the Z-axis, as usual, and $\varphi$ the angle rotated about the z-axis from the x-axis:

$\langle r|H_0|\Psi\rangle = E_0\langle r|\Psi\rangle\rangle = E_0 Y_\ell^m(\theta,\varphi)$

$(A\ell(\ell+1)\hbar^2 + B(m\hbar)^2)Y_\ell^m(\theta,\varphi) = E_{0\ell}Y_\ell^m(\theta,\varphi)$

$E_{\ell,m}^0 = (A\ell(\ell+1) + Bm^2)\hbar^2$ are the eigenvalues and

$\langle r|\Psi\rangle = Y_\ell^m(\theta,\varphi)$ are the eigenstates

| l | m | $E_{\ell,m}/\hbar^2$ | degeneracy |
|---|---|---|---|
| 0 | 0 | 0 | 1 |
| 1 | 0 | 2A | 1 |
| 1 | ±1 | (2A+B) | 2 |
| 2 | 0 | 6A | 1 |
| 2 | ±1 | (6A+B) | 2 |
| 2 | ±2 | (6A+4B) | 2 |

(b) $E = E_{\ell,m}^0 + \langle\Psi|H_1|\Psi\rangle + \cdots$

For $\ell = 1$ energy states there are 3:

$\ell = 1, m = 0$: $\int d\Omega\, Y_\ell^m C L_y^2 Y_{\ell,m} = ?$

$L_+ = L_x + iL_y \qquad\qquad L_- = L_x - iL_y$

$L_1 - L_- = 2iL_y \rightarrow L_y = \frac{1}{2i}(L_+ - L_-)$

$L_\pm Y_{\ell,m} = \sqrt{\ell(\ell+1) - m(m \pm 1)} Y_{\ell,m+1}$

For $\ell = 1, m = 0$: $L_y^2 = -\frac{1}{4}(L_+ - L_-) = -\frac{1}{4}(L_+^2 + L_-^2 - L_+L_- - L_-L)$

$Y_{\ell,m}$ *is rep. by* $Y_{\ell,m}$

$\langle \Psi_{1,0} | L_y^2 | \Psi_{1,0} \rangle = \frac{1}{4} \langle \Psi_{1,0} | L_+L_- + L_-L_+ | \Psi_{1,0} \rangle$

$= \frac{1}{4} \left[ \langle \Psi_{1,0} | L_+\sqrt{2-0} | \Psi_{1,-1} \rangle + \langle \Psi_{1,0} | L_-\sqrt{2-0} | \Psi_{1,-1} \rangle \right]$

$= \frac{1}{4} \left[ \sqrt{2}\sqrt{2-0} + \sqrt{2}\sqrt{2-0} \right] = 1$

$$\boxed{E_{1,0}^{(1)} = C}$$

$\langle \Psi_{1,1} | L_+L_- + L_-L_+ | \Psi_{1,1} \rangle = \langle \Psi_{1,1} | L_+\sqrt{2-0} | \Psi_{1,0} \rangle = 2$

$$\boxed{E_{1,1}^{(1)} = \frac{1}{2} C}$$

$\langle \Psi_{1,1} | L_+L_- + L_-L_+ | \Psi_{1,-1} \rangle = \langle \Psi_{1,1} | L_-\sqrt{2} | \Psi_{1,0} \rangle$

$$\boxed{E_{1,-1}^{(1)} = \frac{1}{2} C}$$

### 4.4.3 Semi classical approximation, a.k.a., the WKB approximation

From the classical description of light wave propagation it is known that a great deal of reflection occurs only when the wavelength $\lambda$ changes by a large fraction of itself within the distance of a wavelength:

$$\delta\lambda = \frac{\partial \lambda}{\partial x} \delta x$$

Set $\delta x = \lambda$

$$|\delta\lambda| = \left| \frac{\partial \lambda}{\partial x} \lambda \right| \ll \lambda \Rightarrow \left| \frac{\partial \lambda}{\partial x} \right| \ll 1$$

When extended to electron waves in de Broglie's description $\left( \lambda = {}^{h}/_{p} \right)$ this becomes

$$\left| \frac{\partial \lambda}{\partial x} \right| = \left| \frac{h}{p^2} \frac{\partial p}{\partial x} \right| \ll 1$$

For non-relativistic motion, $p^2 = 2m(E - V)$, and this becomes

$$\frac{\lambda \left| \frac{\partial V}{\partial x} \right|}{2(E - V)} \ll 1$$

So the absence of a reflected wave is that the potential energy be a slowly changing function of position, and $E - V$ not be to small.

Motion of a wave packet in this approximation then follows that of a particle (since Schrodinger's equation for the wave function leads to Newton's equations of motion in the classical limit) and we have a classical limit.

If $V(x)$ is slowly varying then $p(x) = \sqrt{2m(E-V)}$ is slowly varying and the familiar plane wave solutions to Schro's equation for free motion are approximately detained:

$\Psi = \exp\left(\frac{ipx}{\hbar}\right) \quad ; \quad p = \sqrt{2m(E-V)}$

The general, then, the semiclassical theory may be described in terms of a wavefunction.

$\Psi = \exp\left(\frac{iS}{\hbar}\right)$ (S may be complex in general)

Where $\left|\frac{\hbar}{p^2}\frac{\partial p}{\partial x}\right| \ll 1.$

Using the substitution $\Psi(\vec{r}) = exp\left[\frac{i}{\hbar}S(\vec{r})\right]$ the time-independent Schrodinger equation becomes:

$$(\nabla S)^2 - i\hbar\nabla^2 S - 2m(E-V) = 0.$$

The "Semi-classical approximation" consists in writing $S(\vec{r})$ as a power series in $\hbar$:

$$S = S_0 + \frac{\hbar}{i}S_1 + \left(\frac{\hbar}{i}\right)S_2 + \cdots$$

Then, the regular procedure of equating powers of $\hbar$ gives, for the one dimensional case:

$$(S_0')^2 = 2m(E-V)$$
$$2S_1' = -S_0''/S_0'$$
$$2S_2' = -\left(S_1'' + S_1'^2\right)/S_0'$$

Consider $(S_0')^2 = 2m(E-V)$, then

$$S_0 = \int \pm\sqrt{2m(E-V(x))}dx$$

and let $p(x) = \sqrt{2m(E-V(x))}$ . The zero-order approximation gives:

$$\Psi(x) = Aexp\left[\frac{i}{\hbar}\int p(x)dx\right] + Bexp\left[\frac{-i}{\hbar}\int p(x)dx\right]$$

The approximation is good if:

$$(\nabla S_0)^2 \gg i\hbar\nabla^2 S_0 \rightarrow \left|\frac{\hbar S_0''}{(S_0')^2}\right| \ll 1$$

Let

$$\bar{\lambda} = {}^{\lambda}/_{2\pi}, \lambda = 2\pi\hbar/p(x) \rightarrow \left|\frac{d\bar{\lambda}}{dx}\right| \ll 1$$

Approximate expressions are only valid for sufficiently large values of p(x), or 2m(E-V). Thus, near the classical turning points where p=0 (E=V(x)), the approximation is no longer valid.

The standard "*WKB* Approximation" consists of taking the first two terms of the expansion. So consider next:

$$S_1 = -\frac{1}{2}\ln p(x) + const$$

In the WKB Approximation:

$$\Psi = \frac{A}{\sqrt{p(x)}} e^{\frac{i}{\hbar}\int p dx} + \frac{B}{\sqrt{p(x)}} e^{-\frac{i}{\hbar}\int p dx}$$

checking the validity of WKB: $\hbar|S_2| \ll 1$ we have:

$$-2S_0'S_2' = \frac{1}{4}\left(\frac{S_0''}{S_0'}\right)^2 + \frac{1}{2}\frac{(S_0'')^2}{(S_0')} - \frac{1}{2}\frac{S_0''}{S_0'} = \frac{3}{4}\left(\frac{S_0''}{S_0'}\right)^2 - \frac{1}{2}\frac{S_0''}{S_0'}.$$

So, $\hbar S_2' = \mp\left(\frac{1}{4}\lambda' - \frac{1}{8}\int\frac{\lambda^2}{\lambda}dx\right)$ (where $\bar{\lambda}$ has been replaced with $\lambda$ for convenience).

Thus, $\hbar|S_2| \ll 1$ is satisfied if $|\lambda'| \ll 1$ once again.

When $E < V(x)$ we simply have:

$$\Psi = \frac{A'}{\sqrt{|p|}} e^{-\frac{1}{\hbar}\int|p|dx} + \frac{B'}{\sqrt{|p|}} e^{\frac{1}{\hbar}\int|p|dx}$$

WKB solutions are connected across turning points using nontrivial complex analysis reasoning. In the end, the relations are simply:

For E-crossing junction at increasing potential:

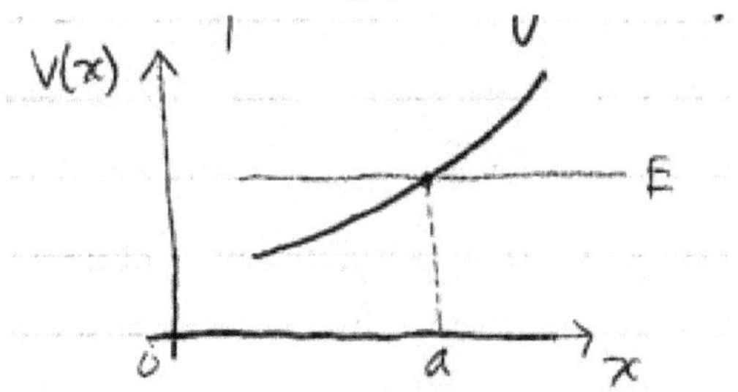

We have:

$$\Psi_1 = \begin{Bmatrix} \frac{1}{\sqrt{p}} \sin\left(\frac{1}{\hbar}\int_x^a p dx + \frac{\pi}{4}\right) for\ x < a \\ \frac{1}{2\sqrt{|p|}} \exp\left(-\frac{1}{\hbar}\int_x^a |p| dx\right) for\ x > a \end{Bmatrix}$$

While for E-crossing junction at decreasing potential:

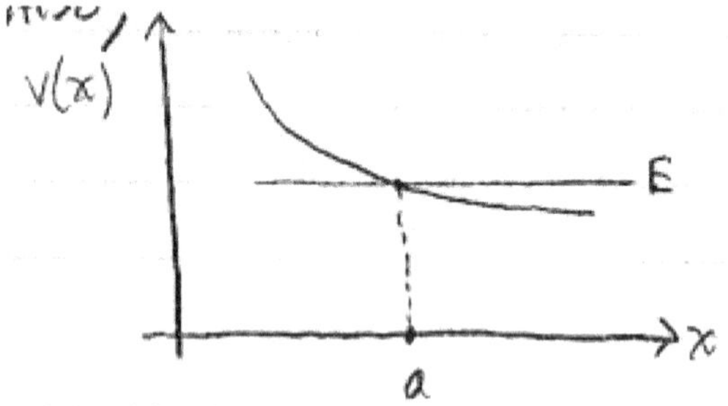


$$\Psi_2 = \begin{Bmatrix} \frac{1}{\sqrt{p}} \sin\left(\frac{1}{\hbar}\int_x^a p dx + \frac{\pi}{4}\right) for\ x > a \\ \frac{1}{2\sqrt{|p|}} \exp\left(-\frac{1}{\hbar}\int_x^a |p| dx\right) for\ x < a \end{Bmatrix}$$

Note that the Wronskian of two linearly independent Solutions to the Schrodinger equation (a 2nd order equation) with the same value of E does not depend on x:

$$W(\Psi_1, \tilde{\Psi}_1) = \begin{vmatrix} \Psi_1 & \Psi'_1 \\ \tilde{\Psi}_1 & \tilde{\Psi}'_1 \end{vmatrix} = const.$$

Using this relation we can find the other WKB solution from the one above or some combination thereof.

**A particle is in a potential well V(x)**

Consider a particle in a well:

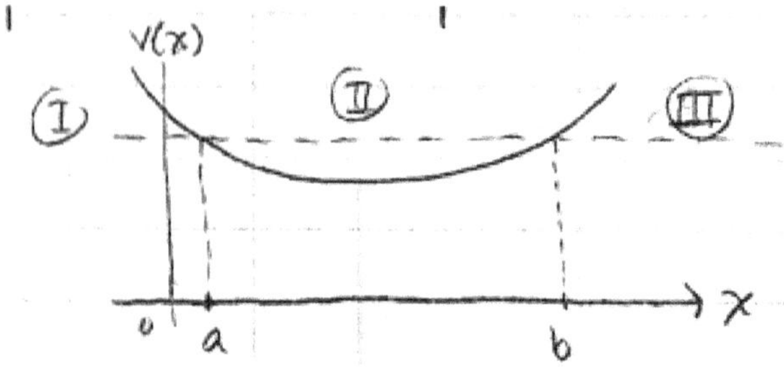


Suppose that a and b are the only turning points. In the WKB approximation, we want the solutions with the appropriate matching (exponential decay) in regions (I) and (III) to conclude:

$$\begin{cases} \Psi_1 = \frac{A}{\sqrt{p}} \sin\left(\frac{1}{\hbar}\int_x^b pdx + \frac{\pi}{4}\right), a < x < b \;\; matches\ to\ right\ x > b \\ \Psi_2 = \frac{B}{\sqrt{p}} \sin\left(\frac{1}{\hbar}\int_a^x pdx + \frac{\pi}{4}\right), a < x < b \;\; matches\ to\ left\ x < a \end{cases}$$

Equating these solutions and their derivatives at a point x in the interval (a,b):

$$Asin\left(\frac{1}{\hbar}\int_x^b pdx + \frac{\pi}{4}\right) - Bsin\left(\frac{1}{\hbar}\int_a^x pdx + \frac{\pi}{4}\right) = 0$$

and

$$A\left[\frac{\sqrt{p}}{\hbar}\cos\left(\frac{1}{\hbar}\int_a^x pdx + \frac{\pi}{4}\right) - \frac{1}{2}\frac{p'}{\sqrt{p^3}}\sin\left(\frac{1}{\hbar}\int_a^x pdx + \frac{\pi}{4}\right)\right] +$$
$$B\left[\frac{\sqrt{p}}{\hbar}\cos\left(\frac{1}{\hbar}\int_a^x pdx + \frac{\pi}{4}\right) - \frac{1}{2}\frac{p'}{\sqrt{p^3}}\sin\left(\frac{1}{\hbar}\int_a^x pdx + \frac{\pi}{4}\right)\right] = 0$$

and using $\left|p'/\sqrt{p^3}\right| \ll \sqrt{p}/\hbar$ . Also, for the system of equations above we want the determinant to vanish, i.e.

$$\sin\left(\frac{1}{\hbar}\int_a^b pdx + \frac{\pi}{2}\right) = 0 \quad \rightarrow \quad \frac{1}{\hbar}\int_a^b pdx + \frac{\pi}{2} = (n+1)\pi$$

Thus, WBK gives the familiar relation:

$$\oint pdx = 2\pi\hbar\left(n + \frac{1}{2}\right)$$

which is the Bohr-Sommerfeld quantization rule of the old quantum theory.

So, in the WKB approximation, we can determine the energy levels for a particle in a potential by an equation directly related to the Bohr-Sommerfeld quantization rule:

$$\int_a^b p(x)dx = \int_a^b \sqrt{2m(E-V)}dx = \pi\hbar\left(n + \frac{1}{2}\right)$$

#### 4.4.4 Application of Semiclassical analysis to shell collapse

Take $\hbar = 1$. Consider $H = \sqrt{p^2 + m^2} - \frac{m^2}{2r}$. For a family of factor orderings in the neighborhood of $= -i\frac{\partial}{\partial x}$ , $p^2 = -\frac{\partial^2}{\partial x,}$ , the semiclassical approximation allows algebraic manipulation of the operator relation for H:

$$\left(H + \frac{m^2}{2r}\right)\Psi = \sqrt{p^2 + m^2}\Psi$$

$$\left(E+\frac{m^2}{2r}\right)\Psi = m\left(1+\frac{1}{2}\left(\frac{p}{m}\right)^2 - \ldots\right)\Psi = m\left(1+\frac{1}{2}\left(\frac{1}{m^2}\Psi''\right) - \ldots\right)$$

Take $\Psi = e^{iS}$ and use $(S')^2 \gg S'', etc.$

$$= m\left(1+\frac{1}{2}\left(\frac{1}{m^2}[iS'' - (S')^2]\right) - \ldots\right)\Psi$$

$$\simeq m\left(1+\frac{1}{2}\left(\frac{1}{m^2}\left[\frac{1}{i}S'\right]^2\right) - \ldots\right)\Psi$$

$$\simeq m\left(1+\frac{1}{2}\left(\frac{1}{m}\frac{1}{i}|\Psi'|^2\right) - \ldots\right)\Psi$$

$$\left(E+\frac{m^2}{2r}\right) \simeq m\sqrt{1+\left(\frac{1}{im}|\Psi'|\right)^2}$$

$$\left(E+\frac{m^2}{2r}\right)^2 \simeq (m^2 + (-i|\Psi'|)^2)$$

$$\left(E+\frac{m^2}{2r}\right)^2\Psi \simeq \left(m^2 + \left(-i\frac{\partial}{\partial x}\right)^2\right)\Psi$$

This shows the equivalence of the $\sqrt{p^2+m^2}$ and Klein-Gordon type equation's in the semiclassical theory. Analysis of the energy eigenstates for the "relativistic Coulomb problem":

$$\left[\left(E+\frac{m^2}{2r}\right)^2 + \nabla^2 - m^2\right]\Psi(r) = 0,$$

Reduces to the radial equation for the non-relativistic Coulomb problem.

In the semiclassical approximation only the large "n" eigenvalues have wavefunctions with $(S')^2 \gg \hbar S''$, etc., i.e. where the main support of the wavefunction has slowly varying ${\partial v}/{\partial x}$ because its far from the singularity. So, the large "n" eigenvalues spectrum is expected to be valid on the semiclassical approximation, for the $\sqrt{\ldots}$ theory (aside from and overall shift in the spectrum).

Consider $H = \sqrt{p^2+m^2} - \frac{m^2}{2r} + M_-(1 - r^{-1}[\sqrt{p^2+m^2} - p])$ . This can be manipulated to obtain ($H\Psi = E\Psi$ for energy eigenvalues) :

$$\left(1-\frac{2M_-}{r}\right)p^2 + \left((E-M_-)+\frac{m^2}{2r}\right)\left(\frac{2M_-}{r}\right)p = \propto + \beta r^{-1} + \gamma r^{-2}$$

Let

$\propto = (E + M_-)^2 - m^2$

$\beta = (E + M_-)m^2$

$\gamma = m^2\left(\left(\frac{m}{2}\right)^2 - M_-^2\right)$

Dividing by $\left(1 - \frac{2M_-}{r}\right)$ this is rewritten:

$$p^2 + \left(1 - \frac{2M_-}{r}\right)^{-1}\left((E - M_-) + \frac{m^2}{2r}\right)\left(\frac{2M_-}{r}\right)p$$

$$= \left(1 - \frac{2M_-}{r}\right)^{-1}(\propto + \beta r^{-1} + \gamma r^{-2})$$

Now, a convenient choice of operator ordering is used to eliminate the "p" term;

$$p = \frac{i}{f(r)}\frac{\partial}{\partial r}f(r)$$

$$p^2 = -\frac{1}{f}\frac{\partial^2}{\partial r^2}f$$

Write $p^2 + \left(1 - \frac{2M_-}{r}\right)^{-1}\left[(E - M_-) + \frac{m^2}{2r}\right]\left(\frac{2M_-}{r}\right)p = p^2 + g(r)p$

$$(p^2 + g(r)p\,)\Psi(r) = \left[-\frac{1}{f}\frac{\partial^2}{\partial r^2}(f\Psi) - \frac{ig}{f}\frac{\partial}{\partial r}(f\Psi)\right]$$

$$= -\left[\frac{\partial^2\Psi}{\partial r^2} + \frac{2}{f}\frac{\partial f}{\partial r}\frac{\partial \Psi}{\partial r} + \frac{\Psi}{f}\frac{\partial^2 f}{\partial r^2} + \frac{ig}{f}\frac{\partial f}{\partial r}\Psi + ig\frac{\partial \Psi}{\partial r}\right]$$

$$= -\left(\frac{\partial^2\Psi}{\partial r^2} + \frac{2}{r}\frac{\partial \Psi}{\partial r}\right) - \frac{\partial \Psi}{\partial r}\left[\frac{-2}{r} + \frac{2}{f}\frac{\partial f}{\partial r} + ig\right] - \left(\frac{1}{f}\frac{\partial^2 f}{\partial r^2} + \frac{ig}{f}\frac{\partial f}{\partial r}\right)\Psi$$

Denote $\hat{p}_f$ and $\hat{p}_f^2$ as the operator representatives with the $f(r)$ terms, and denote $\nabla^2$ for the familiar radial term of the spherical Laplacian: $\nabla^2 = \frac{1}{r}\frac{\partial^2}{\partial r^2}r$. Now,

$$(P_f^2 + gP_f)\Psi(r)$$

$$= -\nabla^2\Psi - \left(\frac{\partial \Psi}{\partial r}\right)\left[-\frac{2}{r} + \frac{2}{f}\frac{\partial f}{\partial r} + ig\right]$$

$$- \left(\frac{1}{f}\frac{\partial^2 f}{\partial r^2} + \frac{ig}{f}\frac{\partial f}{\partial r}\right)\Psi$$

Choose $f$ such that there is no $\frac{\partial \Psi}{\partial r}$ term.

$$\boxed{\frac{1}{f}\frac{\partial f}{\partial r} = \frac{1}{r} - \frac{i}{2}g}$$

$$\frac{\partial^2 f}{\partial r^2} = \frac{\partial}{\partial r}\left(\frac{f}{r} - \frac{i}{2} fg\right) = \frac{1}{r}\frac{\partial f}{\partial r} - \frac{f}{r^2} - \frac{i}{2}\frac{\partial f}{\partial r} g - \frac{i}{2} f \frac{\partial g}{\partial r}$$

$$\frac{1}{f}\frac{\partial^2 f}{\partial r^2} = \frac{1}{r}\left(\frac{1}{f}\frac{\partial f}{\partial r}\right) - \frac{1}{r^2} - \frac{i}{2} g\left(\frac{1}{f}\frac{\partial f}{\partial r}\right) - \frac{i}{2}\frac{\partial g}{\partial r} = -i\frac{g}{r} - \frac{1}{4} g^4 - \frac{i}{2}\frac{\partial g}{\partial r}$$

$$(P_f^2 + gP_f)\Psi(r) = -\nabla^2\Psi - \left(-i\frac{g}{r} - \frac{1}{4} g^2 - \frac{i}{2}\frac{\partial g}{\partial r} + i\frac{g}{r} + \frac{1}{2} g^2\right)\Psi$$

$$= -\nabla^2\Psi + \left(\frac{i}{2}\frac{\partial g}{\partial r} - \frac{1}{4} g^2\right)\Psi$$

$$(P_f^2 + gP_f)\Psi = \left(1 - \frac{2M_-}{r}\right)^{-1}(\propto + \beta r^{-1} + \gamma r^{-2})\Psi$$

$$-\nabla^2\Psi = \left(\frac{1}{4} g^2 - \frac{i}{2}\frac{\partial g}{\partial r}\right)\Psi + \left(1 - \frac{2M_-}{r}\right)^{-1}(\propto + \beta r^{-1} + \gamma r^{-2})\Psi$$

### 4.4.5 Quantum theory of scattering by a potential

Collision phenomena are studied between various particles to identify the forces between those particles. Let's focus on elastic scattering to begin with, and make the following assumptions:
(1) colliding particles have no spin
(2) elastic scattering
(3) assume a "thin target" such that we can neglect multiple scattering processes.
(4) neglect coherence between scattered waves – i.e., assume that the spread of wave packets is small compared with the distance between collision centers (e.g., atoms).
(5) assume that the interaction between the colliding particles can be described by their in-line distance vector, and so reduces to a single particle in a central potential analysis as studied previously.

***Definition of scattering Cross section***

Consider a scattering experiment in which we count the number $dn$ of particles scattered into solid angle $d\Omega(\theta, \varphi)$. We start by assuming that $dn$ is proportional to $d\Omega$ and the incident flux $F_i$:

$$dn = F_i \sigma(\theta, \varphi) d\Omega,$$

where $\sigma(\theta, \varphi)$ has the dimensions of a surface, so is usually referred to as the differential scattering cross-section in the direction $(\theta, \varphi)$. In scattering experiments it is usually convenient to work with cross-section in units of 'barns':

$$1\, barn = 10^{-24} cm^2.$$

Note that the total scattering cross section is simply:

$$\sigma = \int \sigma(\theta, \varphi)\, d\Omega$$

The above definitions for cross-section are carried over to conditions of inelastic scattering as defined.

**Scattering by an arbitrary potential**
Let's describe the time evolution of a particle in an arbitrary potential. The stationary states for such a particle will have the form:

$$\psi(\vec{r}, t) = \varphi(\vec{r}) e^{-iEt/\hbar},$$

where,

$$\left[-\frac{\hbar^2}{2\mu}\Delta + V(\vec{r})\right]\varphi(\vec{r}) = E\varphi(\vec{r}).$$

Let $E = \frac{\hbar^2 k^2}{2\mu}$, $V(\vec{r}) = \frac{\hbar^2}{2\mu} U(\vec{r})$, and assume $V(\vec{r})$ decreases faster than $1/r$:

$$[\Delta + k^2 - U(\vec{r})]\varphi(\vec{r}) = 0.$$

So far, we still have a very general form of solution possible (each k has an infinite number of solutions). Recall that stationary state solutions represent a "probability fluid" in steady flow, where notions of asymptotic behavior are well-defined. Specifically, for large negative (and positive) time the particle is no longer interacting with the potential (too far away) thus it is "free" and should have the form of a plane wave oriented along whatever axis was connecting the particle to the center of the potential. From the potential there will be the scattered wave with radial dependence $e^{ikr}/r$. Together these will make the 'stationary scattering" states, where we have the form:

$$v_k^{(diff)}(r) \sim f_k(\theta, \varphi)\frac{e^{ikr}}{r} + e^{ikz}$$

From the prior definition $dn = F_i \sigma(\theta, \varphi) d\Omega$, with incident $e^{ikz}$ 'current' and scattered $f_k(\theta, \varphi)\frac{e^{ikr}}{r}$ 'current' then gives the relation:

$$\sigma(\theta, \varphi) = |\, f_k(\theta, \varphi)|^2.$$

Note that global conservation on the total number of particles requires destructive interference between the plane and forward scattered waves.

Let's consider collisions in the context of potential scattering. We have the Schrodinger equation $(\Delta + K^2)\,\Psi(\vec{r}) = U(\vec{r})\,\Psi(\vec{r})$ where $K^2 = 2mE/\hbar^2$ $and$ $U(\vec{r}) = 2mV(\vec{r})/\hbar^2$. The solution with the appropriate boundary conditions is:

$$\Psi_k^+(\vec{r}) = \varphi_k(\vec{r}) - \frac{1}{4\pi}\int \frac{\exp\left(ik|\vec{r}-\vec{r'}|\right)}{|\vec{r}-\vec{r'}|} U(\vec{r}')\,\Psi_K^+(\vec{r}')d\vec{r}'$$

Where $\varphi_k(\vec{r}) = \exp\left(i\vec{k}\cdot\vec{r}\right)$ $satisfies$ $(\Delta + k)\varphi_k(r) = 0$. At large distances:

$$\Psi_K^+(\vec{r}) \sim \varphi_k(\vec{r}) + A(E,\theta,\varphi)\frac{e^{ikr}}{r} \quad for\; r \gg r_0$$

Where $A(E,\theta,\varphi) = -\frac{1}{4\pi} < \varphi_{k'}|u|\Psi_k^+ >$, $\varphi_k(\vec{r}) = e^{i\vec{k}\cdot\vec{r}}$, $\vec{k}' = k\vec{r}/r$ and $A$ is called the scattering *amplitude*, for which:

$$d\sigma(E,\theta,\varphi) = |A(E,\theta,\varphi)|^2 d\Omega.$$

Born amplitude:

$$A^{(B)} = -\frac{1}{4\pi} < \varphi_{k'}|u|\varphi_k >$$

Thus,

$$A^{(B)} = -\frac{1}{4\pi}\int e^{-i\vec{k}\cdot\vec{r}}U(\vec{r})\,e^{i\vec{k}\cdot\vec{r}}d\vec{r}$$

Let $\vec{q} = \left(\vec{k} - \vec{k}'\right)$ :

$$A^{(B)}(\vec{q}) = -\frac{1}{4\pi}\int e^{i\vec{q}\cdot\vec{r}}U(\vec{r})d\vec{r}$$

Define the form factor as: $F(\vec{q}) = \frac{A^{(B)}(\vec{q})}{A^{(B)}(0)}$ . This characterizes the interference between waves scattered in different volume elements of the scattering field.

Consider that mutual elastic scattering of two particles with masses $m_1$ and $m_2$ and a potential energy of interaction $V(\vec{r}_2 - \vec{r}_1)$ can be expressed in a center of mass system which gives the scattering of a hypothetical single particle of mass $m=m_1m_2/(m_1+m_2)$ moving in a potential $V(\vec{r})$ fixed at the origin. Relations between scattering angle $(\theta,\varphi)$ and those of particle 1 and 2 in the lab system are then (using scattering units $(\theta,\varphi)$ ):

$$\tan\theta_1 = \frac{\sin\theta}{\gamma + \cos\theta}, \qquad \gamma = \frac{m_1}{m_2}, \qquad \theta_2 = \frac{1}{2}(\pi - \theta), \qquad \varphi_1 = \varphi,$$

$$\varphi_2 = \varphi + \pi$$

**Method of partial waves**

For a central potential $V(\vec{r}) = V(r)$ the direction of the momentum of the incident particle constitutes an axis of symmetry of the problem. Choosing the polar axis in this direction:

$$\Psi_k^+(r,\theta) = \sum_{l=0}^{\infty} a_l \frac{R_{l,k}(r)}{r} P_l(\cos\theta) \qquad a_e = i^l(2l+1)e^{i\delta_l(k)}$$

$R_{l,k}(r)$ are solutions of:

$$\frac{dR_{E,l}}{dr^2} + \frac{2m}{\hbar^2}\left[E - \left(V(r) + \frac{l(l+1)\hbar^2}{2mr^2}\right)\right] R_{E,l} = 0$$

$$R_{E,l}(0) = 0$$

(from radial wave. Equation part of Schrodinger Equation) and have the following asymptotic behavior.

$$R_{E,l}(r) \underset{r\to\infty}{\sim} \frac{1}{k}\sin\left(kr - \frac{l\pi}{2} - \delta_l(k)\right)$$

Asymptotic behavior corresponds to the" normalization":

$$\int_0^{\infty} R_{l,k}(r) R_{l,k'}(r) = \frac{\pi}{2k^2}(k-k')$$

For large r if it is convenient to write the "outgoing" parts of the partial waves in the form:

$$e^{2i\delta_l(k)} \varphi_{l,k}^+(r,\theta)$$

where $\varphi_{l,k}^+(r,\theta)$ is the outgoing part of the l-state wavefunction for the limiting case of a zero scattering potential.

The effect of the non-zero potential then appears in the phase shift factor, and thus the scattering due to a central potential can be represented as a unitary transformation of "free" outgoing partial waves by a scattering operator, which, in the above angular momentum representation, takes the form of a diagonal "scattering matrix" whose elements are:

$$S_{ll'} = \exp[2i\delta_l(k)]\delta_{ll'}$$

The various functions associated with elastic scattering are thus

(a) Phase shift $\delta_l(k)$

(b) Eigenvalues of the following scattering matrix $S_l(k) = exp[2i\delta_l(k)]$

(c) Potential wave amplitudes

$$A_l(E) = \frac{1}{2ik}(S_l(k) - 1) = \frac{1}{k} e^{i\delta_l(k)} \sin\delta_l(k)$$

(d) Scattering amplitude

$$A(E,\theta) = \sum_{l=0}^{\infty} (2l+1) A_l(E) P_l(\cos\theta)$$

(e) Differential cross section

$$d\sigma(E,\theta) = |A(E,\theta)|^2 d\Omega$$

(f) Total cross section

$$\sigma(E) = \sum_{l=0}^{\infty} \sigma^l(E) = \frac{4\pi}{k^2}\sum_{l=0}^{\infty}(2l+1)sin^2\delta_l(k) = \frac{4\pi}{K} I_m A(E,0)$$

When inelastic channels are present:

$$A(E,\theta) = \frac{1}{2ik}\sum_{l=0}^{\infty}(2l+1)\left(e^{2in_l-1}\right)P_l(\cos\theta)$$

$$\sigma_{el}(E) = \frac{\pi}{k^2}\sum_{l=0}^{\infty}(2l+1)|1-C_l|^2, \quad C_l = \exp(2in_l)$$

$$\sigma_{inel}(E) = \frac{\pi}{k^2}\sum_{l=0}^{\infty}(2l+1)(l-1|C_e|^2$$

$$\sigma_{tot}(E) = \sigma_{el}(E) + \sigma_{inel}(E)$$

Conditions of validity of the Born approximation:

$$\Psi_k^+(\vec{r}) = \varphi_k(\vec{r}) - \frac{1}{4\pi}\int \frac{e^{ik|\vec{r}-\vec{r}'|}}{|\vec{r}-\vec{r}'|}U(\vec{r}')\varphi_k(\vec{r}')d\vec{r}' + \cdots$$

$$|\varphi_k(\vec{r})| \gg \left|\frac{1}{4\pi}\int \frac{e^{ik|\vec{r}-\vec{r}'|}}{|\vec{r}-\vec{r}'|}U(\vec{r}')\varphi_k(\vec{r}')d\vec{r}'\right|$$

Since $U_{max}$ is uniquely at r=0:

$$\frac{1}{4\pi}\left|\int \frac{U(\vec{r}')}{\vec{r}'} e^{ikr'+\vec{k}\cdot\vec{r}'|}\varphi_k(\vec{r}')d\vec{r}'\right| \ll 1$$

If the K.E of the incident particle is sufficiently small, $kr_0 \ll 1$, and

$$\frac{1}{4\pi}\int \frac{u(\vec{r}')}{\vec{r}'}d\vec{r}' \gg 1$$

$$\frac{2m}{4\pi\hbar^2}\left|\frac{V(\vec{r}')}{\vec{r}'}d\vec{r}'\right| \ll 1$$

Let

$$k = \sqrt{\frac{2mE}{\hbar^2}} \quad and \quad \bar{V} = \frac{1}{4\pi r_0^2}|\frac{V(r)}{r}d\vec{r}$$

Thus

$$\frac{2mr_0^2}{\hbar^2}\bar{V} \ll 1$$

$$\bar{V} \ll \frac{\hbar^2}{2mr_0^2}$$

where $\frac{\hbar^2}{2mr_0^2}$ is the uncertainty in KE of a particle localized in a region of linear dimensions $r_0$. Thus, our condition requires, in a low energy range, that the potential energy should be much smaller than the kinetic energy of the particles.

If $V(\vec{r}) = V(r)$, then, by choosing the direction of $\vec{k}$ as the polar axis and performing the integration over angular variables, we obtain the condition of validity of the Born approximation. for potentials with spherical symmetry:

$$\frac{1}{4\pi}\left|\int_r \frac{U(\vec{r})}{r} e^{i(kr+kr\cos\theta)}\, r^2 dr d(-\cos\theta) d\varphi\right| \ll 1$$

$$\frac{1}{4\pi}\left|\int_r U(r)\, e^{2ikr} - 1\, dr(2\pi)\right| \ll 1$$

Thus

$$\frac{1}{2kr_0^2}\left|\int_0^\infty V(r)[\, e^{2ikr} - 1]dr\right| \ll \frac{\hbar^2}{2mr_0^2}$$

for validity for Born approximation.

For $k_0 r \gg 1$.

$$\frac{1}{2kr_0^2}\left|\int_0^\infty V(r)dr\right| \ll \frac{\hbar^2}{2mr_0^2}$$

thus, Born approximation more useful at low KE. For low energies $(kr_0 \ll 1)$ we once again arrive at:

$$\frac{1}{4\pi r_0^2}\left|\int \frac{V(r)}{r} dr\right| \ll \frac{\hbar^2}{2mr_0^2}.$$

The possibility of applying the Born approximation. to certain spherically symmetric potentials:

For

$$V(r) = V_0 \exp\left(-\frac{r}{r_0}\right)$$

We have

$$\int_0^\infty V_0 \exp\left(-\frac{r}{r_0}\right)[\, e^{2ikr} - 1]dr = -\frac{2ikr_0^2 V_0}{2kr_0 - 1}$$

Thus,

$$\frac{2ikr_0^2V_0}{\sqrt{4k^2{r_0}^2+1}} \ll \frac{k\hbar^2}{m}$$

or

$$2mr_0^2V_0 \ll \hbar^2\sqrt{1+4k^2r_0^2}$$

So, for

$$kr_0 \ll 1 \quad (low\ energies) \quad 2mr_0^2V_0 \ll \hbar^2$$

and

$$kr_0 \gg 1 \quad (high\ energies) \quad mr_0^2V_0 \ll k\hbar^2$$

Now consider

$$V(r) = \frac{A}{r}\exp(-ar), a = {}^1/_{r_0}$$

Need to evaluate:

$$I(a) = \int_0^\infty e^{-ar}\{e^{2ikr} - 1\}\frac{dr}{r}$$

$$\frac{dI}{da} = -\int_0^\infty e^{-ar}\{e^{2ikr} - 1\}\frac{dr}{r} = \frac{1}{a} - \frac{1}{a-2ik}$$

So,

$$I(a) = lna - \ln(a-2ik) + C = -ln(1-2ikr_0) + C$$
$$= -ln\sqrt{1+4k^2r_0^2} + itan^{-1}(2kr_0) + C$$

For $r_0 = 0,\ \ I = 0 \ \rightarrow C = 0$. So, $mA\left[ln\sqrt{1+4k^2r_0^2} + \left(tan^{-1}(2Kr_0)\right)^2\right]^{1/2} \ll k\hbar^2$

For the "spherical potential well"

$$V(r) = -V_0 \quad for\ r < r_0\ and = 0\ \ for \quad r > r_0.$$

Then,

$$\frac{m}{h\hbar^2}\left|\int_0^{r_0} V_0(e^{2ikr}-1)dr\right| \ll 1 \quad \rightarrow \quad \frac{mV_0}{h\hbar^2}\left|\frac{e^{2ikr}|^{r_0}}{2ik} - r_0\right| \ll 1$$

$$\frac{mV_0}{h\hbar^2}\left|\frac{e^{2ikr_0}}{2ik} - r_0\right| = \frac{mV_0}{h\hbar^2}\left|\frac{\sin(Kr_0)}{k} - r_0\, e^{-ikr_0}\right|$$
$$= \frac{mV_0}{k^2\hbar^2}|\sin(kr_0) - kr_0\{\cos(kr_0) - isin\,(kr_0)\}|$$
$$= \frac{mV_0}{k^2\hbar^2}|\sin^2(kr_0) - kr_0\{kr_0 - \sin(2kr_0)\}|^{\frac{1}{2}} \ll 1$$

Let's consider the differential cross-section in the Born approximation for spherically symmetric potential:

$$d\sigma = |A|^2 d\Omega$$

and

$$A^{(B)} = -\frac{1}{4\pi} < \varphi_k|u|\varphi_k > \quad where\ \vec{q} = \vec{k} - \vec{k}'$$

$$= -\frac{1}{4\pi}\int e^{+i\vec{q}\cdot\vec{r}}u(\vec{r})d\vec{r} \quad and\ q = 2ksin\,\frac{\theta}{2}$$

If $u(\vec{F})$ has spherical symmetry:

$$A^{(B)} = \frac{1}{4\pi}\int e^{+iqrcos\,\theta}u(\vec{r})d(-\cos\theta)\, r^2 dr d\varphi$$

$$= -\frac{1}{2}\int u(\vec{r})rdr\left\{\frac{e^{+iqr} - e^{-iqr}}{iq}\right\}$$

Thus

$$A^{(B)} = -\frac{1}{q}\int_0^\infty ru(r)\sin(qr)\,dr$$

where $u = \frac{2m}{\hbar^2}\ V(r)$ and we have $V(r) = \frac{A}{r}\exp\left(-\frac{r}{r_0}\right)$:

$A^{(B)} = \frac{1}{q}\int_0^\infty Ae^{-r/r_0}\left\{\frac{e^{iqr}-e^{-iqr}}{2i}\right\}dr\left(\frac{2m}{\hbar^2}\right)$

$= \frac{A}{2i}\left\{\left\{e^{r(iq-\frac{1}{r_0}} - e^{r(-iq-\frac{1}{r_0})}\right\}dr\right\}\left(\frac{2m}{\hbar^2}\right)\frac{1}{q}$

$= \frac{A}{2iq}\left[\frac{-1}{iq-\frac{1}{r_0}} - \frac{-1}{iq-\frac{1}{r_0}}\right] = \frac{A}{2iq}\left[\frac{-1}{iq-\frac{1}{r_0}} - \frac{-1}{iq-\frac{1}{r_0}}\right]\left(\frac{2m}{\hbar^2}\right)$

$= \frac{Ar_0}{2iq}\left\{\frac{2iqr_0}{1+q^2r_0^2}\right\}\left(\frac{2m}{\hbar^2}\right) \qquad 1 - \left(2Ksin\,\frac{\theta}{2}\right)^2$

Thus

$$\frac{d\sigma}{d\Omega} = A^2r_0^2\left(\frac{1}{1 + 4r_0^2k^2sin^2\,\frac{\theta}{2}}\right)^2\left(\frac{2m}{\hbar^2}\right)^2 \quad since\ \vec{p} = \hbar\vec{k}$$

$$\frac{d\sigma}{d\Omega} = \left(\frac{2mA}{\left(\frac{\hbar}{r}\right)^2 + 4P^2sin^2\,\frac{\theta}{2}}\right) \qquad \begin{array}{l} p^2 = \hbar^2k^2 \\ p^2 = 2mE \end{array}$$

The Coulomb potential has $V(r) - Z_1Z_2\,e^2/r\,, so\ if\ r_0 \to \infty$:
As $r_0 \to \infty$ the screening disappears, and let $A = Z_1Z_2e^2$, then,

$$\frac{d\sigma}{d\Omega} = \left(\frac{mZ_1Z_2e^2}{2p^2 sin^2\frac{\theta}{2}}\right)^2 = \left(\frac{Z_1Z_2e^2}{4E sin^2\frac{\theta}{2}}\right)^2$$

This is the Rutherford formula. Although obtained in the Born approximation, it is valid for all energies. Note the following features:

(a) $\frac{d\sigma}{d\Omega}$ depends only on the absolute value if the charges

(b) The angular distribution is independent of energy

(c) For a given angle $\frac{d\sigma}{d\Omega} \propto \frac{1}{E^2}$

(d) The total cross-section is infinite

If the potential is an even function of r, then

$$A^{(B)} = -\frac{1}{q}\int_0^\infty rU(r)\sin(qr)\,dr = \frac{1}{q}\int_{-\infty}^\infty rU(r)\,e^{iqr}\,dr$$

So, for a Gaussian potential:

$$A^{(B)} = \frac{1}{q}\frac{V_0}{2i}\left(\frac{2m}{\hbar^2}\right)\int_{-\infty}^\infty re^{-r^2/2r_0^2}\,e^{iqr}\,dr$$

Thus,

$$A^{(B)} = \frac{V_0 m}{qi\hbar^2}\int_{-\infty}^\infty rexp\left\{\left(\frac{r}{r_2 r_0} - i\alpha\right)^2\right\}dr$$

$$A^{(B)} = \frac{mV_0}{qi\hbar^2}\int_{-\infty}^\infty (\sqrt{2}r_0u + iqr_0^2)e^{-u^2}(\sqrt{2}r_0)du\,e^{-\alpha^2}$$

$$= \frac{-mV_0\sqrt{2\pi}r_0^3}{r^2}e^{-\frac{1}{2}r_0^2 1^2}$$

Thus

$$\frac{d\sigma}{d\Omega} = |A| = 2\pi\left(\frac{mr_0^3V_0}{\hbar^2}\right)^2 \exp\left(-4k^2r_0^2 sin^2\frac{\theta}{2}\right)$$

For spherical potential well potential:

$$V(r) = \begin{cases} -V_0 & r < r_0 \\ 0 & r > r_0 \end{cases}$$

$$A^{(B)} = -\frac{1}{q}\int_0^\infty ru(r)\sin(qr)\,dr = \frac{-2m}{q\hbar^2}\int_0^\infty rV(r)\sin(qr)dr,$$

Where $u = \frac{2mV}{\hbar^2}$. Thus,

$$\frac{d\sigma}{d\Omega} = 4\left(\frac{mr_0^3 V_0}{\hbar^2}\right)\frac{(\sin(qr_0) - qr_0\cos(qr_0))^2}{(qr_0)^6}$$

where $q = 2ksin\frac{\theta}{2}$.

A common feature of the scattering differential cross-sections studied above (except for Coulomb) is that they become isotropic at low energies ($kr_0 \ll 1$). This fact is characteristic of all potentials which have a finite range.

**To find the total cross section**

Consider the Yukawa differential cross-section:

Yukawa:

$$\frac{d\sigma}{d\Omega} = \left(\frac{2mA/\hbar^2}{\left(\frac{1}{r_0}\right)^2 + q^2}\right)^2$$

where $p^2 = \hbar^2 k^2$, $q = 2ksin\frac{\theta}{2}$, $0 < q < 2k$:

$$d\Omega = \sin\theta\, d\theta d\varphi = 2\sin\frac{\theta}{2}\cos\frac{\theta}{2} d\theta d\varphi = \frac{1}{k^2} q dq d\varphi$$

$$d\sigma = \frac{2\pi}{k^2} q dq \left(\frac{2mA/\hbar^2}{\left(\frac{1}{r_0}\right)^2 + q^2}\right)^2$$

Thus,

$$\sigma = \frac{2\pi}{k^2}\left(\frac{2mA}{\hbar^2}\right)^2 \int_0^{2k} \frac{q dq}{\left[\left(\frac{1}{r_0}\right)^2 + q^2\right]^2}$$

And we get:

$$\sigma = \pi\left(\frac{4mAr_0^2}{\hbar^2}\right)\left(\frac{1}{1 + 4(kr_0)^2}\right)$$

**Identical particles in scattering**

The wavefunction of a system of identical particles has to be symmetrical (for bosons) or anti-symmetrical (for fermions) with respect to the exchange of any two particles.

***Elastic scattering cross-section for two spinless bosons***

Consider a CMS system $\Psi(\vec{R},\vec{r},\vec{t}) = -\Psi(\vec{R},-\vec{r},\vec{t})$ (symmetrical). If the particles could be distinguished:

$$d\sigma_1(\theta,\varphi) = |A(\theta,\varphi)|^2 d\Omega$$

And

$$d\sigma_2(\theta,\varphi) = |A(\pi-\theta,\varphi+\pi)|^2 d\Omega$$

Since they can't be distinguished $d\sigma$ *and* $d\sigma_2$ cannot be measured separately and we have:

$$d\sigma(\theta,\varphi) = d\sigma_1(\theta,\varphi) + d\sigma_2(\theta,\varphi).$$

Note, if we retain the definition of the total cross-section as being the number of particles scattered out of the incident beam per unit incident flux, we have

$$\sigma = \frac{1}{2}\int d\sigma(\theta,\varphi)$$

Since the symmetrization is valid in the asymptotic region, it implies a corresponding symmetrization of the scattering amplitude:

$$\hat{A}(\theta,\varphi) = \frac{1}{\sqrt{2}}[A(\theta,\varphi) + A(\pi-\theta),\varphi+\pi]$$

So,

$$d\sigma(\theta,\varphi) = 2|\hat{A}(\theta,\varphi)|^2 d\Omega = |A(\theta,\varphi) + +A(\pi-\theta,\varphi+\pi)|^2 d\Omega$$

And an "exchange effect" can lead to substantially greater scattering: consider A independent of $\varphi,\ and\ \theta = {}^{\pi}/_{2} \rightarrow$

$d\sigma\ increased\ by\ factor\ of\ 4.$

***Differential cross-section for proton-proton scattering***

Total spin is a constant of the motion for the problem at hand. Note that the interactions of the two protons in the singlet and triplet spin states can be different. The p-p wavefunction must be anti-symmetrical w.r.t exchange. A triplet state is symmetric under a permutation of spin variables, and hence it must be anti-symmetrical under a permutation of coordinate variables:

$$\vec{A}_t(\theta) = \frac{1}{\sqrt{2}}[A_t(\theta) - A_t(\pi-\theta)]$$

$$d\sigma_t(\theta) = 2|\vec{A}_t(\theta)|^2 d\Omega = |\vec{A}_t(\theta) - A_t(\pi-\theta)|^2 d\Omega$$

A singlet state, on the other hand, is antisymmetric under spin so must be symmetric in coordinate representation:

$$\hat{A}_s(\theta) = \frac{1}{\sqrt{2}}[A_s(\theta) + A_s(\pi-\theta)]$$

$$d\sigma_s(\theta) = 2|\vec{A}_s(\theta)|^2 d\Omega = |\vec{A}_s(\theta) + A_s(\pi-\theta)|^2 d\Omega$$

For an unpolarized beam there are three independent triplet states and one singlet state:

$$d\sigma = \frac{3}{4}d\sigma_t + \frac{1}{4}d\sigma_s$$

If the forces between the particles are spin-independent $A_s = A_t = A(\theta)$.

Coulomb scattering amplitude with unpolarized fermions gives the Mott formula.

As unperturbed wavefunctions we could choose the simultaneous eigenfunctions of $\vec{\ell}^2, \vec{S}^2, \ell_2$ and $S_2$, but then would have to use perturbation theory for the degenerate levels. We can avoid this difficulty by using the simultaneous eigenfunction of the operator $\vec{\ell}, \vec{j}^2$ and $j_2$. Since $\vec{\ell}^2, \vec{j}^2,$ *and* $j_2$ commute with H', it follows that only states with the same quantum numbers can contribute to the perturbation of an unperturbed state. Thus non-degenerate perturbation theory can be applied:

$$< H' > = \langle \frac{\hbar^2 e^2}{2m^2c^2}\frac{1}{r^2}\vec{\ell}\cdot\vec{s}\rangle$$

$$\langle J^2\rangle = j(j+1) = l(l+1) + s(s+1) + 2\langle\vec{\ell}\cdot\vec{s}\rangle$$

$$\langle\frac{1}{r^3}\rangle = \frac{1}{n^3(\ell+I)\left(\ell+\frac{1}{2}\right)\ell}\left(\frac{me^2}{\hbar^2}\right)^3$$

So,

$$\langle H'\rangle = \frac{me^4}{\hbar^2}\left(\frac{e^2}{\hbar c}\right)^2\frac{j(j+1)-\ell(\ell+1)-s(s+1)}{4\pi^3\ell\left(\ell+\frac{1}{2}\right)(\ell+1)}$$

**4.5 Perturbation Theory for linear operators (Kato Rellich Theorem)**
A brief overview of the Kato-Rellich theorem will now be give, First appearing in Kato's solution to the Hydrogen Analysis in 1951 [78,79], it is a central tool in many perturbative analysis. The methods will be carefully applied in the Dust shell collapse quantization in Chapter 7, and the methodology also appears in Book 7 [21] in one of the methods for computing alpha.

**Definition of "A-bounded":** Let A : D(A) → H be a self adjoint operator and let B : D(B) → H be symmetric. We say that B is A-bounded with bound a if D(A) ⊂ D(B) and if there exists b so that

$$||Bf|| \leq a||Af|| + b||f||$$

for all f ∈ D(A).

**Kato-Rellich Theorem:** Let A : D(A) → H be a self adjoint operator and B with A-bounded with bound a < 1. Then A + B : D(A) → H is self adjoint.

**Proof.** A is self adjoint if and only if Ran(A ± iI) = H. More generally, A is self adjoint if and only if Ran(A ± iμI) = H for some real number μ. Now on D(A) consider the operator A + B, which is defined because D(A) ⊂ D(B).

If A is self-adjoint then it satisfies:

$$\|(A + \mu i I)f\|^2 = \|Af\|^2 + \mu^2\|f\|^2$$

Dropping the identity operator notation and using $v = (A + i\mu)f$, we then have:

$$\|v\|^2 = \|A(A + i\mu)^{-1}v\|^2 + \mu^2\|(A + i\mu)^{-1}v\|^2 \;\; \forall\, v \in H$$

Thus,

$$\|A(A + i\mu)^{-1}\| \leq 1, \quad \|(A + i\mu)^{-1}\| \leq 1/|\mu|.$$

Now, using the A-boundedness property, there exists an a such that:

$$\|B(A + i\mu)^{-1}f\| \leq a\|A(A + i\mu)^{-1}f\| + b\|(A + i\mu)^{-1}f\|$$

and with tha above bounds:

$$\|B(A + i\mu)^{-1}f\| \leq (a + \frac{b}{|\mu|})\|f\|$$

Now choose $\mu$ such that $\|B(A + i\mu)^{-1}f\| = \frac{1}{2}(1 + a)\|f\|$ to have the operator

$$1 + B(A + i\mu)^{-1}$$

be invertible. But this implies that the operator

$$A + B + i\mu$$

is bijective, which then implies $A + B$ is self-adjoint on D(A).

**Schrodinger Operator with Coulomb Potential**

Consider a Hilbert Space $H = L^2(\mathbb{R}^3)$. Stripping dimensional parameters the Schrodinger operator is simply the negative Laplacian operator: $P_0 = -\nabla^2 = -\Delta$ (which is self-adjoint with $D(-\nabla^2) = L^2(\mathbb{R}^3)$. Let's write the potential as: $V(x) = \gamma/|x|$, where $\gamma$ is a real number.

It is very easy to see that the Coulomb potential is a potential of this type. Split the potential into two pieces, the part living inside the unit ball and the part outside. Clearly 1/|x| is bounded by 1 outside the unit ball and inside integrates to 4π. The question lingers whether we have found the only self adjoint extension. The following theorem takes care of that. Theorem: Let A : D(A) → H be an essentially self adjoint operator and assume that B is symmetric and A bounded with bound a < 1, i.e., D(A) ⊂ D(B) and Bf ≤ aAf + bf for all f ∈ D(A) where b is some constant. Then A + B on D(A) is also essentially self adjoint and $\overline{A+B} = \bar{A} + \bar{B}$.

## 4.6 Measurement, Wave-collapse, and Objective Reduction

From the Schrodinger equation we have a description for deterministic evolution of a linear superposition of states. From the Dirac axiomatic approach measurement is defined to be a process involving a classical apparatus and a quantum observable (a self-adjoint operator), where measurement of an eigenvalue places the wavefunction in that eigenstate (to the exclusion of others in the superposition description). This means that the measurement event involves a 'wave collapse' that is instantaneous non-locally, thus superluminal. This gives rise to the classic EPR paradox.

### *EPR Paradox*

Consider a state for two systems that are correlated, e.g., such that they don't simplify to a tensor product state. Due to conservation laws, the state of system (2) will depend on measurements of system (1), even when those measurements are performed when systems (1) and (2) are far apart. This allows for correlated events to occur that seem to indicate communication that can be faster than the speed of light. Since the correlated events cannot be controlled, however, we have correlation events that can occur faster than the speed of light, but not 'communication' (information transfer). Nonetheless, such dominance of quantum theory over even special relativity lends further credence to the idea that geometry (including the General Relativity Manifold) is 'apparatus' when it comes to the quantum theory. Thus, General Relativity sets the stage (with choice of time), for the quantum fields to be described in Book 5 [15], where Quantum field Theory will be examined in both flat (including standard model) and curved spacetime.

### *The Measurement Problem*

So, measurement takes us from a superposition of many states, to one state or a set of fewer states, instantaneously. How to make sense of the of this transition, other than simply accepting it, is what is known as the measurement problem. Sometime the measurement problem is made not-so-radical in that it is described as a superposition of states that is reduced to another superposition of states ('peaked' around the observation indicated). If those superpositions are like wave-packets, comprised of an infinite number of contributions in the superposition, then the collapse marks a transition from an infinity of states to another infinity of states (which is perhaps more mathematically palatable). Consider, for example, a tensor product state, $|\varphi(1) > \oplus |\chi(2) >$. This can be considered to represent the simple juxtaposition of two systems, one in the state $|\varphi(1) >$ and the other in the state $|\chi(2) >$. For a tensor state stated thus, the two systems are said to be uncorrelated. (the results of the two types of measurements, bearing either on one system or on the other, that correspond to two independent random variables). When we can't form a tensor product state, we have a more general state for the system which generally reflects the existence of correlations between systems (1) and (2). Now consider that the measurement only 'collapses' the $|\varphi(1) >$ part, this would be an example of a partial collapse. Even so, when describing particle production or scattering, the discrete transfer of events forces the realization that there really is a wave-collapse to "one thing". So, accepting that there really is a wave-collapse, or superposition reduction, in the theory, can we understand what drives this reduction process?

**Wave-collapse Interpretations**

***Everett Many Worlds Model*** – no collapse, just branchings in observer perspectives.

***Subjective Collapse Models*** – we (conscious beings), basically, 'will' our reality.

***Ghirardi–Rimini–Weber (GRW) Model*** – proposes that wave function collapse happens spontaneously as part of the dynamics [80]. Predictions resulting from this theory in known variants indicate behavior not observed.

***Standard Objective Collapse Models*** – proposes to modify the Schrodinger Equation to have nonlinear terms that are stochastic such that macroscopic decoherence is seen and wave-collapse occurs when the source of those nonlinearities terms becomes sufficiently 'strong'. In Penrose's objective reduction, the collapse is tied to gravitational stress with the threshold to trigger wave-collapse to be one graviton [81].

***Emanation Objective Collapse Model*** – Similar to Penrose's objective collapse, but the trigger event is not via a nonlinear term in the Schrodinger equation (it remains unchanged), but by the magnitude of such a term seen as the strain factor between a current emanation of c-geometry/q-field and a adiabatically revised such emanation (with strain relaxed to zero). As with Penrose, the trigger event would involve the production of a one graviton change. Note, however, that the term 'graviton' is suggestive of a quantum gravity theory when no such is assumed, only a first-order 'quantum gravity' theory (already known to exist) that merely correlates the geometry's Einstein tensor with the quantum matter-radiation stress-energy tensor (according to Einstein's relation).

## 4.7 Exercises

### *Exercise 4.1*

A. An electron is prepared with spin $\frac{\hbar}{2}u$ where u is a unit vector.

1) If the component of spin is measured in a direction orthogonal to u, what are the possible outcomes and what is the probability of each?.
2) Suppose, more generally, that the component of the spin of the initial electron is measured along a direction v, at an angle $\theta$ to u. what are the possible outcomes and what is the probability of each?

B. Consider a box of length $2l$, with a partition at the center (walls and partition are impenetrable barriers). A single particle of mass m is in the left half of the box in the lowest energy state. The partition is the suddenly removed. What is the most likely state of the enlarged box for the particle to be in and what is the probability that it is in that state?

Useful information

$$\sigma_x = \begin{pmatrix} 0 & 1 \\ 1 & 0 \end{pmatrix}, \sigma_y = \begin{pmatrix} 0 & -i \\ i & 0 \end{pmatrix}, \sigma_z = \begin{pmatrix} 1 & 0 \\ 0 & -1 \end{pmatrix}$$

$$\int_0^{x/2} \sin nu \sin 2u = -\frac{2\sin\frac{n\pi}{2}}{n^2-4}, \qquad n \neq 2$$

$$\sin A \sin B = \frac{1}{2}\,[cos(A-B) - \cos(A+B)]$$

*Exercise 4.2*

The Hamiltonian of a one- dimensional system is $H_0 + V$ where $V$ is a small time dependent perturbation. If the state $\varphi$ of the system at time t is represented by

$$\varphi(x,t) = \sum_{n=0}^{\infty} a_n(t)\, u_n(x) \exp(-E_n t/\hbar)$$

Where the $u_n$ are the normalized eigenfunctions of $H_0$ with energies . Obtain the set of differential equations satisfied by the coefficients $a_n(t)$.

If

$$H_0 = p^2/2m + \frac{1}{2}\, m\omega_0^2 x^2$$

And

$$V = \frac{1}{2}\, Am\omega_0^2 x^2 \exp(-\,t/\tau) \text{ for } t > 0,$$

where $m, \omega_0,$ $A$ and $\tau$ are constants, find an expression correct to leading order in A for the transition probability from the ground state of the unperturbed system at t=0 to the second excited state at time t=+∞. The ground state and second excited states of an oscillator with frequency $\omega$ and their corresponding energies are:

$$u_0 = \left(\frac{\alpha}{\pi}\right)^{\frac{1}{4}} \exp\left(-\frac{1}{2}\alpha x^2\right); E_0 = \frac{1}{2}\hbar\omega$$

$$u_2 = \left(\frac{\alpha}{\pi}\right)^{\frac{1}{4}} \exp\left(-\frac{1}{2}\alpha x^2\right);\ \ E_2 = \frac{5}{2}\hbar\omega$$

where

$$\alpha = \frac{m\omega}{\hbar},$$

and

$$\left(\frac{\alpha}{\pi}\right)^{\frac{1}{2}} \int_{-\infty}^{\infty} (\alpha x^2)^n \exp(-\alpha x^2)dx = \begin{cases} 1 & (n=0) \\ \frac{1}{2} & (n=1) \\ \frac{3}{4} & (n=2) \end{cases}$$

*Exercise 4.3*

A system of two identical spin ½ fermions is described by the potential

$$V = \frac{1}{2}\mu\omega^2(\vec{r}_1 - \vec{r}_2)^2 + \alpha\vec{s}_1\cdot\vec{s}_2\ \ ;\ \ (0 < \alpha < \omega),$$

where $\vec{r}_1, \vec{s}_1$ are the position and spin operators respectively for the i-th fermion.

(i) Determine the energies of the system and their degeneracies.

(ii) A uniform magnetic field $\vec{B}$ is switched on. Describe what happens to the two lowest energy levels found on part (i). Are the degeneracies removed completely? Assume that the gyromagnetic ratio of the fermions equals 2.

***Exercise 4.4***

A non-relativistic particle with mass m and energy E>0 is (elastically) scattered from a spherically symmetric potential:

$$V(r) - \begin{cases} -V_o \;\; for \;\; 0 \leq r < a & (V_o > 0) \\ -V_1 \;\; for \;\; a \leq r < b & (V_o > V_1 > 0) \\ 0 \;\; for \;\; b \leq r \end{cases}$$

Calculate the phase-shift of the S-wave after the scattering.

***Exercise 4.5***

The Hamiltonian for a two-dimensional simple harmonic oscillator is

$$H = \hbar\omega\left(a_x^+ a_x + a_y^+ a_y + 1\right)$$

where

$$a_x = \sqrt{\frac{m\omega}{2\hbar}}\left(X + \frac{i}{m\omega}P_X\right), \quad a_y = \sqrt{\frac{m\omega}{2\hbar}}\left(Y + \frac{i}{m\omega}P_y\right)$$

with X and Y being position operators and P the momentum operator.

After a rotation by an angle $\theta$,

$$X \to X' = X\cos\theta + Y\sin\theta$$
$$Y \to Y' = -X\sin\theta + Y\cos\theta$$

i) Write the new operators (i.e. after the rotation) $a_x'$ and $a_y'$ in terms of $a_x$ and $a_y$.

ii) Denote $G = a_y^+ a_x - a_x^+ a_y$. Show that $a_x' = e^{\theta G} a_x e^{-\theta G}$ and $a_y' = e^{\theta G} a_y e^{-\theta G}$. (Hint: derive the differential equations for $a_x'$ and $a_y'$ is the transformation $a_x \to a_x{}'$ a unitary one?)

iii) Express the ground and first excited states of the rotated Hamiltonian in terms of the eigenstates of the unrotated H.

*Exercise 4.6*

A yellow line in the spectrum of sodium resulting from the transition ${}^2P_{3/2} \rightarrow {}^2S_{1/2}$ has a wave length of 5890 $\dot{A}$. In a magnetic field of 10,000 gauss, into how many lines will this line split? Find the wavelength of each line and describe its state of polarization when viewed normal to the field. (The ratio of the Bohr magneton to the product of Planck's constant and the velocity of light is $\mu_0/hc = 4.67(10)^{-5} cm^{-1}/gauss$) . The g factor is given by

$$g = 1 + \frac{J(J+1) + S(S+1) - L(L+1)}{2J(J+1)}$$

*Exercise 4.7*

Consider a particle of mass m in the following potential, which vanishes everywhere except on a sphere of radius b:

$$V(r) = -\frac{\hbar^2 \lambda}{27\mathcal{H}} \delta(r-b); \ \lambda > 0$$

(a) Derive the eigenvalue condition for s-wave bound states.
(b) Obtain the s-wave phase shift for scattering from this potential.

*Exercise 4.8*

A spinless particle of mass $\mu$ and charge q moves in the (x,y) plane under the influence of a constant magnetic field B pointing in the z-direction. Use the vector potential.

$$A_x = -By \ , \qquad A_y = A_z = 0$$

(a) Derive the equations of motion for x, y, $p_x$, $p_y$ in the Heisenberg picture.
(b) Show that the solution of the equations of motion is:
$x = A + B\cos\omega t + C\sin\omega t$
$y = D + B\sin\omega t - C\cos\omega t$
Where A, B, C, D are time-independent operators and $\omega$ is a c-number constant.
(c) Relate A, B, C, D to initial values of x, y, , $p_x$, $p_y$ and show that $(B, \mu\omega C)$ and $(D, -\mu\omega A)$are independent pairs of conjugate variables.
(d) Express the Hamiltonian in terms of A, B, C, D and use the result of (c) to find its eigenvalues.
(e) What is the degree of degeneracy?

# Chapter 5 Path Integral Quantization

In this chapter the path integral method and path integral quantization will be described. As the name suggests, a path integral is an integral over spaces of paths (in quantum mechanics) or over spaces of fields (in quantum field theory). If the notion of path integral still sounds mathematically vague, not well-defined in fact, then you would be correct. The process of making the path integral well-defined, in a specific application, or in a more general formulation, is a lengthy history, but the end result will be that it is well-defined, and its utility will be critical in many ways (including the only known path to renormalization for the electroweak part of the theory in the standard model).

Path integrals are infinite dimensional integrals, also known as functional integrals or field integrals, and can be defined in terms of extensions of finite-dimensional integrals. How this definition or extension is accomplished can vary greatly. At a high level we must decide if our path integral implementation shall be deterministic (Feynman Path Integral) or probabilistic (Weiner Path Integral). Intermediate between the two are the Gaussian Path Integrals, where the solution space of Gaussian Integrals is the fundamental representation of the propagator, explicitly solvable, providing the basis for a generator for the theory of N-point functions (in quantum field theory), or of a partition function (in statistical mechanical theory), where the latter is undertaken according to a deterministic or probabilistic interpretation of the theory.

In what follows we start with an overview that begins with Dirac's 1933 paper [60] and Feynman's 1948 paper [62], and describes the path to well-definedness. The underlying notion of highly oscillatory integrals, however, goes back to the inception of the calculus and classical mechanics (1700's Laplace [23], for more details see Book 1 [16]). We will then work with Feynman's path integral defined in terms of classical paths, whose action defines a phase associated with that path, and where a sum on all paths (integral) then defines the path integral. Intuitively, paths will typically add out of phase so the path integral will effectively select for stationary phase solutions. At zeroth order (semiclassical analysis Section 4.4.3) this will give the classical solution, at higher order it will give the Schrodinger's equation, thus we recover standard quantum mechanics from this seemingly odd formulation (just as can be done in classical mechanics for the Euler-Lagrange equations). Next we will review the Green's function solution provided by the path integral formulation for quantum mechanics. This will then be generalized to its quantum field theory version that will be shown to be well-defined. In Book 5 [15] the Green's function will provide the basis for a perturbative and renormalization analysis that will give the modern field theory results, e.g., quantum electrodynamics, that have provided the experiments confirmed with the highest precision of any tests of theory known (freakishly good in fact, as in 16 decimal places of agreement).

In analysis of bound states, path integrals are weak. It took decades, for example, for a solution to the Hydrogen atom to be accomplished using path integrals [9,12,107]. And, even then, the solution required generalizing to path integrals for a curved spacetime and use of analytic time (many of the same methods deployed to make the path integral formulation well-defined). For scattering and perturbative analysis in general, path integrals are strong (already being summation based and explicitly semigroup compatible). Even so, quantum mechanics methods, often in the Heisenberg or Interaction Picture with canonical quantization, provide a clear analysis for such problems. So why bother with path integrals? The answer is you don't need to in quantum mechanics. But, in quantum field theory, the only tractable way to proceed in many situations will be with path integral formulations, and in some cases, such as electroweak renormalization, we are only able to complete the theory in terms of path integral derivations. It all comes down to the Green's function. In quantum field theory many problems reduce to knowing the vacuum expectation of the time-ordered product of Heisenberg field operators (as is shown in Book5 [15]), i.e., what is $\langle 0|T\hat{\varphi}(x_1)\hat{\varphi}(x_2) \dots \hat{\varphi}(x_n)|0\rangle$ ? By use of Wick's theorem (see Book5 [15]), we will see that such n-point functions reduce to a sum on products of 2-point functions. Thus, the core problem reduces to solving a Green's function analysis for the 2-point function. so this derivation is shown in both quantum mechanics and quantum field theory settings in what follows.

Although the Overview, Feynman Path Integral derivation, and Green's function analysis that follows is mainly geared to prepare for those interested in quantum field theory (Book 5 [15]), the special nature of the path integral formulation will also be revealed in two ways: (1) the formalism naturally provides a generating functional of Green's functions that is analytically related to a partition function, suggesting that time is fundamentally analytic (possibly with periodic boundary conditions related to inverse temperature, see Book 6 [20] and App. D for details); (2) the formalism is also obtained in Emanator Theory (Book 7 [21], also a synopsis in App. D) when emanator projection is to a maximally analytic domain.

**5.1 Overview**

In 1933 Dirac proposed that the propagator of quantum mechanics could be argued to correspond with exp $\left(\frac{iS}{\hbar}\right)$ [60]. In 1948 Feynman developed this further, simply appending the Dirac paper to his PhD thesis, and continuing with what would be his foundational paper on path integrals [62], where the fundamental object became:

$$\int e^{\left(\frac{i}{\hbar}\right)S[b,a]}.$$

Note that the integral expression is written with no measure, as this is part of what we must decide how to implement at the outset. Making the notion of path integral well-defined can reduced to the following categories [82]:

(1) The sequential approach, makes use of the Trotter product formula (see App. A for math review) and is what is used in Feynman's 1948 description. If we write the free Hamiltonian operator as: $H = -\frac{\hbar^2}{2m}\Delta$ and the potential operator as $V$ (simply the multiplication operator for factor V acting on $L^2(\mathbb{R}^d)$ ), we then get wavefunction solutions in the form:

$$\psi(t) = \lim_{n\to\infty}\left(e^{-\frac{i\,t}{\hbar n}V}\,e^{-\frac{i\,t}{\hbar n}H}\right)^n \psi(0)$$

In the Green's function analysis that follows we will see that the free Hamiltonian operator satisfies:

$$e^{-\frac{i\,t}{\hbar n}H} = \left(\frac{2\pi i\hbar t}{mn}\right)^{-d/2} e^{-\frac{i}{\hbar}\frac{m}{2t/n}|x-y|^2}$$

which allows us to write the wavefunction solution as:

$$\psi(t,x) =$$

$$\lim_{n\to\infty}\left(\frac{2\pi i\hbar t}{mn}\right)^{-dn/2}\int_{\mathbb{R}^{nd}} \exp\left(-\frac{i}{\hbar}\sum_{j=1}^{n}\frac{t}{n}\left[\frac{m}{2}\frac{\left(x_j-x_{j-1}\right)^2}{\left(\frac{t}{n}\right)^2}\right]\right)\psi(x_0)dx_0 \dots dx_{n-1}$$

From this form, we can arrive at a rigorous definition of Feynman integration in two ways:

(i) Approximate the paths by piecewise linear paths as done by Feynman and since formalized [12,83,84] (derivation of this will be shown shortly).

(ii) Generalize he Trotter formula to semigroups (limits now 'strong'):

$$\lim_{n\to\infty} (F(t/n))^n = \exp(tF'(0))$$

from which a rigorous definition is possible [85].

.

(2) An alternate formulation to have a rigorous definition involves analytic continuation in the physical time parameter (Euclideanization if rotated to pure imaginary). Mathematically this not only results in something well defined, it also provides a versatile formalism to address many problems. Physically, however, the meaning of shifting to imaginary time is unclear. Consider the Schrodinger equation and shift to pure imaginary time, the equation then becomes the heat equation:

$$-\frac{\partial}{\partial t}u(t,x) = -\frac{1}{2}\Delta_x u(t,x) + V(x)u(t,x)$$

The solutions are given in terms of Weiner integrals, the specific form of a solution is known as the Feynman-Kac formula:

$$u(t,x) = \int \exp\left(-\int_0^t V(\omega(s)+x)ds\right)u(0,\ \omega(t)+x)\ dW(\omega)$$

which is well-defined in general.

(3) An alternate approach involving Wiener integrals has also been developed by Daubechies and Klauder [86], which gives a well-defined formalism for Hamiltonians with polynomial position and momentum terms that can be carried over to systems with spin.

(4) The white noise approach is possible if the path integral functional is written as the T-transform of a unique Hida distribution [87]. This construction process is widely applicable (requires analyticity), most especially to the time-dependent harmonic oscillator, which is foundational for path-integral bases field theory.

(5) The Parseval Identity approach to arrive at a (well-defined) Fresnel Integral [88,89]. Provides a detailed method for stationary phase analysis when working in infinite dimensions. Regarding the latter, we could just proceed directly with a generalization of classic oscillatory integrals to the imaginary form (the Fresnel Integrals).

(6) Infinite dimensional oscillatory integrals as a generalization to the Parseval approach, grounded in work starting with Laplace, it provides

general applicability, most notably to phase functions up to degree 4, the latter covering the important cases of quantum field theory to be considered [90,91].

A quick review of the sequential approach to obtain the Feynman configuration space path integral will now be given.

## 5.2 Feynman Path Integral derivation

Feynman and Hibbs [7]:

> "The probability $P(a, b)$ to go from a point $x_a$ at the time $t_a$ to the point $x_b$ at $t_b$ is the absolute square $P(a, b) = |K(b, a)|^2$ of an amplitude $K(b, a)$ to go from a to b."

***The Feynman Path Integral Hypothesis***

The amplitude $K(b, a)$, to go from a to b, shall be the sum over the phase contributions accrued for every path that goes from a to b, where the phase contribution on a given path is taken to be proportional to the action $S[b, a]$ to traverse that path. The functional integral on paths $x(t)$ involves the functional differential '$\mathcal{D}x(t)$':

$$K(b, a) = \frac{1}{A} \int e^{\left(\frac{i}{\hbar}\right) S[b,a]} \mathcal{D}x(t)$$

Where the Action for a quantum system with a classical correspondence is given, it shall satisfy the usual definition in terms of the classical Lagrangian of the system:

$$S[b, a] = \int_{t_a}^{t_b} L(\dot{x}, x, t) dt.$$

In an attempt to formalize the definition of a path integral, Feynman considers an explicit sum over paths for the free particle[7]:

> "The sum over paths is defined as a limit, in which at first the path is specified by giving only its coordinate x at a large number of specified times separated by very small intervals $\varepsilon$. The path sum is then an integral over all these specific coordinates. Then to achieve the correct measure, the limit is taken as $\varepsilon$ approaches 0."

Thus,

$$K(b, a) = \lim_{\varepsilon \to 0} \int e^{\left(\frac{i}{\hbar}\right) S[b,a]} \frac{dx_1}{A} \frac{dx_2}{A} \dots \dots \frac{dx_{N-1}}{A}.$$

Notice that this direct approach involves an explicit time-slicing, where all paths have a monotonically increasing t parameter.

A lengthy derivation involving Gaussian integrals then gives the answer:

$$K(b,a) = \left[\frac{2\pi i\hbar(t_b - t_a)}{m}\right]^{-1/2} exp\left\{\frac{im(x_b - x_a)^2}{2\hbar(t_b - t_a)}\right\}.$$

The analysis leading to the above result is only well-founded (based on a Quantum Mechanical formulation based on a Hilbert space, etc.) for a restricted class of Lagrangians (of which the free particle Lagrangian is one [92]). In what follows an explicit derivation will be given without explicit use of time-slicing and as such will be more amenable to generalization.

Starting with the functional integral description:

$$K(b,a) = \frac{1}{A}\int e^{\left(\frac{i}{\hbar}\right)\int_{ta}^{t_b} L(\dot{x},x,t)dt}\, \mathcal{D}x(t)$$

Let's start by considering classical path $\overline{x}(t)$, where the functional path considered is:

$$x = \overline{x}(t) + y,$$

and we have:

$$S[x(t)] = S[\overline{x}(t) + y(t)]\,.$$

Note that in the phase integral $S[\overline{x}(t) + y(t)] \cong S_{Classical}[\overline{x}(t)] + S[y(t)]$ since elements first order terms in $y(t)$ will have their phases cancelled (by terms with $-y(t)$ ). We, thus, have:

$$K(b,a) = e^{\left(\frac{i}{\hbar}\right)S_{Cl}[b,a]}\int_0^0 exp\left\{\frac{i}{\hbar}\int_{t_a}^{t_b}\frac{1}{2}m\dot{y}^2 dt\right\}\mathcal{D}y(t),$$

where all paths $y(t)$ begin and end at $y = 0$, thus there is no spatial dependance in the integral, only a functional dependence on $t_a$ and $t_b$.

For the classical action we have:

$$S_{Cl}[b,a] = \int_{t_a}^{t_b}\frac{1}{2}m\left(\frac{dx}{dt}\right)^2 dt.$$

Since the classical path simply satisfies:

$$\frac{dx}{dt} = \frac{x_b - x_a}{t_b - t_a}$$

So,

$$S_{Cl}[b,a] = \frac{1}{2}m\frac{(x_b - x_a)^2}{(t_b - t_a)}$$

Furthermore, since the classical path has a $\frac{dx}{dt}$ dependence that only depends on $\Delta x$ and $\Delta t$ then such will be the case for $S_{Cl}$ and, correspondingly, $\mathrm{F}(t_a, t_b) = \mathrm{F}(t_b - t_a)$. Thus,

$$K(b,a) = F(t_a - t_b)\, exp\left(\frac{im(x_b - x_a)^2}{2\hbar(t_b - t_a)}\right)$$

Also, we know (from the Chapman-Kolmogorov relation):

$$K(b,a) = \int_{x_c}^{\square} k(b,c)k(c,a)dx_c$$

So,

$$F(t_a - t_b)exp\left(\frac{im\left(x_y - x_a\right)^2}{2\hbar(t_a - t_b)}\right) = F(t_a - t_c)\mathrm{F}(t_c - t) \cdot I$$

where

$$I = \int_{-\infty}^{\infty} exp\left\{\frac{im}{2\hbar}\left[\frac{(x_b - x_c)^2}{(t_b - t_c)} + \frac{(x_c - x_a)^2}{(t_c - t_a)}\right]\right\}dx_c$$

After regrouping as a Gaussian integral, we get:

$$I = \sqrt{\frac{2\pi\hbar(t_b - t_c)(t_c - t_a)}{(-im)(t_b - t_a)}}\; exp\left(\frac{im}{2\hbar}\left[\frac{(x_b - x_a)^2}{(t_b - t_a)}\right]\right)$$

So,

$$F(t_b - t_a) = F(t_b - t_c)F(t_c - t_a)\sqrt{\frac{2\pi\hbar(t_b - t_c)(t_c - t_a)}{(-im)(t_b - t_a)}}$$

Substituting $\mathrm{F}(t_b - t_a) = \sqrt{\frac{m}{2\pi i\hbar(t_b - t_a)}}\mathrm{f}(t_b - t_a)$ we get

$$\mathrm{f}(t + s) = f(t)f(s) \rightarrow \; f(t) = e^{at}$$

If we consider a very small increment of time we find that $f(\varepsilon) = \left(\frac{2\pi i\hbar\varepsilon}{m}\right)^{-1/2}$:

Consider

$$\Psi(x_2,t_2) = \int_{-\infty}^{\infty} K(x_2,t_2;x_1,t_1)\Psi(x_1,t_1)dx_1$$

For a short time integral,

$$\Psi(x,t+\varepsilon) = \int_{-\infty}^{\infty} \frac{1}{A} exp\left[\varepsilon\frac{i}{\hbar}\left[\frac{m}{2}\right]\left[\frac{x-y}{\varepsilon}\right]^2\right]\Psi(y,t)dy$$

$$= \int_{-\infty}^{\infty} \frac{1}{A} e^{im\eta^2/2\hbar\varepsilon}\ \Psi(x+\eta,t)d\eta$$

Expanding in $\varepsilon$ and equating both sides:

$$\Psi(x,t) + \varepsilon\frac{\partial\Psi}{\partial t} = \int_{-\infty}^{\infty} \frac{1}{A} e^{im\eta^2/2\hbar\varepsilon}\left[\Psi(x,t) + \eta\frac{\partial\Psi}{\partial x} + \frac{1}{2}\eta^2\frac{\partial^2\Psi}{\partial x^2}\right]d\eta$$

Taking the leading terms from both sides we get

$$1 = \int_{-\infty}^{\infty} \frac{1}{A} e^{im\eta^2/2\hbar\varepsilon} d\eta \rightarrow A = \left(\frac{2\pi i\hbar\varepsilon}{m}\right)^{1/2}$$

So, the nonrelativistic propagator is:

$$K(b,a) = \sqrt{\frac{m}{2\pi i\hbar(t_b - t_a)}}\ exp\left(\frac{im(x_b - x_a)^2}{2\hbar(t_b - t_a)}\right)$$

So far we've shown how the path integral formalism can explicitly give rise to known quantum solutions for the free particle case. Feynman was able to extend this result to the 4-D Klein Gordon equation by extending to a 5-D formalism where the K-G equations takes the form of a free particle Lagrangian [93]).

Starting with the Klein Gordon equation:

$$\left(i\frac{\partial}{\partial x_\mu}\right)^2 \Psi = m^2\Psi$$

Define

$$\Psi = \int_{-\infty}^{\infty} exp\left(-\frac{1}{2}im^2u\right)\varphi(x,u)du$$

Then,

$$\left[\left(i\frac{\partial}{\partial x_\mu}\right)^2 - m^2\right]\Psi = \int_{-\infty}^{\infty}\left[\left(i\frac{\partial}{\partial x_\mu}\right)^2 - m^2\right] exp\left(-\frac{1}{2}im^2u\right)\varphi(x,u)du$$

$$= \int_{-\infty}^{\infty}\left\{\left(i\frac{\partial}{\partial x_\mu}\right)^2\varphi - 2i\varphi\frac{\partial}{\partial u}\right\} exp\left(-\frac{1}{2}im^2u\right)du$$

Since $\frac{\partial\Psi}{\partial u} = \left[exp\left(-\frac{1}{2}im^2u\right)\varphi(x,u)\right]_{-\infty}^{\infty} = 0$ (if $\varphi(x,u)$ *bounded*)
we then have:

$$\int_{-\infty}^{\infty} exp\left(-\frac{1}{2}im^2u\right)\frac{\partial\varphi}{\partial u}du = -\int_{-\infty}^{\infty}\left(-\frac{1}{2}im^2\right)exp\left(-\frac{1}{2}im^2u\right)\varphi du$$

Or

$$\int_{-\infty}^{\infty} exp\left(-\frac{1}{2}im^2u\right)\left[\left(i\frac{\partial}{\partial x_\mu}\right)^2 + 2i\left(\frac{\partial}{\partial u}\right)\right]\varphi(x,u)\,du = 0$$

Thus,

$$i\frac{\partial\varphi}{\partial u} = -\frac{1}{2}\left(i\frac{\partial}{\partial x_\mu}\right)^2\varphi$$

Thus, any parabolic partial differential equation can be approached by means of Feynman's method and if it is hyperbolic then the same can be done by considering an extra parameter,
Feynman presents an argument in [62] which shows how

$$\Psi(x_{k+1}, t+\varepsilon) = \int exp\left[\frac{i}{\hbar}S(x_{k+1,}x_k)\right]\Psi(x_k,t)\frac{dx_k}{A},$$

along the lines discussed, and how this generalizes and agrees with the Schrödinger equation. Further use of the $exp\left(-\frac{1}{2}im^2u\right)$ 'regularizer' is made in Feynman Phys. Rev 76, pg. 749 (1949) [94]. These and other methods to formally extend and undergird the path integral formulation will run into further complication, especially when considering the nonrelativistic/noncovariant nature of a particular time slicing. What becomes evident is that a path integral formulation exists that is *generative* such that it generates the correct equations for the quantum system. Much like how the statistical mechanics partition function of a system is generative of a systems thermodynamic potentials. The source of the quantum mechanical path integral formulation as a generative theory is explained in Book 7 of the Series [21], where emanator theory is proposed as giving rise to the quantum theory with its generative expression, along with predictions of the value of alpha and the structure of the Standard Model.

The Path Integral method will be of profound importance when considering quantum field theory in Book 5 [15], where it will provide a pathway to a solution for Quantum electrodynamics, QED (that will produce the best agreement with experimental results observed in physics).

**Propagator for the Schrodinger equation**
In electromagnetism we have Maxwell's equations for the differential point of view, and we have Huygen's principle for the global point of view. The Quantum analogue of Huygen's principle is:

$$\Psi(\vec{r}_2, t_2) = \int d^3r_1\, K(\vec{r}_2, t_2; \vec{r}_1, t_1)\, \Psi(\vec{r}_1, t_1) \qquad (t_2 > t_1) \;\; (or\ K = 0\ for\ t_2 < t_1),$$

where $K$ is called the propagator.

***Proof of the existence of the propagator***
Recall the evolution operator

$$|\Psi(t_2)> = U(t_2, t_1)|\Psi(t_1)\,, \qquad where\ \Psi(\vec{t}_2, t_2) = <\vec{r}_2|\Psi(t_2)>.$$

Thus,

$$\Psi(\vec{r}_2, t_2) = \int d^3r_1 <\vec{r}_2|U(t_2, t_1)|\vec{r}_1><\vec{r}_1|\Psi(t_1)>$$
$$= \int d^3r_1 <\vec{r}_2|U(t_2, t_1)|\vec{r}_1>\Psi(\vec{r}_1, t_1)$$

Thus,

$$K(2,1) = K(\vec{r}_2, t_2; \vec{r}_1, t_1) = <\vec{r}_2|U(t_2, t_1)|\vec{r}_1> \theta(t_2 - t_1).$$

Known as the retarded propagator, it specifies a unique Green's function. The physical interpretation of $K(2,1)$ is that it represents the probability amplitude that the particle, starting from the point $\vec{r}_1$ at $t_1$, will arrive at $\vec{r}_2$ at $t_2$. If H doesn't depend explicitly on time and we consider its eigenstate:

$$H|\varphi_n> = E_n|\varphi_n>,$$

then,

$$U(t_2, t_1) = e^{-iH(t_2,t_1)/\hbar} = e^{-iH(t_2,t_1)/\hbar}\sum_n |\varphi_n><\varphi_n|$$
$$= \sum_n e^{-iE_n(t_2,t_1)/\hbar}|\varphi_n><\varphi_n|$$

Thus,

$$K(2,1) = \sum_n e^{-iE_n(t_2,t_1)/\hbar} <\vec{r}_2|\varphi_n><\varphi_n|\vec{r}_1> \theta(t_2 - t_1),$$

or

$$K(2,1) = \theta(t_2 - t_1) \sum_n \varphi_n^{\square}(\vec{r}_2)\varphi_n^*(\vec{r}_1)\, e^{-iE_n(t_2-t_1)/\hbar}$$

Now, if $\varphi_n(\vec{r}_2)e^{-iE_n t_2/\hbar}$ is a solution of the Schrodinger equation :

$$\left\{i\hbar\frac{\partial}{\partial t_2} - H\left(\vec{r}_2,\frac{\hbar}{i}\nabla_2\right)\right\}\varphi_n(\vec{r}_2)e^{-iE_2 t_2/\hbar} = 0$$

So,

$$\left\{i\hbar\frac{\partial}{\partial t_2} - H\left(\vec{r}_2,\frac{\hbar}{i}\vec{\nabla}_2\right)\right\}K(\vec{r}_2,t_2;\vec{r}_1,t_1) = i\hbar\delta(t_2 - t_1)\sum_n \varphi_n^{\square}(\vec{r}_2)\varphi_n^*(\vec{r}_1)$$

$$\left[i\hbar\frac{\partial}{\partial t_2} - H\left(\vec{r}_2,\frac{\hbar}{i}\nabla_2\right)\right]K(2,1) = i\hbar\delta(t_2 - t_1)\delta(\vec{r}_2 - \vec{r}_1),$$

with boundary condition. $K(2,1) = 0$ $if$ $t_2 < t_1$.

## Lagrangian formulation of quantum mechanics

Consider two pts $(\vec{r}_1, t_1), (\vec{r}_2, t_2)$. Choose N intermediate times $t_\alpha$, $(i = 1, 2, \dots N)$:

$$t_1 < t_{\alpha 1} < t_{\alpha 2} < \cdots < t_{\alpha n} < t_2$$

For each $t_{\alpha i}$ choose a $r_{\alpha i}$. As $N \to \infty$ we get a function $\vec{r}(t)$ (which we shall assume to be continuous)such that: $\vec{r}(t_1) = \vec{r}_1$ and $\vec{r}(t_2) = \vec{r}_2$. Now,

$$U(t_2, t_1) = U(t_2, t_{\alpha N})U(t_{\alpha N}, t_{\alpha N-1}) \dots U(t_{\alpha 2}, t_{\alpha 1})U(t_{\alpha 1}, t_1).$$

Now use $\{|r>\}$ rep and use closure relations $\alpha_N$ times:

$K(2,1)$

$$= \int d^3 r_{\alpha N} \int d^3 r_{\alpha N-1} \dots \int d^3 r_{\alpha 2} \int d^3 r_{\alpha 1}\, K(2, \alpha_N)K(\alpha_N, a_{N-1}) \dots K(\alpha_2, \alpha_1)K$$

As $N \to \infty$: $K(2, \alpha_N)K(\alpha_N, a_{N-1})x \dots K(\alpha_2, 1)K(\alpha_1, 1)$ becomes the probability amplitude of the particle following a given path between 1 and 2. $K(2,1)$ is then the integral which corresponds to the coherent superposition of the amplitudes associated with all possible space-time paths starting from 1 and ending at 2.

**Feynman's Postulates**

Thus far we've seen the propagator formulation of Schrodinger's equation and how this is equivalent to a "sum over paths". The propagator formalism has delineated a spacetime formulation of the postulates of quantum mechanics, and thus is more general (e.g., we have a direct relativistic generalization since already in a spacetime formulation). In the Feynman approach we take the propagator object $K(2,1)$ as fundamental and define it directly as the probability amplitude for a particle to go from $(\vec{r}_1, t_1)$ to $(\vec{r}_2, t_2)$ according to the following rules:

(i) $K(2,1)$ is the sum of an infinity of partial amplitudes, one for each of the spacetime paths connecting $(\vec{r}_1, t_1)$ with $(\vec{r}_2, t_2)$.

(ii) The partial amplitude $K_\Gamma(2,1)$ associated with one of these paths $(\Gamma)$ is given by the classical Action (this is the scenario where such exists), in terms of the classical Lagrangian, according to:

$$S_\Gamma = \int_\Gamma^{\square} \mathcal{L}(\vec{r}, \vec{p}, t)\, dt$$

where

$$K_\Gamma(2,1) = N e^{\frac{i}{\hbar} S_\Gamma},$$

and N is a normalization constant.

Schrodinger's equation follows from the two postulates above. The Canonical commutation relations for $\vec{R}$ *and* $\vec{P}$ are the same (static). Thus, the above postulates permit a formulation of Quantum Mechanics which is different from that of Schrodinger where the time evolution (the 6th Postulate) was given by explicit time reference: $i\hbar \frac{d}{dt} | \Psi(t) > = H(t) | \Psi(t) >$, it is here given in the Action formulation and bundled into the classical Action. This will afford the Feynman approach greater flexibility via clearer separation of the classical/apparatus parts and the quantum parts of the system. Feynman's postulates, in the classical limit, give Hamilton's principle of least action. Feynman's approach generalizes to system with quantum Actions for which there is no classical counterpart (spin angular momentum) and for any classical system which that has a variational formulation even if there is (seemingly) no mechanical aspect (field descriptions).

The disadvantage of the Feynman approach is often attributed to its complexity -- a summation over an infinite number of paths. This disadvantage will be worsened in the case of generalization to quantum field theory, to the point of being ill-defined. Any 'repair' to the theory to make well-defined and connect with a canonical formulation, or experiment, is in effect defining the Feynman theory as providing a 'generative' formulation, where the representation of the experimental system then governs how that formulation will then generate the local quantum theory appropriate to the experiment. This will actually be seen to be in agreement with the thermal quantum theory that results when generalizing to complex time in the propagator (Book 6 [20]), where the partition function will result, with known generative properties to define the entire thermodynamics of the system. So, if the entire theory, other than specific of representation and coordinate system, etc., can be defined in the generative Feynman formulation, is there an even deeper layer of the theory that is generative of the Feynman formulation with the known standard models of both particle physics and cosmology? The answer, is apparently yes, and the theory is referred to as emanator theory (emanator instead of propagator) in Book 7 [21], where it is described in detail.

**Green's function analysis**

In this section we consider how to calculate the quantum field theory n-point function denoted by:

$$\langle 0|T\hat{\varphi}(x_1)\hat{\varphi}(x_2)\dots\hat{\varphi}(x_n)|0\rangle.$$

We begin by consideration of the quantum mechanical equivalent in terms of position operators at time t (not field operators at spacetime coordinate $x_1$):

$$G^{(n)}(t_1,t_2\dots t_n)=\langle 0|T\hat{q}(t_1)\hat{q}(t_2)\dots\hat{q}(t_n)|0\rangle.$$

We will also see that the n-point Green's function can be written as a sum over 2-point Green's functions. Furthermore, since we have the position operators it is convenient to work in the Heisenberg position representation. Thus, our key calculation reduces to evaluating the 2-point Green's function:

$$G^{(2)}(t_1,t_2)=\langle q',t|T\hat{q}(t_1)\hat{q}(t_2)|q,0\rangle.$$

Note that if we remove the "two points" we recover the Feynman propagator

$$K=\langle q',t|q,0\rangle=\int_{q,0}^{q',t}\mathcal{D}q e^{iS}.$$

So, the focus of the Green's function analysis is on evaluating $\langle q',t|T\hat{q}(t_1)\hat{q}(t_2)|q,0\rangle$, let's start with the case $t_1>t_2$:

$$\langle q',t|T\hat{q}(t_1)\hat{q}(t_2)|q,0\rangle=\langle q',t|\hat{q}(t_1)\hat{q}(t_2)|q,0\rangle$$

and

$$\langle q',t|\hat{q}(t_1)\hat{q}(t_2)|q,0\rangle$$
$$= \int dq_1 dq_2 \langle q',t|q_1,t_1\rangle\langle q_1,t_1|\hat{q}(t_1)\ \hat{q}(t_2)|q_2,t_2\rangle\langle q_2,t_2|q,0\rangle$$
$$= \int dq_1 dq_2 q_1 q_2 \langle q',t|q_1,t_1\rangle\langle q_1,t_1|q_2,t_2\rangle\langle q_2,t_2|q,0\rangle$$

Let's shift the Feynman propagators to integral form:

$$\langle q',t|\hat{q}(t_1)\hat{q}(t_2)|q,0\rangle$$
$$= \int dq_1 dq_2 q_1 q_2 \int_{q_1,t_1}^{q',t} \mathcal{D}q e^{iS} \int_{q_2,t_2}^{q_1,t_1} \mathcal{D}q e^{iS} \int_{q,0}^{q_2,t_2} \mathcal{D}q e^{iS}$$

from which it then manifest that (for $t_1 > t_2$):

$$\langle q',t|\hat{q}(t_1)\hat{q}(t_2)|q,0\rangle = \int_{q,0}^{q',t} \mathcal{D}q q_1(t_1) q_2(t_2) e^{iS}.$$

If we repeat the above analysis for $t_1 < t_2$ we get the same result, so the path integral conveniently has one expression for the time-ordered product:

$$\langle q',t|T\hat{q}(t_1)\hat{q}(t_2)|q,0\rangle = \int_{q,0}^{q',t} \mathcal{D}q q_1(t_1) q_2(t_2) e^{iS}.$$

Let's now shift back to an evaluation in terms of vacuum-to-vacuum elements. For this we shift to the time parameter having a small imaginary phase. If we then evolve to large negative imaginary time, all Hamiltonian eigenstates of $\langle q',t|$ and $|q,0\rangle$ will be dominated by the ground state. Thus the following proportionality result must exist:

$$\langle q',t|q,-t\rangle \propto \langle \emptyset,t|\emptyset,-t\rangle$$

where an important notational shift has occurred: '$\emptyset$' references the vacuum state and the time parameter is now in a symmetric form that is standard. We can thus, write:

$$\langle \emptyset,t|\emptyset,-t\rangle \propto \langle q',t|q,-t\rangle = \int_{q,-t}^{q',t} \mathcal{D}q e^{iS}.$$

A similar shift can be done with factors of $\hat{q}(t_1)\hat{q}(t_2) \ldots \hat{q}(t_n)$ present in the integrand, so we have:

$$\langle \emptyset,t|T\hat{q}(t_1)\hat{q}(t_2) \ldots \hat{q}(t_n)|\emptyset,-t\rangle \propto \int_{q,-t}^{q',t} \mathcal{D}q q_1(t_1) q_2(t_2) \ldots q_n(t_n) e^{iS}.$$

We now eliminate the proportionality, and cancel the unknown phase term associated with the large negative imaginary time relation, by dividing it out (as manifest in the Feynman propagator), where the vacuum is now written $\langle \emptyset|$ :

$$G^{(n)}(t_1, t_2 \dots t_n) = \langle \emptyset | T\hat{q}(t_1)\hat{q}(t_2) \dots \hat{q}(t_n) | \emptyset \rangle$$
$$= \frac{\langle \emptyset, t | T\hat{q}(t_1)\hat{q}(t_2) \dots \hat{q}(t_n) | \emptyset, -t \rangle}{\langle \emptyset, t | \emptyset, -t \rangle}$$

Thus,

$$G^{(n)}(t_1, t_2 \dots t_n) = \frac{\int_{q,-t}^{q',t} \mathcal{D}q q_1(t_1) q_2(t_2) \dots q_n(t_n) e^{iS}}{\int_{q,-t}^{q',t} \mathcal{D}q e^{iS}}$$

The method for solving this is well known in quantum field theory and statistical field theory. We begin by defining a generating functional for Green's functions:

$$Z[J] = \frac{\int \mathcal{D}q e^{i(S + \int J(t)q(t)dt)}}{\int \mathcal{D}q e^{iS}} = \frac{\langle \emptyset | \emptyset \rangle_J}{\langle \emptyset | \emptyset \rangle_{J=0}}$$

from which we see that:

$$\left( \frac{1}{i} \frac{\delta}{\delta J(t_1)} \cdots \frac{1}{i} \frac{\delta}{\delta J(t_n)} Z[J] \right) \bigg|_{J=0} = \frac{\int \mathcal{D}q q_1(t_1) q_2(t_2) \dots q_n(t_n) e^{iS}}{\int \mathcal{D}q e^{iS}}$$
$$= G^{(n)}(t_1, t_2 \dots t_n)$$

The case of where the Action S is that of a harmonic oscillator is of special interest for quantum field theory generalization, so let's examine that case and evaluate $Z[J]$. The numerator of $Z[J]$ is:

$$N = \int \mathcal{D}q e^{i\int \left[ \frac{1}{2} m\dot{q}^2 - \frac{1}{2} m\omega^2 q^2 + J(t)q(t) \right] dt}.$$

For the Action $S = \frac{1}{2} m\dot{q}^2 - \frac{1}{2} m\omega^2 q^2 + J(t)q(t)$ let's denote the classical solutions by $q_c(t)$, we then have:

$$N \propto e^{iS[q_c]}.$$

Making use of the fact that $q_c$ satisfies the equations of motion, we can simply to:

$$S[q_c] = \frac{1}{2} \int J(t) q_c(t) dt.$$

The classical solution can be written in terms of the standard Green's function as:

$$\left( \frac{d^2}{dt^2} + \omega^2 \right) G(t, t') = -i\delta(t - t'),$$

where

$$q_c(t) = -i \int J(t') G(t, t') dt'$$

Thus, we can show for the harmonic oscillator:

$$Z[J] = exp\left(\frac{1}{2}\int J(t')G(t,t')J(t)dt'dt\right)$$

For the Green's function solution in momentum space we see that:

$$G(t,t') = \int \frac{dk}{2\pi} \frac{i}{k^2 - \omega^2} e^{-ik(t-t')} dt$$

As is the Green's function is ambiguous since it has poles on the axis of integration and we must decide how to do this. Fortunately our prescription from the outset made use of an asymptotic analyticity relation that involved a small negative imaginary part in the time definition. similarly here, to have $G$ go to zero as $t \to \infty$ we need the (precursor to Feynman propagator) prescription:

$$G(t,t') = \int \frac{dk}{2\pi} \frac{i}{k^2 - \omega^2 + i\epsilon} e^{-ik(t-t')} dt.$$

Note the care taken in understanding and specifying the pole structure…., the pole structure will identify the physical part of propagator in renormalization scheme.

Let's now consider the free Klein-Gordon scalar field. What results is an exact parallel of the preceding analysis. We now have a scalar field $\varphi$ which has a classical solution that satisfies the Klein Gordon equation:

$$(\partial^2 + m^2)\varphi_c(x) = J(x)$$

and the Klein-Gordon Green's function is defined by:

$$(\partial^2 + m^2)\Delta_F(x,x') = -i\delta^4(x - x').$$

Thus,

$$\varphi_c(x) = i\int d^4x J(x')\Delta_F(x,x')$$

so,

$$Z = exp\left(\frac{1}{2}\int d^4x d^4x' J(x')\Delta_F(x,x')J(x)\right).$$

Solving the Green's function in 4-momentum space, adopting the same pole prescription as previously, we then have the Feynman propagator:

$$\Delta_F(x,x') = \int \frac{d^4k}{(2\pi)^4} \frac{i}{k^2 - m^2 + i\epsilon} e^{-ik(x-x')}.$$

# Chapter 6. Relativistic Quantum Mechanics

## 6.1 The Klein-Gordon Equation

The non-relativistic Schrodinger equation was obtained using the d'Alembert procedure. Let's revisit this process and then use it to get the relativistic equation, known as the Klein-Gordon equation.

### The Schrodinger equation

Recall the classical equation for energy (Hamiltonian) for non-relativistic motion:

$$E = \frac{1}{2m}p^2 + V$$

Using the d'Alembert operator substitutions we get:

$$\left(i\hbar\frac{\partial}{\partial t}\right)\psi = \left[\frac{1}{2m}(-i\hbar\nabla)^2 + V\right]\psi$$

Thus,

$$i\hbar\frac{\partial}{\partial t}\psi = \left[-\frac{\hbar^2}{2m}\nabla^2 + V\right]\psi,$$

which is Schrodinger's equation.

***The 1-dimensional Schrodinger equation***

$$i\hbar\frac{\partial}{\partial t}\psi = \left[-\frac{\hbar^2}{2m}\frac{\partial^2}{\partial x^2} + V(x)\right]\psi,$$

If we let $t \to i\tau$, we get the diffusion equation with constant diffusion coefficient. For 1-dimensional situation with a potential we have the general 1-dimensional transient heat transfer equation.

***The 1-dimensional transient Heat transfer Equation***

$$\hbar\frac{\partial}{\partial t}\psi = \left[-\frac{\hbar^2}{2m}\nabla^2 + V(x)\right]\psi,$$

There is evidently a fundamental relation between Quantum Mechanics and Statistical Mechanics / Thermodynamics in terms of complex time. This relation is even more profound in the path integral formulation of quantum field theory, where the pure complex time switch (Wick rotation) links the fundamental quantum propagator for the system's dynamics to the system's partition function for the system's thermodynamics.

**The Klein-Gordon Equation**

Let's now consider the expression for energy in the context of the energy-momentum 4-vector for a particle:

$$p^{\mu} = \left(\frac{E}{c}, \vec{p}\right) \rightarrow \; p^2 = E^2 - \vec{p}^2 = m^2$$

So, with the d'Alembert substitution and setting $\hbar = c = 1$:

$$\left[\frac{\partial^2}{\partial t^2} - \nabla^2\right]\psi + m^2\psi = 0 \quad \rightarrow \quad (\square + m^2)\psi = 0,$$

which is the Klein-Gordon equation. Note that interpreted as a single-particle equation, it can have negative energy states, so problems manifest from the start.

**6.2 The Dirac Equation**

The Dirac Equation (1934 [29]) specializes to spin ½ particles. The Klein-Gordon equation, as just noted, is an expression of the relation between energy, momentum, and mass for a particle, so already applies to the spin ½ particles. The Dirac equation is a relativistic description for a particle (spin ½ particles specifically) that is more direct in that it is a direct expression of Lorentz Invariance in the context of the spin ½ field representation (also known as spinor fields).

To understand Dirac's derivation of the Dirac Equation from Lorentz Invariance we must first mathematically understand:
(1) The Standard rotation group in 3-dimensions, O(3). From the Group representation shift to the Generator/Semigroup representation.
(2) Repeat for SU(2). Show that SU(2) provides a double cover of O(3).
(3) Repeat (2) for $SL(2, C) \approx SU(2) \otimes SU(2)$.
(4) Obtain the 4-spinor irreducible representation of the Lorentz Group extended by parity

***(1) The Standard rotation group in 3-dimensions, O(3)***

Consider

$$\begin{pmatrix} x' \\ y' \\ z' \end{pmatrix} = R \begin{pmatrix} x \\ y \\ z \end{pmatrix}$$

where $R$ is a rotation matrix corresponding to an element of O(3). For rotation about the z-axis by angle $\theta$ we have:

$$R_z(\theta) = \begin{pmatrix} \cos\theta & \sin\theta & 0 \\ -\sin\theta & \cos\theta & 0 \\ 0 & 0 & 1 \end{pmatrix}.$$

Similarly for $R_x(\phi)$ and $R_y(\psi)$. The generator of the rotation transformation is given by:

$$J_z = \frac{1}{i}\frac{dR_z(\theta)}{d\theta}\bigg|_{\theta=0} = \begin{pmatrix} 0 & -i & 0 \\ i & 0 & 0 \\ 0 & 0 & 0 \end{pmatrix}$$

Similarly for $J_x$ and $J_y$. We now recover the rotation in terms of its generator:

$$R_z(\theta) = \lim_{N\to\infty}\{R_z(\theta/N)\}^N = \lim_{N\to\infty}\{1 + iJ_z(\theta/N)\}^N = e^{iJ_z\theta}$$

Similarly for $R_x(\phi)$ and $R_y(\psi)$, and we now have a form for rotation about any axis $\boldsymbol{n}$ through angle $\theta$:

$$R_{\boldsymbol{n}}(\theta) = e^{iJ\cdot n\,\theta}.$$

***(2) Repeat SU(2). Show that SU(2) provides a double cover of O(3)***

For extensive details on the various transformation groups and their generators, see Book 2 [17]. The correspondence between SU(2) transformation U and O(3) transformation R is as follows for rotation about the z-axis:

$$U = \begin{pmatrix} e^{i\theta/2} & 0 \\ 0 & e^{-i\theta/2} \end{pmatrix} \leftrightarrow R_z(\theta) = \begin{pmatrix} \cos\theta & \sin\theta & 0 \\ -\sin\theta & \cos\theta & 0 \\ 0 & 0 & 1 \end{pmatrix}$$

Or, in terms of Pauli matrices:

$$U = e^{i\sigma_z\theta/2} \quad \leftrightarrow \quad R = e^{iJ_z\theta}$$

Similarly for rotations a bout the x- and y-axis. The general relation then becomes:

$$U = e^{i\sigma\theta/2} = \cos\theta/2 + i\boldsymbol{\sigma}\cdot\boldsymbol{n}\sin\theta \quad \leftrightarrow \quad R = e^{iJ\cdot\theta}.$$

The double cover is evident in that the same rotation $\theta$ carries through to an angle $\theta/2$ for spinors and $\theta$ for vectors.

***(3) Repeat (2) for $SL(2,C) \approx SU(2) \otimes SU(2)$***

Let's examine the pure x-direction boost part of the Lorentz Group first. Recall that a Lorentz transformation for a boost in the x-direction by relative speed v, relates the spacetime coordinates of two inertial frames according to:

$$\begin{pmatrix} t' \\ x' \\ y' \\ z' \end{pmatrix} = \begin{pmatrix} \cosh\phi & \sinh\phi & 0 & 0 \\ \sinh\phi & \cosh\phi & 0 & 0 \\ 0 & 0 & 1 & 0 \\ 0 & 0 & 0 & 1 \end{pmatrix}\begin{pmatrix} t \\ x \\ y \\ z \end{pmatrix} = B_x(\phi)\begin{pmatrix} t \\ x \\ y \\ z \end{pmatrix}$$

where

$$\cosh\phi = \gamma = \left(1-\frac{v^2}{c^2}\right)^{-1/2} \quad ; \quad \sinh\phi = \beta\gamma = \left(\frac{v}{c}\right)\left(1-\frac{v^2}{c^2}\right)^{-1/2}.$$

For the generator of x-direction boosts:

$$K_x = \frac{1}{i}\frac{\partial B_x(\phi)}{\partial \phi}\bigg|_{\phi=0} = -i\begin{pmatrix} 0 & 1 & 0 & 0 \\ 1 & 0 & 0 & 0 \\ 0 & 0 & 0 & 0 \\ 0 & 0 & 0 & 0 \end{pmatrix}$$

Similarly for the other Boosts. The rotations appear as 3x3 embedded versions:

$$J_z = -i\begin{pmatrix} 0 & 0 & 0 & 0 \\ 0 & 0 & 1 & 0 \\ 0 & -1 & 0 & 0 \\ 0 & 0 & 0 & 0 \end{pmatrix}$$

and similarly for the other rotations. If we examine the commutations relations to ensure group closure we find there is a problem:

$$[K_x, K_y] = -iJ_z \text{ (plus cyclic permutations)}$$
$$[J_x, K_x] = 0 \text{ (same for y , z)}$$
$$[J_x, K_y] = iK_z \text{ (plus cyclic permutations)}$$

Thus, the pure Lorentz boosts and rotations do not form a group. Consider the groupings:

$$A = \frac{1}{2}(J + iK), \quad B = \frac{1}{2}(J - iK).$$

We now have closure on the $A's$ and $B's$ separately:

$$[A_x, A_y] = iA_z \text{ (plus cyclic permutations)}$$
$$[A_i, B_j] = 0, \quad i, j = x, y, z$$
$$[B_x, B_y] = iB_z \text{ (plus cyclic permutations)}$$

Note how the A's and B's separately each define an SU(2) group under the transformation. Thus,

$$\boldsymbol{SL(2, C) \approx SU(2) \otimes SU(2)}$$

Also, for each SU(2) we refer to a spinor of that type, thus, there are two spinor types that are in inequivalent representations. We can write one spinor type that transforms as:

$$\xi \to \exp\left(\frac{i\sigma}{2}\cdot(\theta - i\phi)\right)\xi$$

while the other spinor type transforms as:

$$\eta \to \exp\left(\frac{i\sigma}{2}\cdot(\theta + i\phi)\right)\eta$$

***(4) Obtain the 4-spinor irreducible representation of the Lorentz Group extended by parity***

Under parity, which changes the velocity to negative velocity, we have

$$\xi \to \eta, \quad \eta \to \xi$$

Thus, if we extend the pure Lorentz transformations to include parity we will have group closure on the entirety. To do this, however, we can't have the 2-spinors free to transform separately, instead they are now grouped as a 4-spinor (the internal transformation of which will comprise the Dirac equation). The 4-spinor will be an irreducible representation of the Lorentz group extended by parity. Thus:

$$\psi = \begin{pmatrix} \xi \\ \eta \end{pmatrix} \to \begin{pmatrix} \exp\left(\frac{i\sigma}{2} \cdot (\theta - i\phi)\right) & 0 \\ 0 & \exp\left(\frac{i\sigma}{2} \cdot (\theta + i\phi)\right) \end{pmatrix} \begin{pmatrix} \xi \\ \eta \end{pmatrix}$$

Let's focus on the case where $\theta = 0$ and let $\xi = \phi_R$ and $\eta = \phi_L$. This transformation will then entail:

$$\phi_R \to \exp\left(\frac{1}{2}\sigma \cdot \phi\right)\phi_R = [\cosh \phi/2 + \sigma \cdot n \sinh \phi/2]\phi_R$$

Taking the original spinor to be at rest, we then have the relations in terms of $\gamma$

$$\cosh \phi = \gamma = \left(1 - \frac{v^2}{c^2}\right)^{-1/2}; \quad \sinh \phi = \beta\gamma = \left(\frac{v}{c}\right)\left(1 - \frac{v^2}{c^2}\right)^{-1/2}.$$

Thus, we can write:

$$\phi_R(\mathbf{p}) = \left[\left(\frac{\gamma + 1}{2}\right)^{1/2} + \sigma \cdot p\left(\frac{\gamma - 1}{2}\right)^{1/2}\right]\phi_R(0)$$

Since $\gamma = E/m$ in $c = 1$ units being used, this can be written:

$$\phi_R(\mathbf{p}) = \frac{E + m + \sigma \cdot p}{[2m(E + m)]^{1/2}}\phi_R(\mathbf{0})$$

and similarly

$$\phi_L(\mathbf{p}) = \frac{E + m - \sigma \cdot p}{[2m(E + m)]^{1/2}}\phi_L(\mathbf{0})$$

Since $\phi_R(\mathbf{0}) = \phi_L(\mathbf{0})$ for a particle at rest, we can regroup our two relations as:

$$\begin{pmatrix} -m & E + \sigma \cdot p \\ E - \sigma \cdot p & -m \end{pmatrix}\begin{pmatrix} \phi_R(\mathbf{p}) \\ \phi_L(\mathbf{p}) \end{pmatrix}.$$

Introducing the 4x4 gamma matrices (in terms of the 2x2 Pauli matrices):

$$\gamma^0 = \begin{pmatrix} 0 & 1 \\ 1 & 0 \end{pmatrix}, \quad \gamma^i = \begin{pmatrix} 0 & -\sigma^i \\ \sigma^i & 0 \end{pmatrix}$$

and also denoting $E = p^0$, we have:

$$(\gamma^\mu p_\mu - m)\psi = 0,$$

the Dirac equation.

Now, suppose we just guess a 4-spinor form that must be Lorentz Invariant, first order, and have mass, and it must be an equation that is 'squarable' to arrive at exactly the second order Klein-Gordon equation to be consistent. Starting with $(\gamma^\mu p_\mu - m)\psi = 0$ as the form needed, the only unknown would be the gamma matrices. But the consistency requires the constraints to be such that

$$\frac{1}{2}(\gamma^\mu\gamma^\nu - \gamma^\nu\gamma^\mu) = 2g^{\mu\nu}$$

which recovers the gammas as defined above.

Notice that

$$\psi^\dagger\psi = \phi_R{}^\dagger\phi_R + \phi_L{}^\dagger\phi_L$$

which is not invariant. To get an invariant we need

$$\bar{\psi} = \psi^\dagger\gamma^0$$

then

$$\bar{\psi}\psi = \phi_L{}^\dagger\phi_R + \phi_R{}^\dagger\phi_L$$

which is invariant (including invariance under parity).

A special gamma matrix can be defined from the others:

$$\gamma^5 = i\gamma^0\gamma^1\gamma^2\gamma^3$$

If we consider the transformation of $\bar{\psi}\gamma^5\psi$ we find:

$$\bar{\psi}\gamma^5\psi = \phi_L{}^\dagger\phi_R - \phi_R{}^\dagger\phi_L$$

will b invariant under proper Lorentz transformation but not under parity, i.e., $\bar{\psi}\gamma^5\psi$ is a pseudoscalar (while the fully invariant $\bar{\psi}\psi$ transforms as a scalar). Thus, the 4-spinor (partial) invariants possible with gamma matrices is more than that familiar from simple vectors:

$\bar{\psi}\psi$ is scalar,
$\bar{\psi}\gamma^5\psi$ is pseudoscalar,
$\bar{\psi}\gamma^\mu\psi$ is vector,
$\bar{\psi}\gamma^\mu\gamma^5\psi$ is axial vector,
$\bar{\psi}(\gamma^\mu\gamma^\nu - \gamma^\nu\gamma^\mu)\psi$ is an antisymmetric tensor.

# Chapter 7. Theoretical Foundations Revisited

## 7.1 Theoretical foundations revisited

Quantum Theory seems arbitrary, at first, so it's natural to ask if it could have been done any other way… Or, being weird, could it be even weirder? To coin a term from [95], are any other quantum theories possible from "theory-space"? Remarkably, the answer is NO. Although arrived at in many different ways with many different formulations, quantum theory is shown to be equivalent in these formulations, with the notion of a complex-valued wavefunction typical, such as $\psi(x,t)$ for a one-dimensional wavefunction (counting in spatial degrees of freedom, with one time dimensions counted separate). The wavefunction gives us non-deterministic information about the outcomes of measurement on the system according to a probability of observing a particular configuration of interest, where $P(x) \propto |\psi(x,t)|^2$ (Born interpretation of wavefunction 1927 [28]). So let's start with

(1) $\psi$ is a complex-valued function; and
(2) The probability derives from a 2-norm on $\psi$: $|\psi(x,t)|^2$

Can either of these structures be altered in the theory or its interpretation? In Book 7 of the Series [21], a discussion of propagatable constructs, mathematically, is provided, including the emanator construct that parallels the quantum propagator construct. Here, and in Section 4.7, we focus on what variances in standard quantum theory (with above constructs and dynamics from Schrödinger's equation) are possible and where possible, if they give rise to a different quantum theory (they won't). The larger question of generalization of such constructs with dynamics merely from the multiplication operation in a higher-order hypercomplex mathematics is discussed in Book 7 [21] (where it results in Emanator Theory). Note, however, that in the p-Norm generalizations considered, and soon to be excluded for $p \neq 2$, there is mentioned by [95] the possibility of manual normalization: given a state $|\psi\rangle = \sum_x a_x |x\rangle$, the probability of outcome $|x\rangle$ is given by $|a_x|^p / \sum_y |a_y|^p$. In such a manual (decoupled) normalization, there is no longer a need to keep $|\psi\rangle$ itself normalized. This will allow the possibility of local normalization according to the local action of 'A' any invertible matrix. For quantum theories in terms of wavefunctions this variation would result in superluminal signaling and proof of NP=P, and limit elements that are zero-divisors [95], so can be excluded, and the quantum theory will remain focused on p=2 in what follows. The manual normalization described above is mentioned, however, because this is precisely the mechanism for propagation ('emanation') in emanator theory, but the manual normalization is actually with p=2! However, the freedom of local normalization with local action, and sum-on-paths normalization overall, is then employed to arrive at emanation theory's simple (multiplicative) propagation step (see App. D for brief description and Book 7 [21] for details).

So, as regards the p-Norm generalization, we see that even Emanator Theory will restrict to p=2. This is no surprise if it is recalled that the properties of a sum on squares are fundamental in many areas: Pythagorean theorem, Fermat's Last Theorem, and least-squares regression in mathematics, and adding noise in quadrature in physics. A more formal mathematical explanation for why p=2 is strictly needed is that it is the only norm that permits non-trivial norm-preserving linear maps (shown in [95]). From Gleason's theorem [96] it also noted that in a Hilbert space of dimension of 3 or higher, use of $p \neq 2$ will require contextuality, or violate some other basic attribute of the theory [97,98].

As for the argument for complex amplitudes, this is made by first generalizing the classical deFinetti relation to the quantum deFinetti relation [99]. Starting from the quantum deFinetti relation (which gives classical Bayesian reasoning), we find that simple power counting shows that only complex amplitudes are possible. Consider $f(n)$ the number of real parameters to describe an n-dimensional mixed state (an $n \times n$ matrix). For real amplitudes: $f(n) = n(n+1)/2$; for complex: $f(n) = n^2$; and for quaternionic: $f(n) = 2n^2 - n$. Thus, only complex $f(n)$is compatible with the norm-preserving constraint: $f(n_A n_B) = f(n_A)f(n_B)$. So, from simple power counting we see that anything other than complex amplitudes will require other more contrived constraints to make the theory work.

Thus, roughly speaking, the wavefunction and related formalism is uniquely singled out. This will be comforting as we move through the usual quantum mechanics representations in terms of such wavefunctions. The significance of this convenience is lost, however, if not considered in the context of Emanator theory. So a brief description of Emanator Theory is given in App. D. There it will be shown that certain (maximal perturbation) parameters and structures of the standard model of particle physics have their origin in Emanator theory. The Emanator theory emanation process is hypothesized to have an association with a projected quantum process, undergoing quantum propagation, that inherits alpha and the standard model gauge group. What remained undetermined in this process was why the projection of the propagatable construct (to parallel the emanator propagator construct as a projection) should reduce to the form of wavefunction and norm as discussed – this is now clarified – the theory in terms of wavefunction and 2-norm is the only one allowed.

### 7.2 Lorentz invariance in General Settings

A now obscure paper, published in 1917, describes Lorentz invariance for complex bi-quaternions [100] (modern notation [101,102]). Given the importance of the Lorentz Invariance in generating Dirac's equation, as just shown, the question arises as to whether a generalized Lorentz Invariant formulation might lead to a generalized physical theory. This is one of the pathways to seeing the generalized emanator theory that is described in detail in Book7 [21].

***Notation***

We can write the coordinate transformation under Lorentz transform as

$$x^{\mu'} = \Lambda^{\mu'}{}_{\upsilon} x^{\upsilon} \text{ ,}$$

where $x^{\mu'}$ is a contravariant coordinate, and if the 4-vector length is invariant, then the metric is invariant:

$$\eta = \Lambda^T \eta \Lambda.$$

Any 4-vector entity $z^{\upsilon}$ that transforms as $x^{\upsilon}$, is, similarly, called contravariant. This is to distinguish it from covariant 4-vectors that transform as:

$$z'_{\mu} = (\Lambda^{-1})^{\upsilon}{}_{\mu} z_{\upsilon}.$$

Note that for rotation transformations in 3D we have $R^{-1} = R^T$, so the '3-vectors' transform identically, and there is no differentiation of contravariant and covariant vectors. For the 4D Lorentz transform, however:

$$\eta = \Lambda^T \eta \Lambda \quad \rightarrow \quad \Lambda^{-1} = \eta^{-1} \Lambda^T \eta$$

and $\Lambda^{-1}$ and $\Lambda^T$ are not the same.

#### 7.2.1 Lorentz-invariant Spinor Solutions in addition to Vector solutions [103]

Consider that a mathematical subtlety has arisen at this juncture that is not due to the novel three boosts introduced by the 4D Lorentz transformation, but due to the three (standard) rotations. (This discussion originally appeared in Book7 [21].) This is because the rotation group in 3D, SO(3), admits a double cover from the group SU(2), indicating two types of solutions. We find that there is a new type of Lorentz invariant description, other than the 4-vectors, known as a spinor. The easiest way to show that Lorentz invariance extends to the new mathematical object is to rewrite a 4-vector as a 2x2 Hermitian matrix, where $V^a$ with components $(V^0, V^1, V^2, V^3)$ is written as the 1-1 mapping as

$$\psi(V^a) = V^{AA'} = \begin{pmatrix} V^{00'} & V^{01'} \\ V^{10'} & V^{11'} \end{pmatrix} = \frac{1}{\sqrt{2}} \begin{pmatrix} V^0 + V^3 & V^1 + iV^2 \\ V^1 - iV^2 & V^0 - V^3 \end{pmatrix}$$

where the length of the 4-vector is related to the determinant of the matrix:

$$\det[\psi(V^a)] = \frac{1}{2}\eta_{ab} V^a V^b.$$

Thus, there is a map from SL(2,C) to the Lorentz transforms where there is a 2-1 isomorphism. This can be seen to be directly related to the SU(2) double cover (2-1 mapping) on SO(3) as mentioned previously (we consider the Lorentz transform leaving the time-like component of a chosen orthonormal tetrad invariant, what remains is the 3D rotation group SO(3) ). This is not simply a useful mathematical relation. We've identified the invariance under Lorentz transform as a fundamental element of the theory (giving rise to special relativity, etc.). *If we see this then conveyed to an extended set of spinorial solutions, in addition to vectorial, then new forms of energy/matter are indicated for the spinorial solutions. This is precisely what is observed: matter is spinorial (spin* $\pm 1/2$*) and (force) fields are vectorial.*

**7.2.2 Lorentz invariance via complex biquaternions [100-102]**

Let's now consider a similar process involving transformational invariance but instead of encoding the Lorentz transform in the form of matrix transformation invariance let's use elements of the Cayley algebras instead. Specifically, consider the following transformation:

$$q' = aqa_c^*, \quad where \quad aa_c = 1,$$

where $a$ is a (unitary) complex bi-quaternion: $H(\mathbb{C}) \times H(\mathbb{C})$, and $q = (ct, ix, iy, iz)$ (note this notation is for $q = (ReH_1, iImH_1, iReH_2, iImH_2)$ ). The $q' = (ct', ix', iy', iz')$ that results will correspond to a proper orthochronous Lorentz transform [100-102].

This is a remarkable result, but is there more at higher dimensionality/complexity? Consider that a complex bi-quaternion is isomorphic to a complex octonion which is isomorphic to a sedenion, thus a theory for sedenion emanation is indicated from this result as far back as 1917. The 'halting condition' on the generalization in 1917 seems to be that octonions are the highest-order of the division algebras that have inverses defined (which is necessary to have $aa_c = 1$ be defined). But they have already extended past octonions since these are complex octonions ≅ chiral sedenions, so how are they guaranteed to have $aa_c = 1$ be defined? This is possible for the sedenions, as shown in [104], if restricted to be unit norm. So now we have our answer based on the results from [104] – we can go one complexation order higher:

$$Q' = AQA_c^*, \;\; where \;\; AA_c = 1,$$

where $A$ is a (unitary) bi-complex bi-quaternion: $H(\mathbb{C} \times \mathbb{C}) \times H(\mathbb{C} \times \mathbb{C})$, which is isomorphic to a unitary quaternionic bi-quaternion, $H(\mathbb{H}) \times H(\mathbb{H})$ , and $Q = (ct, ix, iy, iz)$.

where $Q = (ReH_1, iImH_1, iReH_2, iImH_2)$ as before, except $H_1 = H \times \mathbb{H}$ not $H_1 = H \times \mathbb{C}$

The $Q' = (ct', ix', iy', iz')$ that results will again correspond to a proper orthochronous Lorentz transform. Again, how do we know the critical operation $AA_c = 1$ can always be satisfied? Previously we saw that a complex bi-quaternion was equivalent to a sedenion (in the sense described in [104]), thus a bi-complex bi-quaternion is isomorphic to a complex chiral sedenion, which is isomorphic to a trigintaduonion. This is precisely the construct examined in [104], so a generalization of the 1917 result to unitary quaternionic bi-quaternions appears possible. As shown in [104], however, there is no higher order construct. This latter form (on doubly-chiral trigintaduonions) establishes the Emanator as Lorentz Invariant explicitly. To consider the 1917 relation (involving sedenions) directly in the analysis, we shall simply expand the trigintaduonion to a sedenion and operate at that level in the analysis.

### 7.2.3 Lorentz invariance via complex biquaternions using Split Octonion Algebra

Maximum information flow occurs for octonions when split, because additional structure can occur, such as zero divisors. Maximum information flow with a propagate-able mathematical structure is thus a chiral trigintaduonion of the form

$$T_{chiral}^{(k)} = \begin{cases} ((O_{split}, \alpha), \beta) \\ ((\alpha, O_{split}), \beta) \\ (\beta, (O_{split}, \alpha)) \\ (\beta, (\alpha, O_{split})) \end{cases}.$$

Point-like matter appears in the theory when the zero-divisors in the propagation are managed by removal according to maximal analytic domain (the maximum flow hypothesis will occur in a maximal domain). Thus, it is found that the zero-divisor general formulation of the maximal propagator is indicated by maximum information flow and indicates point-like matter. This formulation involves the split-Cayley algebras. Thus, the chiral trigintaduonion is a chiral bi-sedenion as originally indicated [104]. The chirality at sedenion-level is:

$$S_{split,L} = \left(O_{split}, O_{real}\right) \; or \; S_{split,R} = \left(O_{real}, O_{split}\right).$$

If we are working with a chain of products of chiral emanator trigintaduonions $T^{(1)}_{chiral}$ this will reduce to a chain of products of $((O_{split}, \alpha), \beta)$ and upon using the unit-norm constraint to eliminate $\beta$ as an independent variable, we have a product of $(O_{split}{}', \alpha')$. So, dropping the primes, we have products:

$$(O_{split}, \alpha)_1 \times (O_{split}, \alpha)_2 \times \ldots$$

Since $O_{split} \cong H \times H$ and $(O, \alpha) \cong O(\mathbb{C})$, we have

$$(H(\mathbb{C}), H(\mathbb{C}))_1 \times (H(\mathbb{C}), H(\mathbb{C}))_2 \times \ldots$$

Recall in the background section that if we wanted to consider the standard equivalence class of invariants for unitary complex bi-quaternion operations, then we call

$$(H(\mathbb{C}), H(\mathbb{C}))_1 = a_c^*, \;\; aa_c = 1, \;\; |a| = 1,$$

and have propagation according to:

$$qa_c^* \rightarrow q_{new}$$

and invariance according to

$$q' = aqa_c^*$$

where invariance for $q = (ct, ix, iy, iz)$ is realized such that the $q' = (ct', ix', iy', iz')$ that results will correspond to a proper orthochronous Lorentz transform. The maximal information hypothesis underlying the proposed Emanator theory, thus indicates Lorentz invariant point-like matter.

**7.3 Spin-statistics**

First, to clarify, when we speak of a relation between spin and statistics we are actually speaking of a deeper relation between spin and commutation rules, where the commutation rules in turn, give rise to Fermi-Dirac or Bose-Einstein statistics. We will see in Book 5 [15], that canonical quantization for a simple quantum field theory is equivalent to second quantization, which is equivalent to a collection of harmonic oscillator quantizations (Section 4.3) where the creation and annihilation operators satisfy either (Bose) commutation rules:

$$[a_n, a_m] = [a_n{}^{\dagger}, a_m{}^{\dagger}] = 0, \qquad [a_n, a_m{}^{\dagger}] = \delta_{nm},$$

or (Jordan-Wigner) anticommutation rules:

$$\{a_n, a_m\} = \{a_n{}^\dagger, a_m{}^\dagger\} = 0, \qquad \{a_n, a_m{}^\dagger\} = \delta_{nm}.$$

To arrive at Pauli's spin-statistics relation, Pauli imposed two simple constraints on the system: (1) causality; and (2) Total system energy is non-negative. By causality, Pauli required observables outside each others light-cones, e.g., spatially separated, to not have any influence on each other, thus commute (or anticommute), so no delta term above. In Book3 [18] we see that non-negativity of system energy is not only required globally, but also locally according to various formulations (positive energy condition), and it so happens that spinorial matter on a manifold induces the positive energy condition by its existence [71]. So the constraints are already required by current understanding of the theory. With the two constraints indicated, Pauli was then able to prove [70] that quantization of integral spin particles with anticommutation rules will violate causality, and quantization of half-integral spin particles with commutation rules will violate non-negativity. This leaves only the standard spin-statistics relations as experimentally observed.

More rigorous derivations of the spin-statistics relations would be obtained in later years [72,73], with significant clarification from Wightman's theorem allowing the expectation of field products to be analytically continued to all separations [74].

## 7.4 The Euclideanized Quantum Propagator and the Classical Partition Function

Indications of a connection between Quantum Mechanics and Statistical Mechanics (likewise, quantum field theory and statistical field theory) have occurred in prior examples, let's now explore this in detail. Time consisting of a real parameter that labels causal events is familiar and is part of the dynamical description in a special role from the earliest classical mechanical descriptions. From special relativity we learn time has different forms (parametrizations) and in the context of spacetime, is simply another coordinate variable (albeit with differently signed signature). Once demoted to being a coordinate variable, even part of the time (no pun intended), we can ask about complex time in that context. The metric signature will change from Lorentzian to Euclidean if we switch to pure imaginary time, for example, where all of the Euclidean-based sums will be well-defined. In essence, we will find that the connection between Quantum Mechanics and Statistical Mechanics is that they share the same analytic time, one referencing the real-part of time in a standard dynamical context (Quantum Mechanics) while the other references the imaginary part of time in a standard equilibrium thermodynamics description of the system. The analyticity of time overall will be maximally extended in the manner that gives rise to the Feynman propagator. (which will embed the proper causality into the theory).

As usual, when a convenient mathematical connection is used a lot (Planck's original introduction of his constant and quantization [19]; or use of Vector potential in electromagnetism, proven to be real by the Aharonov-Bohm Effect [75]) we should consider what it means if analytic time really 'exists'. And it probably does, a supporting reason for this is that bound state descriptions, such as the solution for the hydrogen atom, so easy in standard Schrodinger or Heisenberg analysis, if attempted with the path integral formalism that works everywhere else (when other methods don't), is found to only work describing bound states if in curved spacetime with analytic time. So analytic time is needed to fully utilize the path integral formalism. What does it mean to have complexified time? From the Unruh analysis of the accelerated observer [15], one possibility is that this indicates a universal thermality according to acceleration. In the analysis, imaginary time is periodic and that periodicity defines the inverse temperature of the system. what results for the accelerated observer (in a standard quantum vacuum) is the appearance of a thermal flux from direction accelerating away from, with thermal spectrum at temperature:

$$T = \frac{\hbar a}{2\pi c k_B},$$

where $\hbar$ is Planck's constant, $a$ is the acceleration, $c$ is the speed of light, and $k_B$ is Boltzmann's constant. For an acceleration of $1 m/s^2$ we have a temperature $T = 4.06 \times 10^{-21} K$. The strongest acceleration we usually feel is due to Earth, 9.8 $m/s^2$, so we are bathed in a thermal bath at temperature $T = 3.98 \times 10^{-20} K$ from this effect (so drowned out by the CMBR). In Cosmological and Black Hole analysis complexified time provides a bridge to the thermodynamics of the system, which allows a thermal quantum gravity solution to be described (but not a quantum gravity solution, especially since such a solution might not exist). Once we've adjusted to the notion of analytic time, its natural to ask for the maximal analytic extension and if there is more than one choice (there is) the choice consistent with other aspects of the physics (such as causality) can be adopted at this juncture, and that is what is done with the aforementioned Feynman propagator. Analytic time provides a more complete and interconnected physical description, but it also constrains time and other aspects even more than before. So we now need to know why we have a theory with analytic time in addition to the odd local gauge group that is a product group of U(1), SU(2), and SU(3), and the set of 19 (or 22?) constants, etc. To have a deeper understanding of time, and the rest, requires an enveloping formalism to the structured formalism currently seen, and that is what is explored in Book 7 [21], with a synopsis provided in App. D.

### 7.4.1 The Propagator with Complexified Time

From the first appearance of the Schrodinger equation it was noted that changing time to imaginary time, $t \to -i\tau$, would give the diffusion equation. Let's now consider complexification of time when working with the propagator.

#### 7.4.1.1 Direct Substitution

Recall that the propagator is based on the unitary evolution operator that is defined by the Hamiltonian for the system, or by the Action on paths:

$$K(t,q,q') = \langle q|e^{itH/\hbar}|q'\rangle = \int_{\chi(0)=q'}^{\chi(t)=q} \mathcal{D}\chi e^{iS[\chi]/\hbar},$$

with paths parametrized by $\chi(t)$, that start at $q'$ and end at $q$. Let's now shift from $t \to i\tau$ in this context:

$$K_E(\tau,q,q') = \langle q|e^{-\tau H/\hbar}|q'\rangle = \int_{\chi(0)=q'}^{\chi(\tau)=q} \mathcal{D}\chi e^{-S_E[\chi]/\hbar},$$

where now the integrand is real and well-defined. The partition function is then simply given by:

$$\int dqK_E(\tau,q,q') = tr[e^{-\tau H/\hbar}]$$

where the temperature is $T = \hbar/\tau k_B = 1/\beta$. In standard notation ($k_B = 1$) using for partition function $Z(\beta)$ and free energy $F(\beta)$ we then have:

$$Z(\beta) = tr\left[e^{-\beta H}\right] = e^{-\beta F} = \oint_{\chi(0)=\chi(\tau)} \mathcal{D}\chi e^{-S_E[\chi]/\hbar}$$

Note that to have the correspondence with the definition of partition function we identified the ends of the paths, or equivalently, we've shifted to integration on periodic paths with period $\tau = \hbar\beta$. Let's now apply this to the harmonic oscillator fundamental case to see if it makes sense. For the Euclideanized harmonic oscillator, we have:

$$K_E(\hbar\beta, q, q) = \sqrt{\frac{m\omega}{2\pi\hbar\sinh(\hbar\omega\beta)}}\exp\left[-\frac{2m\omega q^2}{\hbar}\frac{\sinh^2(\hbar\omega\beta/2)}{\sinh(\hbar\omega\beta)}\right]$$

$$Z(\beta) = \int dqK_E(\tau, q, q') = \frac{1}{2\sinh\left(\frac{\hbar\omega\beta}{2}\right)} = \sum_n \exp\{-\beta E_n\},$$

where

$$E_n = \hbar\omega\left(n + \frac{1}{2}\right).$$

For low temperature, $\beta \to \infty$, and $F(\beta) \to E_0 = \frac{1}{2}\hbar\omega$, as expected. This simple substitution works out, so let's now consider complex time in more detail. In particular, is time analytic?

#### 7.4.1.2 Full analyticity – Wick rotation

We've seen that the direct substitution $t \to -i\tau$ provides interesting connections. If we arrive at this change more formally in terms of analytic time, we will have better understanding. First, as regards a Lorentzian spacetime with integrations (the Action) described as integrals in real time, it was observed that complexified time being analytic allowed for a change in the contour of integration, effectively a rotation from the real axis by 90 degrees about the origin to turn the integration into an integral along the imaginary axis, i.e., we've achieved $t \to -i\tau$ by way of a "Wick rotation". There are different ways to do this in the context of the quantum propagator, however, but only one encodes the causal structure consistently, the Feynman propagator.

Before continuing with analysis of the Feynman propagator, consider the significance of analyticity and the Wick rotation is not only that a 1-dimensional time and (N-1)-dimensional spatial dynamics problem can be turned into a N-dimensional statics problem, allowing for Euclideanized path integrals that are convergent and well-defined to obtain system integral solutions (that can be analytically carried back to the Lorentzian system representation, thereby making the path integrals well-defined via analyticity). We can also take this in the other direction, an intractable N-dimensional statics problem might be more easily solvable as a (N-1)-dimensional dynamics problem.

#### 7.4.1.3 Green's Function – Feynman Propagator – Choice of analytic extension

Let's follow the discussion of analyticity given by [105] and consider the Feynman propagator given in their notation by:

$$G_F(x^\alpha, t; x'^\alpha, t') = \begin{cases} i\sum_n \frac{\psi_n(x^\alpha)\psi_n(x'^\alpha)}{2\omega_n}\exp[-i\omega_n(t-t')], & t > t' \\ i\sum_n \frac{\psi_n(x^\alpha)\psi_n(x'^\alpha)}{2\omega_n}\exp[i\omega_n(t-t')], & t < t' \end{cases}$$

To effect the substitution by $t \to -i\tau$ by analytically extending off the real line, there must be a rotation as indicated in Figure 1, which requires analyticity in quadrants Two and Four as shown in shade.

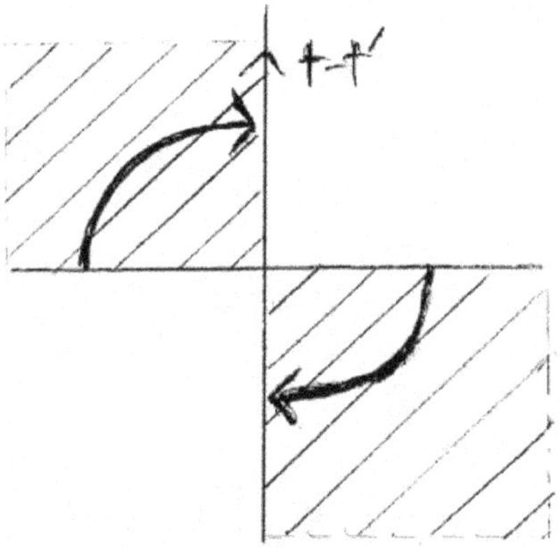


Here's the new Euclideanized Green's function $G_E$ (or Euclideanized Feynman propagator):

$$G_E(x^\alpha, t; x'^\alpha, t') = \begin{cases} i\sum_n \frac{\psi_n(x^\alpha)\psi_n(x'^\alpha)}{2\omega_n}\exp[-\omega_n(\tau-\tau')], & \tau > \tau' \\ i\sum_n \frac{\psi_n(x^\alpha)\psi_n(x'^\alpha)}{2\omega_n}\exp[\omega_n(\tau-\tau')], & \tau < \tau' \end{cases}$$

which is well-defined and unique given the fall-off condition to zero for large$|\tau - \tau'|$.

The choice of shaded regions allows the "Wick rotation" shown, and this convention for analytic extension is usually captured in the mathematics (e.g., not diagrammatically as shown) by introduction of a small imaginary part to $\omega_n$, or in momentum representation (taking a Fourier Transform), to $p^2 - m^2$. Let's shift to the standard Feynman propagator form with this in mind. Staring with the standard Green's function to the Klein Gordon equation we have the standard solution

$$G(x, y) = \frac{1}{(2\pi)^4}\int d^4p \frac{e^{-ip(x-y)}}{p^2 - m^2 \pm i\varepsilon}$$

where the $\pm i\varepsilon$ denotes various choice of deformation to the integration contour to have a well-defined propagator, and these solutions are different. We want the Feynman propagator convention allowing the "Wick rotation" indicated, thus, we want:

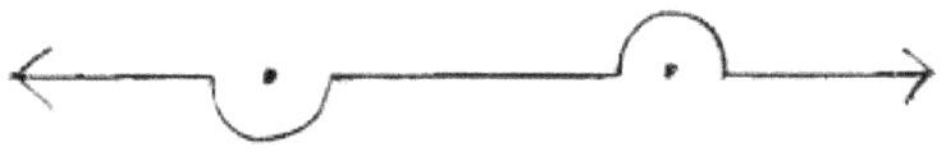

This is equivalent to the definition in terms of the limit as the $i\varepsilon$ contour deformation goes to zero:

$$G_F(x, y) = \lim_{\varepsilon \to 0} \frac{1}{(2\pi)^4} \int d^4p \frac{e^{-ip(x-y)}}{p^2 - m^2 + i\varepsilon}.$$

**7.4.1.4 Thermal Green's Function**

Let's now consider a system at a temperature $T = 1/\beta$ and get the associated Thermal Green's Function. If we have a temperature and using the grand canonical ensemble formalism (see [20]) we can write the expectation value of any operator $A$ as:

$$\langle A \rangle_\beta = \frac{Tr[\exp(-\beta H)\, A]}{Tr[\exp(-\beta H)]}$$

The thermal Green's function would then be:

$$G_T(x, y) = i\langle T\varphi(x)\varphi(y)\rangle_\beta.$$

The solution has the same terms as before, but now has a Bose-Einstein statistics contribution:

$$G_T\left(x^\alpha, t; x'^\alpha, t'\right) = \left\{ \begin{array}{ll} i\sum\limits_n \dfrac{\psi_n(x^\alpha)\psi_n\left(x'^\alpha\right)}{2\omega_n} \left\{ \begin{array}{c} (1 + n_B)\exp[-i\omega_n(t - t')] \\ + n_B \exp[i\omega_n(t - t')] \end{array} \right\}, & t > t' \\ i\sum\limits_n \dfrac{\psi_n(x^\alpha)\psi_n\left(x'^\alpha\right)}{2\omega_n} \left\{ \begin{array}{c} (1 + n_B)\exp[i\omega_n(t - t')] \\ + n_B \exp[-i\omega_n(t - t')] \end{array} \right\}, & t < t' \end{array} \right\}$$

where

$$n_B = \frac{1}{(\exp(\omega_n \beta) - 1)}$$

As before, we can analytically continue to obtain the more manageable Euclideanized form, and this will correspond to the analytic continuation under conditions requiring periodicity in imaginary time with period $\beta$:

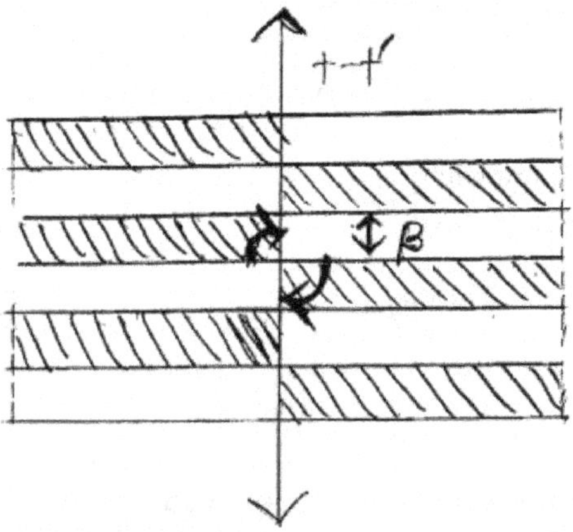


This will be useful in describing the Black holes thermodynamics analysis that is used in [20, 125,126].

### 7.4.2 Thermal Quantum Field Theory

#### 7.4.2.1 Complexification of time arrives at thermal state and partition function

If we have a quantum system that is in equilibrium with its environment, and where nontrivial exchange is possible in energy and particles with that environment, the 'Euclideanization' of the spacetime described above brings us to a Grand-Canonical Ensemble formulation for the thermal state given by:

$$\Phi_{\beta,\mu} = \frac{e^{-\beta(H-\mu N)}}{Z}, \quad Z = Tr\left[e^{-\beta(H-\mu N)}\right]$$

where $H$ is the Hamiltonian operator of the system, $N$ is the number operator, $\mu$ is the chemical potential, and $Z$ is the partition function (from which the entire system thermodynamics can be derived).

#### 7.4.2.2 Non-interacting particles: Bose-Einstein and Fermi-Dirac statistics

Suppose we have non-interacting particles with energy and number density given by (harmonic oscillator form):

$$H = \sum_n \varepsilon_n a_n{}^{\dagger} a_n, \quad N = \sum_n a_n{}^{\dagger} a_n$$

where $\{a_n{}^{\dagger}, a_n\}$ are the standard creation and annihilation operators, which for Bosons will obey the standard commutation relations:

$$[a_j, a_k{}^{\dagger}] = \delta_{jk}, \qquad [a_j, a_k] = 0, \qquad [a_j{}^{\dagger}, a_k{}^{\dagger}] = 0$$

The mean number of Bosons molecules in mode $n$ is then $n_B$:

$$n_B = Tr\left[a_n{}^\dagger a_n \Phi_{\beta,\mu}\right] = \frac{1}{\exp(\beta[\varepsilon_n - \mu]) - 1},$$

where $\varepsilon_n > \mu$ is required to have a positive mode number.

For Fermions the commutator relations above get replaced with anticommutators, and we now have:

$$n_F = Tr\left[a_n{}^\dagger a_n \Phi_{\beta,\mu}\right] = \frac{1}{\exp(\beta[\varepsilon_n - \mu]) + 1},$$

there is no longer a problem $\varepsilon_n < \mu$ in this situation to stay consistent with positive occupation numbers (we have antimatter).

### 7.4.3 Laws of Thermodynamics are recovered

#### 7.4.3.1 Definition of Heat Bath

Let's consider a total system Hamiltonian $H$ consisting of the system Hamiltonian $H_S$ and of a heat bath Hamiltonian $H_B$, and an interaction Hamiltonian between System and Heat Bath, $H_{SB}$. Thus:

$$H = H_S + H_B + H_{SB}.$$

The Hamiltonian gives the unitary time evolution operator as

$$U(t) = T\{\exp -i \int_0^t dt' H(t')\}$$
$$\equiv \lim_{\delta t \to 0} e^{-i\delta t H(t)} e^{-i\delta t H(t-\delta t)} \ldots e^{-i\delta t H(\delta t)} e^{-i\delta t H(0)}$$

(the starting point form many path integral formulations). Instead of a wavefunction, with mixing, and with Euclideanization, we have a system density matrix that is acted on by the unitary evolution operator:

$$\rho(t) = U(t)\rho(0)U^\dagger(t), \quad \frac{\partial \rho(t)}{\partial t} = -i[H(t), \rho(t)].$$

Let's consider the mean energy change in the heat reservoir:

$$\frac{\partial \langle H_B(t) \rangle}{\partial t} = \frac{\partial}{\partial t}\langle H_B(t) - \mu N \rangle + \mu \frac{\partial}{\partial t}\langle N \rangle = F + P,$$

where $F = \frac{\partial}{\partial t}\langle H_B(t) - \mu N \rangle$ is the heat flow that enters the reservoir and $P = \mu \frac{\partial}{\partial t}\langle N \rangle$ is the power that enters the reservoir. The separation into $F$ and $P$ parts is indicated by the definition of entropy in the next section.

#### 7.4.3.2 Entropy

Recall the Boltzmann entropy formula in terms of system probabilities (Shannon Entropy) is:

$$S = -k_B p \ln p,$$

here we take $k_B = 1$ and switch to the density matrix formulation, to get the standard von Neumann entropy form relevant to our situation:

$$S = -Tr\{\rho \ln \rho\}$$

and we then have:

$$\partial_t S = -\partial_t Tr\{\rho \ln \rho\} = 0$$

under unitary evolution. Now suppose we split the density matrix into a non-thermal system part (non-system parts traced out) and a thermal system part $\Phi_{\beta,\mu}$ as indicated previously (thereby ignoring reservoir information). Denote the effective description by $\rho_S \otimes \Phi_{\beta,\mu}$ . The entropy difference (divergence) that describes the difference between the full density matrix and the non-thermal state density matrix, the information loss by such an approximation, is then $(\rho||\rho_S)\otimes\Phi_{\beta,\mu}$ and the entropy of this can be written

$$S[(\rho||\rho_S)\otimes\Phi_{\beta,\mu}] = S[\rho_S] - S[\rho] + \beta\langle H_B - \mu N\rangle + \ln Z$$

Thus (showing $k_B$):

$$\partial_t S[(\rho||\rho_S)\otimes\Phi_{\beta,\mu}] = \partial_t S[\rho_S] + \frac{F}{k_B T}.$$

This allows us to write:

$$\partial_t S[\Phi_{\beta,\mu}] = \beta\partial_t\langle H_B - \mu N\rangle = \frac{F}{k_B T}$$

**7.4.3.3 First Law of Thermodynamics**

Let's start with the implications of number conservation. Global number conservation gives:

$$[N_S + N_B, H] = 0,$$

while local number conservation gives:

$$[N_B, H_B] = 0, \quad [N_S, H_S] = 0.$$

Taken together we then know that the thermal coupling terms satisfies:

$$[N_S + N_B, H_{SB}] = 0,$$

which means that particles are either in the system or in the bath (the coupling does not change their total number). If there were a coupling number, this would be the same as the change in that number being zero. There is a coupling energy term, however, so let's make change in the coupling energy over time equal to zero. This means that:

$$\partial_t\langle H_{SB}\rangle = i\langle[H_S + H_B, H_{SB}]\rangle = 0.$$

In terms of the $F, P$ variables introduced earlier:

$$F = i\langle[H_S - \mu N_S, H_{SB}]\rangle, \qquad P = i\mu\langle[N_S, H_{SB}]\rangle,$$

and

$$\partial_t\langle H_S\rangle = \langle\partial_t H_S\rangle - (F + P),$$

where $\langle\partial_t H_S\rangle$ is power entering the system, and $F, P$ describe energy flows leaving the system. Thus, the first law of thermodynamics is shown for the system.

#### 7.4.3.4 Second Law of Thermodynamics

Let's return to the product notation for the effective density of states given our (typical) missing information of a system: $\rho_S \otimes \Phi_{\beta,\mu}$. Let's now consider the real system to have precisely this tensor product form as initial state:

$$\rho(0) = \rho_S(0) \otimes \Phi_{\beta,\mu}.$$

We can now write change in entropy as:

$$\Delta S(t) = S\left[(\rho(t) || \rho_S(t)) \otimes \Phi_{\beta,\mu}\right] = S[\rho_S(t)] - S[\rho(t)] + \frac{Q}{k_B T}$$

where heat is

$$Q = Tr\{(H_B - \mu N)\rho(t)\} - Tr\{(H_B - \mu N)\rho(0)\}.$$

Note that the expression above for change in entropy is in terms of a relative entropy, which is always positive of zero. Thus,

$$\Delta S(t) \geq 0,$$

the Second Law of Thermodynamics.

# Chapter 8. General Relativistic Shell Collapse Quantization

## 8.1 Dust Shell Collapse in full General Relativity

In classical general relativity, every three-manifold occurs as the spatial topology of a globally hyperbolic vacuum spacetime. In a canonical approach to quantum gravity, the spatial topology is frozen, and one can ask for ground states corresponding to each topology. (Even in a theory that permits topology change, topologies threaded by electric of magnetic charge (flux) in source-free Einstein-Maxwell theory (or in higher-dimensional gravity with Kaluza-Klein asymptotic behavior) cannot evolve to Euclidean space. If there is a nonsingular quantum theory of such a system, it must allow a ground state with nonzero asymptotic charge and non-Euclidean topology. Topological geons with half-integral angular momentum in a quantum theory of gravity would similarly be unable to settle down to Euclidean topology.

Thus, spherically symmetric minisuperspaces also provide simple models for the quantization of geometries with non-Euclidean topology (such as topological geons). The spatial topologies consistent with spherical symmetry and asymptotic flatness are R3, the wormhole S2xR of the extended Schwarzschild geometry with two asymptopias, and the RP3 geon, a manifold with a single asymptopia obtained by removing a point from the compact manifold RP3. This last manifold is the space acquired from an extended Schwarzschild geometry by identifying diametrically opposite points on an U + V = constant slice, with U and V the usual Kruskal null coordinates (see Book 3 [18]).

For pure Einstein gravity in four spacetime dimensions, spherically symmetric minisuperspace has been considered by several authors [108-117]. For extensions to related theories, including spherically symmetric Einstein-Maxwell theory and lower-dimensional dilatonic theories, see [116,118-126]. For discussions within the Euclidean context, see for example [127-136] and the references therein. In the present analysis we add to spherically symmetric Einstein gravity a single degree of freedom by introducing a thin dust shell of (strictly) positive rest mass. A shell of vanishing rest mass has been considered recently by Kraus and Wilczek [137,138], following an earlier minisuperspace treatment of a bubble wall by Fischler et al . [139]. The work in [137] also contained an essentially complete derivation of a Hamiltonian for a shell with positive rest mass, using a gauge with flat spatial sections outside the shell. However, the resulting Hamiltonian is a complicated function of the shell radius and the conjugate momentum, determined via the solution to a transcendental equation that cannot be explicitly inverted.

In this analysis we first (re)derive the Hamiltonian action of Kraus and Wilczek [137] for a massive shell in a simpler way, directly computing the symplectic product on a reduced phase space to obtain the momentum conjugate to the shell radius. We then introduce a formalism for reparametrizing time in a Hamiltonian theory with a two-dimensional phase space. Applying this formalism to the shell Hamiltonian, we redefine the coordinate time to coincide with the proper time of the shell, recovering a Hamiltonian that can be given in terms of elementary functions [140]. An alternative choice of the coordinate time yields a Hamiltonian that generalizes to our self-gravitating shell the familiar Hamiltonian of a spherical test shell in Minkowski space.

The proper time Hamiltonian for a shell on a wormhole spacetime was considered previously by Berezin et al [140]. For a flat geometry interior to the shell, our proper time Hamiltonian has been considered classically in [141] and quantum mechanically in [142]. For a at geometry interior to the shell, a super-Hamiltonian quantization obtained via yet another time reparameterization has been considered in [143].

The formalism so far describes a shell in a three-geometry with Euclidean topology or wormhole topology. However, this formalism is easily adapted to a shell in a space with topology RP3. The covering space has wormhole topology, and by lifting shell and geometry to the covering space, one obtains a left-right symmetric geometry with two symmetrically placed shells.

The paper concludes with a discussion of the prospects for quantization. Quantization of the vacuum case is revisited to emphasize choices that lead to discrete or continuous mass spectra. The additional degree of freedom provided by the shell does not appear to qualitatively alter these choices.

A disadvantage of using the shell to model an additional degree of freedom is that it brings the pathology of the relativistic Coulomb problem: a Hamiltonian that is not bounded below when the shell has rest mass larger than the Planck mass.

Adding a spherically symmetric scalar field would be a more realistic and more difficult way to model the gravitational degrees of freedom that are ruled out by spherical symmetry. A discussion of the difficulties involved is given in [144,145]. For a discussion in the context of a dilatonic black hole, see [146].

Latin indices will denoter abstract spacetime indices, as in Wald.

### 8.1.1 Lagrangian formulation

Consider the quantization of a spherically symmetric geometry whose classical evolution is described by Einstein's theory of General Relativity. The goal is to better understand the properties of "Quantum Gravity." Spherically symmetric and asymptotically flat spacetime, with no matter, has no dynamics. This assumes, however, that no further restrictions are placed on the canonical 3-geometry information on the 3-dimensional hypersurfaces. Otherwise, dynamical structure can be introduced through a variety of reductions. If hypersurface slicing restrictions (gauge choice) are only effected by boundary conditions (such as asymptotically flat) and the presence of matter, then a simple yet non-trivial dynamical theory might involve a spherically symmetric, infinitesimally thin, shell of dust matter. Such is studied here. (Note, in Chapter 8 the equation numbers are those used in the original PhD thesis so may be out of sequence in places, but the text references to those equation numbers are still valid.)

The general spherically symmetric four-dimensional metric is

$$(1)\ ds^2 = \sum_{\alpha,\beta=t,r} g_{\alpha\beta}\,(t,r)dx^\alpha dx^\beta + R(t,r)^2\,(d\theta^2 + \sin^2\theta d\theta^2).$$

For canonical quantization, the spherically symmetric ADM form for the metric is used (which casts General Relativity into canonical form):

$$(2)\ ds^2 = -N^2\,dt^2 + \Lambda^2(dr + N^r dt)^2 + \ R^2 d\Omega^2,$$

Where $d\Omega^2$ is the metric on the unit two-sphere, and N, $N^r, \Lambda$ and R are functions of $t$ and $r$ only. N and $N^r$ describe the lapse and shift between hypersurfaces in the foliation, $\Lambda$ is $ds/dr$, and R is transverse radius. We shall work in natural units, $\hbar = c = G = 1$.

Einstein's equations, $G_{\mu\nu} = 8\pi T_{\mu\nu}$, can be obtained from variation of the action

(3) $S[\Lambda, R, N, N^r] = \frac{1}{16\pi} \int d^4x \sqrt{-{}^{(4)}g}\,\left({}^{(4)}R\right) + S_M +$ (boundary terms)

Where ${}^{(4)}R$ is the four dimensional Ricci scalar, $S_M$ is the matter action, the boundary terms are obtained from the restriction of hypersurface 3-geometry data to asymptotically flat, and the terms to be varied are given square brackets.

For the metric in (2) we can write:

(4a) $\mathcal{L} = \mathcal{L} + \partial_\mu \alpha^\mu = \frac{1}{4\sin\theta} \sqrt{-{}^{(4)}g}\,\left({}^{(4)}R\right),$

(4b) ${}^{(1)}\mathcal{L} = \frac{1}{2}\Lambda N - \frac{1}{2}\Lambda N^{-1}(DR)^2 + \frac{1}{2}N\Lambda^{-1}(R')^2 + RR'(N')\Lambda^{-1}$ $-RN^{-1}(DR)(\dot{\Lambda} - (\Lambda N^r)')$

(4c) $\alpha^t = \frac{1}{2}R^2 N\left(\dot{\Lambda} - (\Lambda N^r)'\right) + R\Lambda N^{-1}(DR),$

(4d) $\alpha^r = -\frac{1}{2}R^2\left(N'\Lambda^{-1} + N^r N^{-1}(\dot{\Lambda} - (\Lambda N^r)'\right) -$ $R\Lambda N^r N^{-1}(DR) - RN\Lambda^{-1}R'.$

The notation $D \equiv \partial_t - N^r\partial_r$ has been adopted for convenience. Since boundary terms are to be considered later, the surface terms contribution from $\partial_\mu \alpha^\mu$ is simply dropped from the action.
$\left(\text{Note that } ' \text{ represents } \frac{d}{dr} \text{ and } \cdot \text{ represents } \frac{d}{dt}.\right)$

The derivation for the Lagrangian description for a shell of dust is given in Section X. The resulting action is

(5a) $S_M = -\mu \int_{shell} d^3\Sigma,$

Where $d^3\Sigma$ is the volume element of the shell's trajectory in spacetime. Such an action may well have been guessed, and then verified, as it has the minimal geometric structure expected of a variational formulation. Written in terms of the metric variables:

$$(5b)\ S_M = -4\pi\mu \int \hat{R}^2\, d\tau = -m \int dt \sqrt{\hat{N}t^2 - \hat{L}^2\left(\dot{\hat{r}} + \hat{N}^r\right)^2},$$

Where the proper time difference for radial shell motion,

$$d\tau^2|_{shell} = \hat{N}^{t^2} dt^2 - \hat{L}^2\left(d\hat{r} + \hat{N}^r dt\right)^2,$$

has been used, as well as the $m = 4\pi\hat{R}^2\mu$ conserved mass relation. The hatted variables denote evaluation at the shell radial coordinate $\hat{r}$.

A Legendre transformation is now performed to obtain a Hamiltonian form for shell Action. For the gravitational part of the action:

$$(6a)\ P_R = \frac{\delta\mathcal{L}}{\delta\dot{R}} = N^{-1}\left(\Lambda\left(\dot{R} - R'N^r\right) + R\left(\dot{\Lambda} - (\Lambda N^r)'\right)\right),$$

$$(6b)\ P_\Lambda = \frac{\delta\mathcal{L}}{\delta\dot{L}} = -N^{-1}R\left(\dot{R} - R'N^r\right),$$

While for the matter part of the action:

$$(6c)\ \hat{p} = \frac{\partial\mathcal{L}_M}{\partial\dot{\hat{r}}} = \frac{m\hat{L}^2\left(\dot{\hat{r}} + \hat{N}^r\right)}{\sqrt{\hat{N}^{t^2} - \hat{L}^2\left(\dot{\hat{r}} + \hat{N}^r\right)^2}}$$

Inverting these equations in terms of the coordinate velocities, $\dot{R}, \dot{L},$ and $\dot{\hat{r}}$, which can be done explicitly, with substitution into the Lagrangian action then leads to the Hamiltonian action:

$$(7a)\ S[\Lambda, R, P_\Lambda, P_R; N, N^r] = \int dt \int_{-\infty}^{\infty} dr\left(P_\Lambda\dot{\Lambda} + P_R\dot{R} - NH - N^r H_r\right) + \int dt\hat{p}\dot{\hat{r}} + \text{(boundary terms)},$$

where the super-Hamiltonian constraint H, and the radial super momentum constraint $H_r$ are given by

$$H = \int dr(N\mathcal{H} + N^r\mathcal{H}_r),$$

$$\mathcal{H} = \frac{\Lambda P_\Lambda^2}{2R^2} - \frac{P_\Lambda P_R}{R} - \frac{1}{2}\left[\left(\frac{2RR'}{\Lambda}\right)' - \frac{(R')^2}{\Lambda} - \Lambda\right]$$

$$\mathcal{H}_r = R'P_R - \Lambda P_\Lambda'$$

where we have considered spherically symmetric spacetimes,

$$ds^2 = -N^2(r,t)dt^2\Lambda^2(r,t)(drt\,N^r(r,t)dt)^2 + R^2(r,t)d\Omega^2.$$

The condition of asymptotic flatness on fully extended hypersurfaces leads to the addition of boundary terms to the Action:

$$S[\Lambda, P_\Lambda R, P_R; N, N^r] = \int dt \int_{-\infty}^{\infty} dr\left(P_\Lambda\dot{\Lambda} + P_R\dot{R} - N\mathcal{H} - N^r\mathcal{H}_r\right) - \int dt\,(N_+M_+ + N_-M_-),$$

where $N_\pm$ are prescribed functions of $t$ label (not to be varied), and $M_\pm$ are the $ADM$ energies.

We then consider spherically symmetric spacetimes with an infinitesimally thin matter shell:

$$S = \int dtp\dot{\hat{r}} + \int dtdr\left[P_\Lambda\dot{\Lambda} + P_R\dot{R} - N\mathcal{H} - N^r\mathcal{H}_r\right] - \int dt(N_+M_+ + M_-M_-)$$

Where $\mathcal{H} = \mathcal{H}_{Grav'} + \mathcal{H}_{shell}$ and $\mathcal{H}^r = \mathcal{H}^r_{Grav} + \mathcal{H}^t_{shell'}$ *and* $\mathcal{H}_{Grav}$ and $\mathcal{H}^r_{Grav}$ are the $\mathcal{H}$ and $\mathcal{H}^r$ above, while:

$$\mathcal{H}_{shell} = \sqrt{\left(p/\hat{\Lambda}\right)^2 + m^2}\,\delta(r-\hat{r})$$

$$\mathcal{H}^r_{shell} = -p\delta(r-\hat{r}).$$

.

Following the approach of Kraus and Wilczek, initially, the shell contribution to the action will be re-expressed via a reconstruction. The reconstruction permits the shell to be described solely in terms of the geometrics to either side; and in terms of the junction condition ties between these geometrics.

**Geometric reconstruction of shell description**
**I. Impose constraints away from shell, write dS**

Since the Poisson bracket of $\mathcal{H}_t$ and $\mathcal{H}_r$ close with no central charge, they are first class. If a first class system is solved for all momenta, $p_i - \underline{P_i}(x) = 0$, then first class implies:

$$\left[P_i - \underline{P_i}(x), P_j - \underline{P_j}(x)\right] = 0,$$

$\partial_i \underline{P_j} - \partial_j \underline{P_i} = 0 \; \forall x \Rightarrow \underline{P_i} = \partial_i F(x)$for some $F$.

$L = P_i \dot{x}^i = (\partial_i \; F)\dot{x}^i = \frac{d}{dt} F(x(t)),$

Hence,

$$S = \int_{t_0}^{t} dt' \, P_i(x(t))\dot{x}^i(t') \; = \int_{x_0}^{x} P_i \; dx^i, \quad \rightarrow \qquad ds = P_i dx^i,$$

Thus, for the shell-geometry action:

$$ds = pd\hat{r} + \int dr(P_R \delta R + \; P_\Lambda \delta \Lambda) - (N_+ M_> + N_- M_<)$$

**II. Consider the bulk contribution to dS (i.e. not shell or surface terms)**

$$(dS)_{bulk} = \int_{-\infty}^{\hat{r}-\epsilon} dr \, (P_R \; \delta R + \; P_\Lambda \delta \Lambda) + \int_{\hat{r}+\epsilon}^{+\infty} dr \, (P_R \; \delta R + \; P_\Lambda \delta \Lambda)$$

Integrate $(dS)_{bulk}$ along a path in the constraint surface for which only $\Lambda$ is changing $\left(\hat{r}, P, R, \hat{\Lambda} \text{ are held fixed}\right)$ until $P_R = 0 = P_\Lambda$. Then, keeping $P_R = 0 = P_\Lambda$, integrate along a path to any desired R, L geometry. The latter integral, in the constraint surface, contributes nothing. With any R, L data posited initially, obtain thereby an expression for the bulk contribution to the action for unitary R, L. The reference R, L data posited initially is fixed and so may be dropped in the variational formulation – thus no artifact of this reconstruction remains:

$$S' = \int (dS)_{bulk} = \int_{-\infty}^{\hat{r}-\epsilon} dr \int \delta \Lambda R \sqrt{(R'/\Lambda)^2 - 1 + 2M_</R}$$

$$+ \int_{\hat{r}+\epsilon}^{+\infty} dr \int \delta \Lambda R \sqrt{(R'/\Lambda)^2 - 1 + 2M_</R}$$

$$= \int_{-\infty}^{\hat{r}-\epsilon} \left[ R\Lambda \sqrt{(R'/\Lambda)^2 - 1 + 2M_</R} \right.$$

$$\left. + RR' log \left| \frac{(R'/\Lambda) - \sqrt{(R'/\Lambda)^2 - 1 + 2M_</R}}{|1 - 2M_</R|} \right| \right]$$

$$+ \int_{\hat{r}+\epsilon}^{+\infty} \underbrace{\left[R\Lambda\sqrt{(R'/\Lambda)^2 - 1 + 2M_</R} + RR' \log\left|\frac{(R'/\Lambda)-\sqrt{(R'/\Lambda)^2-1+2M_>/R}}{|1-2M_>/R|}\right|\right]}_{Call\ this\ F(R,R',\Lambda,\mathcal{M})}$$

Consider now the general variation of the bulk action S':

$$S' = \int_{-\infty}^{\hat{r}-\epsilon} + \int_{\hat{r}+\epsilon}^{\infty} dr\, F(R, R', \Lambda, \mathcal{M})$$

$$dS' = \int_{-\infty}^{\hat{r}-\epsilon} + \int_{\hat{r}+\epsilon}^{\infty} dr \left[\left(\frac{\partial F}{\partial R} - \frac{\partial}{\partial r}\left(\frac{\partial F}{\partial R'}\right)\right)\delta R + \frac{\partial F}{\partial \Lambda}\delta\Lambda\right]$$
$$+ \left(\frac{\partial F}{\partial R'}\delta R\right)\Big|_{-\infty}^{\hat{r}-\epsilon} \Big|_{\hat{r}+\epsilon}^{+\infty} + \int_{-\infty}^{\hat{r}-\epsilon} + \int_{\hat{r}+\epsilon}^{\infty} dr\left(\frac{\partial F}{\partial M}\right) d\mathcal{M}$$

Using the expression for the momenta obtained from imposing the constraints:

$$dS' = \int_{-\infty}^{\hat{r}-\epsilon} + \int_{\hat{r}+\epsilon}^{\infty} dr[P_R\delta R + P_\Lambda\delta\Lambda] - \left(\frac{\partial F}{\partial R}\delta R\right)\Big|_{\hat{r}-\epsilon}^{\hat{r}+\epsilon}$$
$$+ \int_{-\infty}^{\hat{r}-\epsilon} + \int_{\hat{r}+\epsilon}^{\infty} dr \left(\frac{\partial F}{\partial M}\right)\partial\mathcal{M}$$

Since the correct action is supposed to yield

$$\frac{\delta S}{\delta R} = P_R \quad , \quad \frac{\delta S}{\delta \Lambda} = P_\Lambda$$

The bulk contribution to the action that includes shell has variation:

$$dS = dS' + \left(\frac{\partial F}{\partial R}\delta R\right)\Big|_{\hat{r}-\epsilon}^{\hat{r}+\epsilon} - \int dr\left(\frac{\partial F}{\partial \mathcal{M}}\right) d\mathcal{M},$$

Including the boundary terms from the asymptotic regions, the reconstructed action is then

$$S = \int_{-\infty}^{\hat{r}-\epsilon} + \int_{\hat{r}+\epsilon}^{+\infty} dr\left\{R\Lambda\sqrt{(R'/\Lambda)^2 - 1 + 2\mathcal{M}/R} + \right.$$
$$\left. RR'\log\left|\frac{(R'/\Lambda)-\sqrt{(R'/\Lambda)^2-1+2\mathcal{M}/R}}{\sqrt{|1-2\mathcal{M}/R\,|}}\right|\right\} - \int_{-\infty}^{\hat{r}-\epsilon} + \int_{\hat{r}+\epsilon}^{\infty} dt dr\left(\frac{\partial F}{D\mathcal{M}}\right)\dot{\mathcal{M}} +$$
$$\int dt\left(\frac{\partial F}{DR}\dot{\hat{R}}\right)\Big|_{\hat{r}-\epsilon}^{\hat{r}+\epsilon} - \int dt(N_+M_> + N_-M_<),$$

In a form with constraints imposed and shell described via geometry and junction conditions. As a check:

$$\frac{dS}{dt} = p\dot{r} + \int_{-\infty}^{\infty} dr\left[P_R R' + P_\Lambda \dot{\Lambda}\right] - (N_+ M_> + N_- M_<),$$

In agreement with the $\frac{dS}{dt}$ taken with direct substitution of constraints.

The junction conditions are now incorporated into the Action. This is done in a general way, with no gauge choice necessary, by a careful limit argument on the geometry in the neighborhood of the shell. (An alternate derivation is given in [147], where gauge choice is necessary, but same result.) Depending on whether the shell is actually embedded in the right or left geometry, there are two limit procedures:

If the shell is embedded in the left hand side (LHS) geometry the RHS limit argument is:

$R'$ is freely specified for $-\infty < r \leq \hat{r} - \epsilon - \mu$ and $\hat{r} + \epsilon + \mu \leq r < \infty$. For LHS limit take

$$R'_< = \lim_{\mu \to 0} R'(\hat{r} - \epsilon - \mu) = R'(\hat{r} - \epsilon),$$

While only a smooth transition is imposed on $R'$ between $\hat{r} + \epsilon$ and $\hat{r} + \epsilon + \mu$, $R'_> \neq R'(\hat{r} + \epsilon)$, such that the $\mu \to 0^+$ limit describes a discontinuity. The expression for the action is thus:

$$S = \int_{-\infty}^{\hat{r}-\epsilon-\mu} + \int_{\hat{r}-\epsilon-\mu}^{\hat{r}-t} + \int_{\hat{r}+t}^{\hat{r}+\epsilon+\mu} + \int_{\hat{r}+\epsilon+\mu}^{\infty} dr\, \{R\Lambda\sqrt{(R'/\Lambda)^2 - 1 + 2\mathcal{M}/R} + RR' \log\left|\frac{(R'/\Lambda) - \sqrt{(R'/\Lambda)^2 + \frac{2\mathcal{M}}{R}}}{\sqrt{\left|1 - \frac{2\mathcal{M}}{R}\right|}}\right|\}$$

$$+\left\{\text{similar} \int dt dr \left(\frac{\partial F}{\partial M}\right)\dot{\mathcal{M}}\right\} + \int dt \left[\frac{\partial F}{\partial R}\right]_{\hat{r}-\epsilon}^{\hat{r}+\epsilon} - \int dt (N_+ M_> + N_- M_<),$$

where

$$F = \{R\Lambda\sqrt{(R'/\Lambda)^2 - 1 + 2\mathcal{M}/R} + RR' \log\left|\frac{(R'/\Lambda) - \sqrt{(R'/\Lambda)^2 + \frac{2\mathcal{M}}{R}}}{\sqrt{\left|1 - \frac{2\mathcal{M}}{R}\right|}}\right|\}$$

Since $R'$ is at worst a step function, the $\mu \to 0$ limits may be taken in the action and the $\mu$ interval integrals contribute zero. Thus $R'$ is freely specified and $R' = R'_{\gtrless}$ on the RHS and LHS of the shell. The junction conditions are then incorporated at the level of the Lagrangian calculation from $L = dS/dt$:

$$L = \frac{dS}{dt} = \dot{\hat{r}}\left[\hat{R}\hat{\Lambda}\left(\sqrt{<} - \sqrt{>}\right) + \hat{R}R'_{<}\log|<| - \hat{R}R'_{>}\log|>|\right]$$

$$+\hat{R}R\log\left|\frac{(R'_{+\epsilon}/\Lambda) - \sqrt{"(+\epsilon)"}}{\sqrt{1-2M_{<}/R}}\right| - \hat{R}R\log\left|\frac{(R'_{-\epsilon}/\Lambda) - \sqrt{"-\epsilon"}}{\sqrt{1-2M_{<}/R}}\right|$$

$$-(N_{+}M_{>} + N_{-}M_{<})$$

$$+\int_{\substack{-\infty\to\hat{r}-\epsilon-\mu \\ \hat{r}+\epsilon+\mu\to+\infty}} dr\, \underbrace{\left\{\frac{\partial F}{\partial R}\dot{R} + \frac{\partial F}{\partial R'}\left(\dot{R}'\right) + \frac{\partial F}{\partial \Lambda}\dot{\Lambda} + \frac{\partial F}{\partial \mathcal{M}}\dot{\mathcal{M}}\right\}}_{P_R\dot{R} + P_\Lambda\dot{\Lambda} + \left(\frac{\partial F}{\partial R'}\dot{R}\right)'} -$$

$$\int_{\substack{-\infty\to\hat{r}-\epsilon-\mu \\ \hat{r}+\epsilon+\mu\to+\infty}} dr\left(\frac{\partial F}{\partial \mathcal{M}}\right)\dot{\mathcal{M}}$$

$$L = \dot{\hat{r}}\hat{R}\hat{\Lambda}\left[\sqrt{<} - \sqrt{>}\right]$$

$$+\dot{\hat{R}}\hat{R}\log\left|\frac{(R'_{+\epsilon}/\Lambda) - \sqrt{" + \epsilon"}}{\sqrt{1-2M_{>}/R}}\right| - \dot{R}\log\left|\frac{(R'_{>}/\Lambda) - \sqrt{>}}{\sqrt{1-2M_{>}/R}}\right| - \dot{\hat{r}}\hat{R}R'_{>}\log|>|$$

$$-\dot{\hat{R}}\hat{R}\log\left|\frac{(R'_{+\epsilon}/\Lambda) - \sqrt{" + \epsilon"}}{\sqrt{1-2M_{<}/R}}\right| - \dot{R}\log\left|\frac{(R'_{<}/\Lambda) - \sqrt{<}}{\sqrt{1-2M_{<}/R}}\right| - \dot{\hat{r}}\hat{R}R'_{<}\log|<|$$

$$+\int dr\left\{P_R\dot{R} + P_\Lambda\dot{\Lambda}\right\} - (N_{+}M_{+} + N_{-}M_{-})$$

where,

$\sqrt{<} = \sqrt{\left(R'_{<}/\hat{\Lambda}\right)^2 - 1 + 2M_{<}/R}$, similarly for $\sqrt{>}$, $\sqrt{" \pm \epsilon"}$

$\log|<| = \log\left|\frac{(R'_{<}/\hat{\Lambda}) - \sqrt{<}}{\sqrt{|1-2M_{<}/R|}}\right|$, similarly for $\log|>|$

$R'_{\pm\epsilon} = R'(\hat{r} \pm \epsilon)$

$\dot{\hat{R}}(r,t) = \dot{R}(\hat{r},t) + \dot{\hat{r}}R'(\hat{r},t)$

For RHS limits $R'_{<} = R'_{-\epsilon}$ :

$$L = \dot{\hat{r}}\hat{R}\hat{\Lambda}\left[\sqrt{<} - \sqrt{>}\right] + \dot{\hat{R}}\hat{R}\log\left|\frac{(R'_{+\epsilon}/\Lambda) - \sqrt{"+\epsilon"}}{(R'_{>}/\Lambda) - \sqrt{>}}\right| +$$

$$\int_{\substack{-\infty\to\hat{r}-\epsilon \\ \hat{r}+\epsilon\to\infty}} dr\left[P_R\dot{R}\,\tau\,P_\Lambda\dot{\Lambda}\right] - (N_{+}M_{+} + N_{-}M_{-})$$

From the j.c.'s:

$$R'_{+\epsilon} = R'_{-\epsilon} - \frac{1}{\hat{R}}\sqrt{p^2 + m^2\hat{\Lambda}^2} = R'_{<} - \frac{1}{\hat{R}}\sqrt{p^2 + m^2\hat{\Lambda}^2}$$

$$\sqrt{"+\epsilon"} = \sqrt{"-\epsilon"} - \frac{P}{\hat{R}\hat{\Lambda}} = \sqrt{<} - \frac{P}{\hat{R}\hat{\Lambda}}$$

$$L = \dot{\hat{r}}\hat{R}\hat{\Lambda}\left[\sqrt{<} - \sqrt{>}\right] + \dot{\hat{R}}\hat{R}\log\left|\frac{(R'_{<}/\Lambda) - \sqrt{<} + \left(\frac{P}{\hat{R}\hat{\Lambda}} - \frac{1}{\hat{R}\hat{\Lambda}}\right)\sqrt{p^2 + m^2\hat{\Lambda}^2}}{(R'_{<}/\Lambda) - \sqrt{>}}\right|$$
$$+ \int dr\left[P_R\dot{R} + P_\Lambda\dot{\Lambda}\right] - (N_+M_> + N_-M_<)$$

A static gauge is now chosen in the vicinity of the shell. To be more precise, let $R = r\ \ and\ \ \Lambda = 1$ for $r_< < \hat{r} < r_>$ :

$$L = \dot{\hat{r}}\left[\sqrt{2M_< r} - \sqrt{2M_> r} - \dot{\hat{r}}r\log\left|\frac{\left(1 - \sqrt{2M_>/r}\right)}{\left(1 - \sqrt{2M_>/r}\right) - r^{-1}\left(\sqrt{p^2 + m^2 - p}\right)}\right|\right]$$
$$+ \int_{-\infty}^{r_<} dr\left[P_R\dot{R} + P_\Lambda\dot{\Lambda}\right] + \int_{r_>}^{\infty} dr\left[P_R\dot{R} + P_\Lambda\dot{\Lambda}\right] - (N_+M_> + N_-M_<)$$

And $M_{\lessgtr}$ are related by the junction conditions according to:

$$M_> - M_< = \sqrt{p^2 + m^2} - \frac{m^2}{2r} - p\sqrt{\frac{2M_<}{r}}$$

From here drop hats: $\dot{\hat{r}} \equiv \dot{r}$ :

$$P_< = \frac{\partial L}{\partial \dot{r}} = \left[\sqrt{2M_< r} - \sqrt{2M_> r} - r\log\left|\frac{\left(1 - \sqrt{2M_>/r}\right)}{\left(1 - \sqrt{2M_</r}\right) - r^{-1}\left(\sqrt{p^2 + m^2 - p}\right)}\right|\right]$$

Choose $N_\pm = \pm 1$:

$$\boxed{L = \dot{r}\,p_< - (M_> - M_<) + \int_{-\infty}^{r_<} + \int_{r_>}^{+\infty} dr\left[P_R\dot{R} + P_\Lambda\dot{\Lambda}\right],}$$

Where the constraint have been imposed throughout.

Reverting to the form of an action:

$$S = \int L dt = \int [\dot{r}\,P_< - (M_> - M_<)]\,dt + \int_{-\infty}^{r_<} + \int_{r_>}^{+\infty} dr\left[P_R\dot{R} + P_\Lambda\dot{\Lambda}\right],$$

Enforcing the constraints in the radial integrals in terms of the usual Lagrange multipliers of "Lapse" and "Shift";

$$S = \int [\dot{r}\,P_< - (M_> - M_<)]\,dt + \int_{\gtrless} drdt\left[P_R\dot{R} + P_\Lambda\dot{\Lambda} - NH - N^rH_r\right]$$

From this form the radial integrals are reconstituted via a canonical transformation to the Kuchař variables [148-150].

In order for the Kuchař-type transformation to be canonical the difference of the Liouville forms must be an exact form:

$$\int_{-\infty}^{\infty} dr\,(P_\Lambda\,\delta\Lambda + P_R\delta R) = \int_{-\infty}^{\infty} dr\,(P_M\,\delta M + P_R\delta R) + \delta\omega$$

Direct calculation reveals that (Kuchar [148-150]):

$$\begin{aligned} P_\Lambda\,\delta\Lambda + P_R\delta R - P_M\,\delta M - P_R\delta R \\ = \delta\left(\Lambda P_\Lambda + \frac{1}{2}RR'\ln\left|\frac{RR' - \Lambda P_\Lambda}{RR' + \Lambda P_\Lambda}\right|\right) \\ + \left(\frac{1}{2}R\delta R'\ln\left|\frac{RR' - \Lambda P_\Lambda}{RR' + \Lambda P_\Lambda}\right|\right)'. \end{aligned}$$

So, at issue in whether the transformation is canonical is whether.

$$\left(\frac{1}{2}RR'\ln\left|\frac{RR' - \Lambda P_\Lambda}{RR' + \Lambda P_\Lambda}\right|\right)\Bigg|_{r_<}^{r_>} = 0$$

By continuity, a choice gauge is imposed at specified $r_{\gtrless}$ such that $R = r$, i.e. $\dot{R} = 0$, so each of the above boundary terms is zero individually (as they need to be).

$$(P_\Lambda = R\sqrt{(R'/L)^2 - 1 + 2\mathcal{M}/R} \Rightarrow (P_\Lambda)(r = r_>) = \sqrt{2M_> r_>}$$

thus no singular term multiplies $\dot{R} = 0$.

A reduction in the canonical variables away from the shell region of gauge fixing (and prior reduction) is now done in the manner of Kuchar. As with Kuchar, embedding degrees of freedom has now been properly incorporation such that $M_{\gtrless}(t)$ are dynamic variables:

$$\int_{\gtrless} drdt\left[P_R\dot{R} + P_\Lambda\dot{\Lambda} - NH - N^r H_r\right]$$

$$\Rightarrow \int_{\gtrless} drdt\left[P_M\dot{M} + P_R\dot{R} - N^M M' - N^R P_R\right]$$

With reduction: $M'(t,r) = 0 \Rightarrow M(t,r) = M(t)$, the action now reads:

$$S = \int [\acute{r}P_< - (M_> - M_<)]dt \int dt\,(p_>\dot{M}_> + p_<\dot{M}_<)$$

where

$$P_> = \int_{r_>}^{\infty} dr\,P_M(r)\,, \quad P_< = \int_{-\infty}^{r_<} dr\,P_M\,(r).$$

So,

$$S = \int \left( P_> \dot{M}_> + P_< \dot{M}_< + P_< \dot{r} - (M_> - M_<) \right) dt,$$

And $M_> - M_<$ is related to $P_<$ and $r$ through the implicit relations:

$$P_< = \Bigg[ \sqrt{2M_< r} - \sqrt{2M_> r} - r \log \left| \frac{\left(1 - \sqrt{2M_>/r}\right)}{\left(1 - \sqrt{2M_>/r}\right) - r^{-1}\left(\sqrt{(p^2 + m^2 - p)}\right)} \right| \Bigg]$$

$$M_> - M_< = \sqrt{p^2 + m^2} - \frac{m^2}{2r} - p\sqrt{\frac{2M_<}{r}}.$$

The implicit relations may be expressed in a slightly different form if instead one takes for LHS limits $R'_> = R'_{+\epsilon}$ and at RHS $R'_< \neq R'_{-\epsilon}$

$$P_< = \Bigg[ \sqrt{2M_< r} - \sqrt{2M_> r} - r \log \left| \frac{\left(1 - \sqrt{2M_>/r}\right) + \frac{1}{r}\left(\sqrt{(p^2 + m^2 - p)}\right)}{\left(1 - \sqrt{2M_>/r}\right)} \right| \Bigg]$$

$$M_> - M_< = \sqrt{p^2 + m^2} + \frac{m^2}{2r} - p\sqrt{\frac{2M_>}{r}}.$$

The latter relations are convenient when showing that the positive energy condition is satisfied, the former is convenient in directly obtaining the equations of motion for the shell. They can be seen to be directly related by interchanging $M_>$ and $M_<$ and changing the sign of the action.

Clearly the reduced Hamiltonian is $H = M_> - M_<$. If the second set of implicit relations is used (and assume $M_>(r, P_<, M_<)$)

$$\dot{r} = \frac{\partial H}{\partial P_<} = \frac{\partial M_>}{\partial p_<} = \left(\frac{\partial M_>}{\partial p}\right)\left(\frac{\partial p}{\partial P_<}\right) = \left(\frac{\partial M_>}{\partial p}\right)\left[\frac{\partial P_<}{\partial p} + \frac{\partial P_<}{\partial M_>}\frac{\partial M_<}{\partial p}\right]^{-1}$$

$$\dot{r} = \frac{P}{\sqrt{p^2 + m^2}} - \sqrt{\frac{2M_>}{r}} \quad \text{results.}$$

Thus $\left(\frac{P}{m}\right) = \eta_1 \frac{\left(\dot{r} + \sqrt{\frac{2M_>}{r}}\right)}{\sqrt{1 - \left(\dot{r} + \sqrt{\frac{2M_>}{r}}\right)^2}}$ , $\eta_1 = \pm 1$ ,

And the Hamiltonian is:

$$H = \eta_1 \frac{m\left(1 - \left(\dot{r} + \sqrt{\frac{2M_>}{r}}\right)\sqrt{\frac{2M_>}{r}}\right)}{1 - \left(\dot{r} + \sqrt{\frac{2M_>}{r}}\right)^2} + \frac{m^2}{2r}$$

For time like shell trajectory of positive energy the first term in the Hamiltonian is positive, since $m^2/2r$ is clearly positive for $0 < r < \infty$ we get that $H \geq 0$ . The positive energy condition is satisfied (positive $M_< \Rightarrow$ positive $M_>$ ). If similar consideration is made for the shell trajectory viewed from a LHS embedding this result doesn't necessarily hold. However, the junction conditions in conjunction with the requirement of time like shell trajectory in both LHS and RHS embeddings ensures that positive energy is satisfied.

Instead of RHS static gauge "time", one might prefer proper time, and intuitively one might expect this to yield a simpler analysis. Starting with the implicit expression for the Hamiltonian above $\dot{r} = \dot{r}(r, P_<)$ , and the parameter space restrictions showing agreement with the positive mass conjecture, let's convert now to a proper time description.

In the time reparameterization that follows the implicit Hamiltonian is first considered as a conserved energy functional on static-gauge time shell trajectories. With:

$$E(r, \dot{r}) = \frac{m\left(1 - \left(\dot{r} + \sqrt{\frac{2M_>}{r}}\right)\sqrt{\frac{2M_>}{r}}\right)}{\sqrt{1 - \left(\dot{r} + \sqrt{\frac{2M_>}{r}}\right)^2}} + \frac{m^2}{2r}$$

And

$$\dot{r} = \frac{\dot{\boldsymbol{r}}\left(1 - \frac{2M_>}{r}\right)}{\sqrt{\frac{2M_>}{r}} \pm \sqrt{\left(1 - \frac{2M_>}{r}\right) + \boldsymbol{\dot{r}}^2}}$$

(positive sign so $\dot{r} \simeq \dot{\boldsymbol{r}}$ as $r \to \infty$, where bold is the new time derivative 'dot'). Which is the trajectory relation in terms of the RHS metric to proper time frame:

$$E(r, \dot{\boldsymbol{r}}) = m\sqrt{\left(1 - \frac{2M_>}{r}\right) + \dot{\boldsymbol{r}}^2} + \left(\frac{m^2}{2r}\right) = M_> - M_<.$$

This expression is recursively defined, but it can be given as a direct analytic relation for $M_>$:

$$E(r, \dot{\boldsymbol{r}}) = M_> - M_< = m\sqrt{\left(1 - \frac{2M_>}{r}\right) + \dot{\boldsymbol{r}}^2} - \left(\frac{m^2}{2r}\right)$$

This is now converted to a true Hamiltonian form:

$$H(r, p) = \mathrm{m}\cosh\left(\frac{p}{m}\right) - \left(\frac{mM_<}{r}\right)\exp\left(\frac{-p}{m}\right) - \left(\frac{m^2}{2r}\right)$$

The Action is now

$$S = \int (P_> \dot{\boldsymbol{M}}_> + P_< \dot{\boldsymbol{M}}_< + p\dot{\boldsymbol{r}} - [M_>(r, P, M_<) - M_<])dt$$

$$M_> - M_< = H(r, p)$$

Instead of integration over static-gauge time label (K + W gauge [137,138]), the proper-time label is chosen to match the re-expressed energy functional $E(r, \dot{r})$ (where 'dot' denotes time-derivative w.r.t proper time), which is given in terms of proper-time.

$$S_{shell} = \int [p\dot{r} - E(r, \dot{r})]\, dt = \int \left[p\dot{r} - E(r, \dot{r})\left(\frac{\partial t}{\partial \tau}\right)\right] d\tau$$

$$= \int [p\dot{r} - E(r, \dot{r})N^{-1}]\, d\tau$$

where $\frac{\partial r}{\partial t} dt \Rightarrow \frac{\partial r}{\partial \tau} d\tau$ and $N = \left(\frac{\partial t}{\partial \tau}\right)^{-1}$.

The explicit time dependence in the new energy functional is eliminated by a lapse that corresponds to shell proper time labeling: $N_\pm = \pm N$.

$$S_{shell} = \int [p\dot{r} - E(r, \dot{r})]\, d\tau = \int [p\dot{r} - H(r, p)]\, d\tau$$

This is not a canonical transformation, but the dynamical description is the same. Note, also, that the total lapse at $\pm\infty$ is now finite as it is the shell proper time lapse. This type of transformation is possible due to the arbitrariness of the lapse (choice of time). Taking the dot operation to mean proper time derivative:

$$S = \int \left[P_> \dot{M}_> + P_< \dot{M}_< + p\dot{r} - H(r,p)\right] dt$$

The choice of time is implemented through specification of $N_\pm(t)$. For vacuum there is some redundancy as $N(t) = N_+(t) + N_-(t)$ is how the $N_\pm$ occur – there essentially being one choice of time for the reduced theory, and there is one canonical pair of variables.

In the theory with shell, the dynamics is simplest (no explicit time dependence) when shell proper time is chosen.

In the theory with throat, a canonical transformation is possible starting from the action with asymptotic killing time to one in terms of throat trajectory variables. At a guess, it would seen that for geodesic trajectories a proper time description is possible in two ways: (1) via a canonical transformation where the new canonical momentum is taken to be the proper time of the geodesic or (2) via a choice of geodesic proper time for $N_\pm(t)$; specialism to throat, $M$ taken as a dependent variable of $\{a, \acute{a}\}$ through a hypersurface restriction to Novikov gauge at the throat (the latter construction is first implemented in the Lagraugian description). Either way, if throat proper time is chosen, the canonical variables describe the throat (or geodesic).

In the theory with shell and throat no universal time choice present itself. This is clear if the coupling term in $N_+ M_>(r, P, \mu_<)$ is considered:

$$S = \int \left( \left[P_> \dot{M}_> \tau\ p\dot{r} - N_+ M_>\right] dt_+ + \int \left[P_< \dot{M}_< - N_- M_<\right] \right) dt_-$$

$M_>$ is taken as dependent variable so its equations of motion are trivial: $\dot{M}_> = 0, \dot{P}_> = -N_+$ (and $N_+$ is chosen to give shell proper time). For $M_< : \dot{M}_< = 0, \dot{P}_< = -N_- - N_+ \left(mr^{-1} e^{-\frac{P}{m}}\right)$. The dynamics for the throat (or shell) will not be simple in this approach and the canonical transformation for the throat is no longer time independent.

(7b) $H = H_{shell} + H_{Grav}$

$$H_{Grav} = \frac{\Lambda P_\Lambda^2}{2R^2} - \frac{P_\Lambda P_R}{R} + \left(\frac{RR'}{\Lambda}\right)' - \frac{R'^2}{2\Lambda} - \frac{\Lambda}{2},$$

$$H_{shell} = \sqrt{\left(\hat{p}/\hat{\Lambda}\right)^2 + m^2}\ \delta(r - \hat{r}),$$

(7c) $H^r = H^r_{shell} + H^r_{Grav}$ ,
$H^r_{Grav} = R'P_R - \Lambda P'_\Lambda$,
$H^r_{shell} = \hat{p}\delta(r - \hat{r})$

The boundary terms are determined by the fall-off behavior of the canonical variables $\Lambda, R, P_\Lambda$ and $P_R$ required to satisfy asymptotic flatness [148-151]:

$$\Lambda(t,r) = 1 + M_\pm(t)|r|^{-1} + O^\infty\left(|r|^{-(1+\epsilon)}\right),$$
$$R(t,r) = |r| + O^\infty(|r|^{-\epsilon}),$$
$$P_\Lambda(t,r) = O^\infty(|r|^{-\epsilon}),$$
$$P_R(t,r) = O^\infty\left(|r|^{-(1+\epsilon)}\right).$$

For each of the two asymptotically flat regions (primordial wormhole topology) there are boundary contributions

$$S_{boundary} = -\int dt\left(N_+(t)E_+(t) + N_-(t)E_-(t)\right)$$

The boundary terms are the product of a lapse, $N_\pm(t)$, and the ADM energy, $E_\pm(t) = M_\pm(t)$. In what follows we adopt $N_\pm = \pm 1$ and $M_+ \equiv M_>, M_- \equiv M_<$.

The constraints $H = 0 = H^r$ are imposed to reduce the theory to the physical degrees of freedom. This is done prior to quantization. If no further gauge restriction is imposed away from shell, it is possible to reduce the theory via a reconstruction program [137-139]. The reconstruction program for the reduced shell/geometry action proceeds in four steps:

(i) Consider the variation of the action from the shell, making use of the properties of first class constraints, in a convenient gauge.
(ii) Generalize the analysis away from the shell to arbitrary gauge while accounting for included variations due to the junction conditions that describe the shell.
(iii) Choose a limit gauge on R near the shell such that $R'$ effectively becomes freely specifiable, obtain the partially reduced Lagrangian and impose the junction conditions.

(iv) A final choice gauge for L and R is made in the vicinity of the shell, otherwise still arbitrary. The choice made near the shell is that of Kraus +Wilcek: $L = 1, R = r$.

When the reduction outlined above is complete, the following Lagrangian formulation is obtained (dropping hats):

$$(10)\ L = \dot{r}P_< - M_> + M_<,$$
$$P_< = P_<(r,p) = \sqrt{2M_< r} - \sqrt{2M_> r}$$
$$- r\log\left|\left[\left(1 - \sqrt{2M_>/r}\right) + \frac{1}{r}\left(\sqrt{p^2 + m^2 - p}\right)\right]\right.$$
$$\left./\left[1 - \sqrt{2M_</r}\right]\right|$$

And where the junction conditions defines $M_>$ in terms of $\{r,p\}$ :

$$(11)\ M_> + M_< = \sqrt{p^2 + m^2} + \frac{m^2}{2r} - p\sqrt{\frac{2M_>}{r}} = E(r,p);$$

Here $M_> = M_>(r,p;\ M_<)$ and $M_<$ is taken to be a configuration space variable.

If the Hamiltonian is taken as $H(r,p) = M_>(r,p)$ (regrouping the configuration variable $M_<$):

$$H(r,p) = \mathrm{mcosh}\left({}^{p}/_{m}\right) - \frac{m^2}{2r} + M_<\left(1 - \left(\frac{m}{r}\right)e^{-p/m}\right).$$

If the Hamiltonian is rewritten in terms of dimensionless variables:

$\bar{r} = mr$ $\qquad \alpha = \frac{1}{2}m^2$

$\bar{p} = p/m$

$\bar{\tau} = m\tau$ $\qquad \beta = \left(\frac{M_<}{m}\right)$

$\bar{H} = H/m$

$$\bar{H} = \cosh(\bar{p}) - \frac{\alpha}{\bar{r}} + \beta\left(1 - \frac{2\alpha}{\bar{r}}\ e^{-\bar{p}}\right).$$

The over-bars are dropped to obtain the form considered in the operator analysis:

$$(H - \beta) = \cosh(p) - 2\alpha\beta r^{-1}e^{-p} - \alpha r^{-1}$$

From the relatives $\dot{r} = \frac{\partial H}{\partial p}, \dot{p} = \frac{\partial H}{\partial r}$, the first integral is obtained:

$$\frac{d}{dt}\left[\sqrt{\left(1 - \frac{2M_<}{r}\right) + \dot{r}^2} - \frac{m}{2r}\right] = 0$$

$$\left(\text{here 'dot' denotes } \frac{d}{d\tau}\right)$$

$$E = \text{const} = m\sqrt{\dot{r}^2 + \left(1 - \frac{2M_<}{r}\right)} - \frac{m^2}{2r}.$$

This agrees with the junction condition analysis. Furthermore;
$\dot{r}^2 + V_{eff}(r) = 0,$
$V_{eff}(r)$ is monotonically increasing on $\mathbb{R}^+$ and $V_{eff}(\infty) = 1 - \left(\frac{E}{m}\right)^2$.

There are three types of classical solution:

(1) $E \geq m, \dot{r} > 0$. The shell starts at $r = 0$ at finite $\tau_0$ and approaches $r = \infty$ with $\tau \to \infty$ and $\dot{r} = \sqrt{-V_{eff}}$
(2) Time reverse of (1)
(3) $E < m$. Shell starts at $r = 0, \dot{r} = \infty$, reaching a turning point at $V_{eff}(r_0) = 0$, and falling back to $r = 0, \dot{r} = -\infty$.

Integration of constraints across the shell yields the junction conditions (j.c.'s):

$$P_\Lambda(\hat{r} + \epsilon) - P_\Lambda(\hat{r} - \epsilon) = -P_\Lambda/\hat{\Lambda},$$
$$R'(\hat{r} + \epsilon) - R'(\hat{r} - \epsilon) = -\frac{1}{\hat{R}}\sqrt{p^2 + m^2\hat{\Lambda}^2}.$$

Grouping the constraints:

$$0 = \frac{R'}{\Lambda}\mathcal{H} + \frac{P_\Lambda}{R\Lambda}\mathcal{H}_r = -\mathcal{M}' + \frac{\hat{R}'}{\hat{\Lambda}}\mathcal{H}_{shell} + \frac{\hat{P}_\Lambda}{\hat{R}\hat{\Lambda}}\mathcal{H}^r_{shell},$$
$$\mathcal{M} = \frac{P_\Lambda^2}{2R} + \frac{R}{2} - \frac{RR'^2}{\partial\Lambda^2}.$$

Away from the shell $[\mathcal{M}(r,t)] = 0$, i.e., $\mathcal{M}(r,t) = M(t)$. We write

$$\mathcal{M}(t) = M_<(t) \qquad r < \hat{r},$$
$$\mathcal{M}(t) = M_>(t) \qquad r > \hat{r},$$

And $M_< \neq M_>$ due to the presence of the shell.

Away from the shell, solution of the constraints provides the following relations on the momenta:

$$P_\Lambda = R\sqrt{(R'/\Lambda)^2 - 1 + 2M_>/R}\,;\; P_R = \frac{\Lambda}{R'}P'_\Lambda \qquad r > \hat{r},$$
$$P_\Lambda = R\sqrt{(R'/\Lambda)^2 - 1 + 2M_</R}\,;\; P_R = \frac{\Lambda}{R'}P'_\Lambda \qquad r < \hat{r}.$$

The constraints may be grouped to obtain the following form:

(8a) $$O = \mathcal{M}' + \frac{\hat{R}\prime}{\hat{L}} H_{shell} + \frac{\hat{\pi}_L}{\hat{R}\hat{L}} H^r_{shell}$$
$$\mathcal{M} = \left(\frac{P_\Lambda^2}{2R} + \frac{R}{2} + \frac{RR'^2}{2\Lambda^2}\right).$$

Thus, away from the shell, $H_{shell} = 0 = H^r_{shell}$, and $\mathcal{M}$ = constant. The following convention is adopted to be consistent with the boundary conditions:

(8b) $$\mathcal{M} = \begin{cases} M_< & r < \hat{r}, \\ M_> & r > \hat{r}. \end{cases}$$

Regrouping the $\mathcal{M}$ relation and also using the $\mathcal{H}^r_{Grav} = 0$ supermomentum constraint, the momenta away from the shell are given by

(8c) $$P_\Lambda = R\sqrt{(R'/\Lambda)^2 - 1 + 2\mathcal{M}/R}\;;\; P_R = \frac{\Lambda}{R'} P'_\Lambda.$$

Coordinates are chosen such that $\Lambda$and $R$ are continuous across the shell and $P_R$ and $P_\Lambda$ are free of singularities there. Integration of the constraints across the shell then yields

(9a) $P_\Lambda(\hat{r} + \epsilon) - P_\Lambda(\hat{r} - \epsilon) = \hat{p}/\hat{L}$,
(9b) $R'(\hat{r} + \epsilon) - R'(\hat{r} - \epsilon) = -\frac{1}{R}\sqrt{\hat{p}^2 + m^2\hat{L}^2}$

The Lagrangian is now manipulated into a more familiar Hamiltonian form.

$L(r, \dot{r}; p) = \dot{r}\, P_{<}\,(r, p) - E(r, p).$

From the Euler-Lagrange equation for $p$: $\frac{\partial L}{\partial p} = \dot{r}\,\frac{\partial P_{<}}{\partial p} - \frac{\partial E}{\partial P} = 0.$

Solving for $\frac{\partial L}{\partial p} = 0 \Rightarrow \dot{r} = \left(\frac{p}{\sqrt{p^2 + m^2}}\right) - \sqrt{\frac{2M_{<}}{r}}$, which inverts:

$$\frac{p}{m} = \frac{\eta\left(\dot{r} + \sqrt{\frac{2M_{<}}{r}}\right)}{\sqrt{1 - \left(\dot{r} + \sqrt{\frac{2M_{<}}{r}}\right)^2}}, \eta = \pm 1.$$

Thus, $p = p\,(r, \dot{r})$ and substitution into $P_{<}$ yields a relation $P_{<} = P_{<}(r, \dot{r})$ that is not analytically invertible. This is inconvenient since $P_{<}$ is the conjugate momentum for $r$:

$$\frac{\partial L}{\partial \dot{r}} = P_{<} + \frac{\partial p}{\partial r}\left(\dot{r}\,\frac{\partial P_{<}}{\partial p} - \frac{\partial E}{\partial P}\right) = P_{<}.$$

Thus, the Legendre transformation runs into complications. However, the energy functional $E\big(r, p\,(r, \dot{r})\big)$ is not in the desired form to begin with. Instead, an energy functional on dynamical paths with proper time parameterization is what is desired:

$$E\big(r, p\,(r, \dot{r})\big) \equiv E(r, \dot{r}) = \frac{\eta m\left(1 - \left(\dot{r} + \sqrt{\frac{2M_{<}}{r}}\right)\sqrt{\frac{2M_{<}}{r}}\right)}{\sqrt{1 - \left(\dot{r} + \sqrt{\frac{2M_{<}}{r}}\right)^2}} - \frac{m^2}{2r},$$

And the switch to proper time is governed by the metrical relation

$$\dot{r} = \left(\frac{d\tau}{dt}\right)\dot{r} = \left[\frac{\left(1 - \frac{2M_{<}}{r}\right)}{\sqrt{\frac{2M_{<}}{r}}\,\dot{r} \pm \sqrt{\left(1 - \frac{2M_{<}}{r}\right) + \dot{r}^2}}\right]\dot{r},$$

Where $\dot{r} = \frac{dr}{d\tau}$. Upon direct substitution:

$$E\big(r, \dot{r}\,(r, \dot{r})\big) \equiv E(r, \dot{r}) = \eta m \sqrt{\left(1 - \frac{2M_{<}}{r}\right) + \dot{r}^2} - \frac{m^2}{2r}, \quad \eta = \pm 1$$

From the $E(r, \dot{r})$ energy functional on proper-time paths, a Hamiltonian is defines up to certain canonical transformations:

$$E(r, \dot{r}) \equiv H\big(r, p(r, \dot{r})\big) = \dot{r}p - L(r, \dot{r}) = \dot{r}^2 \frac{\partial}{\partial \dot{r}}\left(\frac{L(r, \dot{r})}{\dot{r}}\right).$$

Here $p = \frac{\partial L}{\partial \dot{r}}$ is a new canonical momentum and

$$L(r, \dot{r}) = \dot{r}\left\{\int_{<}^{\dot{r}} dx \left(\frac{E(r, x)}{x^2}\right) + f(r)\right\}$$

$$p(r, \dot{r}) = \int_{<}^{\dot{r}} dx \left(\frac{E(r, x)}{x^2}\right) + \frac{E(r, \dot{r})}{\dot{r}} + f(r)$$

$$(r > 2M_{<}): \int_{<}^{\dot{r}} dx \, \frac{\sqrt{a^2 + x^2}}{x^2} = \left[\frac{-\sqrt{a^2 - x^2}}{x} + \ln\left(x + \sqrt{x^2 - a^2}\right)\right]_{<}^{\dot{r}}$$

$$(r > 2M_{<}): \int_{<}^{\dot{r}} dx \, \frac{\sqrt{x^2 + a^2}}{x^2} = \left[\frac{-\sqrt{x^2 - a^2}}{x} + \ln\left(x + \sqrt{x^2 - a^2}\right)\right]_{<}^{\dot{r}}$$

For the $r > 2M_{<}$ case $x > a$ is required to have a real timelike trajectory.

$$p = \eta m \ln\left(\dot{r} + \sqrt{\dot{r}^2 + \left(1 - \frac{2M_{<}}{r}\right)} - \eta m \ln\left(c + \sqrt{c^2 + \left(1 - \frac{2M_{<}}{r}\right)}\right)\right)$$

$$-\frac{1}{c}\left(\frac{m^2}{2r}\right) + \frac{\sqrt{c^2 - a^2}}{c} + f(r).$$

The $\eta = +1$ case is chosen, and $f(r)$ and $c$ chosen so that

$$p = m \ln\left(\dot{r} + \sqrt{\left(1 - \frac{2M_{<}}{r}\right) + \dot{r}^2}\right)$$

Inverting this,

$$\dot{r} = \sinh(p/m) + \left(\frac{M_{<}}{r}\right) e^{-p/m}$$

### 8.1.2 Newtonian Limit Tests

***Newtonian Limit #1: slow moving particle in a weak gravitational field***

Metric convention: $(1,-1,-1,-1)$, Straumann [152].

Weak: $g_{\mu v} = \eta_{\mu v} + h_{\mu v}, |h_{\mu v}| \ll 1$

Slow: $\frac{dx^0}{ds} \simeq 1, \left|\frac{dx^i}{ds}\right| \ll \left|\frac{dx^0}{ds}\right|$

$$\frac{d^2x^i}{dt^2} \simeq \frac{d^2x^i}{ds^2} = -\Gamma^i_{\alpha\beta}\,\frac{dx^\alpha}{ds}\,\frac{dx^\beta}{ds} \simeq -\Gamma^i_{00}$$

Recall $\Gamma^\mu_{\alpha\beta} = \frac{1}{2}\,g^{\mu v}\left(g_{\alpha v,\beta} + g_{\beta v,\alpha} - g_{\alpha\beta,v}\right)$

$$\Gamma^i_{00} \simeq \frac{1}{2}h_{00,1} - \underbrace{h_{0i,0}}$$

Drop by assuming grav., field to be stationary

So,

$$\frac{d^2\vec{x}}{dt^2} = \frac{1}{2}\bar{V}h_{00} \Rightarrow \ddot{\vec{x}} = \nabla\varphi \quad \text{if } h_{00} = 2\varphi + const.$$

Since $\varphi$ and $h_{00}$ vanish far from all masses, $const. = 0$, and

$$g_{00} = 1 + 2\varphi$$

***Newtonian limit #2***

Metric convention: $(-1,1,1,1)$, Weinberg [153]

Slow particle in weak stationary gravitational field:

$$\left|\frac{d\vec{x}}{d\tau} \ll \frac{dt}{d\tau}\right|, \qquad g_{\alpha\beta} = \eta_{\alpha\beta} + h_{\alpha\beta}\;, \;\; |h_{\alpha\beta}| \ll 1$$

So, $\frac{d^2x^\eta}{d\tau^2} = -\Gamma^\eta_{\mu v}\,\frac{dx^\mu}{d\tau}\,\frac{dx^v}{d\tau} \simeq \Gamma^\lambda_{00}\,\left(\frac{dt}{d\tau}\right)^2$

$$\Gamma^\sigma_{\lambda\mu} = \frac{1}{2}g^{v\sigma}\left\{\frac{\partial g_{\mu v}}{\partial x^\lambda} + \frac{\partial g_{\lambda v}}{\partial x^\mu} - \frac{\partial g_{\mu\lambda}}{\partial x^v}\right\}$$

↓ stationary

$$\Gamma^\sigma_{00} \cong \frac{1}{2}\eta^{v\sigma}\left\{\frac{\partial h_{0v}}{\partial x^0} + \frac{\partial h_{0v}}{\partial x^0} - \frac{\partial h_{00}}{\partial x^v}\right\} \simeq +\frac{1}{2}\,\eta^{i\sigma}\,\frac{\partial h_{00}}{\partial x^i}$$

$$\Gamma^i_{00} \simeq \frac{1}{2}\nabla h_{00}, \text{ other } \Gamma'^s \equiv 0$$

$$\left.\begin{array}{l} \frac{d^2\vec{x}}{d\tau^2} = \frac{1}{2}\left(\frac{dt}{d\tau}\right)^2 \nabla h_{00} \\ \frac{d^2 t}{d\tau^2} = 0 \Rightarrow \left(\frac{dt}{d\tau}\right) = const \end{array}\right\} \frac{d^2\vec{x}}{dt^2} = \frac{1}{2}\nabla h_{00} = -\nabla\varphi \text{ when Newtonian } \varphi = -GM/r.$$

As before $h_{00} = -2\varphi + const$

$$g_{00} = -(1 + 2\varphi)$$

***Newtonian limit as discussed in MTW [154]***

If $\Gamma^j_{00} \simeq \frac{1}{2}\nabla h_{00} = -\nabla_j\varphi$ and all other $\Gamma^\alpha_{\beta\gamma}$ vanish, then using

$$R^\alpha_{\beta\gamma\delta} = \Gamma^\alpha_{\beta\delta,\gamma} - \Gamma^\alpha_{\beta\gamma,\delta} + \underbrace{\Gamma^\alpha_{\mu\gamma}\Gamma^\mu_{\beta\delta} - \Gamma^\alpha_{\mu\delta}\Gamma^\mu_{\beta\gamma}}_{drop\ out}$$

$$R^i_{0j0} = \Gamma^i_{00,j} = -\frac{\partial^2\varphi}{\partial x^i \partial x^j} \text{ all other } R'^s \text{ vanish}$$

$$\underbrace{R_{00} = -\nabla^2\varphi}_{\text{All other } \acute{R}_{\mu\nu} \text{ vanish}} \quad \rightarrow \quad -\nabla^2\varphi = 4\pi\rho \text{ Newtonian theory}$$

In this instance $\underline{\mathcal{H}_G \simeq -\nabla^2\varphi}$

See MTW Ex. 18.3 for inclusion of matter, also section 19.3. The Newtonian limit is now verified in a number of settings, so let's proceed with the equations of motion.

**8.2 Equation of motion from Lagrangian**

One of the forms obtained to describe dust hell collapse is:

$$L = \dot{r}\left[\sqrt{2Mr} - \sqrt{2M_+ r} - \eta\dot{r}r\log\left|\frac{\sqrt{r} - \eta\sqrt{2M_+}}{\sqrt{r} - \eta\sqrt{2M}}\right| - M_+\right]$$

$$= \dot{r}\left(\frac{\partial L}{\partial \dot{r}}\right) - M_+ \;\; = \dot{r}\,p_< - M_+.$$

The variational parameters are: $\{r, \dot{r}; p, \dot{p}\}$, for which the equations of motion are:

$$\frac{\partial L}{\partial r} - \frac{d}{dt}\left(\frac{\partial L}{\partial \dot{r}}\right) = 0 \quad \text{and} \quad \frac{\partial L}{\partial p} - \frac{d}{dt}\left(\frac{\partial L}{\partial \dot{p}}\right) = 0$$

Since L is first order in $\dot{r}$, the $\{r, \dot{r}\}$ eqn of motion simplifies substantially:

$$L = \dot{r}\,\alpha(r,p) - M_+(r,p)$$
$$\frac{\partial L}{\partial r} = \dot{r}\frac{\partial \alpha}{\partial r} - \frac{dM_+}{\partial r}$$
$$\frac{d}{dt}\left(\frac{dL}{\partial \dot{r}}\right) = \frac{d}{dt}(\alpha) = \frac{d\alpha}{\partial r}\dot{r} + \frac{\partial \alpha}{\partial p}\dot{p}$$

So, $\boxed{\frac{dM_+}{\partial r} - \frac{\partial \alpha}{\partial p}\dot{p} = 0}$

Since there is no $\dot{p}$ term the other eqn of motion is:

$$\boxed{\dot{r}\frac{\partial P_<}{\partial p} - \frac{\partial M_+}{\partial p} = 0}$$

The latter equation is all that is needed, so now to simplify.

$$P_< = \sqrt{2Mr} - \sqrt{2M_+r} - \eta\dot{r}r\log\left|\frac{\sqrt{r} - \eta\sqrt{2M_+}}{\sqrt{r} - \eta\sqrt{2M_-}}\right|$$

$$M_+(r,p): \qquad p = \frac{M_+ - M}{\eta - \sqrt{2M_+/r}}$$

Now to solve $\dot{r}\frac{\partial p_<}{\partial p} - \frac{\partial M_+}{\partial p} = 0$

$$\frac{\partial p_<}{\partial p} = \frac{\partial p_<}{\partial M_+}\frac{\partial M_+}{\partial p}, \qquad \text{so} \qquad \frac{\partial M_+}{\partial p}\left[\dot{r}\frac{\partial p_<}{\partial M_+} - 1\right] = 0$$

$$\boxed{\dot{r} = \left(\frac{\partial p_<}{\partial M_+}\right)^{-1}}$$

$$\frac{\partial p_<}{\partial M_+} = -\sqrt{\frac{r}{2M_+}} - \eta r\left\{\frac{-\eta/\sqrt{2M_+}}{\sqrt{r} - \eta\sqrt{2M_+}}\right\} = -\sqrt{\frac{r}{2M_+}} + \sqrt{\frac{r}{2M_+}}\left\{\frac{1}{1-\eta\sqrt{\frac{2M_+}{r}}}\right\} =$$

$$\frac{\eta\sqrt{\frac{2M_+}{r}}}{1-\eta\sqrt{\frac{2M_+}{r}}}\sqrt{\frac{r}{2M_+}}$$

$$\dot{r} = \frac{1-\eta\sqrt{\frac{2M_+}{r}}}{\eta} = -\sqrt{\frac{2M_+}{r}} + \eta$$

These trajectories are the null geodesics of the RHS metric:

$$ds^2 = -dt^2 + \left( dr \pm \sqrt{\frac{2M}{r}} dt \right)^2 + r^2 d\Omega^2$$

And correspond to choosing the + sign preceding $\sqrt{\frac{2M}{r}}$ in the metric. (then $\eta = -1$ has no horizon structure while $\eta = +1$ does.)

Prior to choice of gauge, the Lagrangian is expressed by:

$$L = \frac{ds}{dt} = \dot{\hat{r}} \hat{R} \hat{L} \left[ \sqrt{(R'_{<}/L)^2 - 1 + 2M/\hat{R}} - \sqrt{\left(R'_{>}/\hat{L}\right)^2 - 1 + 2M/\hat{R}} \right.$$

$$-\dot{\hat{R}} \hat{R} \log \left| \frac{R\prime(\hat{r}-\epsilon)/\hat{L} - \sqrt{(R\prime(\hat{r}-\epsilon)/\hat{L})^2 - 1 + 2M/\hat{R}}}{R'_{<}/\hat{L} - \sqrt{\left(R'_{<}/\hat{L}\right)^2 - 1 + 2M/\hat{R}}} \right| + \int_0^\infty dr \left[ \pi_R \dot{R} + \pi_L \hat{L} \right] - M_+$$

In reducing to this form preparation has been made to account for the junction condition via $R'(\hat{r} - \epsilon)$ terms. What this means when the K+W [137] gauge is chosen "across the shell" is that the junction conditions reduce to

$$R'(\hat{r} - \epsilon) = 1 + \frac{1}{r} \sqrt{p^2 + m^2}$$

$$\sqrt{\left(R'(\hat{r} - \epsilon)/\hat{L}\right)^2 - 1 + 2M/\hat{R}} = \sqrt{2M_+/\hat{R}} + \frac{1}{r} p.$$

What must be remembered at this juncture, in order to interpret "$\frac{d}{dt}$", is the limit process that defined the Lagrangian above, what this limit entailed was a description for L and R in the range $\hat{r} \leq r \leq \infty$ while L, R were detailed for $r_{min} < r < \hat{r}$ by the limit process. Thus "$\frac{d}{dt}$" is given in terms of the RHS gauge with $M_+$.

With the above definition of "$\frac{d}{dt}$" in mind it may prove most convenient in the $m \neq 0, \underline{M_< = 0}$ verification to first work with the $R'(\hat{r} + \epsilon)$ replacement in order that "$\frac{d}{dt}$" can be defined relative to the interior – flat – metric. In this instance the Lagrangian is simply:

$$L = \dot{\hat{r}} \hat{R} \hat{L} \left[ \sqrt{<} - \sqrt{>} \right] + \dot{\hat{R}} \hat{R} \log \left| \frac{R\prime(\hat{r}-\epsilon)/\hat{L} - \sqrt{(R\prime(\hat{r}-\epsilon)/\hat{L})^2 - 1 + 2M_+/R}}{R'_{>}/\hat{L} - \sqrt{\left(R'_{>}/\hat{L}\right)^2 - 1 + 2M_+/R}} \right| +$$

$$\int_0^\infty dr \left[ \pi_R \dot{R} + \pi_L \dot{L} \right] - M_+$$

The junction conditions now yield:

$$L = \dot{\hat{r}}\hat{R}\hat{L}\left[\sqrt{<} - \sqrt{>}\right] + \dot{\hat{R}}\hat{R}\log\left|\frac{(R'_{<}/\hat{L}) - \frac{1}{\hat{R}\hat{L}}\sqrt{p^2+m^2L^2} - \sqrt{<} + \frac{p}{RL}}{(R'_{>}/\hat{L}) - \sqrt{>}}\right| + \int_0^\infty dr\left[\pi_R \acute{R} + \pi_L \acute{L}\right] - M_+$$

Now the gauge $L = 1, R = r$ is chosen (drop hats, let $M = 0$ redefine $M_+ = M$)

$$L = \dot{r}r\left[-\sqrt{2M/r}\right] + \dot{r}r\log\left|\frac{1 - \frac{1}{r}\left(\sqrt{p^2+m^2} - p\right)}{1 - \sqrt{2M/r}}\right| - M,$$

And "$M$" is defined by:

$$1 = R'(\hat{r} + \epsilon) + \frac{1}{r}\sqrt{p^2 + m^2}$$

$$0 = \pm\sqrt{\left(R'(\hat{r} + \epsilon)\right)^2 - 1 + 2M/r} + \frac{1}{r}p$$

And $\frac{d}{d\dagger}$ is defined in terms of the metric on LHS, which is flat:

$$ds^2 = -dt^2 + dr^2 + r^2\,d\Omega^2$$
$$d\tau^2 = dt^2 - dr^2 = (1 - \dot{r}^2)dt^2$$

So,

$$L = \dot{r}P_c\big(r, p, M(r,p)\big) - M(r,p)$$

Consider $\frac{\partial}{\partial p} - \frac{\partial L}{\partial t}\left(\frac{\partial L}{\partial p}\right) = 0 \Rightarrow \dot{r}\left[\left(\frac{\partial p_c}{\partial p}\right) + \left(\frac{\partial p_c}{\partial M}\right)\left(\frac{\partial M}{\partial p}\right)\right] - \left(\frac{\partial M}{\partial p}\right) = 0$

$$\dot{r}\left\{\left(\frac{\partial p_c}{\partial p}\right) + \left(\frac{\partial p_c}{\partial M}\right)\left(\frac{\partial M}{\partial p}\right)\right\} - \left(\frac{\partial M}{\partial p}\right) = 0$$

The $\left(\frac{\partial M}{\partial p}\right)$ won't be cancelled as in the $m = 0$ case, so the expression for $M(r,p)$ must be considered. The junction conditions give:

$$1 = R'(\hat{r} + \epsilon) + \frac{1}{r}\sqrt{p^2 + m^2}$$

$$0 = \pm\sqrt{\left(R'(\hat{r} + \epsilon)\right)^2 + 2M/r} + \frac{1}{r}p$$

Thus,

$$(R' + \epsilon)^2 - 1 + 2M/r = \left(\frac{p}{r}\right)^2$$

$$\left(1-\frac{1}{r}\sqrt{p^2+m^2}\right)^2-1+2M/r=\left(\frac{p}{r}\right)^2$$
$$1-\frac{2}{r}\sqrt{p^2+m^2}+\frac{p^2}{r^2}+\frac{m^2}{r^2}-1+2M/r=\left(\frac{p}{r}\right)^2$$
$$\sqrt{p^2+m^2}=\frac{m^2}{2r}+M$$
So,

$$\boxed{M=\sqrt{p^2+m^2}=\frac{m^2}{2r}} \text{ and } \boxed{\frac{\partial M}{\partial p}=\frac{p}{\sqrt{p^2+m^2}}}$$

$$P_c=-\sqrt{2M/r}+r\log\left|\frac{1-\frac{1}{r}\left(\sqrt{p^2+m^2}-p\right)}{1-\sqrt{2M/r}}\right|$$

$$\frac{\partial P_c}{\partial M}=-\sqrt{\frac{r}{2M}}-r\left[\frac{-\sqrt{\frac{r}{2Mr}}}{1-\sqrt{\frac{2M}{r}}}\right]=\sqrt{\frac{r}{2M}}\left[\frac{-\sqrt{\frac{2M}{r}}}{1-\sqrt{\frac{2M}{r}}}\right]=\frac{1}{1-\sqrt{\frac{2M}{1}}}$$

$$\boxed{\frac{\partial P_c}{\partial M}=\left(1-\sqrt{\frac{2M}{1}}\right)^{-1}}$$

$$\left(\frac{\partial P_c}{\partial p}\right)=r\left[\frac{-\frac{1}{r}\left(\frac{p}{\sqrt{p^2+m^2}}-1\right)}{1-\frac{1}{r}\left(\sqrt{p^2+m^2}-p\right)}\right]$$

$$\dot{r}=\left(\frac{\partial M}{\partial p}\right)\left[\frac{\partial P_c}{\partial p}+\frac{\partial P_c}{\partial M}\frac{\partial M}{\partial p}\right]^{-1}=\left[\frac{\partial P_c}{\partial p}\frac{\partial p}{\partial M}+\frac{\partial P_c}{\partial M}\right]^{-1}$$

$$\left(\frac{\partial P_c}{\partial p}\right)\left(\frac{\partial p}{\partial M}\right)=\frac{-\left(1-\frac{\sqrt{p^2+m^2}}{P}\right)}{\left(1-\frac{1}{r}\left(\sqrt{p^2+m^2}-p\right)\right)}=\frac{\sqrt{p^2+m^2}-p}{p-\frac{p}{r}\left(\sqrt{p^2+m^2}-p\right)}$$

$$\left(\frac{\partial P_c}{\partial M}\right)=\left(1-\sqrt{\frac{2M}{r}}\right)^{-1}$$

$$\dot{r}=\left[\frac{\sqrt{p^2+m^2}-p}{p\left(1-\frac{1}{r}\left(\sqrt{p^2+m^2}-p\right)\right)}+\frac{1}{1-\sqrt{\frac{2M}{r}}}\right]^{1-}=\left[\left(\frac{\sqrt{p^2+m^2}}{p}\right)\left(\frac{1-{}^{p}\!/\!\sqrt{p^2+m^2}}{1-\frac{1}{r}\left(\sqrt{p^2+m^2}-p\right)}\right)+\right.$$
$$\left.\frac{1}{1-\sqrt{\frac{2M}{r}}}\right]^{-1}$$

$$= \left(\frac{\sqrt{p^2+m^2}}{p}\right)^{-1}\left[\frac{1-{}^{p}\!/\!\sqrt{p^2+m^2}}{1-\frac{1}{r}\left(\sqrt{p^2+m^2}-p\right)}+\frac{1}{1-\sqrt{\frac{2M}{r}}}\frac{p}{\sqrt{p^2+m^2}}\right]^{-1}$$

Consider $\frac{d}{d\tau}r = \frac{dr}{dt}\cdot\frac{dt}{d\tau}$ where $d\tau^2 = (1-\dot{r}^2)dt^2$:

$$\dot{\boldsymbol{r}} = \frac{\dot{r}}{\sqrt{1-\dot{r}^2}}$$

$$1+\dot{\boldsymbol{r}}^2 = \frac{\dot{r}^2}{1-\dot{r}^2} = \frac{1}{1-\dot{r}^2} \Longrightarrow \frac{1}{1-\dot{r}^2} = 1-\dot{r}^2 \Rightarrow \dot{r}^2 = 1-\frac{1}{1+\dot{\boldsymbol{r}}^2} = \frac{\dot{\boldsymbol{r}}^2}{1+\dot{\boldsymbol{r}}^2}$$

$$\dot{r} = \frac{\dot{\boldsymbol{r}}}{\sqrt{1+\dot{\boldsymbol{r}}^2}} = \frac{m\dot{\boldsymbol{r}}}{\sqrt{m^2+(m\dot{\boldsymbol{r}})^2}}$$

So, large bracket is simply unity.

$$L = \dot{r}P_c\big(r,p,M(r,p)\big) - M(r,p)$$

$$\frac{\partial L}{\partial r} = \dot{r}\left(\frac{\partial p_c}{\partial r}+\frac{\partial p_c}{\partial M}\frac{\partial M}{\partial r}\right)-\frac{\partial M}{\partial r}$$

$$\frac{d}{dt}\left(\frac{\partial L}{\partial \dot{r}}\right) = \frac{d}{dt}(P_c) = \left(\frac{\partial p_c}{\partial r}+\frac{\partial p_c}{\partial M}\frac{\partial M}{\partial r}\right)\dot{r}+\left(\frac{\partial p_c}{\partial p}+\frac{\partial p_c}{\partial M}\frac{\partial M}{\partial p}\right)\dot{p}$$

$$\frac{\partial L}{\partial r}-\frac{d}{dt}\left(\frac{\partial L}{\partial \dot{r}}\right) = -\frac{\partial M}{\partial r}-\left(\frac{\partial p_c}{\partial p}+\frac{\partial p_c}{\partial M}\frac{\partial M}{\partial p}\right)\dot{p} = 0$$

$$\dot{p} = -\left(\frac{\partial M}{\partial r}\right)\left(\frac{\partial p_c}{\partial p}+\frac{\partial p_c}{\partial M}\frac{\partial M}{\partial p}\right)^{-1}$$

$$M = \sqrt{p^2+m^2}-\frac{m^2}{2r} \Longrightarrow \left(\frac{\partial M}{\partial r}\right) = \frac{m^2}{2r^2}$$

$$P_c = -\sqrt{2Mr} - r\log\left|1-\frac{\sqrt{2M}}{r}\right| + r\log\left|1-\frac{1}{r}\left(\sqrt{p^2+m^2}-p\right)\right|$$

$$\left(\frac{\partial p_c}{\partial M}\right) = \left(1-\frac{\sqrt{2M}}{r}\right)^{-1}$$

$$\left(\frac{\partial p_c}{\partial p}\right) = \frac{1-P/\sqrt{p^2+m^2}}{1-\frac{1}{r}\left(\sqrt{p^2+m^2}-p\right)}$$

$$\left(\frac{\partial M}{\partial p}\right) = \frac{p}{\sqrt{p^2+m^2}}$$

$$\dot{p} = -\left(\frac{m^2}{2r^2}\right)\left[\frac{1-P/\sqrt{p^2+m^2}}{1-\frac{1}{r}\left(\sqrt{p^2+m^2}-p\right)}+\frac{1}{1-\frac{\sqrt{2M}}{r}}\frac{p}{\sqrt{p^2+m^2}}\right]^{-1}$$

So,

$$\dot{p} = -\left(\frac{m^2}{2r^2}\right)\left[\dot{r}\left(\frac{\sqrt{p^2+m^2}}{p}\right)\right]$$

Consider $M = \sqrt{p^2+m^2} - \frac{m^2}{2r^2} \Rightarrow \dot{M} = \frac{P\dot{P}}{\sqrt{p^2+m^2}} + \frac{m^2}{2r^2}\dot{r} = 0$

This indicates that $\dot{M} = 0$, as expected.

<u>The Hamiltonian form of the derivation</u>:

$L = \acute{r}\,P_c - (M_> - M_<)\,;\; P_c = (r, p, M_>, M_<)$, an explicit analytic form
$p = p(r, P_c)$, an implicit analytic form
$M_> = M_>(r, p)$, explicit
$M_< =$ parameter like $m$

$$H = M_> - M_< = \sqrt{p^2+m^2} - \frac{m^2}{2r} - p\sqrt{\frac{2M}{r}}$$

$P_c = \left[\sqrt{2M_< r} - \sqrt{2M_> r}\right] -$
$r\ln\left|\left[\left(1-\sqrt{2M_>/r}\right) + r^{-1}\left(\sqrt{p^2+m^2} - p\right)\right]/\left[1-\sqrt{2M_</r}\right]\right|$

$\dot{r} = \frac{\partial H}{\partial P_c} = \frac{\partial M_>}{\partial P_c} \Rightarrow$ if $m = 0$, then done, $P_c = P_c\,(r, M_>)$

$$= \left(\frac{\partial M_>}{\partial p}\right)\frac{\partial p}{\partial p_c} = \left(\frac{\partial M_>}{\partial p}\right)\left[\frac{\partial p_<}{\partial p} + \frac{\partial p_<}{\partial M_>}\frac{\partial M_>}{\partial p}\right]^{-1}$$

From here the analysis is like that in the Lagrangian calculation, the result is:

$\dot{r} = \frac{p}{\sqrt{p^2+m^2}} - \sqrt{\frac{2M_>}{r}}\,;\quad p = p(r, P_<), M_> = M_>(r, P_<)$ (both implicitly defined)
Inverting:

$$\left(\frac{p}{m}\right) = \frac{\eta\left(\dot{r} + \sqrt{\frac{2M_>}{r}}\right)}{\sqrt{1-\left(\dot{r} + \sqrt{\frac{2M_>}{r}}\right)^2}}$$

The Hamiltonian is then expressed as:

$$H = \eta\, m \frac{\left(1-\left(\dot{r}+\sqrt{\frac{2M_>}{r}}\right)\sqrt{\frac{2M_>}{r}}\right)}{\sqrt{1-\left(\dot{r}+\sqrt{\frac{2M_>}{r}}\right)^2}} + \frac{m^2}{2r}$$

Since $M_> = H + M_<$ this is a recursively defined expression. Expanding it leads to a 6th –order polynomial expression in H, for which there is generally not an explicit form of the solution.

$$H(r,P_0) = H\big(r,\dot{r}(r,P_0)\big) = m\sqrt{\left(1-\frac{2M_<}{r}\right)+\dot{r}^2} - \frac{m^2}{2r} = m\left(\exp\left(\frac{P_0}{m}\right) - \dot{r}\right) - \frac{m^2}{2r}$$

$$= \left(m\cosh\left(\frac{P_0}{m}\right) - \frac{m^2}{2r}\right) - \left(\frac{mM_<}{r}\right)\exp\left(\frac{-P_0}{m}\right)$$

When $M_< \to 0$ this result agrees with Hajicek –where it is claimed to be uniquely determined from $H(r,\dot{r})$.

$H(r,P_0)$ is a Hamiltonian with explicit analytic form, as is the pseudo-Hamiltonian $H(r,\dot{r})$. The $H(r,\dot{r})$ form is implicitly defined via a sixth-order polynomial root while $H(r,P_c)$ is implicitly defined in terms of algebraic and logarithmic terms.

What is desirable is a direct expression of the Hamiltonian from the Lagrangian of K+W [137] instead of solving for $\hat{r}$ from the equation of motion and "bootstrapping" for the equation of motion and the Hamiltonian related to it. In this effort the substitutions that led to the proper time expression seem crucial in obtaining an explicit expression of the Hamiltonian. This motivates starting from the minisuperspace Lagrangian that expresses a proper time labeling of slices.

Recall that this is a minisuperspace analysis on spherically symmetric asymptotically flat spacetimes with a shell of dust. The constraints that are imposed away from the shell require that the hypersurface describing the system at a given instant be a spacelike slice of Kruskal spacetime. The integration of the constraints across the shell permits calculation of the required junction conditions between the right hand side (RHS) and left hand side (LHS) shell regions.

An analysis by Kraus and Wilczek describes the action for the dust system. From this action a succinct expression for the Lagrangian is given. The Lagrangian given is found by limit arguments and makes no further restriction on gauge choice than that already required in the vicinity of the shell (such that no pathologies arise). If the limit argument is applied on the LHS of the shell the Lagrangian takes the following form:

$$L = \dot{\hat{r}}\hat{R}\hat{L}\left[\sqrt{\left(R'_{<}/\hat{L}\right)^2 - 1 + 2M_{<}/\hat{R}} - \sqrt{\left(R'_{>}/\hat{L}\right)^2 - 1 + 2M_{<}/\hat{R}}\right]$$

$$-\dot{\hat{R}}\hat{R}\log\left|\frac{R'\left(\hat{r}-\epsilon\right)/\hat{L} - \sqrt{\left(R'\left(\hat{r}-\epsilon\right)/\hat{L}\right)^2 - 1 + 2M_{<}/\hat{R}}}{R'_{<}/\hat{L} - \sqrt{\left(R'_{<}/\hat{L}\right)^2 - 1 + 2M_{<}/\hat{R}}}\right|$$

$$+\int_{LHS}^{RHS} dr\left[\pi_R\dot{R} + \pi_L\dot{L}\right] - \underline{M_{>} + M_{<}}$$

The hatted variables are evaluated at the shell. The symbols $>$ and $<$ denote evaluation on the RHS and LHS respectively. $R'_{>}$ is defined such that $R'_{>} = R'(\hat{r}+\epsilon)$, where $\epsilon$ is a limit variable used to describe the junction conditions. Those junction conditions, for massive dust shell, are given by

$$R'(\hat{r}-\epsilon) = R'(\hat{r}+\epsilon) + \frac{1}{\hat{R}}\sqrt{p^2 + m^2\hat{L}^2}$$

$$\sqrt{\left(R'\left(\hat{r}+\epsilon\right)/\hat{L}\right)^2 - 1 + 2M_{<}/\hat{R}}$$
$$= \sqrt{\left(R'\left(\hat{r}+\epsilon\right)/\hat{L}\right)^2 - 1 + 2M_{>}/\hat{R}} + \frac{P}{\hat{R}\hat{L}}.$$

The expression for $R'$ close the LHS of the shell, but sufficiently far away that it be freely specifiable, is given by $R'_{<}$ . it is this term, $R'_{<}$, that requires that a limit argument be applied to obtain the Lagrangian as given.

To examine the Lagrangian further a choice of gauge is made. The gauge is chosen such that (i) there is no contribution from the proper integral,

(ii) it be well defined at the shell, and (iii) that the slicing indicated maintains the topology of interest. For this purpose the following is considered:

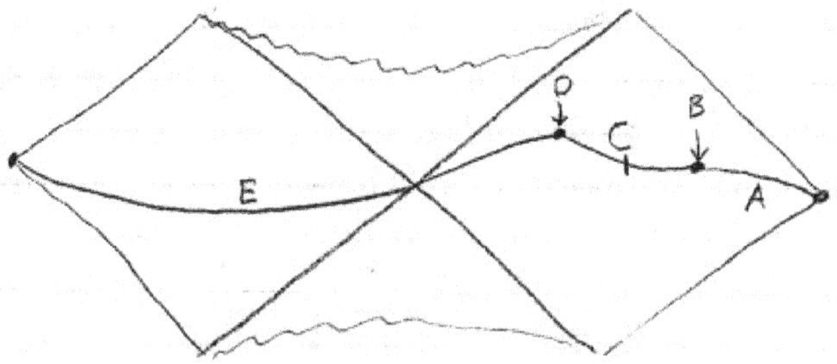


The A and E sections of the slice are sections of a Schwarzschild slicing: $\pi_{L,R} = 0$. Kruskal section could just as well have been used, a minor alteration that might be indicated if zero lapse at the bifurcate horizon 2-sphere presents difficulties.

The infinitesimal B and D sections have any gauge that smoothly connects the A,E gauge to the C gauge and that maintains $\acute{\lambda}, \acute{R} = 0$.

The C section has a gauge compatible with the junction conditions across the shell. For this purpose the gauge of Krause and Wilczek is chosen: $L = 1, R = r$.

The Lagrangian in the gauge described is then (dropping hats):

$$L = \dot{r}\left[\sqrt{2M_< r} - \sqrt{2M_> r}\right] - \dot{r} r \log\left|\frac{\left(1-\sqrt{2M_>/r}\right) + \frac{1}{r}\left(\sqrt{p^2+m^2}-p\right)}{\left(1-\sqrt{2M_</r}\right)}\right| - M_> + M_< = \dot{r}P_c - M_> + M_<$$

Where

$$P_c = \left[\sqrt{2M_< r} - \sqrt{2M_> r}\right] - r\log\left|\left[\left(1-\sqrt{2M_>/r}\right) + \frac{1}{r}\left(\sqrt{p^2+m^2}-p\right)\right]\Big/\left[\left(1-\sqrt{2M_>/r}\right)\right]\right|$$

and where the junction conditions define $M_>$ in terms of $\{r, p\}$:

$$M_> + M_< = \sqrt{p^2+m^2} + \frac{m^2}{2r} - P\sqrt{\frac{2M_>}{r}},$$

Here $M_> = M_<(r, p)$ and $M_<$ is taken to be a parameter (like m).

The variational parameters in the Lagrangian are $\{r, \dot{r}; p, \dot{p}\}$, However, as there is no $\dot{p}$ dependence, the equation of motion for p is a constraint. Such an instance is familiar from the analysis of the Maxwell Field in Coulomb gauge. In that instance $\dot{A}_0$ does not occur in the Lagrangian, leaving $A_0$ to enter as a Lagrange multiplier. The equation of motion for $A_0$ then yields Gauss Law. In the problem at hand p enters in much more complicated fashion than linearly. Nonetheless, its equation of motion can be expressed simply, and it relates $p = p(r, \dot{r}, M_>)$:

$$0 = \frac{\partial L}{\partial p} = \dot{r}\left[\frac{\partial p_<}{\partial p} + \frac{\partial p_c}{\partial M_>}\frac{\partial M_>}{\partial p}\right] - \frac{\partial M_>}{\partial p} = 0,$$

We have:

$$\begin{cases} \left(\frac{\partial M_>}{\partial p}\right) = \left(\frac{p}{\sqrt{p^2+m^2}} - \sqrt{\frac{2M_>}{r}}\right)\left(1 + \frac{p}{r}\sqrt{\frac{r}{2M_>}}\right)^{-1} \\ \left(\frac{\partial p_c}{\partial p}\right) = \left[\frac{\left(1-P/\sqrt{p^2+m^2}\right)}{\left(1-\sqrt{2M_>/r}\right)+\frac{1}{r}\left(\sqrt{p^2+m^2}-p\right)}\right] \\ \left(\frac{\partial p_c}{\partial M_>}\right) = \left[\frac{\left(1-\sqrt{p^2+m^2}-p\right)/\sqrt{2M_>r}}{\left(1-\sqrt{2M_>/r}\right)+\frac{1}{r}\left(\sqrt{p^2+m^2}-p\right)}\right] \end{cases}$$

So

$$\dot{r} = \left(\frac{p}{\sqrt{p^2+m^2}} - \sqrt{\frac{2M_>}{r}}\right)$$

The equation for $\dot{r}$ can be inverted to give

$$\left(\frac{p}{m}\right) = \pm\frac{\left(\dot{r} + \sqrt{\frac{2M_>}{r}}\right)}{\sqrt{1 - \left(\dot{r} + \sqrt{\frac{2M_>}{r}}\right)^2}}$$

This expression for p is recursively defined as $M_> = M_<(r, p)$. Expansion to explicit form gives an expression for p as the root of an eighth order polynomial. If $p(r, \dot{r}, M_>)$ above is substituted into the $M_> - M_<$ expression resulting from the junction conditions, the resulting expression is

$$M_> - M_< = \frac{\pm m\left(\dot{r} + \sqrt{\frac{2M_>}{r}}\right)}{\sqrt{1 - \left(\dot{r} + \sqrt{\frac{2M_>}{r}}\right)^2}} \left(\frac{1}{\left(\dot{r} + \sqrt{\frac{2M_>}{r}}\right)} - \sqrt{\frac{2M_>}{r}}\right).$$

This recursive, upon expansion, yields a sixth order polynomial in $(M_> - M_<)$, so still no convenient form.

If now, p as a function (implicit) of $r$ and $\dot{r}$ is replaced throughout the Lagrangian $L = L(r, \dot{r}, p(r, \dot{r}))$, the resulting functional on $\{r, \dot{r}\}$ can be Legendre Transformed to Hamiltonian form:

$$\frac{\partial L(r,\dot{r})}{\partial \dot{r}} \equiv \frac{\partial L(r,\dot{r},p(r,\dot{r}))}{\partial \dot{r}} = P_c + \dot{r}\frac{\partial P_c}{\partial p}\frac{\partial p}{\partial \dot{r}} - \frac{\partial (M_> - M_<)}{\partial p}\frac{\partial p}{\partial \dot{r}}$$
$$= P_c + \left(\frac{\partial L}{\partial p}\right) = P_c,$$

Where now $P_c(r,\dot{r}) \equiv P_c(r, p(r,\dot{r}))$ and $M_>(r,\dot{r}) \equiv M_>(r, p(r,\dot{r}))$ are replaced in the Lagrangian expression:

$$L(r,\dot{r}) = r\dot{P}_c(r,\dot{r}) - H(r,\dot{r}), \text{ and } H(r,\dot{r}) = M_>(r,\dot{r}) - M_<.$$

The expression for the conserved energy functional, $H(r,\dot{r})$, could just as well have been obtained directly from

$$L = \dot{r}P_c(r,p) - (M_>(r,p) - M_<) = \dot{r}P_c(r,p) - H(r,p)$$

If the variable p were interpreted as merely parametrizing the dependence between $P_c$ and $H$. In that case,

$$\dot{r} = \frac{\partial H}{\partial P_c} = \left(\frac{\partial M_>}{\partial p}\right)\left(\frac{\partial P_c}{\partial p}\right)^{-1} = \left(\frac{p}{\sqrt{p^2 + m^2}} - \sqrt{\frac{2M_>}{r}}\right),$$

The same result as before.

Whichever way it is obtained, the energy functional $H(r,\dot{r})$ is now re-expressed using the proper time at the shell instead of the RHS shell time (implicitly chosen via the limit construction that gave the Kraus +Wilczek Lagrangian). The times are related by their respective line elements:

$$d\tau^2 = dt^2 - \left(dr + \sqrt{\frac{2M_>}{r}}dt\right)^2,$$

$$\left(\frac{dt}{d\tau}\right) = \frac{\dot{r}\sqrt{\frac{2M_>}{r}} \pm \sqrt{\left(1 - \frac{2M_>}{r}\right) + \dot{r}^2}}{\left(1 - \frac{2M_>}{r}\right)},$$

And,

$$\dot{r} = \frac{\dot{r}\left(1 - \frac{2M_>}{r}\right)}{\sqrt{\frac{2M_>}{r}}\dot{r} \pm \sqrt{\left(1 - \frac{2M_>}{r}\right) + \dot{r}^2}}, \qquad \text{where } \dot{r} = \frac{dr}{d\tau}.$$

Substitution into energy functional, $H(r,\dot{r}) = H\big(r,\dot{r}(r,\dot{r},M_>)\big)$, then yields

$$H\big(r,\dot{r}(r,\dot{r},M_>)\big) = M_>\big(r,\dot{r}(r,\dot{r},M_>)\big) - M_<$$
$$= \pm m\sqrt{\left(1 - \frac{2M_>}{r}\right) + \dot{r}^2 + \frac{m^2}{2r}}$$

The expression for $H = M_> - M_<$ is still recursively defined through its $M_>$ functional dependence. With the $\dot{r} = (r,\dot{r},M_>)$ substitution, however, the recursive expression can be manipulated into an explicit form that gives H as a function of $r,\dot{r},m,$ and $M_<$ (which is just a parameter like m):

$$H = \big(r,\dot{r}(r,\dot{r},M_>)\big) = m\sqrt{\left(1 - \frac{2M_<}{r}\right) + \dot{r}^2 - \left(\frac{m^2}{2r}\right)}$$

If this energy functional is now considered as acting on paths given by $r$ and $\dot{r}$, it may be uniquely related to a Hamiltonian on proper time:

$$H(r,\dot{r}) = m\sqrt{\left(1 - \frac{2M_<}{r}\right) + \dot{r}^2 - \left(\frac{m^2}{2r}\right)} = \dot{r}P_0 - L(r,\dot{r})\,, P_0 = \left(\frac{\partial L}{\partial \dot{r}}\right)$$
$$= \dot{r}^2 \frac{\partial}{\partial \dot{r}}\left(\frac{L}{\dot{r}}\right)$$

$$L(r,\dot{r}) = \dot{r}\left\{\int_c^{\dot{r}} dx\left(\frac{H(r,x)}{x^2}\right) + f(c)\right\}$$

The function $f(c)$ is chosen such that $L(r,\dot{r}) = \dot{r}^2\frac{\partial}{\partial \dot{r}}\left(\frac{L}{\dot{r}}\right)$. Furthermore,

$$P_0 = \left(\frac{\partial L}{\partial \dot{r}}\right) = \int_c^{\dot{r}} dx\left(\frac{H(r,x)}{x^2}\right) + \frac{H(r,\dot{r})}{\dot{r}} - \left(\frac{H(r,c)}{c}\right) + f(c),$$

Where the parameter c is chosen such that $P_0 = (L + H)/\dot{r}$, i.e. $H(r, c) = 0$. This process appears to uniquely specify $L(r, \dot{r})$ as it is invertible. The resulting expression for $L(r, \dot{r})$ is:

$$L(r, \dot{r}) = -H(r, \dot{r}) + \dot{r}\left\{m \ln\left(\dot{r} + \sqrt{\left(1 - \frac{2M_<}{r}\right) + \dot{r}^2}\right)\right\}$$

From the equation for $L(r, \dot{r})$ the expression $P_0$ is obtained (either by inspection or by evaluation of $\frac{\partial L}{\partial p}$ that also serves as a check):

$$P_0 = m \ln\left(\dot{r} + \sqrt{\left(1 - \frac{2M_<}{r}\right) + \dot{r}^2}\right)$$

Inverting the $P_0$ equation to get $\dot{r}$:

$$\dot{r} = \frac{1}{2}\left(\exp\left(\frac{P_0}{m}\right) - \exp\left(\frac{-P_0}{m}\right)\right) + \frac{1}{2}\left(\frac{2M_<}{r}\right)\exp\left(\frac{P_0}{m}\right)$$

Substitution into $H(r, \dot{r})$ then yields the desired (explicit) expression for the Hamiltonian:

$$H(r, P_0) = m \cosh\left(\frac{P_0}{m}\right) - \frac{m^2}{2r} - \left(\frac{mM_<}{r}\right)\exp\left(\frac{-P_0}{m}\right).$$

When $M_< \to 0$ this result agrees with Hajicek (where it is claimed to be uniquely determined from $H(r, \dot{r})$ also).

The expression for the Hamiltonian of the system has been obtained by way of the Lagrangian and making use of a novel time reparameterization method. In App. C an alternate, direct, derivation for the Hamiltonian is given.

## 8.3 Time Reparameterization

### 8.3.1 The short derivation

Consider 1-dimensional Hamiltonian $h = h(q, p)$. The inverse of the Legendre transformation with canonical momentum p to a velocity v is $V(q, p) = {}^{\partial h}/_{\partial p}$.

Reparametrize time *t* in the manner

$$dT = N dt, N(t) = N\big((q/t), v(t)\big)$$

$$V(T) = \frac{d}{dt} q(t)$$

is related to $V(t)$ by

$$V = N(q,v)V$$

In the Hamiltonian framework, the canonical momentum associated with $(q,v)$ is given by

$$\frac{\partial P}{\partial p} = N(q,p) \text{ where } N(q,p) := N\big(q, v(q,p)\big)$$

where $p = \int^{P} N(q,p')dp'$, where solutions differ by canonical transformation of the form $P \to P + f(q)$.

Consider a free relativistic particle, parameterized by Minkowski time $t$, and described by the Hamiltonian

$$h = (p^2 + m^2)^{1/2}$$

With $v = \frac{\partial h}{dp} = \frac{p}{(p^2+m^2)^{1/2}}$. If we adopt new time T to be proper time $(T = \tau)$, then

$$N = (1 - V^2)^{1/2} = \frac{m}{(p^2 + m^2)^{1/2}},$$

$$P = \int N dp = m\sinh^{-1}\frac{p}{m} \Rightarrow p = m\sinh\frac{P}{m},$$

And we obtain the familiar $H = m\cosh\frac{P}{m}$. For the dust shell we have

$$h = \sqrt{P(r,p)^2 + m^2} - \frac{m^2}{2r} - P(r,p)\sqrt{\frac{2M_-}{r}}$$

$$\dot{r} = \frac{\partial H}{\partial p} = \frac{P}{\sqrt{p^2 + m^2}} - \sqrt{\frac{2M_-}{r}}$$

$$P = \frac{m\left(\dot{r} + \sqrt{\frac{2M_-}{r}}\right)}{\left[1 - \left(\dot{r} + \sqrt{\frac{2M_-}{r}}\right)^2\right]^{1/2}}$$

$$\tilde{h} = M_+ = m\frac{1 - \left(\dot{r} + \sqrt{\frac{2M_-}{r}}\right)\sqrt{\frac{2M_-}{r}}}{\left[1 - \left(\dot{r} + \sqrt{\frac{2M_-}{r}}\right)^2\right]^{1/2}} - \frac{m^2}{2r} + M_-$$

$$V = \frac{dr}{d\tau} = \frac{1}{N}\dot{r}\ , N = \left(\frac{dt}{d\tau}\right)^{-1}$$

$$d\tau^2 = dt^2 - \left( dr + \sqrt{\frac{2M_-}{r}} dt \right)^2$$

Thus

$$\dot{r} = NV = \frac{\left(1 - \frac{2M_-}{r}\right)}{\sqrt{\frac{2M_-}{r} \pm \left[1 + \left(1 - \frac{2M_-}{r}\right) V^{-2}\right]^{1/2}}}$$

$$\tilde{H}(r, V) = \pm m \left[V^2 + 1 - \frac{2M_-}{r}\right]^{1/2} - \frac{m^2}{2r}$$

$$\frac{\partial P}{\partial V} = \frac{1}{V} \frac{\partial H}{\partial V} = m \left[V^2 + 1 - \frac{2M_-}{r}\right]^{1/2}$$

$$P = m\log V + \sqrt{\left(1 - \frac{2M_-}{r}\right) + V^2}$$

$$V = \sinh(P/m) + \left(\frac{M_-}{r}\right) \exp(-P/m)$$

$$H(r, P) = m \cosh {}^{P}/_{m} - \frac{m^2}{2r} + M_- \left(2 - \left(\frac{m}{r}\right) \exp\left(\frac{-p}{m}\right)\right)$$

***Starting directly the Hamiltonian energy functional***
$\tilde{h}(q, v) = h(q, P(q, v))$ one can obtain $p(q, v)$ via

$$\frac{\partial p}{\partial v} = \frac{1}{v} \frac{\partial \tilde{h}}{\partial v} \quad \left(\text{using } p(q, v) = \frac{\partial \ell}{\partial v}\right)$$

Now with $V = N(q, V)V$ one can write

$$\tilde{H}(q, V) = \tilde{h}(q, N(q, V)V)$$

For which $\frac{\partial P}{\partial V} = \frac{1}{V} \frac{\partial \tilde{H}}{\partial V}$

For each choice of P, we recover a Hamiltonian $(q, P)$ :

$$H(q, P) = \tilde{H}\left(q, V(q, P)\right)$$

**8.3.2 General formalism for a one-dimensional Hamiltonian**
(Note, in Chapter 8 the equation numbers are those used in the original PhD thesis so may be out of sequence in places, but the text references to those equation numbers are still valid.)

Consider a 1-dimensional Hamiltonian of the form:

$$h = h(q, p). \tag{4.32}$$

The inverse of the Legendre transformation relating the canonical momentum p to a velocity v is given by

$$v(q, p) = \frac{\partial h}{\partial p}. \tag{4.33}$$

That is, solutions q(t); p(t) to Hamilton's equations correspond to dynamical solutions q(t); v(t) to the Euler-Lagrange equations associated with the Lagrangian

$$\ell(q, v) = p(q, v)v - h(q, p(q, v)), \tag{4.34}$$

with

$$v(t) = \frac{d}{dt}q(t). \tag{4.35}$$

The Hamiltonian, expressed as a function on velocity space, will be denoted by h tilda:

$$\begin{aligned} \tilde{h}(q, v) &= h(q, p(q, v)), \\ &= p(q, v)v - \ell(q, v). \end{aligned} \tag{4.36}$$

If one reparametrizes the time t in the manner

$$dT = Ndt, \quad N(t) = N(q(t), v(t)), \tag{4.37}$$

the corresponding velocity, V (T) = d dT q(T), is related to v(t) by

$$v = N(q, v)V. \tag{4.38}$$

In the Hamiltonian framework, the canonical momentum P associated with (q; V ) is given by

$$\frac{\partial P}{\partial p} = N(q, p), \tag{4.39}$$

where

$$N(q, p) := N(q, v(q, p)). \tag{4.40}$$

That is, with a new symplectic form dP ^ dq replacing dp ^ dq, and with the Hamiltonian written in terms of P and q,

$$H(q, P) = h(q, p(q, P)), \tag{4.41}$$

each dynamical path q(t); p(t) of the Hamiltonian h coincides with a reparametrized path q(T); P(T) of H.

equation (4.39) has a family of solutions,

$$P = \int^{p} N(q, p')dp', \tag{4.42}$$

corresponding to a choice of initial point. The solutions differ from one another by canonical transformations of the form,

$$P \to P + f(q). \tag{4.43}$$

Given H(q; P), one can recover the reparametrized velocity V from the relation

$$V = \frac{\partial H}{\partial P}. \tag{4.44}$$

Consider, for example, the motion of a free relativistic particle, parametrized by a Minkowski time t, and described by the Hamiltonian

$$h = \left(p^2 + m^2\right)^{\frac{1}{2}}, \tag{4.45}$$

with

$$v = \frac{\partial h}{\partial p} = \frac{p}{(p^2 + m^2)^{\frac{1}{2}}}. \tag{4.46}$$

If we adopt, as a new time parameter, the proper time along each path, then T is related to Minkowski time by equation (4.37) with

$$N - \left(1 - v^2\right)^{\frac{1}{2}} = \frac{m}{(p^2 + m^2)^{\frac{1}{2}}}. \tag{4.47}$$

We then have

$$P = \int N dp = m \sinh^{-1} \frac{p}{m}, \tag{4.48}$$

with inverse

$$p = m \sinh \frac{P}{m}. \tag{4.49}$$

Writing H(P; q) = h(q; p(P; q)), with h given by equation (4.45), we obtain the familiar form

$$H = m \cosh \frac{P}{m}. \tag{4.50}$$

If one starts directly from the Hamiltonian on velocity space, from equations (4.45) and (4.46), the momentum p(q; v) is given by the relation

$$\frac{\partial p}{\partial v} = \frac{1}{v} \frac{\partial h}{\partial v}, \tag{4.51}$$

By writing the change in velocity corresponding to a reparameterization of time in the manner,

$$v = N(q, V)V, \tag{4.52}$$

one can re-express the energy as a function of q and V :

$$\tilde{H}(q, V) = \tilde{h}(q, N(q, V)V). \tag{4.53}$$

The corresponding canonical momentum P(q; V ) is again given by

$$\frac{\partial P}{\partial V} = \frac{1}{V} \frac{\partial \tilde{H}}{\partial V}, \tag{4.54}$$

or

$$P = \int dV \frac{1}{V} \frac{\partial \tilde{H}}{\partial V}. \tag{4.55}$$

As in equation (4.43), P is determined only up to a canonical transformation of the form P(q; V ) →P(q; V ) + f(q). For each choice of P, we recover a Hamiltonian H(q; P) by inverting P(q; V ) to obtain

$$H(q, P) = \tilde{H}(q, V(q, P)). \tag{4.56}$$

### 8.3.3 Proper-time parametrization of a self-gravitating dust shell

(Note, in Chapter 8 the equation numbers are those used in the original PhD thesis so may be out of sequence in places, but the text references to those equation numbers are still valid.)

The formalism presented in the preceding section is now used to obtain a simple Hamiltonian that describes a self-gravitating, spherically symmetric shell of dust, using proper time along the shell's trajectory. We begin with the Lagrangian expressed via equations (4.30) and (4.31):

$$\ell = \dot{r}p - h, \quad h = M_+ + M_- = \sqrt{\hat{p}^2 + m^2} + \frac{m^2}{2r} - \hat{p}\sqrt{\frac{2M_+}{r}} + 2M_- \tag{4.57}$$

where p is the canonical momentum associated with r, and $\hat{p}$ is the radial component of the shell's momentum in an orthonormal frame. The time indicated in r is the static-gauge time of the left-hand-side geometry. Here M- is a free parameter (not a dynamical variable), while the Hamiltonian h (and M+= h + M-) is determined by equation (4.57). The relation between $\dot{r}$ and $(\hat{p}, r)$, equation (4.30) has inverse

$$\hat{p} = m\frac{\left(\dot{\hat{r}} + \sqrt{\frac{2M_+}{r}}\right)}{\left[1 - \left(\dot{\hat{r}} + \sqrt{\frac{2M_+}{r}}\right)^2\right]^{1/2}}. \tag{4.58}$$

$$\tilde{h} = M_+ + M_- = m\frac{1 - \left(\dot{\hat{r}} + \sqrt{\frac{2M_+}{r}}\right)\sqrt{\frac{2M_+}{r}}}{\left[1 - \left(\dot{\hat{r}} + \sqrt{\frac{2M_+}{r}}\right)^2\right]^{1/2}} + \frac{m^2}{2r} + 2M_-. \tag{4.59}$$

Let us now reparametrize the time along each trajectory, replacing t by the proper time $\tau$, with

$$V = \frac{d\hat{r}}{d\tau} = \frac{1}{N}\dot{r}, \quad N = \left(\frac{dt}{d\tau}\right)^{-1}. \tag{4.60}$$

Using the 3-metric on the shell's time-like world-sheet, we have

$$d\tau^2 = dt^2 - \left(d\hat{r} + \sqrt{\frac{2M_+}{\hat{r}}}dt\right)^2. \tag{4.61}$$

implying for the function N the form

$$\frac{1}{N} = \frac{\sqrt{\frac{2M_+}{\hat{r}}} \pm \left[1 + \left(1 - \frac{2M_+}{\hat{r}}\right)V^{-2}\right]^{1/2}}{1 - \frac{2M_+}{\hat{r}}}V. \tag{4.62}$$

Then

$$v \equiv \dot{r} = NV = \frac{1 - \frac{2M_+}{r}}{\sqrt{\frac{2M_+}{r}} \pm \left[1 + \left(1 - \frac{2M_+}{r}\right)V^{-2}\right]^{1/2}}. \tag{4.63}$$

By substituting this expression for $\dot{r}$ into equation (4.59) and simplifying, we obtain h in terms of V:

$$\begin{aligned}\tilde{H}(r,V) &= \tilde{h}(r,v(r,V)) \\ &= \pm m\left[V^2+1-\frac{2M_+}{r}\right]^{\frac{1}{2}}+\frac{m^2}{2r}+2M_-. \end{aligned}\tag{4.64}$$

This is easily inverted to obtain an expression dependent on M- :

$$\begin{aligned}\tilde{H}(r,V) &= \tilde{h}(r,v(r,V)) \\ &= \pm m\left[V^2+1-\frac{2M_-}{r}\right]^{\frac{1}{2}}-\frac{m^2}{2r}+2M_-. \end{aligned}\tag{4.65}$$

Using equation (4.54), we then have

$$\frac{\partial P}{\partial V}=\frac{1}{V}\frac{\partial H}{\partial V}=\pm m\left[V^2+1-\frac{2M_-}{r}\right]^{-\frac{1}{2}}. \tag{4.66}$$

The canonical momentum P is given in terms of the r-component V of the shell's 4-velocity by

$$P=m\int\frac{du}{\left[u^2+1-\frac{2M_-}{r}\right]^{\frac{1}{2}}}=m\log V+\sqrt{\left(1-\frac{2M_-}{r}\right)+V^2}, \tag{4.67}$$

with inverse

$$V=\sinh(P/m)+\left(\frac{M_-}{r}\right)\exp(-P/m). \tag{4.68}$$

With equation (4.68) substituted into the relation given by equation (4.65) for h, we find after some calculation a simple expression for the Hamiltonian H(r; P), namely

$$H(r,P)=m\cosh P/m-\frac{m^2}{2r}+M_-(2-(m/r)\exp(-P/m)). \tag{4.69}$$

**8.3.4 An alternative parametrization of dust-shell time**

(Note, in Chapter 8 the equation numbers are those used in the original PhD thesis so may be out of sequence in places, but the text references to those equation numbers are still valid.)

A different time reparameterization leads to a Hamiltonian for the dust shell that is analogous to the Minkowski-time Hamiltonian

$$H=\sqrt{p^2+m^2} \tag{4.70}$$

for a free spherical dust shell in at space. Starting from the proper-time Hamiltonian (4.69), we set

$$N = \frac{m}{\sqrt{p^2 + m^2}}, \tag{4.71}$$

with p the momentum associated with the new time. equation (4.39),

$$\frac{\partial P}{\partial p} = N(q,p) = \frac{m}{\sqrt{P^2 + m^2}}, \tag{4.72}$$

has solution

$$p = m \sinh \frac{P}{m}. \tag{4.73}$$

Then

$$\cosh \frac{P}{m} = \frac{\sqrt{P^2 + m^2}}{m}, \tag{4.74}$$

and we have

$$H = \sqrt{p^2 + m^2} - \frac{m^2}{2r} + M_{-}[2 - \frac{1}{r}(\sqrt{p^2 + m^2} - p)]. \tag{4.75}$$

When the interior mass M- vanishes, this is the Hamiltonian used by Kuchar et. al. [148-150].

### 8.3.5 Hamiltonians for the RP3 geon

As noted earlier, an asymptotically at, spherically symmetric spacetime M; g with a single asymptotic region can have the spatial topology of the manifold $\Sigma = RP^3/\{\infty\}$, where $/\{\infty\}$ denotes mod or removal of point at infinity. We will refer to such a spacetime, or to an asymptotically at initial data set with this topology, as an $RP^3$-geon. Its covering space has the wormhole topology of the extended Schwarzschild geometry.

One can obtain a spherically symmetric vacuum spacetime with this topology from extended Schwarzschild, $\bar{M}$, $\bar{g}$, by picking a Kruskal chart $(U; V; \theta; \phi)$, with U and V the Kruskal null coordinates. Surfaces $\bar{\Sigma}$ of constant Kruskal time T = U+V have a throat, a symmetric two-sphere of minimum radius, and they are symmetric under the involution

$$I : (R, \theta, \phi) \to (-R, \pi - \theta, \phi + \pi), \tag{4.76}$$

that identities points on opposite sides of the throat. The spacetime M; g is the quotient space M = $\bar{M}/I$, with g and $\bar{g}$ locally isometric. Each slice

Σ of the spacetime $\overline{M}$ , has as a throat a copy of RP2. The spacetime M; g outside the throat's history is indistinguishable from half of the extended vacuum Schwarzschild geometry $\overline{M}$, $\bar{g}$.

An initial data set with a dust shell on Σ, satisfying the constraint equations, is characterized by the circumferential radius of the dust r, a conjugate momentum P, and a mass M describing the geometry interior to the shell. The data can be lifted to a locally identical initial data set with two dust shells on $\bar{\Sigma}$ in which the two dust shells lie on opposite sides of the throat, each at circumferential radius r. Outside the shells, the Schwarzschild masses have the same value $M_{-} = M_{+}$, and between the shells is a Schwarzschild geometry with a smaller mass M. From equation (4.69) the Hamiltonian $\overline{H}$ for the two-shell system has the proper-time form,

$$\bar{H} = M_{-} + M_{+} = 2m\cosh P/m - 2\frac{m^2}{2r} + 2[M_{-}(2 - (m/r)\exp(-P/m))]. \quad (4.77)$$

The corresponding Hamiltonian for the single shell on the RP3-geon manifold Σ is then

$$H = m\cosh P/m - \frac{m^2}{2r} + M_{-}(2 - (m/r)\exp(-P/m)), \qquad (4.78)$$

identical in form to equation (4.69). In terms of the alternative time-parametrization of equation (4.75), the system has Hamiltonian

$$H = \sqrt{p^2 + m^2} - \frac{m^2}{2r} + M - [1 - \frac{1}{r}(\sqrt{p^2 + m^2} - p)]. \qquad (4.79)$$

### 8.4 Quantization Overview

We first consider the vacuum case before turning to the quantization of the shell + geometry. This will be an overview, with the lengthy details placed in App. C.

#### 8.4.1 Mass spectrum of spherically symmetric vacuum wormholes and RP3 geons

One can include in the asymptotic conditions for a wormhole geometry, a particular value $M_{-}$ = M for the mass at the left asymptopia.

If this is done, then, once the constraint equations are solved, the right-side mass is determined, M+ = $M_{-}$ = M, and there is no remaining degree of freedom. The system is trivial, and its mass, a c-number by construction, is not quantized. If, on the other hand, one retains $M_{-}$ = M as a dynamical variable, then, after the constraints are satisfied, the Hamiltonian is the mass

$$H = M_- + M_+ = 2M, \tag{4.80}$$

and the symplectic structure is $dP_M \wedge dM$. If, as Kuchar does, one takes M as the configuration space variable, then, in a Schrodinger representation the operator M acts in the manner

$$\hat{M}\Psi(M) = M\Psi(M), \tag{4.81}$$

and M has continuous spectrum.

One can, however, replace the canonical variables M and $P_M$ by any other canonical pair. In particular, we can use as a configuration space variable the throat radius, a, of a symmetric slicing of the Schwarzschild geometry. That is, each solution to the constraint equations is specified by a and its conjugate momentum $p_a$. With the choice N = 1, $\dot{a}$ is the derivative of the throat radius with respect to the throat's proper time. Written in terms of a and $\dot{a}$, the mass is

$$M = \frac{1}{2}(a\dot{a}^2 + a). \tag{4.82}$$

A canonical transformation from M, $P_M$ to a, $p_a$ gives

$$H = M = \frac{1}{2}\left(\frac{p_a^2}{a} + a\right), \tag{4.83}$$

where

$$p_a = sgn(p)\sqrt{2Ma - a^2}. \tag{4.84}$$

For obvious choices of measure on the configuration space R+ associated with a, factor orderings that make $\hat{M}$ self-adjoint yield a spectrum that, like the spectrum of the nonrelativistic Coulomb Hamiltonian has discrete and continuous parts.

The continuous spectrum that Kuchar obtains for the mass of this system can be obtained in an analogous way for any dynamical system in which the Hamiltonian is not explicitly time-dependent. That is, for any function H with nonvanishing gradient on phase space, one can find a local canonical chart of the form

$$H, q_2, \cdots, q_n, p_H, p_2, \cdots, p_n, \tag{4.85}$$

a chart in which H is one of the canonical variables. If, as Kuchar does, one takes a Schrodinger representation with Hilbert space $L_2(R_+) \otimes \mathcal{H}$, with $\mathcal{H}$ a Hilbert space for the remaining q's, then $\widehat{H}$ acts as a multiplication operator,

$$\hat{H}\psi(H, q_2, \cdots q_n) = H\psi(H, q_2, \cdots, q_n), \tag{4.86}$$

and its spectrum is $R_+$.

The dynamics of the RP3-geon geometry are described by the formally identical Hamiltonian (4.84), for the throat radius with a proper-time parametrization, and the quantizations and corresponding mass spectra are identical. Only the interpretation is altered.

**8.4.2 Discussion of quantization of geometry with dust-shell**

To do the quantization of the geometry with dust shell, consider the proper-time Hamiltonian (4.69).

When $M_- = 0$, the shell encloses a flat interior with trivial topology, and the Hamiltonian (4.69) takes the form corresponding to a relativistic particles in a Coulomb potential

$$H(\hat{r}, P) = m \cosh\frac{P}{m} - \frac{m^2}{2\hat{r}}, \tag{4.87}$$

discussed by Hajicek [142]. One can adopt a Schrodinger representation corresponding to configuration-space variable $\hat{r} \in R_+$ and Hilbert space

$$\mathcal{H} = L_2(\mathrm{R}_+, r^\alpha d\hat{r}). \tag{4.88}$$

With the factor ordering

$$\widehat{\cosh\frac{P}{m}} = \lim_{N\to\infty} \hat{r}^{\frac{\alpha}{2}} \sum_{n=0}^{\infty} \frac{(-1)^n}{(2n)!} \Delta^n \hat{r}^{\frac{\alpha}{2}}, \tag{4.89}$$

where $\Delta = \partial_{\hat{r}}^2$, $\widehat{H}$ is self-adjoint with domain [142]

$$D(\hat{H}) = \left\{ f^{(2n)}(0) = 0,\ f^{(n)} \in L_2,\ for all\, n \right\}. \tag{4.90}$$

For m < 1.9 Planck mass H is bounded from below, and its spectrum, like that of the nonrelativistic Coulomb problem, has discrete and continuous parts.

With specified $M_- > 0$, the shell Hamiltonian has the form (4.69). Again we expect a spectrum that is partly discrete and bounded below for small m and we have:

$$\hat{H} = m\cosh\widehat{\frac{P}{m}} - \frac{m^2}{2\hat{r}} + 2M_- - mM_-\hat{r}^{-\frac{1}{2}}\exp(\widehat{-P}/m)\hat{r}^{-\frac{1}{2}} \tag{4.91}$$

When $M_-$ is dynamical, the parameter $M_-$ in $\widehat{H}$ is replaced by an operator $\widehat{M}_-$ One can effectively separate variables by first considering the eigenvalue equation for $\widehat{M}_-$,

$$\widehat{M}_-\psi = M_-\psi$$

For each eigenspace of $\widehat{M}_-$, the shell Hamiltonian then has the form (4.91) with c-number $M_-$. Again by a choice of factor ordering (or of time) one can obtain either a continuous or discrete spectrum for $\widehat{M}_{\square}$. That is, as in the vacuum case, if one chooses a representation $\psi(\hat{r}, M_-)$, with $\widehat{M}_-\psi = M_-\psi$, the spectrum of $\widehat{M}_-$ is continuous on $L_2(R_+, dM_-$; and by choosing variables $\alpha, p_\alpha$ canonically related to $M_-, P_{M_-}$ by equation (4.83) and (4.84), we can recover the form

$$\hat{M}_- = \frac{1}{2}\left(\frac{\hat{P}_\alpha^2}{\alpha} + \alpha\right). \tag{4.92}$$

Note, however, that as presented, this canonical transformation does not have the physical interpretation it had in the vacuum case. One cannot immediately relate $\alpha$ and $p_\alpha$ to the radius a of the throat and its conjugate momentum for a proper-time parametrization of the throat: Had we replaced $M_-$ by its expression in terms of $a$ and $\dot{a}$ in the action prior to a reparameterization of time, the corresponding momentum would not satisfy equation (4.84); and it is not clear to us how one could recast the resulting action in a form that uses proper time of the throat at the throat and proper time of the shell at the shell.

Finally, as noted earlier, the Hamiltonian (4.78) for a spherically symmetric RP3-geon geometry with a dust shell is formally identical to that of a wormhole geometry with shell. In the case of a geon, however, the interior mass is not the mass of an asymptotic region, and one does not have the option of treating it as a parameter characterizing the asymptotics rather than a dynamical variable. This is essentially a distinction of interpretation, because one is free to restrict consideration to an eigensubspace of the interior mass.

### 8.4.3 Quantization Technical Overview

***For the dust shell we have***

$$h = \sqrt{P(r,p)^2 + m^2} - \frac{m^2}{2r} - P(r,p)\sqrt{\frac{2M_-}{r}}$$

$$\dot{r} = \frac{\partial H}{\partial p} = \frac{P}{\sqrt{P^2 + m^2}} - \sqrt{\frac{2M_-}{r}}$$

$$P = \frac{m\left(\dot{r} + \sqrt{\frac{2M_-}{r}}\right)}{\left[1 - \left(\dot{r} + \sqrt{\frac{2M_-}{r}}\right)^2\right]^{1/2}}$$

***Switch to dimensionless variables and consider quantization***

$$\hat{H} = \widehat{\cosh(p)} - \frac{2 \propto \widehat{\beta}}{r} e^{-p} - \frac{\hat{\propto}}{r} + \beta$$

Consider just $\cosh(p)$, among its possible representations there are

(i) $\widehat{\cosh(p)} = \frac{1}{2}\left(\widehat{e^P} + \widehat{e^{-P}}\right), e^{tp} f = \lim_{n\to\infty}\left(1 - \frac{tp}{n}\right)^{-n} f$;
and the maximal domain is where the limit is defined such that it is and $L^2$ function

(ii) $\widehat{\cosh(p)} = \lim_{n\to\infty} \sum_{n=0}^{N} \frac{(-1)^n}{(2n)!}\left(\frac{d}{dr}\right)^{2n}$, and the maximal domain is where the $N - \mathit{limit}$ is an $L^2$ function.

Choose the closure of case (ii): $\underline{(RP^3)}$

$\widehat{\cosh(p)} \equiv \bar{T}$ $\qquad$ $\underline{L^2(r, dr)}$

$T \equiv \lim_{n\to\infty} T_N$ (weak limits sufficient)

$$T_N f \equiv \sum_{n=0}^{N} \frac{(-1)^n}{(2n)!} (\Delta_0^{\propto})^n f\,, \qquad \Delta_0^{\propto} = \left(\frac{d}{dr}\right)^2$$

$$D(T_N) = D((\Delta_0^{\propto})^N) = \{f \in D((\Delta_0^{\propto})^{N-1}) | (\Delta_0^{\propto})^{N-1} f \in D(\Delta_0^{\propto})|\}$$

$$D(\Delta_0^{\propto}) = \{f \in H^2 \,|f(0^+)\sin(\propto) + f'(0^+)\cos(\propto) = 0,\ \propto \in [0,\pi]\},$$

$$D(T) = D_0 \left\{ f \in \bigcap_{N=0}^{\infty} D(T_N) \,\middle|\, \lim_{n\to\infty} T_N f \text{ exists in } L^2 \right\}$$

Claim: $\Delta_0^{\propto}$ is a one parameter family of self-adjoint operates for $\Delta_0^{\propto} = \left(\frac{d}{dr}\right)^2$ with domain parameterized by

$$D(\Delta_0^{\propto}) = \{f \in L^2, \Delta_0 f \in L^2 \,|f(0^+)\sin(\propto) + f'(0^+)\cos(\propto) \equiv F_{\propto}(0^+) = 0, \propto \epsilon[0,\pi]\}$$
$$= \{f \epsilon H^2 | F_{\propto}(0^+) = 0\}$$

Claim: $T_N$ is self-adjoint

Claim: $\bar{T}$ is self-adjoint

Proof:
Lemma 1: $-\Delta_0$ is non-negative
Proof: $(\delta, -\Delta_0 f) = \int_0^\infty |\delta'|^2 \, dr \geq 0$
Remark: since $-\Delta_0$ is non-negative and so it can be used generate a semi-group
$\{e^{t\Delta_0} | t \geq 0\} < \mathcal{L}(L^2)$, where $\mathcal{L}(L^2)$ is the set of linear functionals on L.

Lemma 2: $T_N$ is non-negative
Proof: $(T_N f, f) = \sum_{n=0}^{N} \frac{(-1)^n}{(2n)!} (\Delta_0^n f, f) = \sum_{n=0}^{N} \frac{1}{(2n)!} \, |\nabla' f|^2 \geq |f|^2 \geq 0$

And $|T_N f||f| \geq |f|^2 \geq 0$,
$|T_N f| \geq |f| \geq 0$

The limit of the operate sequence $T_N$ need only be wealthy defined to them obtain $|Tf| \geq |f| \geq 0$ . So, T is non-negative also.

Lemma 3: $A\,(\Delta_0^{\propto}) = \{e^{s\Delta_0} f \,| s\epsilon\, (0\ 1], f \in L^2\}$ is a core for $L^2$
Proof:
Consider $e^{s\Delta_0} f; f \in L^2$:

$$T_N(e^{s\Delta_0} f\,) = \sum_{n=0}^{N} \frac{(-1)^n}{(2n)!} (\Delta_0)^n (e^{s\Delta_0} f)$$

$$|\Delta_0^n e^{s\Delta_0} f| = \left|\left(\Delta_0 e^{\frac{s}{n}\Delta_0}\right)^n f\right|$$

Since $e^{s\Delta_0}$ is a semi-group (an analylic semi-group in fact), the theorem by Hille and Yoside is applicable:

$$\left|\Delta_0 e^{s\Delta_0} f\right| \leq \frac{M}{S}\,|f|,$$

Where $f \in D(e^{s\Delta_0})$ and $M$ is an absolute constraint, since $\left|\left(\Delta_0 e^{\frac{s}{n}\Delta_0}\right)\hat{f}\right| \leq M^1\left(\frac{M}{S}\,|f|\right)$ it possible to write

$$|T_N(e^{s\Delta_0} f)| \leq \sum_{n=0}^{N} \frac{1}{(2n)!}\frac{M^n n^n}{S^n}\,|f| < \sum_{n=0}^{N} \frac{1}{n!}\left(\frac{M}{S}\right)^n |f|$$

With either strong or weak limit defining the operator sequence, it then follows that

$$|T(e^{s\Delta_0} f)| \leq \sum_{n=0}^{\infty}\left|\frac{(-1)^n}{(2n)!}\Delta_0^n e^{s\Delta_0} f\right| < \sum_{n=0}^{\infty}\frac{1}{n!}\left(\frac{M}{S}\right)^n |f|,$$

$$< \exp\left(\frac{M}{S}\right)|f|$$

So, $e^{s\Delta_0} f \in D_0$, i.e., $e^{s\Delta_0}: L^2 \to D_0$. From the well established semigroup properties $A(\Delta_0)$ is clearly dense in $L^2$ : $e^{s\Delta_0} f \to f$ as $s \downarrow 0$

.

So, for $f \in D_0$:

$$\left.\begin{array}{l} e^{s\Delta_0} f \to f \\ Te^{s\Delta_0} f = e^{s\Delta_0} Tf \to Tf \end{array}\right\} \text{ indicating that } A(\Delta_0) \text{ is a core.}$$

Since $A(\Delta_0)$ is dense in $L^2$ and $(\Delta_0) \subset D_0$ , this also proves that $D_0$ is dense.

Lemma 4: The closure of the range of $T$ is $L^2$ ,$\overline{R(\tau)} = L^2$.

Proof:

Define $R(T) = \{g \in L^2 | g = Tf \text{ for some } f\}$

Write $L^2 = \left(\overline{R(\tau)}\right)^1 \oplus \overline{R(\tau)}$ and $D_0 \cap \left(\overline{R(\tau)}\right)^\perp \neq \{0\}$ , then $\exists f \neq 0$ in the intersection (by the denseness of $D_0$ ) which satisfies

$0 = (f, Tf) \geq |f| > 0$

Which is a contradiction. So $\left(\overline{R(\tau)}\right)^\perp = \{0\}$, i.e., $\overline{R(\tau)} = L^2$.

Lemma 5: $T^{-1}$ is non-expansive

Proof: $|Tf| \geq |f| \Rightarrow |T^{-1}g| \leq |g|$ for $g \in R(\tau)$
Since $\overline{R(\tau)} = L^2$ , the limit $g = \lim g_n , g_n \in D_0$ exists.

Lemma 6: the limit $n \to \infty$ $of$ $T^{-1}$ $g_n$ exists.
Proof: $|T^{-1}g_m - T^{-1}g_n| = |T^{-1}(g_m - g_n)| \leq |g_m - g_n|$
Since $|g_m - g_n| \to 0$ as a Cauchy sequence, so also for $T^{-1}g_n$ as a Cauchy sequence, thus lim $T^{-1}g_n$ exists and indicates as extension of $T^{-1}$:

$$\lim_{n\to\infty} T^{-1}g_n \equiv \overline{T}^{-1} g$$

Lemma 7: $\overline{T}^{-1}$ exists
Proof: $\left|\overline{T}^{-1} g - {T_N}^{-1} g\right| = |T^{-1}g - T_N^{-1}g| = |f - T_N^{-1}Tf|$

$$= |T_N^{-1}T_N\mathrm{f} - T_N^{-1}Tf| \leq |T_N f - Tf|$$

$$\to 0 \quad \text{as } N \to \infty,$$

Where the concluding relation follows from the definition of $T$

Lemma 8: $\overline{T}^{-1}$ exists, so $\overline{T}$ exists
Proof:
Take $f_n \in D(T), f_n \to f, Tf_n = g_n \to g$, then

$$f_n \to f \Rightarrow T^{-1}g_n \to f \Rightarrow f = T^{-1}g \Rightarrow \left(\overline{T}^{-1}\right)^{-1} f = g$$

Which implies that $\overline{T}$ exists and is related to $\overline{T}^{-1}$ by the expected relation $\overline{T} \equiv \left(\overline{T}^{-1}\right)^{-1}$.
Since $T_N$ is self-adjoint the symmetry of $T$ is already established. From which:

$$T \text{ symmetric } \Rightarrow \overline{T} \text{ symmetric } \Rightarrow \overline{T}^{-1} \text{ symmetric}$$

The existence and symmetry of $\overline{T}^{-1}$ being established, note that
$D\left(\overline{T}^{-1}\right) = R(\overline{T}) = \overline{R(T)} = L^2$

Since $D\left(\overline{T}^{-1}\right) = L^2$, then $D\left(\left(\overline{T}^{-1}\right)t\right) = L^2$ also, and $\overline{T}^{-1}$ is self-adjoint

<u>Lemma 9</u>: $\overline{T}^{-1}$ self-self-adjoint implies $\overline{T}$ is self-adjoint.
Proof:
$(g, \overline{T}f) = (h, f)$ for $f \in D(\overline{T})$

$$(g,\hat{f}) = \left(h,\overline{T}^{-1}\hat{f}\right) \ for\ \hat{f} = \overline{T}f \in L^2$$

$$\Rightarrow g = \overline{T}^{-1}h \Rightarrow h = \overline{T}g \Rightarrow \overline{T}^t = \overline{T}$$

This proves claim that $\overline{T}$ is self-adjoint. Similar analysis for other cases.

# Chapter 9 Conclusion

We began with the pre-Quantum physics history that leads into Quantum Theory. A thorough description of the modern Schrodinger and Heisenberg theories and applications was then given. Material used in this endeavor derives from lecture notes and material from over a dozen excellent quantum mechanics textbooks [1-14]. After the standard, undergraduate-level, quantum mechanics description, a more mathematically formal construction was made with Dirac's axiomatic description, and more advanced quantum mechanical issues were addressed, such as relativistic quantum mechanics.

The quantum analysis of hydrogen shows state quantization and precisely the spectral oddities observed and captured approximately in various models going back to Balmer – all with a fit involving a constant that is based on the mysterious 'alpha' constant (also known as the fine-structure constant). The Hydrogen analysis indicates the chemical model shell structure (based on fermionic particles), thus all of the standard chemistry rules.

The quantum theory from its Schrödinger formulation is inherently non-relativistic. It is also oddly asymmetric in time in the manner of a diffusion equation (in analytic extension). The diffusion equation form (complex time) can be described in terms of a random-walk process, which, in turn, can be described as a Martingale process – in agreement with Emanator Theory [21]. Also, at the at the low-energy low-velocity extreme, we see behavior in agreement with classical motion[16].

To go to a relativistic theory is the next step [15] and this was successfully undertaken by Dirac with the famous Dirac equation. This equation is first-order in time and space differentials, and describes a wave process. Furthermore, the equation indicates spinorial functions for solutions, thus we now have a true representation of Fermionic matter (all of the elementary particles in the standard model are fermions) for the first time. The Dirac equation, however, breaks down in interpretation, although it does indicate a multiparticle vacuum state and the existence of holes in this vacuum that would be seen as positrons, as observed. The problem is that the Dirac equation, like the Schrodinger equation before it, is (originally) a one-particle wavefunction theory. The quantum field theory formulation will allow for many-particle descriptions, while covariant, and when applied in the specific form of quantum electrodynamics we will get remarkable agreement with various experiments [15].

As many-particle solutions are fully realized in the quantum field theory context, often with black-body thermal distributions, we begin to suspect a thermal aspect to the theory, fundamental, not just derivative of the assemblage of structures and their inter-relation (including spin-statistics relations both kinematically-based in flat spacetime and dynamically-based in curved spacetime [15]). This fundamental thermality is most evident in the complex structure (trivially extended) of the theory at equilibrium where (inverse) temperature is related to the periodicity in complex time. The time-thermality connection is exploited in a full general relativity analysis using Hamiltonian Thermodynamics analysis [20], where more details on the odd complex-structure (complex time with periodicity) thermal aspect of the theory is described.

A minisuperspace formalism is explored in the latter part of the book, where a spherically symmetric geometry with a shell of dust is examined in detail. A spherically symmetric geometry without a dust shell has two dynamical degrees of freedom when the constraint equations are solved. These degrees of freedom may be identified with the system's total mass and momentum or, equivalently, with the position of the center of mass and one additional metric parameter. For extended Schwarzschild topology, if the momentum is fixed in one asymptotic region, the momentum constraint fixes it in the other asymptotic region. If the mass is also fixed in one asymptotic region, the Hamiltonian constraint determines it in the other, and the system has no remaining freedom. Adding a dust shell adds one dynamical degree of freedom in each case. In the minisuperspace analysis for dust shell in extended Schwarzschild the degree of freedom associated with asymptotic momentum is not retained, but the differences that arise from retaining or discarding the freedom associated with the system's total mass will be considered. In the context of a reduced phase-space approach, one can thus choose to regard the minisuperspace Hamiltonian as a function of one or two dynamical variables depending (in the case of Schwarzschild topology) on whether a fixed asymptotic mass is incorporated into the asymptotic conditions on the metric.

The Hamiltonian obtained upon reduction is in a peculiarly complicated form. In order to simplify its expression, and eventual quantization, a new formalism is developed for reparametrizing time in one-dimensional Hamiltonian systems. The formalism then permits reformulation of the reduced Hamiltonian to a simpler (and known) form associated with the proper time of the shell. Quantization of the system is then done in detail. Whether or not the energy spectrum is discrete depends both on the representation and factor ordering, and on the choice of time. Unfortunately this means that we have to put in the answer to get the answer, an overall freedom that is on par with classical apparatus freedom. Even if geometry is apparatus in the quantum measurement sense, however, we can still side-step measurement theory complications (more unknowns at the theoretical level) by shifting to a Hamiltonian thermodynamics analysis (instead of tackling quantum gravity head-on, we tackle thermal quantum gravity) and this is what is done will be done in the next two books in the Series [15,20]).

# Appendix A. Mathematics and Notation of Quantum Mechanics

This appendix contains math notes gathered from the quantum mechanics texts mentioned earlier [1-14], as well as texts by Constantinescu [155] and Cohen-Tannoudji [156,157].

## A.1 Operator Mathematics

***Products of operators***

$(AB)\Psi = A[B\Psi]$

In general $AB \neq BA$ (just consider matrix representation).

***The Commutator***

$[A, B] = AB - BA$ is called the commutator.

***Discrete orthogonal bases in $\mathcal{F}$: $\{u_i(r)\}$***

Consider a countable set of functions of $\mathcal{F}$, labelled by a discrete index i:

The set $\{u_i(r)\}$ is orthonormal if $(u_i, u_j) = \int d^3r\, u_i^* u_j = \delta_{ij}$.
The set constitutes a basis if every $\Psi(\vec{r}) \in \mathcal{F}$ can be expanded in only one way in terms of the $u_i$ :

$$\Psi(\vec{r}) = \sum_i c_i u_i(r).$$

***Components of a wave function in the $\{u_i(r)\}$ basis***

Project out:

$$(u_j, \Psi) = \left(u_j, \sum_i c_i u_i\right) = \sum_i c_i (u_j u_i) = \sum c_i \delta_{ij} = c_j$$

Thus,

$$c_j = (u_j, \Psi) = \int d^3r u_j^*(\vec{r})\, \Psi(\vec{r}).$$

The $\{c_i\}$ represent $\Psi(\vec{r})$ in the $\{u_j(\vec{r})\}$ basis. (The same $\Psi$ can have different components if in two different bases.)

***Expression for the scalar product in terms of the components***

$$\varphi(\vec{r}) = \sum_i b_i u_i \quad , \quad \Psi(r) = \sum_j c_j u_j$$

Consider

$$(\varphi, \Psi) = \sum_{i.j} b_i^* c_j (u_i, u_j) = \sum_{i,j} b_i^* c_j \delta_{ij} = \sum_i b_i^* c_i$$

In particular, $(\Psi, \Psi) = \sum_i |c_i|^2$ (analogous to $\vec{v} \cdot \vec{w} = \sum_{ij}^3 v_i w_i$ for vectors in $R^3$).

***Closure relation***

Expresses the fact that $\{u^i\}$ constitute a basis:

$$\Psi = \sum_i c_i u_i = \sum_i (u_i, \Psi) u_i = \sum_i \left[ \int d^3r' u_1^*(\vec{r'}) \Psi(\vec{r'}) \right] u_i(r)$$

$$= \int d^3r' \Psi(\vec{r'}) \left[ \sum_i u_i(\vec{r'}) u_i^*(\vec{r'}) \right]$$

Thus, $\sum_i u_i(\vec{r'}) u_i^*(r^{*\prime}) = \delta\left(\vec{r} - \vec{r'}\right)$.

**Introduction of "bases" not belonging to $\mathcal{F}$ (or $L^2$ at all)**

***Plane waves (in 1D)***

Recall the advantage of the Fourier transform in analysis:

$\Psi(x) = \frac{1}{\sqrt{2\pi\hbar}} \int_{-\infty}^{\infty} dp\, \overline{\Psi}(p) e^{ipx/\hbar} = \int_{-\infty}^{\infty} dp\, \overline{\Psi}(p) v_p(x)$

$\overline{\Psi}(p) = \frac{1}{\sqrt{2\pi\hbar}} \int_{-\infty}^{\infty} dp\, \Psi(x) e^{-ipx/\hbar} = (v_p, \Psi) = \int_{-\infty}^{\infty} dx v_p^*(x) \Psi(x)$

So, consider the function $v_p(x) = \frac{1}{\sqrt{2\pi\hbar}} e^{ipx/\hbar}$. Now, $\left|v_p(x)\right|^2 = \frac{1}{\sqrt{2\pi\hbar}}$, which diverges upon integration over x, thus $v_p(x) \notin \mathcal{F}$ .

***Parseval's Relation***

Parseval's relation follows from the above with: $(\Psi, \Psi) = \int_{-\infty}^{\infty} dp\, |\overline{\Psi}(p)|^2$ since we have the closure relation $\int_{-\infty}^{\infty} dp\, V_p(x) V_p^*(x') = \delta(x - x')$. The generalization to 3D then directly follows. Thus, $v_{\vec{p}}(\vec{r})$ can be considered to constitute a "continuous basis":

$$i \leftrightarrow \vec{p}$$

$$\sum_i \leftrightarrow \int d^3p$$

$$\delta_{ij} \leftrightarrow \delta(\vec{p} - \vec{p}')$$

***Delta function basis***

Consider the set $\{\varepsilon_{\vec{r}_0}(\vec{r})\}$ where $\varepsilon_{\vec{r}_0}(\vec{r}) = \delta(\vec{r}-\vec{r}_0)$, where $\varepsilon_{\vec{r}_0}(\vec{r}) \notin \mathcal{F}$ . We have:

$$\Psi(\vec{r}) = \int d^3r_0\, \Psi(\vec{r}_0)\delta(\vec{r}-\vec{r}_0) = \int d^3r_0\, \Psi(\overrightarrow{r_0})\varepsilon_{\vec{r}_0}(\vec{r})$$

and

$$\Psi(\overrightarrow{r_0}) = \int d^3r\delta(\vec{r}_0-\vec{r})\,\Psi(\vec{r}) = (\varepsilon_{\vec{r}_0}, \Psi) = \int d^3r\varepsilon_{\vec{r}_0}{}^*(\vec{r})\,\Psi(\vec{r})$$

where, $(\varphi, \Psi) = \int d^3r_0\varphi^*(\vec{r}_0)\,\Psi(\vec{r})$ gives back the definition of scalar product orthogonalization (and closure condition also trivial).

Note: A physical state must always correspond to a square-integrable wave function. The continuous basis $v_p(x)$ and $\varepsilon_{\vec{r}_0}$ are not square-integrable and are only used as intermediaries in calculations.

Consider a general continuous "orthogonal" basis $\{w_\alpha(\vec{r})\}$:

$$(w_\alpha, w_{\alpha'}) = \int d^3r\, w_\alpha^*(\vec{r})\, w_{\alpha'}(\vec{r}) = \delta(\alpha-\alpha')$$

and

$$\int d\alpha\, w_\alpha(\vec{r})\, w_\alpha^*(\vec{r}') = \delta(\vec{r}-\vec{r}').$$

Note: for $\alpha = \alpha', (w_\alpha, w_\alpha')\ diverges,$ thus a continuous basis has $w_\alpha(\vec{r}) \notin \mathcal{F}$.

***Mixed (discrete and continuous) Basis***

$$\text{Orthonormality}\begin{Bmatrix} (u_i, u_j) = \delta_{ij} \\ (\omega_\alpha, \omega_{\alpha'}) = \delta(\alpha-\alpha') \\ (u_i, \omega_\alpha) = 0 \end{Bmatrix}$$

$$\text{Closure } \textstyle\sum_i u_i(\vec{r})u_i^*(\vec{r}') + \int d\alpha\, w_\alpha(\vec{r})w_\alpha^*(\vec{r}') = \delta(\vec{r}-\vec{r}')$$

## A.2 Properties of linear operators

***The Trace***

$TrA = \sum_i < u_i|A|\, u_i >$ which is invariant, independent of chosen basis. If A is an observable, choose a diagonalizing basis, then $TrA = \sum_n g_n a_n$ where the $\{a_n\}$ are the eigenvalues and $\{g_n\}$ are the associated degrees of degeneracy.

$TrAB = TrBA$ and $Tr\ ABC = Tr\ BCA = Tr\ CAB$ (under cyclic permutations).

***Commutator relations***

$$[\mathrm{A}, \mathrm{BC}] = [\mathrm{A}, \mathrm{B}]\mathrm{C} + \mathrm{B}[\mathrm{A}, \mathrm{C}]$$

$$[\mathrm{A}, [\mathrm{B}, \mathrm{C}]] + [\mathrm{B}, [\mathrm{C}, \mathrm{A}]] + [\mathrm{C}, [\mathrm{A}, \mathrm{B}]] = 0$$

$$[\mathrm{A}, \mathrm{B}]^\dagger = [\mathrm{B}^\dagger, \mathrm{A}^\dagger]$$

Thus

$$\text{If } [\mathrm{A}, \mathrm{C}] = [\mathrm{B}, \mathrm{C}] = 0 \text{ and } \mathrm{C} = [\mathrm{A}, \mathrm{B}], \text{then } [\mathrm{A}, \mathrm{B}^n] = n\mathrm{C}\mathrm{B}^{n-1}$$

Thus

$$[\mathrm{A}, \mathrm{F}(\mathrm{B})] = [\mathrm{A}, \mathrm{B}]\mathrm{F}'(\mathrm{B}) \quad and \quad [\mathrm{B}, \mathrm{G}(\mathrm{A})] = [\mathrm{B}, \mathrm{A}]\mathrm{G}'(\mathrm{A})$$

And it follows

$$[\mathrm{Q}, \mathrm{P}] = i\hbar \rightarrow [\mathrm{Q}, \mathrm{F}(\mathrm{P})] = i\hbar\mathrm{F}'(\mathrm{P})$$

***Restriction of an operator to a subspace***

$$P_q = \sum_{i=1}^{q} |\varphi_i >< \varphi_i|$$

is a projection onto a subspace, for which we have the restriction of an operator to that subspace given by:

$$\hat{A}_q = P_q A P_q$$

**Functions of operators**

$A^n$ is defined as n successive applications of $A$. The inverse, $A^{-1}$, is defined by $A^{-1}A = 1$ (if it exists). Any function expressible by a power series $F(Z) = \sum_{n=0}^{\infty} f_n Z^n$ allows us to similarly define a corresponding operator $F(A)$. Consider, for example, $F(A) = e^A$, where

$$e^A = \sum_{n=0}^{\infty} \frac{A^n}{n!} = 1 + A + {A^2}/{2} + \cdots$$

If F(Z) is real function then the $f_n$ are real, and if $A$ is Hermitian, then $F(A)$ is Hermitian. Thus, if $A|\varphi_2 >= a|\varphi a > then\ F(A)|\varphi_a >= F(a)|\varphi_a >$.

Note that if $A$ is in its diagonalized basis, then $F(A)$ is the operator which is represented in the same basis by the diagonal matrix whose elements are $F(a_i)$. Take for example:

$$\sigma_Z = \begin{pmatrix} 1 & 0 \\ 0 & -1 \end{pmatrix} \quad \rightarrow \quad e^{\sigma_Z} = \begin{pmatrix} e & 0 \\ 0 & 1/e \end{pmatrix}.$$

Note that $e^A e^B \neq e^B e^A \neq e^{(A+B)}$ , $unless\ [A,B] = 0$. Consider the following important example -- the potential operator:

$$V(X)|x> = V(x)|x>$$

So,

$$<x|V(X)|\Psi> = V(x)\,\Psi(x)$$

as used in Schrodinger equation.

***Commutators involving functions of operators***

Consider two observables P and Q satisfying $[Q,P] = i\hbar$ (canonical commutation relations for 1 degree of freedom). Using just $[Q,P] = i\hbar$ we have:

$$[Q,P^2] = [Q,P]P + P[Q,P] = 2i\hbar P,$$

which appears to act as a derivative. Let's now show that this generalizes to the relation:

$$[Q,P^n] = i\hbar nP^{n-1}.$$

Let's assume the relation is true and verify iteratively:

$$[Q,P^{n+1}] = [Q,P]P^n + P[Q,P^n] = i\hbar p^n + i\hbar pp^{n-1} = i\hbar(n+1)P^n$$

by recurrence from initial condition $[Q,P^2\ ] = i\hbar P^{n+1}$ it is thus shown. Thus, in general

$$[Q,F(P)] = i\hbar F'(P).$$

Analogously:

$$[P,G(Q)] = -i\hbar G'(Q).$$

Note that $[Q,\Phi(Q,P)]$ is complicated due to ordering problems.

Note: If $[A,C] = [B,C] = 0\quad and\ C = [A,B]$ then we have

$[A,B^2] = [A,B]B + B[A,B] = CB + BC = 2CB$

Which generalizes to

$[A,B^n] = nCB^{n-1}$

Thus

$[A,F(B)] = CF'(B) = [A,B]F'(B)$

$[B,G(A)] = -CG'(A) = [B,A]G'(A)$

Thus, the essential communication properties of canonical variables follows not from $[Q,P] = constant$ but from the general property:

$$\big[Q,[Q,P]\big] = 0 \quad and \quad \big[P,[Q,P]\big] = 0.$$

So, $[Q,P]$ could even remain as an operator as long as it commutes with Q, P.

### A.3 Differentiation of an operator

The above detail on essential commutation properties is revisited to obtain Glauber's Formula where the derivative is also used. The derivative, if it exists, will have the usual form:

$$\frac{dA}{dt} = \lim_{\Delta t \to 0} \frac{A(t+\Delta t) - A(t)}{\Delta t},$$

for the matrix elements of A(t) in an arbitrary basis of t-independent vectors $|u_j>$:

$$< u_i|A|u_j > = A_{ij}$$

Then,

$$\left(\frac{dA}{dt}\right)_{ij} = < u_i \left|\frac{dA}{dt}\right| u_j >.$$

Consider

$$e^{At} = \sum_{n=0}^{\infty} \frac{(At)}{n!} \to \frac{d(e^{At})}{dt} = Ae^{At}$$

Similarly

$$\frac{d}{dt}(e^{At}e^{Bt}) = Ae^{At}e^{Bt} + e^{At}Be^{Bt}.$$

Note that $\frac{d}{dt}e^{A(t)}$ is generally not equal to $\frac{dA}{dt}e^{A(t)}$. In order for this equality to hold $A(t)$ and $d(A(t))/dt$ must commute.

Now to derive Glauber's formula:

$$e^A e^B = e^{A+B} e^{\frac{1}{2}[A+B]}$$

where both A and B are assumed to commute with their commutators. Consider $F(t) = e^{At}e^{Bt}$:

$$\frac{dF}{dt} = Ae^{At}e^{Bt} + e^{At}e^{At}Be^{Bt} = (A + e^{At}Be^{-At})F(t)$$

since

$$[e^{At}, B] = -[B, A]te^{At} = [A, B]te^{At}$$
$$\to e^{At}B = Be^{At} + t[A, B]e^{At}$$

we can write:

$$\frac{dF}{dt} = (A + [Be^{At} + t[A, B]e^{At}]e^{-At})F(t) = (A + B + t[A, B])F(t)$$

Thus,

$$\ln F(t) = (A + B)t + \frac{1}{2}[A, B]t^2 \ + const$$

$$F(t) = \exp\left\{(A+B)t + \frac{1}{2}t^2[A,B]\right\}$$

*Let* $t = 1$:

$$e^A e^B = e^{(A+B)} e^{\frac{1}{2}[A,B]}.$$

**A.4 Schwarz inequality and other relations**

Schwarz inequality can be derived using $< \varphi|\varphi > \geq 0$ with:

$$|\varphi> = |\varphi_1> + \lambda|\varphi_2>, \text{and choose } \lambda = \frac{-< \varphi_2|\varphi_1 >}{< \varphi_2|\varphi_2 >},$$

Then:

$$|< \varphi_1|\varphi_2 >|^2 \leq < \varphi_1|\varphi_1 >< \varphi_2|\varphi_2 >$$

Thus,

Schwartz inequality : $|< u|v >| \ \leq \ \sqrt{< u|u >}\sqrt{< v|v >}$

In 3-space $|\vec{A}\cdot\vec{B}| \leq |\vec{A}||\vec{B}|$

$|(\underline{u},\underline{v})| \leq \|\underline{u}\|^{1/2}\|\underline{v}\|^{1/2} = iff\ \underline{u} = \lambda\underline{v}$ Schwartz inequality

$\underline{\omega} = \underline{u} + \lambda\underline{v}$

$\|\omega\|^2 \geq 0 \rightarrow (\underline{u}+\lambda\underline{v}, \underline{u}+\lambda\underline{v}) \geq 0$

0 only if $\underline{u} = -\lambda\underline{v}$

$0 \leq (\underline{u},\underline{u}) + (\underline{v},\underline{v})|\lambda|^2 + \lambda(\underline{u},\underline{v}) + \lambda^*(\underline{v},\underline{u})$

$0 \leq u^2 + v^2\lambda^2 + [\lambda(\underline{u},\underline{v})] + c.c$

Choose $\lambda = -\frac{(\underline{v},\underline{u})}{\|\underline{u}\|^2} \rightarrow$

$0 \leq u^2 + v^2\left(\frac{(\underline{u},\underline{v})^2}{|v|^2}\right) - \left(\frac{(\underline{u},\underline{v})^2}{|v|^2}\right) - \left(\frac{|\underline{u},\underline{v}|^2}{|v|^2}\right) - (\underline{v},\underline{u})$

$0 \leq u^2 + \frac{|\underline{u},\underline{v}|^2}{|v|^2} \rightarrow u^2v^2 \geq |(\underline{u},\underline{v})|^2$

$|(\underline{u},\underline{v})| \leq \|\underline{u}\|^{1/2}\|\underline{v}\|^{1/2}$

Thus

$|(\underline{u}+\underline{v})| \leq \|\underline{u}\| + \|\underline{v}\|$

$|(\underline{u}+\underline{v})| = (\underline{u}+\underline{v}, \underline{u}+\underline{v}) = \|\underline{u}\|^2 + \|\underline{v}\|^2 + (\underline{u}+\underline{v}) + (\underline{v}+\underline{u})$

$\leq \|\underline{u}\|^2 + \|\underline{v}\|^2 + 2|\|\underline{u}\|\|\underline{v}\||$

$\leq (\|\underline{u}\| + \|\underline{v}\|)^2$

Triangle inequality : $\sqrt{< u+v|u+v >} \leq \sqrt{< u|u >} + \sqrt{< v|v >}$

Translation operator :

$$\Omega(a)\varphi(x) = \varphi(x+a)$$

$$= \sum_{n=0}^{\infty} \frac{a^n}{n!} \frac{d^n}{dx^n} \varphi(x) = \sum_{n=0}^{\infty} \frac{1}{n!} \left(\frac{iap}{\hbar}\right)^n \varphi(x) = e^{\frac{i}{\hbar}ap} \varphi(x)$$

where

$$\Omega(a) = \exp(iap/\hbar)$$

Also recall some operator relations:

$[AB, C] = A[B, C] + [A, C]B$

If $[[A, B], A] = 0$ $then$ $[A^m, B] = mA^{m-1}[A, B]$

$[A, B^n] = \sum_{k=0}^{n-1} B^k [A, B] B^{n-k-1}$

$e^L A e^{-L} = A + [L, A] + \frac{1}{2!}[L, [L, A]] + \frac{1}{3!}[L, [L, A]] + \cdots$

$e^A e^B = e^{A+B} e^{-\frac{1}{2}[A,B]}$ $if$ $[[A, B], A] = [[A, B], B] = 0$

**Complex Wavefunction Example**

Complex numbers are used frequently in quantum mechanics. Let's examine some properties:

(a) Given that $z_1$ and $z_2$ are complex numbers, is $z_1^* \, z_2^* = (z_1 z_2)^*$?

(b) Given that $z_1$ and $z_2$ are complex numbers, is $Re\langle z_1\rangle Re\langle z_2\rangle$ the same as $Re(z_1 z_2)$?

(c) Given that $\psi = (a + ib)e^{-iwt}$, where a, b, and w are real, find $\psi\psi^*$ and $|\psi|$.

(d) Given that $\psi_1(x) = a(x) + ib(x)$ and $\psi_2(x) = e^{i\varphi(x)}\psi_1(x)$
Is $\psi_1^*\psi_1 = \psi_2^*\psi_2$? Is $\psi_1^* \frac{d\psi_1}{dx} = \psi_2^* \frac{d\psi}{dx^2}$?

**Answers**

(a) Let $z_1 = x_1 + iy_1, z_2 = x_2 + iy_2$, then:

$$z_1^* z_2^* = (x_1 - iy_1)(x_2 - iy_2) = x_1x_2 - y_1y_2 - i(x_1y_2 + x_2y_1)$$
$$= [x_1x_2 - y_1y_2 + i(x_1y_2 + x_2y_1)]^* = [(x_1 + iy_1)(x_2 + iy_2)]^* = (z_1z_2)^*$$

(b) $Re(z_1) = x_1$ ; $Re(z_2) = x_2$

$$z_1z_2 = x_1x_2 - y_1y_2 + i(x_1y_2 + x_2y_1)$$
$$Re(z_1z_2) = x_1x_2 - y_1y_2 \neq x_1x_2$$
$$Re(z_1)Re(z_2) \neq Re(z_1z_2)$$

(c) $\psi = (a + ib)e^{-iwt} = (a + ib)(\cos wt - i\sin wt)$

$$= a\cos wt + b\sin wt + i(b\cos wt - a\sin wt)$$

$$\psi^* = a\cos wt + b\sin wt - i(b\cos wt - a\sin wt)$$

$= a(\cos wt + i \sin wt) - ib(\cos wt + i \sin wt)$
$= (a - ib)e^{iwt}$ (the signs of all of the 'i's flip).

$$\psi\psi^* = (a + ib)(a - ib)e^{iwt}e^{-iwt} = a^2 + b^2$$

$$|\psi| = \sqrt{a^2 + b^2} = \sqrt{\psi\psi^*}$$

(d) $\psi_1^*\psi_1 = a^2(x) + b^2(x)$

$$\psi_2^*\psi_2 = \left[e^{-i\varphi(x)}\psi_1^*(x)\right] \cdot \left[e^{-i\varphi(x)}\psi_1(x)\right]$$
$$= \psi_1^*(x)\psi_1(x)$$

$$\psi_2^*\frac{d\psi_2}{dx} = \left[e^{-i\varphi(x)}\psi_1^*(x)\right] \cdot \left[e^{-i\varphi(x)}\frac{d\psi_1}{dx} + \psi_1 e^{-i\varphi(x)} i \frac{d\varphi}{dx}\right]$$
$$= \psi_1^*\frac{d\psi_1}{dx} + \psi_1^*\psi_1 \cdot i\frac{d\varphi(x)}{dx}$$

$\psi_2^*\frac{d\psi_2}{dx} = \psi_1^*\frac{d\psi_1}{dx}$ only if $\psi_1^*\psi_1 i\frac{d\varphi}{dx} = 0$ .

**Trigonometry Relations**

Trigonometry Relations that are useful based on $e^{i\theta} = \cos\theta + i\sin\theta$ .

$$e^{i(\theta_1+\theta_2)} \Rightarrow \begin{cases} \cos(\theta_1 \pm \theta_2) = \cos\theta_1\cos\theta_2 \mp \sin\theta_1\sin\theta_2 \\ \sin(\theta_1 \pm \theta_2) = \sin\theta_1\cos\theta_2 \mp \sin\theta_1\cos\theta_2 \end{cases}$$

$$\Rightarrow \begin{cases} 2\cos\theta_1\cos\theta_2 = \cos(\theta_1 + \theta_2) + \cos(\theta_1 - \theta_2) \\ 2\sin\theta_1\cos\theta_2 = \sin(\theta_1 - \theta_2) + \sin(\theta_1 + \theta_2) \end{cases}$$

$$\Rightarrow \begin{cases} \cos^2\theta = \frac{1}{2}[1 + \cos(2\theta)] \\ \sin 2\theta = 2\sin\theta\cos\theta \end{cases}$$

$(e^{i\theta})^2 \Rightarrow \cos 2\theta = \cos^2\theta - \sin^2\theta$

**A.5. The 1955 Von Neumann Axioms of Quantum Mechanics**

Notes in this section largely derive from von Neumann's text on Quantum Mechanics [6], with added comments. Let's start with the definitions for states and observables:

1. States are represented by (normalized) vectors in $H$, $(\underline{\Psi}, \underline{\Psi}) = 1$ (equivalence class of $\underline{\Psi} \rightarrow e^{i\theta}\,\underline{\Psi}$ exists, e.g., rays in $H$ $\{i.e\ e^{i\theta}\underline{\Psi}\}$ .

2. Measurable (physical) quantities are represented by observables in H, where observables are Hermitian operators acting on vectors of H.
3. $< A >= (\underline{\Psi}, A\underline{\Psi})$ real since A is hermitian.

Under what conditions is the measurement sharp? $(\Delta A)^2 = 0$? Let's repeat a measurement N times $x_i$, $1 = 1, \dots N$, where $\overline{x} = \frac{1}{N}\sum x_i$ and the dispersion of the measurement is

$$(\Delta x)^2 = (x - \overline{x})^2 = \frac{1}{N}\sum_i (x_i - \overline{x})^2$$

If $\Delta x = 0$ we have a sharp measurement.

$$(\Delta A)^2 = 0 =< (A - < A >)^2 >= (\underline{\Psi}, (A - < A >)^2 \underline{\Psi})$$
$$= \left((A - < A >)\underline{\Psi}, \right) = \|(A - < A >)\underline{\Psi}\|^2$$

By properties of norm:

$$\left((A - < A >)\underline{\Psi}\right) = 0 \rightarrow A\underline{\Psi} =< A > \underline{\Psi}$$

Consider $A\underline{u}_n^{(r)} = \alpha_n \underline{u}_n^{(r)}$ $\qquad A = \sum_n \alpha_n P^{(n)}$

$< A >= (\underline{\Psi}, A\underline{\Psi}) = \sum_n \alpha_n (\underline{\Psi}, P^{(n)}\underline{\Psi}) = \sum_n \alpha_n (P^{(n)}\underline{\Psi}, P^{(n)}\underline{\Psi})$

$= \sum_n \alpha_n \|\underline{P}^{(n)}\underline{\Psi}\|^2$ , which has the correct properties for probabilities:

$$\|P^{(n)}\underline{\Psi}\|^2 \geq 0, \qquad \sum_n \|\underline{P}^{(n)}\underline{\Psi}\|^2 = (\underline{\Psi}, \underline{\Psi}) = 1$$

**The only possible outcomes of measuring A are its eigenvalues $\alpha_n$ and the probabilities are $\|\underline{P}^{(n)}\underline{\Psi}\|^2$.**

Generalization continuous observables $(\underline{\Psi}, A\underline{\Psi}) = \int_{-\infty}^{\infty} x d(\underline{\Psi}, E_x\underline{\Psi})$,

$A = \int_{-\infty}^{\infty} x dE_x$

$\int_{-\infty}^{\infty} x d(\underline{\Psi}, E_x\underline{\Psi}) = (\underline{\Psi}, \underline{\Psi}) = 1$

$P(a < x < b) = \int_a^b d\,(\underline{\Psi}, E_x\underline{\Psi})$

Spectrum for the position $Q$: $E_x\,\Psi(x') = \begin{cases} \Psi(x') & x \geq x' \\ 0 & x < x' \end{cases}$

$(\underline{\Psi}, E_x\underline{\Psi}) = \int_{-\infty}^{x} dx'\, \Psi^*(x')\,\Psi^*(x') = \int_{-\infty}^{x} dx \,\|\,\Psi(x')\|^2$

Thus $P(a < x < b) = \int_a^b d\int_{-\infty}^{x} dx' \,\|\,\Psi(x')\|^2 = \int_a^b \|\,\Psi(x)\|^2 dx$

**Representations**

$\underline{u}_n \; (n = 1,2 \dots)$

$(\underline{u}_{n'}, \underline{u}_n) = \delta_{nn}$ $\qquad \sum_n |\underline{u}_n >< \underline{u}_n| = I$

$\underline{\Psi} = \sum_n C_n \underline{u}_n \qquad C_n = (\underline{u}_{n'}, \underline{\Psi})$

$$\underline{\Psi} \to \overline{\Psi} = \begin{pmatrix} \underline{u}_1, \underline{\Psi} \\ \underline{u}_2, \underline{\Psi} \\ , \\ \\ , \\ , \end{pmatrix}$$

$(\underline{\varphi}, \underline{\Psi}) = \overline{\varphi}^* \cdot \overline{\Psi} = \sum_n (\underline{u}_{n'}, \varphi)^* (\underline{u}_{n'}, \underline{\Psi})$

$< \underline{\varphi}, \underline{\Psi} > = < \varphi|\underline{u}_n >< \underline{u}_n|\Psi > \qquad I = \sum_n |\underline{u}_n >< \underline{u}_n|$

$A \to A_{nn'} = (\underline{u}_{n'}, A\underline{u}_n{}')$

$A + B \to A_{nn'} + B_{nn'} \qquad AB \to \sum_{n''} A_{nn'} B_{n''n'}$

$A\underline{\Psi} \to \sum_{n'} A_{nn'} (\underline{u}'_{n'}, \Psi)$

$A_{\underline{u}_n} = \alpha_n \underline{u}_n \to A_{nn'} = \alpha_n \delta_{nn'}$

$$\begin{cases} \underline{\Psi} \to \Psi_{A,n} = (\underline{u}_{n'}, \underline{\Psi}) \\ B \to (B_A)_{nn'} = (\underline{u}_{n'}, B\underline{u}_{n'}) \end{cases}$$

**Dirac transformation theory**

Ƶ is another observable : $Z_{\underline{V}_k} = Ƶ_k \underline{V}_k$

$$\begin{cases} \underline{\Psi} \to \Psi_{Ƶ,k} = (V_{k'}, \underline{\Psi}) \\ B \to (B_Ƶ)_{kk'} = (V_{k'}, B\underline{V}_{k'}) \end{cases}$$

Dirac notation convenient (whenever you have to use the closure relations)

$$\begin{cases} \Psi_{Ƶ,k} =< V_k|\Psi >= \sum_n < V_k >< u_n >< u_n|\Psi > = \sum_n < V_k|u_n > \Psi_{A,n} \\ (B_Ƶ)_{KK'} =< V_k|B|V_{k'} >= \sum_{nn'} < V_k|u_n >< u_n|B|u_n >< u'_n|u'_k > \\ \qquad = \sum_{nn'} < V_k|u_n > (BA)_{nn'} < V_k|u_{k'} > \end{cases}$$

$S_{kn} =< V_k|u_n >$

$SS^+ = S^+S = I$ unitary (a result of closure relation)

$(SS^+)_{KK'} = \sum_n S_{Kn} S^*_{nk'} = \sum_n < V_k|\, u_n >< u_n|V_{k,} >$

$=< V_k|V_{k'} = \delta_{KK'}$

$$\begin{cases} \Psi_Ƶ, k = \sum_n S_{kn} \Psi_{A,n} \\ (B_Ƶ)_{KK'} = \sum_{n'n'} S_{kn} (B_A)_{nn'} S^*_{k'n'} = \sum_{nn'} \sum_{n'n'} S_{kn} (B_A)_{nn'} S^+_{k'n'} \end{cases}$$

$$\begin{cases} \Psi_Ƶ = S\,\Psi_A \\ B_Ƶ = S \cdot B_A \cdot S^+ \qquad S =< V_k|u_n > \end{cases}$$

The core of transformation theory.

Properties independent of representation

(i) Scalar products: $\Psi_Z^+ \cdot \Psi_Z = \Psi_A^+ \cdot S^+ \cdot S\Psi_A = \Psi_A^+\Psi_A$
(ii) Algebraic relations
(iii) Conjugation relations
(iv) Eigenvalues of operators
(v) Traces

**Traces**

$Tr(B_Z) = \sum_K (B_Z)_{KK} = \sum_K \sum_{nn'} S_{kn} (B_A)_{nn'}, S^+_{n'K}$
$= \sum_n (B_A)_{nn}$
$= Tr(B_A)$
$\underline{u}_k \rightarrow \underline{V}_k$
$\mho|\underline{u}_k > \; = |\underline{V}_k > \quad = \mho^+|\underline{V}_k >$
$S_{kn} = (\underline{V}_k, u_n) = (U\underline{u}_{k'}, \underline{u}_n) = (\underline{u}_{k'}, U^+\underline{u}_n)$
So, $S = (U_A^+) = S = (U_A^+)$
Passive, active transformation, in geometry have an analog here:

$$\begin{cases} \Psi_Z = S\Psi_A \\ B_Z = SB_AS^{-1} \quad \text{passive trans} \\ S_{kn} =< u_k|u_n > \end{cases}$$

What if active is not equivalent to passive?
Active transformation:
$|\overline{\Psi}> \; = u^+|\Psi>$
$\overline{B} > \; = u^+Bu$
$\overline{\Psi}_{A,n} =< u_n|\mho^+|\Psi> \; = \sum_{n'} < u_n |U^+|u_{n'} >< u_{n'}|\Psi> =$
$\sum_{n'} S_{nn'} \Psi_{A,n'}$
$(\overline{B})_{nn'} = \sum < u_n |\overline{B}|u_{n'} >= \sum_{kk'} S_{nk} (B_A)_{KK'} S^+_{K'n'}$

Suppose $\hat{A} = u^+Au$
$|\Psi>= u^+|\Psi>$
$[A, B] = C \qquad [\overline{A}, \overline{B}] =?$
$[\overline{A}, \overline{B}] = \overline{AB} - \overline{B}\,\overline{A} = u^+Auu^+Bu - u^+Buu^+Au$
$\qquad = u^+ABu - u^+BAu = u^+[A, B]u = u^+Cu = \overline{c}$
So, $[\overline{A}, \overline{B}] = \overline{C}$ commutation relations preserved.

In this sense the active trans. Is the direct analog of the canonical trans., of the classical theory.

Consider a transformation depending on $\theta$ in a continuous way:

Classically expand by infinitesimals, same here:

$\overline{U}(\theta) = I - \frac{i}{\hbar}\theta G + \cdots$

$\overline{U}(\theta) = I$ , $G =$ "Generator"

$u^+u = I = \left(I - \frac{i}{\hbar}\theta G G^+\right)\left(I - \frac{i}{\hbar}\theta G\right) = I - \frac{i}{\hbar}\theta(G - G^+)$ at 1[st] order

$G - G^+ = 0 \leftarrow$ Hermitian

G Hermitian-usually associated with an observable u cannot be so associated as it is unitary.

$\delta|\Psi>= |\overline{\Psi}> - \left|\Psi>= \left(I + \frac{i}{\hbar}\theta G\right)\right| \Psi> = \frac{i}{\hbar}\theta G|\Psi>$

Thus $\delta|\Psi>= \frac{i}{\hbar}\theta G|\Psi$.

$\delta A = \overline{A} - A = \left(I + \frac{i}{\hbar}\theta G\right)A\left(I - \frac{i}{\hbar}\theta G\right) - A$

$= \frac{i}{\hbar}\theta[GA - AG] = \frac{i}{\hbar}\theta[G, A]$

Thus $\delta A = \frac{\theta}{i\hbar}[A, G]$

Classical Mech: $\delta A = \theta\{A, G\} \rightarrow \frac{I}{i\hbar}[A, G]$ Dirac correspondence

Usual statement $[A, G] = \frac{I}{i\hbar}\{A, G\}$

In limit $\hbar \rightarrow 0$ , $[A, G] \rightarrow 0$

1 – dim system with classical counterpart

$[Q, P] = i\hbar \qquad [Q, Q] = [P, P] = O$

(i) Q (or P) is by itself a complete set of community observables.

(ii) Suppose it were possible to find $Q_{nn'}$, $P_{nn'}$ such that

$\underset{n''}{\Sigma}[Q_{mn''}P_{n''n'} - P_{nn''}Q_{n''n'}] = i\hbar\delta_{nn'}$

Now take trace : Tr$(QP)$-Tr$(PQ)$ = $i\hbar$d, so 0 = d? What is wrong is that the reordering property of the trace is only defined for finite dimensional matrices.

(iii) Q and P are unbounded

A is bounded if $||A\varphi|| \leq C||\varphi||$ (smallest C is called norm)

$[Q, P^n] = i\hbar n P^{n-I}$ $||[Q, P^n]|| = \hbar n||P||^{n-I}$ assuming bounded

$||QP^n - P^nQ|| \leq ||QP^n|| + ||QP^A|| = 2||Q||\ ||P^n||$

$2||Q||\ ||P^n|| \geq \hbar n||P||^{n-I}$

$||Q||\,||P|| \geq \frac{\hbar}{2}n$ for all n, thus its unbound by symmetry of $[Q, P] = i\hbar$ both Q must be unbounded.

Consider

$V(\eta) = e^{i\eta P}$ and $V^+V = I$

$e^{-i\eta P} Q^{i\eta P}$ = Q - $i\eta[P,Q]$+$\frac{(-i\eta)^2}{2}$ $[P,[P,Q]]$ +,,,,,,,= Q - $\hbar\eta$
QV(η) = V(η) $[Q - \hbar\eta]$
Q|∝⟩ = ∝ |∝⟩
QV(η)|∝⟩ = V(η) $[Q - \hbar\eta]$ |∝⟩ = V(η) (∝ $-\hbar\eta$) |∝⟩
If Q has large value ∝ it also has large value ∝ - $\hbar\eta$ for arbitrary η . → Q cannot have any discrete spectrum. Same goes for P,

So we usually state: Q$|x\rangle = x|x\rangle$

**The Dirac Delta function**
We have: $\langle x|x'\rangle = \delta(x - x')$

Products of $\delta's$ can complicate the analysis, etc.
$\int_{-\infty}^{\infty} |x\rangle dx\langle x|$ = I
$\langle x'| \int_{-\infty}^{\infty} |x\rangle dx\langle x|x''\rangle = \int_{-\infty}^{\infty} \delta(x' - x)dx\delta(x - x') = \delta(x' - x) = \langle x|x''\rangle =$
$\langle x|I|x''\rangle$
$\langle x|\varphi\rangle = \varphi(x)$ a function, not a column vector suit x is continuous
$\langle x|Q|\varphi >$ = x$\varphi(x)$

How is the momentum operator represented? Exactly analogous to position representation
$e^{i\propto P/\hbar} |x\rangle = |x-\propto\rangle$
$\langle x|e^{i\propto P/k} |x\rangle = \delta(x' - x+\propto)$
$\frac{\hbar}{i}\frac{d}{d\propto}\langle x|e^{i\propto P/\hbar} x\rangle|_{\propto=0} = \frac{\hbar d}{id\propto}\delta(\square)|_{\propto=0}$
$\langle x'|P|x\rangle = \frac{\hbar}{i}\frac{d}{dx'}\delta(x' - x)$
Recall
$\langle x'|Q|x\rangle = x\delta(x' - x)$

P is off-diagonal:
$\langle x|P|\varphi\rangle = \int \langle x|P|x'\rangle$ dx'$\langle x'|\varphi\rangle$
$= \frac{\hbar}{i}\frac{d}{dx}\int \delta(x - x')dx'\varphi(x')$
$= \frac{\hbar}{i}\frac{d}{dx}\varphi(x)$
Ʊ(∝)$|\varphi\rangle = |\hat{\varphi}\rangle$ Ʊ(∝) AƱ(∝)$^{-1}$ =$\hat{A}$
Consider U(∝) = $e^{-i\propto^P/\hbar}$
$\langle x|\hat{\varphi}\rangle = \left\langle x \middle| e^{-i\propto^P/\hbar} \middle| \varphi \right\rangle = \langle x-\propto|\varphi\rangle$

$x-\propto)$

$) U^{-I}(\propto) = \sum_{n,n'} C_{nn'} Q^n P^{n'}$

$\sum_{n,n'} C_{nn'}[UQU^{-I}][UPU^{-I}]^{n'}$=f$(Q, \propto, P)$

P$|p\rangle$ = p$|p\rangle$ P$|p'\rangle$ = $\delta(\rho - \rho)$

$\langle x|p\rangle$= $(2\pi\hbar)^{-1/2} e^{ipx}$

**Oscillator (Fock) representation**

Suppose a =A $[Q + i\lambda P]$and $a^\dagger = A[Q - i\lambda P]$ (a is not hermitian)

Want $[a, a^\dagger]$ =I

$A^2\ [Q + i\lambda P, Q - i\lambda P] = A^2(2n\ \lambda) = \mathrm{I} \Rightarrow \mathrm{A} = \frac{\mathrm{I}}{\sqrt{2\hbar\lambda}}$

For dimensional consistency $(\lambda) = \frac{[Q]}{[P]} = \frac{\mathrm{I}}{mass.fre.}$

So, $\lambda = \frac{I}{\mu\omega}$, thus

A= $\sqrt{\frac{\mu\omega}{2\hbar}}\left[Q + \frac{i}{\mu\omega} P\right]$

Dimensions of a? $[a]$=$\sqrt{\frac{\mu\ell^2 \dagger^{-I}}{\hbar}}$ = dimensionless

Define $a^\dagger a = N$.

Properties of $N$:

(i) $N$ is non-negative i.e. $(\varphi, N\varphi) \geq 0$

N$(\varphi, \varphi) = (\varphi, a^\dagger a\varphi) = (a\varphi, a\varphi) = \|a\varphi\|^2 \geq 0$

(ii) $[N, a] = [a^\dagger a, a] = [a^\dagger, a]a = -a$

$[N, a^\dagger] = a^\dagger$

$N$ is hermitian

Since N is Hermitian and non-negative:

$N|\lambda\rangle = \lambda|\lambda\rangle$ ; $\lambda \geq 0$

Consider

$N[a|\lambda\rangle] = (aN + [N, a])|\lambda\rangle = \lambda a|\lambda\rangle - a|\lambda\rangle = (\lambda - 1)\ a|\lambda\rangle$

N$[a^\dagger|\lambda\rangle] = (\lambda + 1)\ a^\dagger|\lambda\rangle$

This tells us that (in the N eigenvector space)

a$|\lambda\rangle$ = c$|\lambda - 1\rangle$

$a^+|\lambda\rangle$ = c$|\lambda + 1\rangle$

Repeat application of a ……. eigenvalues decrease; $\lambda,\ \lambda - 1$, …. But this would eventually arrive at $\lambda - n < 0$ unless we have $\lambda = n$ (since non-negative).

So, $\lambda = n$ ( integer) then

$$a|\lambda - n\rangle = 0$$

$$a|0\rangle = 0$$

$a|n\rangle = c|n-1\rangle \Rightarrow |c|^2 = \langle n|a^+a|n\rangle = n$

$a^+|n\rangle = c'|n+1\rangle \Rightarrow |c'|^2 = \langle n|aa^+|n\rangle = \langle n|(N+1)\,|n\rangle = n+I$

*so,*

$a|n\rangle = \sqrt{n}|n-1\rangle$

$a^+|n\rangle = \sqrt{n+1}|n+1\rangle$

Since $N$ is hermitian $\langle n|n'\rangle = \delta_{nn'}$

What is Q and P in Fock rep?

$$Q = \sqrt{\frac{\hbar}{2\mu\omega}}(a + a^+)\ ,\ P = -i\sqrt{\frac{\hbar\mu\omega}{2}}(a - a^+)$$

$$Q_{nn'} = \langle n|Q|n'\rangle = \sqrt{\frac{\hbar}{2\mu\omega}}\left(\sqrt{n'}\ \delta_{n,n'-I} - \sqrt{n'+1}\,\delta_{n,n'+1}\right)$$

$$P_{nn'} = -i\sqrt{\frac{\hbar}{2\mu\omega}}\left(\sqrt{n'}\ \delta_{n,n'-1} - 1\,\delta_{n,n'+1}\right)$$

$$\sum_{n''}(Q_{nn''}P_{n''n'} - P_{nn''}Q_{n''n'}) = i\hbar\delta_{nn'}$$

$[Q,P] = i\hbar$

Change of basis is not trace, between position and number basis.

$S_n(x) = \langle x|n\rangle$

$$S_n(x) = C_n H_n\left(\sqrt{\frac{\mu\omega}{\hbar}}x\right)e^{\frac{-\mu\omega}{2k}x^2}$$

Hermite polynomials a complete basis, any square integral function rep. in terms of H poly's.

Knowing Fock representation

$S_n(x) =?$, $a|0\rangle = 0$

$$\sqrt{\frac{\mu\omega}{2\hbar}}\left(Q + \frac{iP}{\mu\omega}\right)|0\rangle = 0$$

$$\left(x + \frac{i\hbar}{\mu\omega i}\frac{d}{dx}\right)S_0(x) = 0 \quad \Rightarrow 0(x) = C_0 e^{-\frac{\mu\omega}{2\hbar}x^2}$$

$$|n\rangle = \frac{a^\dagger}{\sqrt{n}}|n-1\rangle,,,,, = \frac{(a^\dagger)^n}{\sqrt{n!}}|0\rangle$$

$$S_n(x) = \frac{1}{\sqrt{n!}}\left(\frac{\mu\omega}{2\hbar}\right)^{n/2}\left(x - \frac{\hbar}{\mu\omega}\frac{d}{dx}\right)^n S_0(x)$$

There is also a representation not based on eigenvectors or eigenvalues but on eigenvectors of the a operators: $a|Z\rangle = Z|Z\rangle$ a coherent state. Based on the eigenstates of a non-Hermitian operator:

$$a = \sqrt{\frac{\mu\omega}{2\hbar}}Q + \frac{i}{\mu\omega}P$$

$$N = a^{\dagger}a \ ,N|n\rangle = n|n\rangle \qquad (n = 0,1,2)$$

$$\langle n|a|n'\rangle = \sqrt{n'}\delta_{n,n'} - 1$$

Consider the eigenvalue problem $a|Z\rangle = Z|Z\rangle$, since a is not Hermition, $Z$ is not necessarily real.

Expand $|Z\rangle = \sum_{n=o}^{\propto} C_n|n\rangle$:

$$a|Z\rangle = \sum_{n=I}^{\propto} C_n\sqrt{n}|n-1\rangle = \sum_{n=o}^{\infty} C_{n+1}\sqrt{n+1}|n\rangle$$

want $Z|Z\rangle = \sum_{n=o}^{\infty} ZC_n|n\rangle$

so, $C_{n+I} = \frac{Z}{\sqrt{n+I}}C_n$ a recursion relation is determined for the eigenvalue problem

$$C_n = \frac{Z}{\sqrt{n}}C_{n-I}$$

$$C_1 = ZC_0$$

$$C_2 = \frac{Z}{\sqrt{2}}1 = \frac{Z^2}{\sqrt{2}}C_0$$

$$C_3 = \frac{Z^3}{\sqrt{2.3}}C_0$$

$$\boxed{C_n = \frac{Z^n}{n!}C_0}$$

$$|Z\rangle = C_o \sum_{n=o}^{\infty} \frac{Z^n}{n!}|n\rangle$$

With $C_o$ determined by normalization:

$$|Z'\rangle = C'_o \sum_{n=o}^{\infty} \frac{Z'}{\sqrt{n!}}|n'\rangle$$

$$\langle Z'|Z\rangle = C'_oC_o \sum_{n,n'} \frac{Z'^{n'}}{\sqrt{n'|}} \frac{Z^n}{\sqrt{n|}} \langle n'|n\rangle = C'_oC_o \sum_n \frac{(Z'Z)^n}{n!}$$

Set Z=Z' to find normalization

$$\langle Z|Z\rangle = |C_o|^2 \sum_n \frac{(|Z|^2)^n}{n!} = |C_o|^2 e^{|C_o|^2} = 1$$

Can normalize by $C_o = e^{-|Z|^2/2}$

So,

$$|Z\rangle = e^{-|Z|^2/2} \sum_{n=o}^{\infty} \frac{Z^n}{\sqrt{n!}}|n\rangle$$

If n corresponds to the number of the above is a covalent state.

The states $|Z\rangle$ are orthogonal

$\langle Z'|Z\rangle = e^{-(|Z|^2+|Z|^2)^{2/2}} e^{Z'Z}$

$\frac{1}{2}(|Z|^2 + |Z'|^2)$-$Z'Z = \frac{1}{2}|Z - Z'|^2 + \frac{1}{2}(Z^x Z' - ZZ'^x)$

$\frac{1}{2}|Z - Z'|^2 = \frac{1}{2}(|Z'|^2 + |Z'|^2)$ - $Z'Z'^x - Z^x Z'$

$= \frac{1}{2}|Z - Z'|^2$ -i$\mathrm{Im}(Z'^x Z)$

So, $|\langle Z'|Z\rangle| = e^{-\frac{1}{2}|Z-Z'|^2} \neq 0$, so never orthogonal.

Is there a closure relation? yes:

$\int |Z\rangle \frac{dZdZ^x}{\pi} \langle Z| = 1$ Closure

$dZdZ^x = (ReZ)d(ImZ)$

Proof

$$\int \langle n|Z\rangle \frac{dZdZ^x}{\pi} \langle Z|n'\rangle = \langle n|n'\rangle = \delta_{nn'}$$

$= \int e^{-|Z|^2} \frac{Z^n}{\sqrt{n!}} \frac{(Z^x)^{n'}}{n'!} \frac{dZdZ^x}{\pi}$ write $Z = |Z|e^{i\varphi}$

$= \int e^{-|Z|^2} \frac{|Z|^{2n}}{n!} 2|Z|d|Z|\delta_{nn'}$

$$\int_0^{2\pi} e^{i(n-n')} d\varphi = 2\pi\delta_{n,n'}$$

$= \frac{I}{n!} \int_0^{\infty} e^{-x} n^x dx \delta_{nn'}$

$= \delta_{nn'}$

$|Z\rangle = \int |Z'\rangle \frac{dZdZ^x}{\pi} \langle Z'|Z\rangle \rightarrow$ over complete basis.

$$|\varphi\rangle = \int |Z\rangle \frac{dZdZ^x}{\pi} \langle Z|\varphi\rangle$$

$|\varphi\rangle \rightarrow \langle Z|\varphi\rangle = \varphi(Z)$

Represented by function of a complex variable. The functions are vey well behaved.

$$\varphi(Z) = \langle Z|\varphi\rangle = e^{-|Z|^{2}/2} \sum_n \frac{Z^{*n}}{\sqrt{n!}} \langle n|\varphi\rangle$$

$\equiv e^{-|Z|^{2}/2} \delta(Z)$

$$|\delta(Z)| \leq \sum_n \frac{|Z|^n}{\sqrt{n!}} |\langle n|\varphi\rangle|$$

$|\langle n|\varphi\rangle| \leq \|\varphi\|$

$$|f(Z)| \leq \|\varphi\| \sum_n \frac{|Z|^n}{\sqrt{n!}}$$

So, $\varphi(Z)$ is very well behaved in $Z$

$\langle Z'|a|Z\rangle = Z\langle Z'|Z\rangle$
$\langle Z'|a^\dagger|Z\rangle = (Z')^*\langle Z'|Z\rangle$
If $A(a, a^\dagger) = \sum_{r,s} C_{rs}(a^\dagger)^r a^S$

$$\langle Z'|A|Z\rangle = \sum_{r,s} C_{rs}(Z'^x)^r Z^S \; A(Z, Z'^x)\langle Z'|Z\rangle$$

Consider
$\theta = \cdots + (a^\dagger)^k a^r (a^\dagger)^S a^P + \cdots$
$:\theta: = \cdots + \; (a^\dagger)^k (a^\dagger)^S a^r a^P$ (the normal ordered form)
Consider
$\theta = e^{\tau a^\dagger a}$
Recall $|Z\rangle = e^{-Z^2/2}\sum_n \frac{Z^n}{\sqrt{n!}}|n\rangle$
$\langle Z'|\theta|Z\rangle = e^{-|Z|^2+|Z\prime|^2/2}\sum_{n,n\prime}\frac{(Z\prime x)^{2\prime}}{\sqrt{n'!}}\frac{(2)^n}{\sqrt{n!}} < n'|e^{Za^\dagger a}|n>$
$\langle 2'|\theta|2\rangle = e^{-\left(|2|^2+|2'|^2\right)/2} e^{2'^* 2 e^x}$
$\langle 2'|e^{xa'a}|2\rangle = e^{(e^x - i)2'^* 2}\langle 2'|2\rangle$
Now, what is $\langle 2'|:|e^{\tau a'a}|2\rangle$ ?
Simplify $|e^{xa'a}|2\rangle = e^{xa'a}\langle 2'|2\rangle$
$: e^{\tau a'a}: = \; e^{\ell n(1+\tau)a^\dagger a}$
Suppose $\tau = -1$
$e^{-a'a}: = e^{-\infty a'a} = \delta(a^\dagger a)$

Projects the vacuum!

$$\boxed{: e^{-a'a}: = \; |0\rangle\langle 0|}$$

Recall $Q|x\rangle = x|x\rangle$
$\langle x|Q|\Psi\rangle = x\Psi(x)$
What about $2$
$A|2\rangle = 2|2\rangle$
$\langle 2|\Psi\rangle = \Psi(2)$
$\langle 2|a^\dagger|\Psi\rangle = 2^*\Psi(2)$
$\langle 2|a|\Psi\rangle =$ nontribal some a is nonhermitian

$$\langle 2|a|\Psi\rangle = \int\langle 2|a|2'\rangle\frac{d2'd2^*}{\pi}\langle 2'|\Psi\rangle$$

$$= \int 2' e^{-\left(|2|^2+|2|^2\right)/2 + 2^* 2'}\frac{d2'd2^*}{\pi}\Psi(2')$$

$$\int\left(\frac{2^*}{2}+\frac{\partial}{\partial 2^*}\right)e^{-\left(|2|^2+|2|^2\right)/2+2^*2'}\frac{d2'd2^*}{\pi}\Psi(2')$$

$$= \left(\frac{z^*}{2} + \frac{\partial}{\partial z^*}\right) \int \langle z|z'\rangle \frac{dz' dz^*}{\pi} \langle z'|\Psi\rangle\langle z|\Psi\rangle$$

# Appendix B. Review of Classical Mechanics

Lagrangian for conservative forces discussed $q_{i,}\ 1 = 1, \dots, n$

$L = T(q, \dot{q}) - v(q) \rightarrow eqn. of\ motion\ \frac{d}{dt}\left(\frac{\partial L}{\partial \dot{q}_1}\right) - \frac{\partial L}{\partial q_i} = 0$

Nonconservative forces (no exact differential to give potential field)

With $\vec{F}^{nc}$ how do we modify L?

$$\frac{d}{dt}\left(\frac{\partial L}{\partial \dot{q}_1}\right) - \frac{\partial L}{\partial q_i} = \vec{F}_i^{nc}$$

The exception is: $T(q, \dot{q})$ such that $\vec{F}_i^{nc}$ can be expressed as:

$\vec{F}_i^{nc} = \frac{\partial L}{\partial \dot{q}_1} - \frac{d}{dt}\left(\frac{\partial L}{\partial \dot{q}_1}\right)$

Then, $\frac{d}{dt}\left(\frac{\partial (L+u)}{\partial \dot{q}_1}\right) - \frac{\partial (L+u)}{\partial q_i}$

$L' = L + u$ Lagrangian corresponding to conservative forces

$E - m$ force, i.e. the Lorentz force satisfies the required property.

$\vec{F} = q\vec{E} + \frac{q}{c}\vec{V}x\vec{B}$ $\qquad \vec{E} = -\vec{\nabla}\varphi - \frac{1}{c}\frac{\partial}{\partial t}\vec{A}\ ,\ \vec{B} = \vec{\nabla}x\vec{A}$

$\vec{F}_i = -q\frac{\partial \varphi}{\partial x_i} - \frac{q}{c}\frac{\partial A_i}{\partial t} + \frac{q}{c}\sum_j\left[\dot{x}_j\frac{\partial A_j}{\partial x_i} - \dot{x}_j\frac{\partial A_i}{\partial x_j}\right]$

$F_i = -q\frac{\partial \Psi}{\partial x_i} + \frac{q}{c}\sum_j \dot{x}_j\frac{\partial A_j}{\partial x_i} - \frac{q}{c}\frac{dA_i}{dt}$

$u = -q\varphi + \frac{q}{c}\sum_j A_j\dot{x}_j = -q\,\Psi\varphi + \frac{q}{c}\vec{A}\cdot\dot{\vec{x}}$

$L = T - V - q\varphi + \frac{q}{c}\vec{A}\cdot\dot{x} \quad \frac{q}{c}\vec{A}\cdot\vec{J}$

Hamiltonian Formalism

Based in phase space $\{q_i, p_i\}$

$P_i = \frac{\partial L}{\partial q_i}$

$L(q, \dot{q}) \rightarrow H(q, p)$ Legendre transformation

$dL = \sum_i\left[\frac{\partial L}{\partial q_i}dq_i + \frac{\partial L}{\partial q_i}d\dot{q}_i\right] = \sum_i\left[\frac{\partial L}{\partial q_i}dq_i + \frac{\partial L}{\partial q_i}d\dot{q}_i\right]$

$= \sum_i\left[\frac{\partial L}{\partial q_i}dq_i + d(\dot{q}_i p_i) - \dot{q}_i dp_i\right]$

$H = \sum_i p_i\dot{q}_i - L$ define, then

$dH = \sum_i\left[\dot{q}_i dp_i - \frac{\partial L}{\partial q_i}dq_i\right]$

$\frac{\partial L}{\partial q_i} = \frac{d}{dt}\left(\frac{\partial L}{\partial q_i}\right) = \dot{p}_i$ from equation of motion.

$H = \sum_i p_i \dot{q}_i$; $\dot{q}_i = \frac{\partial H}{\partial \dot{p}_i}$ ; $\dot{p}_i = -\frac{\partial H}{\partial q_i}$; $p_i = \frac{\partial L}{\partial q_i}$

$L = \frac{1}{2}m\left(\dot{r}^2 + r^2\dot{\theta}^2\right) + \frac{k}{r}$

$p_r = m\dot{r}$ , $P_\theta = mr^2\dot{\theta} = |\vec{L}|$

$H = \frac{Pr^2}{2m} + \frac{P_\theta^2}{mr^2}$ ; $H = m\dot{r} + mr^2\dot{\theta} - k$

$H = P_r\dot{r} + P_\theta\dot{\theta} - L = m\dot{r}^2 + mr^2\dot{\theta} - \frac{1}{2}mr^2\dot{\theta}^2 - \frac{k}{r}$

$H = \frac{p_r^2}{2m} + \frac{P_\theta^2}{mr^2} - \frac{k}{r}$

$0 = \dot{P}_\theta$ ; $\dot{P}_r = -\frac{k}{r^2} + mr\dot{\theta}^2$

$\frac{P_r}{m} = \dot{r}$ ; $\frac{P_\theta}{mr^2} = \dot{\theta}$

$\dot{P}_r = -\frac{k}{r^2} + \frac{P_\theta^2}{mr^3}$

As long as $T = \sum_{i,j} T_{ij}(q)\dot{q}_i\dot{q}_j$ is bilinear in the velocity, then

$H = T + V - Energy$

$L = T - V - q\varphi + \frac{q}{c}\vec{A}\cdot\dot{\vec{x}}$

$T = \frac{1}{2}m\sum_i \dot{x}_i^2$

$P_i = \frac{\partial L}{\partial \dot{x}_i} = m\dot{x}_i + \frac{q}{c}\vec{A}_i \leftarrow \vec{x}_i = \frac{1}{m}\left(\vec{P} - \frac{q}{c}\vec{A}\right)$

So, if $H = \sum_i P_i\dot{x}_i - L = \frac{1}{2}m\sum \dot{x}_i^2 - V - q\varphi$ $\quad$ *(the $A_i'S$ cancel!)*

$H = \frac{1}{2}m\Sigma$

$H = \frac{1}{2m}\left(\vec{P} - \frac{q}{c}\vec{A}\right)^2 - V - q\varphi$

So, $\vec{P} \to \vec{P} - \frac{q}{c}\vec{A}$ and $V \to V + q\varphi$

Minimal substitution.

(i) $F(q, Q, t)$ *yields* $P_i = \frac{\partial F}{\partial q_i}$ , $P_i = \frac{\partial F}{\partial Q_i}$ , $K = H + \frac{\partial F}{\partial t}$

(ii) $F(q, P)$

$\sum P_i\acute{q}_i - H = \sum P_i\dot{Q}_i - K + \sum\left(\frac{\partial F}{\partial q_i}\dot{q}_i + \frac{\partial F}{\partial P_i}\dot{P}_i\right) + \frac{\partial F}{\partial t}$

$= -\sum Q_i\dot{P}_i - K + \sum\left(\frac{\partial F}{\partial q_i}\dot{q}_i + \frac{\partial F}{\partial P_i}\dot{P}_i\right) + \frac{\partial F}{\partial t} + \frac{\partial}{\partial t}\frac{\Sigma}{T}(P_i)$

$P_i = \frac{\partial F}{\partial q_i}, Q_i = \frac{\partial F}{\partial P_i}$ , $K = H + \frac{\partial F}{\partial t} + \frac{\Sigma}{T}\frac{\partial}{\partial t}(P_iQ_i)$

Consider $F = \sum_i q_iQ_i$ (using trans. (i) of course)

$P_i = Q_i$ $\quad$ $P_i = -q_i$ $\quad$ $K = H$

Try $F = \sum_i q_i P_i$ Using (ii) of course

$P_i = P_i$ $\quad$ $Q_i = q_i$ identity

Try $F = \sum_i f_i(q)P_i$ (ii)

$Q_i = \frac{\partial F}{\partial P_i} = f_i(q)$ $\qquad$ $P_i = \frac{\partial F}{\partial q_i} = \sum_i \frac{\partial F_i}{\partial q_i} P_j$

Is $\begin{matrix} Q_i = Q_i(q,p,t) \\ P_i = P_i(q,p,t) \end{matrix}$ a canonical transformation?

$$\{Q_i, Q_j\}_{(q,p)} = \sum_K \left( \frac{\partial Q_i}{\partial q_k} \frac{\partial Q_j}{\partial P_k} - \frac{\partial Q_i}{\partial P_k} \frac{\partial Q_j}{\partial q_k} \right) = 0$$

A canonical transformation preserves the Poisson bracket structure.

Conservation + invariance

$q_i \to Q_i = Q_i(q,p,t;\theta)$ some continuous parameter $\theta$

$P_i \to P_i = P_i(q,p,t;\theta)$

$\theta = 0 \to ident. trans.$

$Q_i(\theta = 0) = q_i$

$P_i(\theta = 0) = P_i$

Since continuous consider $\theta = \theta \ll 1$

$Q_i = q_i + \in \frac{\partial Q_i}{\partial \theta} \equiv q_i + \delta q_i$

$P_i = p_i + \in \frac{\partial P_i}{\partial \theta} \equiv p_i + \delta p_i$

$F(q,p) \to Q = \frac{\partial F}{\partial q}$ as before

$F(q,p) = \sum_i q_i p_i + \in G(q,p)$ G being the generator of the continuous set of trans.

$\begin{matrix} Q_i = q_i + \in \frac{\partial G}{\partial p_i} \\ P_i = p_i + \in \frac{\partial G}{\partial q_i} \end{matrix} \to \delta q_i = \in \{p_i, G\}$

So, $\delta f(q,p) = \sum_i \left[ \frac{\partial f}{\partial q_i} \delta q_i + \frac{\partial f}{\partial p_i} \delta p_i \right] = \epsilon \sum_i \left[ \frac{\partial f}{\partial q_i} \frac{\partial G}{\partial p_i} - \frac{\partial f}{\partial p_i} \frac{\partial G}{\partial q_i} \right]$

$$\delta f = \epsilon \{f, G\}$$

if $G$ is the generator. Consider

$$\frac{df}{dt} = \{f, H\} + \frac{df^0}{dt} \quad \to \partial f = \delta t \{f, H\}$$

the Hamilton is the generator of the transformation in time.

Suppose, $Q_i = Q_i(q_i, p_i, \theta)$ etc. no explicit time dependence

So, $\delta H = 0 \to \{H, G\} = 0 \to \frac{dG}{dt} = 0$

Thus, G=const. of motion

Space translation along i-axis

$q_j \to Q_j + a\delta_{ij}$

$F = \sum_j (q_j + a\delta_{ij}) p_i = \sum_j q_j p_i + aP_j ;\ G = P_j$

$\frac{dG}{dt} = 0 \rightarrow P_j = const$

Rotations around the Ƶ axis

$Q_i = \sum_j R_{ij}(\theta) q_j$

$$R(\theta) = \begin{pmatrix} \cos\theta & -\sin\theta & 0 \\ \sin\theta & \cos\theta & 0 \\ C & 0 & 1 \end{pmatrix}$$

$F = \sum_{i,j} R_{ij}(\theta)\, q_i P_i$ small rot. $= I + \in \begin{pmatrix} 0 & -1 & 0 \\ 1 & 0 & 0 \\ 0 & 0 & 0 \end{pmatrix}$

$= \sum_i q_i P_i + \in (-q_2 P_1 + q_1 P_2)$

$G = L_Ƶ$

$\frac{dG}{dt} = 0 \rightarrow L_Ƶ = const$

# C. Dust Shell Collapse Quantization

## C.1 Equation of motion from Hamiltonian

Recovering the massive dust shell equations of motion from the Hamiltonian action.

### 1. ACTION AND EOMS

Recall that the metric is

$$ds^2 = N^2dt^2 + \Lambda^2(dr + N^r dt)^2 + R^2 d\Omega^2 \tag{2.1}$$

The Hamiltonian bulk action is

$$S_\Sigma = \int dt\, P\dot{\hat{r}} + \int dt \int dr\ \left(P_\Lambda\dot{\Lambda} + P_R\dot{R}\right) - NH - N^r H_r, \tag{2.2}$$

Where

$$H = H^g + H^s, \tag{2.3a}$$

$$H_r = H_r^g + H_r^s, \tag{2.3b}$$

With

$$H^g = -R^{-1}P_\Lambda + \frac{1}{2}R^{-1}\Lambda P_\Lambda^2 + \Lambda^{-1}RR'' - \Lambda^{-2}RR'\Lambda' + \frac{1}{2}\Lambda^{-1}R'^2 \frac{1}{2}\Lambda, \tag{2.4a}$$

$$H_r^g = P_R R' - \Lambda P'_\Lambda \tag{2.4b}$$

and

$$H^s = \sqrt{\hat{\Lambda}^{-2}p2 + m^2}\delta(r-\hat{r}), \tag{2.5a}$$

$$H_r^\delta = -P\delta(r-\hat{r}) \tag{2.5b}$$

Formal local variation gives the constraints

$$H = 0 \tag{2.6a}$$

$$H_r = 0 \tag{2.6b}$$

and the dynamical equations

$$\dot{\Lambda} = N(R^{-2}\Lambda P_\Lambda - R^{-1}P_R) + (N^r\Lambda)'. \tag{2.7a}$$

$$\dot{R} = -NR^{-1}P_\Lambda + N^r R' \tag{2.7b}$$

$$\dot{P}_\Lambda = \frac{1}{2} N \left[ -R^{-2}P_\Lambda^2 - (\Lambda^{-1}R')^2 + 1 + \frac{2P^2\delta(r-\hat{r})}{\hat{\Lambda}^3\sqrt{\hat{\Lambda}^{-}p2+m^2}} \right] - \Lambda^{-2}N'RR' + N^rP'_\Lambda \qquad (2.7c)$$

$$\dot{P}_R = N\left[\Lambda R^{-3}P_\Lambda^2 - R^{-2}P_\Lambda P_R(\Lambda^{-1}R')'\right] - (\Lambda^{-1}N'R)' + (N^rP_R)', \qquad (2.7d)$$

$$\dot{\hat{r}} = \frac{\hat{N}p}{\hat{\Lambda}^2\sqrt{\hat{\Lambda}^{-2}p2+m^2}} - \hat{N}^r \qquad (2.7e)$$

$$\dot{p} = \frac{\hat{N}\hat{\Lambda}'p2}{\hat{\Lambda}^2\sqrt{\hat{\Lambda}^{-2}p2+m^2}} - \hat{N}^1\sqrt{\hat{\Lambda}^{-2p2}+m^2} + \mathrm{P}(\widehat{\mathrm{N^r}})'$$

(2.7f)

We shall assume that all the fields are smooth (say $C^2$) functions of r, with the exception that $N', (N^r)', \Lambda', R', P_\Lambda$ *and* $P_R$ may have finite discontinuities at isolated values of r, and that the locations of the discontinuities may be smooth functions of t. with the exception of (2.7f), all the equations (2.6) and (2.7) are then well defined in the distributional sense, containing at worst delta-functions.

[More precisely, the situation is as follows. The constraint equations (2.6) contain explicit delta-functions from the matter contribution and implicit delta-functions in $R''$ *and* $P'_\Lambda$. The right hand sides of (2.7a) and (2.7b) contain at worst finite discontinuities, and the right hand sides of (2.7c) and (2.7d) contain at worst delta-functions; this is consistent with the left hand sides, recalling that the loci of nonsmoothness of $\Lambda, R, P_\Lambda$ *and* $P_R$ may evolve in t.]

The right hand side of (2.7f) is not defined. We shall discuss this below.

Outside the shell, the equations are equivalent to vacuum Einstein. What are they at the shell?

**2. SHELL STRESS-ENERGY**

Some notation. Let $v^a := (1, \dot{\hat{r}}, 0,0)$. The (normalized) shell four-velocity is $u^a = (-v^cv_c)^{-1/2}v^a$ . The right-point unit normal is

$$n_a = \frac{\hat{N}\hat{\Lambda}}{\sqrt{-v^cv_c}}(1, \dot{\hat{r}}, 0,0). \qquad (3.1)$$

Latin indices four-dimensional. The projector to the shell three-surface is $h_{ab} := g_{ab} - n_a n_b$.

For future use, inverting (2.7e) gives

$$P = \frac{m\hat{\Lambda}^2 (\dot{\hat{r}} + \widehat{N^r})}{\sqrt{-v^a v_a}} = m u_r \qquad (3.2)$$

Recall the def of the stress-energy tensor:

$$\delta_g = \frac{1}{2} \int \sqrt{-g} d^4 x T^{ab} \delta(g_{ab}) \qquad (3.3)$$

Taking the variation and using (3.2) we find

$$T^{ab} = \frac{m\sqrt{-v^c v_c}}{4\pi N \Lambda R^2} u^a u^b \delta(r - \hat{r}) \qquad (3.4)$$

From this, the surface stress-energy tensor of the shell is $[MTW, 21.163]$

$$S^{ab} = \frac{m}{4\pi \hat{R}^2} u^a u^b \qquad (3.5)$$

This confirms that the matter is pressureless dust, with surface energy density $m/(4\pi\hat{R}^2)$.

The total rest mass is m.

**3. THE (CORRECT) CLASSICAL DYNAMICS**

Recall that the full content of the Einstein equation at the shell is encoded in Israel's junction conditions

$$-8\pi\left(S_{ab} - \frac{1}{2} h_{ab} S\right) = K_{ab}^{+} - K_{ab}^{-} \qquad (4.1)$$

Where $K_{ab} + h^c{}_a h^d{}_b \nabla_c n_d$ is the extrinsic curvature tensor wrt the right-point normal $n_a$, and the + (-) refers to the right (left) side of the shell.

With the stress-energy tensor (3.5), and with Schwarzschild on each side of the shell, the $\theta\theta$component of (4.1) reads

$$-\frac{m}{\hat{R}} = \epsilon_+ \sqrt{\left(\frac{d\hat{R}}{d\tau}\right)^2 + F_+} - \epsilon_- \sqrt{\left(\frac{d\hat{R}}{d\tau}\right)^2 + F_-} \qquad (4.2)$$

Where $F\pm := 1 - 2M_\pm/\hat{R}$ *and* $M_+$ $(M_-)$ is the right (left) hand side Schwarzschild mass.

$\epsilon_+ = 1$ $(\epsilon_- = 1)$ if, when viewed from the geometry right (left) of the shell, the shell is in the right hand side exterior region of the Kruskal diagram, or if the shell is in the white hole region and moving to the right, or if the shell is in the black hole region and moving to the left. Otherwise $\epsilon_+ = -1$ $(\epsilon_- = 1)$

It can be verified that (4.2) implies

$$(n_a u^b \nabla_b u^a)_+ - (n_a u^b \nabla_b u^a)_- = \frac{m}{\hat{R}^2}, \qquad (4.3a)$$

$$(n_a u^b \nabla_b u^a)_+ - (n_a u^b \nabla_b u^a)_- = 0\,. \qquad (4.3b)$$

(4.3a) is recognized as the projection of (4.1) along $u^a$. Note that the only independent components of (4.1) are the $\theta\theta$ component and the projection along $u^a$.

Upshot: Einstein outside the shell plus the $\theta\theta$ component of (4.1) implies the full dynamics of the shell. One consequence of the dynamics is (4.3b)

**4. RECOVERING THE CORRECT CLASSICAL DYNAMICS**

Now we would like to show that our Hamiltonian equations imply the above dynamics.

Notation: $\Delta(quantity) := (quantity)_+ - (quantity)_-$.

The constraint $H = 0$ at the shell reads

$$\Delta R' = -(\hat{\Lambda}/\hat{R})\sqrt{\hat{\Lambda}^{-2}P^2 + m^2} \qquad (5.1)$$

Using (2.7e) this becomes

$$\Delta R' = \frac{m\hat{N}\hat{\Lambda}}{\hat{R}\sqrt{-v^a v_a}} \qquad (5.2)$$

Using (3.1), (3.5), the Christoffels for (2.1), and the fact that

$$0 = \Delta\dot{R} + \dot{\hat{r}}\Delta R' \qquad (5.3)$$

One finds that (5.2) is precisely the $\theta\theta$ component of (4.1). Hence we do recover enough equations to imply the correct dynamics. To proceed, now need to check that the rest of our equations are consistent with this dynamics.

The constraint $H_r = 0$ at the shell reads

$$\Delta P_{\hat{\Lambda}} = -\frac{\mathcal{P}}{\hat{\Lambda}} \qquad (5.4)$$

Use (2.7e), express $P_\Lambda$ in terms of the velocities, and use (5.3): one gets an equation that is $\hat{R}\left(\dot{\hat{r}} + \hat{N}^r\right)/\hat{N}$ times (5.2). Hence the momentum constraint is consistent.

Equations (2.7a) and (2.7b) contain no delta-function, so they are consistent.

Equations (2.7c) and (2.7d) contain well-defined delta-functions on both sides. The delta-function parts can be readily isolated, and a direct calculation shows that they are consistent. The delta-terms in (2.7c) are simple. The delta-terms in the (2.7d) terms contain both $\Delta\Lambda', \Delta N', \Delta(N^r)', and\ \Delta R'$, but these terms group to $\Delta(R^2 v^a v_a)'$, which vanishes by arguments similar to those below. What remains is (2.7f). The LHS is unambiguous but the RHS is not. How should we understand this equation? On each side of the shell, plugging (3.2) into (2.7f) yields:

$$\frac{d(u_r)}{dt} = \frac{1}{2}\sqrt{-v^c v_c}\, u^a u^b (g_{ab})' \qquad (5.5)$$

Where the noncovariant RHS is only meant to hold in our coordinate system. Use $\sqrt{-v^c v_c} = d\tau/dt, where\ \tau$ is the proper time of the shell, to write this as

$$\frac{d(u_r)}{dt} = \frac{1}{2} u^a u^b (g_{ab})' \qquad (5.6)$$

Still not covariant but true in our coordinate system

We want to covariantize (5.6). Let $f$ denote the vector field $\partial_r$: $f^a = (\partial_r)^a$. The left hand side of (5.6) is $u^b \nabla_b (f_a u^a)$. On the right hand side, substitute $(g_{ab})' = \pounds_f g_{ab}$, which plainly holds in our coordinates. The right hand side becomes then covariant and equals by standard manipulations $u_a u^b \nabla_b f_a$. Writing $f^a = Au^a + Bn^a$ (where necessarily $B \neq 0$ because $f$ is spacelike), the terms proportional to $\nabla_a A$ cancel and (5.6) becomes

$$0 = -Bn_a u^b \nabla_b u^a \qquad (5.7)$$

This is equivalent to the geodesic equation, and it is not satisfied on either side of the shell. However, (4.3b) shows that it is satisfied if RHS is understood as the average over the two sides of the shell.

The Hamiltonian equations produce the correct dynamics, provided the RHS of (2.7f) is interpreted as the average of its values on the two sides of the shell.

***Hamiltonian via Legendre Transformation on conveniently chosen time reparameterization***

Consider

$$L(r,\dot{r};p) = \dot{r}P_c(r,p) - E(r,p)$$

From the Euler-Lagrange equation for $p$: $\frac{\partial L}{\partial p} = \dot{r}\frac{\partial P_C}{\partial p} - \frac{\partial E}{\partial p} = 0$. The conjugate momentum to "r":

$$r := \frac{\partial L}{\partial \dot{r}} = P_c + \frac{\partial p}{\partial \dot{r}}\left(\dot{r}\frac{\partial P_c}{\partial p} - \frac{\partial E}{\partial p}\right) = P_c.$$

Thus, the variable $P_c$ and $E$ are aptly named, where E is the Hamiltonian energy functional:

$$\frac{\partial L}{\partial p} = 0 \Rightarrow \dot{r} = \left(\frac{p}{\sqrt{p^2+m^2}} - \sqrt{\frac{2M_<}{r}}\right)$$

Thus

$$\left(\frac{p}{m}\right) = \frac{\eta\left(\dot{r} + \sqrt{\frac{2M_<}{r}}\right)}{\sqrt{1 - \left(\dot{r} + \sqrt{\frac{2M_<}{r}}\right)^2}}, \eta = \pm 1$$

So, $p = p(r, \dot{r})$ and substitution into $P_c$ yields a relation $P_c = P_c(r, \dot{r})$ that is not analytically invertible. Thus, the Legendre transformation runs into complications.

However, as already discussed, the energy functional $E\big(r, p(r, \dot{r})\big)$ is not in the desired form any way. Instead, energy functional on dynamical paths with proper time parameterization is what is desired.

$$E(r, \dot{r}) = M_> - M_< = \frac{\eta m\left(1 - \left(\dot{r} + \sqrt{\frac{2M_<}{r}}\right)\sqrt{\frac{2M_<}{r}}\right)}{\sqrt{1 - \left(\dot{r} + \sqrt{\frac{2M_<}{r}}\right)^2}} - \frac{m^2}{2r}$$

The switch to proper time is governed by the metrical relation

$$\dot{r}\left(\frac{d\tau}{dt}\right)\dot{r} = \left\{\frac{\left(1 - \frac{2M_<}{r}\right)}{\sqrt{\frac{2M_<}{r}}\,\dot{r} \pm \sqrt{\left(1 - \frac{2M_<}{r}\right) + \dot{r}^2}}\right\}\dot{r}$$

Upon direct substitution:

$$E(r,\dot{r}) = M_> - M_< = \eta m \sqrt{\left(1 - \frac{2M_<}{r}\right) + \dot{r}^2 - \left(\frac{m^2}{2r}\right)}\ ,\eta = \pm 1.$$

From this energy functional on proper-time paths a Hamiltonian is defined up to canonical transformations:

$$E(r,\dot{r}) \equiv H\big(r, P(r,\dot{r})\big) = \dot{r}P - L(r,\dot{r}) = \dot{r}^2 \frac{\partial}{\partial \dot{r}}\left(\frac{L}{\dot{r}}\right)$$

$$L(r,\dot{r}) = \dot{r}\left\{\int_c^{\dot{r}} dx \left(\frac{E(r,x)}{x^2}\right) + f(r)\right\}$$

$$P = \frac{\partial L}{\partial \dot{r}} = \int_c^{\dot{r}} dx \left(\frac{E(r,x)}{x^2}\right) + \frac{E(r,\dot{r})}{\dot{r}} + f(r)$$

$$(r > 2\mu M_<): \int_c^{\dot{r}} dx\, \frac{\sqrt{a^2 + x^2}}{x^2} = \left[-\frac{\sqrt{a^2 + x^2}}{x} + \ln\left(x + \sqrt{a^2 + x^2}\right)\right]_c^{\dot{r}}$$

$$(r > 2M_<): \int_c^{\dot{r}} \frac{\sqrt{a^2 - x^2}}{x^2} = \left[-\frac{\sqrt{a^2 - x^2}}{x} + \ln\left(x + \sqrt{a^2 - x^2}\right)\right]_c^{\dot{r}}$$

(For the $r > 2M_<$ case $x > a$ is required of real timelike trajectories, so the integration need not be complex).

$$P = \eta m \ln\left(\dot{r} + \sqrt{\dot{r}^2 + \left(1 - \frac{2M_<}{r}\right)}\right) - \eta m \ln\left(c + \sqrt{c^2 + \left(1 - \frac{2M_<}{r}\right)}\right)$$
$$-\frac{1}{c}\left(\frac{m^2}{2r}\right) + \frac{\sqrt{c^2 - a^2}}{c} + f(r)$$

$f(r)$ is chosen so that the r-dependent only terms in P are eliminated. This is consisted with $c = a$. For a description where $\dot{r}^2 < a^2$ and $r > 2M_<$ (starting from rest in the exterior), it is more appropriate to take $c = 0$ and then consider the change of $f(r)$ indicated to be just a canonical transformation.

$\eta = +1$ case is also chosen.

$$P = m \ln\left(\dot{r} + \sqrt{\left(1 - \frac{2M_<}{r}\right) + \dot{r}^2}\right)$$

$$\left(e^{p/m} - \dot{r}\right) = \sqrt{\left(1 - \frac{2M_<}{r}\right) + \dot{r}^2}$$

$$e^{p/m} - 2\dot{r}e^{p/m} = \left(1 - \frac{2M_<}{r}\right)$$

$$\dot{r} = -\frac{1}{2}\left(1 - \frac{2M_<}{r}\right)e^{-P/m} + \frac{1}{2}\, e^{P/m}$$
$$= \frac{1}{2}\left(e^{P/m} - e^{-P/m}\right) + \left(\frac{M_<}{r}\right)e^{-P/m}$$
$$= \sinh(P/m) + \left(\frac{M_<}{r}\right)e^{-P/m}$$ (for $r \to \infty$ this agrees with the free motino)

Shift the definition of the Hamiltonian to $(r, p) = M_>(r, p)$, and maintain $\eta = +1$ choice:

$$H(r,p) = M_< + E\left(r, \dot{r}\,(r,p)\right)$$
$$= M_< - \left(\frac{m^2}{2r}\right) + m\sqrt{\left(1 - \frac{2M_<}{r}\right) + \dot{r}^2}$$
$$= M_< - \left(\frac{m^2}{2r}\right) + m\left(e^{P/m} - \dot{r}\right)$$
$$= M_< - \left(\frac{m^2}{2r}\right) + m\left(\cosh(P/m) - \left(\frac{2M_<}{r}\right)e^{-P/m}\right)$$

$$H(r,p) = \left[m\cosh(P/m) - \left(\frac{m^2}{2r}\right)\right] + M_<\left[1 - \left(\frac{m}{r}\right)e^{-P/m}\right]$$

Now to verify that this yields the e.o.m. that is obtained in the junction condition analysis.

$$\dot{r} = \frac{dr}{d\tau}$$

$$\dot{r} = \frac{\partial H}{\partial p} = \sinh(P/m) + \left(\frac{M_<}{r}\right)e^{-P/m}$$

$$\dot{p} = \frac{\partial H}{\partial r} = -\frac{m^2}{2r^2} - \frac{mM_<}{r^2}e^{-P/m}$$

$$\ddot{r} = \cosh(P/m)\,(\dot{P}/m) + \left(\frac{M_<}{r}\right)e^{-P/m}\left(-\frac{\dot{P}}{m}\right) - \left(\frac{M_<}{r^2}\right)\dot{r}e^{-P/m}$$
$$= \left(\cosh(P/m) - \left(\frac{M_<}{r}\right)e^{-P/m}\right)\left(-\frac{m}{2r^2} + \frac{M_<}{r^2}e^{-P/m}\right) - \left(\frac{M_<}{r^2}\right)e^{-P/m}\dot{r}$$

$$= \sqrt{\left(1 - \frac{2M_<}{r}\right) + \dot{r}^2} \left( -\frac{m}{2r^2} + \left(\frac{M_<}{r^2}\right) \frac{1}{\dot{r} + \sqrt{\left(1 - \frac{2M_<}{r}\right) + \dot{r}^2}} \right) -$$

$$\left(\frac{M_<}{r^2}\right) \frac{\dot{r}}{\dot{r} + \sqrt{\left(1 - \frac{2M_<}{r}\right) + \dot{r}^2}}$$

$$= \sqrt{\left(1 - \frac{2M_<}{r}\right) + \dot{r}^2} \left(-\frac{m}{2r^2}\right) - \left(\frac{M_<}{r^2}\right)$$

$$\frac{\left[\ddot{r} + \left(\frac{M_<}{r^2}\right)\right](\dot{r})}{\sqrt{\left(1 - \frac{2M_<}{r}\right) + \dot{r}^2}} = \left(-\frac{m}{2r^2}\right)(\dot{r})$$

$$\frac{d}{dt}\left[\sqrt{\left(1 - \frac{2M_<}{r}\right) + \dot{r}^2} - \frac{m}{2r}\right] = 0$$

$$E = const = m\sqrt{\dot{r}^2 + \left(1 - \frac{2M_<}{r}\right)} - \frac{m^2}{2r}$$

***The Hamiltonian is now rewritten in dimensionless variables***

$\bar{r} = mr$ $\quad\quad$ $\propto = \frac{1}{2}m^2(\ell_{pl}^2)$ $\quad\quad$ $\ell_{pl} = 1$ in units

$\bar{p} = p/m$ $\quad\quad$ where $G = \hbar = c =$

1

$\bar{\tau} = m\tau$ $\quad\quad$ $\beta = \left(\frac{M_<}{m}\right)$ $\quad\quad$ as chosen

$\bar{H} = H/m$

$$\bar{H} = \cosh(\bar{p}) - \frac{\propto}{\bar{r}} + \beta\left(1 - \frac{2^{\propto}}{\bar{r}} e^{-\bar{P}}\right)$$

The bars will now be dropped.

$$\boxed{\bar{H} = \cosh(p) - \frac{\propto}{\bar{r}} + \beta\left(1 - \frac{2^{\propto}}{\bar{r}} e^{-\bar{P}}\right)}$$

From the first integral

$$\dot{\bar{r}}^2 + \left(1 - \frac{2M_< m}{\bar{r}}\right) = \left({}^{E}/_{m} + \frac{m^2}{2\bar{r}}\right)^2$$

$$\dot{\bar{r}}^2 + \left(1 - \frac{2M_< m}{\bar{r}} - \left({}^{E}/_{m} + \frac{m^2}{2\bar{r}}\right)^2\right) = 0$$

$$\dot{\bar{r}}^2 + V_{eff}(\bar{r}) = 0$$

$V_{eff}\ (\bar{r})$ is a monotonically increasing function on $[0, \infty)$ and $V_{eff}(\infty) = 1 - m^{-2}\ E^2$

There are three solutions:

(1) $E \geq m\ , \dot{\bar{r}} > 0$ shell starts at $\bar{r} = 0$ at finite $\tau_-$ and approaches $\bar{r} = \infty$ with $\bar{\tau} = \infty$ and $\dot{\bar{r}} = \sqrt{-V_{eff}(\infty)}$

(2) The reverse of (1)

(3) $E < m$ . Shell starts at $\bar{r} = 0$ with $\tau_-$ and $\dot{\bar{r}} = \infty$ , reaching a turning point at $V_{eff}(\bar{r}_0) = 0$ and falling back to $\bar{r} = 0$ at $\tau_t$ with $\dot{\bar{r}} = -\infty$.

This is qualitatively the same as $M_< = 0$, <u>the dynamics is incomplete</u>.

The Hamiltonian is not bounded from below as $r \to 0$ or as $p \to -\infty$

***Evaluating the Action***

Since Poisson brackets of $H\ and\ H_r$ close with no central charge, the constraints are first class and the shell geometry action has variation

$$dS = pd\hat{r}t \int dr\ (P_R\ \delta R +\ P_\lambda \delta\ \lambda) -\ (N_+ M_+)$$

$$\hookrightarrow \frac{\partial R}{\partial \lambda} d\lambda \qquad\qquad \uparrow q + 1 \text{ (independence of path)}$$

Reconstruction begins with the bulk geometric contribution to $dS$:

$$(dS)_{bulk} = \int_{-\infty}^{\hat{r}-\epsilon} dr(P_R\ \delta R +\ P_\lambda \delta\ \lambda) + \int_{\hat{r}+\epsilon}^{+\infty} dr\ (P_R\ \delta R +\ P_\lambda \delta\ \lambda).$$

Path → only $\lambda$ charging to $P_R = 0 = P_\lambda$ , then to any R, L

$$S_{bulk} = \int_{-\infty}^{\hat{r}-\epsilon} I\ dr + \int_{\hat{r}+\epsilon}^{\infty} Idr$$

$$I = R\ \lambda \sqrt{(R'/\lambda)^2 - 1 + 2M/R}$$

$$+ RR' \log \left| \frac{(R'/\lambda) - \sqrt{(R'/\lambda)^2 - 1 + 2M/R}}{\sqrt{|1 + 2M/R|}} \right|$$

Unrestricted variation of $S_{bulk}$ produces (back sub for momenta)

$$dS_{bulk} = \int_{-\infty}^{\hat{r}-\epsilon} dr[P_R\,\delta R + P_\lambda \delta\,\lambda] + \int_{\hat{r}+\epsilon}^{\infty} dr\,[P_R\,\delta R + P_\lambda \delta\,\lambda]$$

$$-\left(\frac{\partial I}{\partial R}\partial R\right)_{-\infty}^{\hat{r}-\epsilon} - \left(\frac{\partial I}{\partial R}\partial R\right)_{\hat{r}+\epsilon}^{\infty}$$

$$+\int_{-\infty}^{\hat{r}-\epsilon} dr\left(\frac{\partial I}{\partial M}\right)\partial M + \int_{\hat{r}+\epsilon}^{\infty} dr\left(\frac{\partial I}{\partial M}\right)\partial M$$

Correct action should yield $\frac{\delta S}{\delta R} = P_R$ and $\frac{\delta S}{\delta \lambda} = P_\lambda$. The bulk geometric action can be modified to agree with this by adding appropriate "shell" terms:

$$dS_{bulk+shell} = dS_{bulk} + \left(\frac{\partial I}{\partial R}\delta R\right)_{\hat{r}-\epsilon}^{\hat{r}+\epsilon}$$

$$-\int_{-\infty}^{\hat{r}-\epsilon} dr\left(\frac{\partial I}{\partial M}\right)\partial M + \int_{\hat{r}+\epsilon}^{\infty} dr\left(\frac{\partial I}{\partial M}\right)\partial M\,.$$

Including the boundary terms from the asymptotic regions, the reconstructed action is then

$$S = \int_{-\infty}^{\hat{r}-\epsilon} I dr \int_{\hat{r}+\epsilon}^{\infty} I dr + \int d\dagger\left(\frac{\partial I}{\partial R}\,\dot{\hat{R}}\right)_{\hat{r}-\epsilon}^{\hat{r}+\epsilon}$$

$$-\int_{-\infty}^{\hat{r}-\epsilon} dr\int dt\left(\frac{\partial I}{\partial M}\right)\dot{M} - \int_{\hat{r}+\epsilon}^{\infty} dr\int dt\left(\frac{\partial I}{\delta M}\right)\dot{M} - \int dt(N_+ M_+)$$

***Derivation of the Minisuperspace Lagrangian and Hamiltonian.***

We have the action, S, so calculation of $L = \frac{dS}{dt}$ will yield the Lagrangian. However, care must be taken so as to properly account for the junction conditions in what is a solely geometric expression for the action. So consider the following troublesome junction condition.

$$R'(\hat{r}+\epsilon) = R'\,(\hat{r}-\epsilon) - \frac{1}{\hat{R}}\sqrt{P^2 + m^2\,\lambda^2}.$$

Eventually we intend to choose the static gauge $R = r, \lambda = 1$. Let $R'_< = R'(\hat{r}-\epsilon)$ and $R'_> = R'(\hat{r}+\epsilon+\mu)$ and for $r < \hat{r}-\epsilon$ and $r > \hat{r}+\epsilon+\mu$ let $R = r$ as desired.

As $\mu \to 0$ it is possible for a smooth interpolative function $R'$ between $\hat{r} - \epsilon$ and $\hat{r} + \epsilon + \mu$ to remain smooth, with such a gauge choice we can then have

$$R'_< = 1 \text{ and } R'_> = 1 \;\text{ where } R'(\hat{r} - \epsilon) = R'_< \text{ and } R'(\hat{r} + \epsilon) = R'_< - \frac{1}{\hat{R}}\sqrt{P^2 + m^2\,\lambda^2}$$

Due to the equality $R'(\hat{r} + \epsilon) = R'_<$ in this form of the $\mu \to 0$ limit argument the shell is described as "embedded in the LHS geometry." A similar derivation is possible for shell "embedded in the RHS geometry", with equivalent results. Calculation of the Lagrangian now proceeds with

$$S = \left\{ \int_{-\infty}^{\hat{r}-\epsilon} + \int_{\hat{r}+\epsilon}^{\hat{r}+\epsilon+\mu} + \int_{\hat{r}+\epsilon+\mu}^{\infty} \right\} \left( I dr + \int dt \left(\frac{\partial I}{\partial M}\right) \dot{M} \right) + \int dt \left[ \left(\frac{\partial I}{\partial R}\right) \dot{\hat{R}} \right]_{\hat{r}-\epsilon}^{\hat{r}+\epsilon} - \int dt (N_+ M_+)$$

Taking $L = \frac{dS}{dt}$ and then $\mu \to 0$ limits:

$$L = \dot{r}\hat{R}\,\hat{\lambda} \left[ \sqrt{(R'_</\hat{\lambda})^2 - 1 + 2M_-/R} - \sqrt{(R'_>/\hat{\lambda})^2 - 1 + 2M_+/R} \right]$$

$$\lim_{\epsilon \to 0} + \frac{d\hat{R}}{dt}\,\hat{R} \left( \log \frac{(R'(\hat{r}+\epsilon)/\lambda) - \sqrt{(R'(\hat{r}+\epsilon)/\lambda)^2 - 1 + 2M_+/R}}{\sqrt{|1 + 2M_+/R|}} \right)$$

$$- \log \frac{(R'(\hat{r}-\epsilon)/\lambda) - \sqrt{(R'(\hat{r}-\epsilon)/\lambda)^2 - 1 + 2\mu M_-/R}}{\sqrt{|1 - 2M_-/R|}}$$

$$-\hat{R}(\dot{r} + \dot{\hat{r}} R'_>) \log \frac{(R'_>/\lambda) - \sqrt{(R'_>/\lambda)^2 - 1 + 2M_+/R}}{\sqrt{|1 + 2M_+/R|}}$$

$$+ \hat{R}(\dot{r} + \dot{\hat{r}} R'_<) \log \frac{(R'_>/\lambda) - \sqrt{(R'_</\lambda)^2 - 1 + 2M_-/R}}{\sqrt{|1 + 2M_-/R|}}$$

$$+ \int_{-\infty}^{\hat{r}-\epsilon} dr\,(P_R \dot{R} + P_\lambda\,\dot{\lambda}) + \int_{\hat{r}+\epsilon}^{\infty} dr\,(P_R \dot{R} + P_\lambda\,\dot{\lambda}) - (N_+ M_+)$$

The junction conditions with the $\mu \to 0$ limits conventions adopted read:

$$R'(\hat{r} + \epsilon) = R'_< - \frac{1}{\hat{R}}\sqrt{P^2 + m^2\,\hat{\lambda}^2}$$

$$\sqrt{(R'(\hat{r}+\epsilon)/\lambda)^2 - 1 + 2M_+/R} = \sqrt{(R'_</\hat{\lambda})^2 - 1 + 2M_-/\hat{R}} - \frac{P}{\hat{R}\,\hat{\lambda}}$$

The resulting form of the Lagrangian is then

$$L = \dot{\hat{r}}\,\hat{\lambda}\,\hat{R}\left[\sqrt{(R'_{<}/\hat{\lambda})^2 - 1 + 2M_-/R} - \sqrt{(R'_{>}/\lambda)^2 - 1 + 2M_+/\hat{R}}\right]$$

$$+\lim_{\epsilon\to 0}\frac{d\hat{R}}{dt}\hat{R}\log\frac{(R'_{<}/\lambda)-\sqrt{\left(R'_{<}/\hat{\lambda}\right)^2-1+2M_-/\hat{R}}+\left(P/(\hat{R}\hat{\lambda})-(\hat{R}\hat{\lambda})+\sqrt{P^2+m^2\hat{\lambda}^2}\right)}{(R'_{>}/\lambda)-\sqrt{\left(R'_{>}/\hat{\lambda}\right)^2-1+2M_+/\hat{R}}} +$$

$$\int_{-\infty}^{\hat{r}-\epsilon}\left(P_R\dot{R} + P_\lambda\,\dot{\lambda}\right) + \int_{\hat{r}+\epsilon}^{\infty}\left(P_R\dot{R} + P_\lambda\,\dot{\lambda}\right) - (N_+M_+)$$

The static gauge is now adopted: $R = r\,,\ \lambda = 1$

$$L = \dot{\hat{r}}\ \left[\sqrt{2M_-r} - \sqrt{2M_+r}\right]$$

$$+\ \dot{\hat{r}}\,r\log\frac{1-\sqrt{2M_+/r}}{1-\sqrt{2M_-/r}\ - r^{-1}\left(\sqrt{P^2+m^2}-P\right)} - N_+M_+$$

$M_+$ is related to $r$ and $P$, but not $\dot{r}$ , through the junction condition relation $M_+ - M_- = \sqrt{P^2+m^2} - \frac{m^2}{2\hat{r}} - P\sqrt{\frac{2M_-}{r}}$

So,

$$p = \frac{\partial L}{\partial r} = \sqrt{2M_-r} - \sqrt{2M_+r}$$

$$- r\log\left(\frac{1-\sqrt{2M_+/r}}{1-\sqrt{2M_-/r}-\left(\sqrt{P^2+m^2}-P\right)}\right)$$

And

$$L = \dot{r}\,p - M_+$$

Since $L = \dot{r}p - M_+$ it immediately follows that $H = M_+$

$$\dot{r} = \frac{\partial H}{\partial p} = \frac{P}{\sqrt{P^2+m^2}} - \sqrt{\frac{2M_-}{r}}$$

Written in terms of the implicit expression $\dot{r}(r,p)$:

$$H = m\,\frac{\left(1-\left(\dot{r}(r,p)+\sqrt{\frac{2M_-}{r}}\right)\sqrt{\frac{2M_-}{r}}\right)}{\sqrt{1-\left(\dot{r}(r,p)+\sqrt{\frac{2M_-}{r}}\right)^2}} - \frac{m^2}{2r}$$

In agreement with the Lagrangian analysis.

## C.2 Shell Quantization using Proper Time – Full Derivation

### C.2.1 Introduction

Spherically symmetric, asymptotically-flat geometrics are considered for which the only matter present is a spherical dust shell. From a reduced phase space formalism a reduced Hamiltonian is obtained for geometry plus matter upon imposing the gravitational constraints. A time reparametrized to shell proper time then yields the Hamiltonian theory to be studied.

### C.2.2 The Representation of the Hamiltonian Operator

The Hamiltonian describing classical shell motion is:

$$H(r,p) = M_{>} + M_{<} = m\cosh(p/m) - \frac{m^2}{2r} - M_{<}\frac{m}{r}e^{-p/m} + 2M_{-} \quad (1)$$

In terms of dimensionless variables:

$$(H-\beta) = \cosh(p) - \frac{2\alpha\beta}{r}\, e^{-p} - \frac{\alpha}{r} \qquad (2)$$

Upon quantization the above expression is formally taken to be an operator acting on the Hilbert space of state functions describing the shell:

$$(H-\beta) = \cosh(p) - 2\alpha\beta r^{-1}\, e^{-p} - \alpha r^{-1} = H_o - 2\alpha\beta C - \alpha V, \quad (3)$$

Where $H_0 = \cosh(p), C = r^{-1}e^{-p}$, and $V = r^{-1}$. A reasonable representation will first be sought for $H_0$ alone, and its self adjointness determined in a manner expected to be generalizable to the full analysis of H. Among the possible representations for $\cosh(p)$ there are:

$$\cosh(p) = \frac{1}{2}(e^{p} + e^{-p}) \qquad (4)$$

Where

$$e^{tp} f = \lim_{n\to\infty}\left(1 - \frac{tp}{n}\right)^{-n} f \qquad (5)$$

And the maximal domain is where the limit is denied such that it is an $L^2$ function. Alternatively,

$$\cosh(p) = \lim_{N\to\infty} \sum_{n=0}^{N} \frac{(-1)^n}{(2n)!}\left(\frac{d^{2n}}{dr}\right), \qquad (6)$$

And the maximal domain is where the N-limit is an $L^2$ function. Remark: both of the above maximal domains are much larger than that of $\cosh(p) = \cosh(p)\ where\ D(\cosh(p)) = \bigcap_{n=0}^{\infty} D(P^n), D(P^n) = \{f \in D(P^{n-1}) | P^{n-1} f \in D(P)\}$.

The representation chosen will be along the lines of the latter case above:

$$\cosh(p) = \overline{T} \tag{7}$$

$$T = \lim_{N\to\infty} T_N \ strong\ limit \tag{8}$$

$$T_N f = \sum_{n=0}^{N} \frac{(-1)^n}{(2n)!} \Delta_0^n f,\ \ \Delta_0 = \left(\frac{d}{dr}\right)^2 \tag{9}$$

$$D(T_N) = D(\Delta_0^N) = \{f \in D(\Delta_0^{N-1}) | \Delta_0^{N-1} f \in D(\Delta_0)\} \tag{10}$$

$$D(\Delta_0) = \{f \in H^2 | f(0^+)\sin(\alpha) = 0, \alpha \in [0,\pi]\} \tag{11}$$

$$D(T) = D_0 = \left\{f \in \cap_{N-0}^{\infty} D(T_N) \,\middle|\, \lim_{N-\infty} T_N f \in L^2\right\} \tag{12}$$

Remark: in the analysis to follow the strong limits (s-lim) may be taken as weak limits (w-lim) in obtaining the result. It is yet be determined whether this freedom may prove helpful when considering H in its entirely.

### C.2.3 Self-Adjointness of the Free Shell Operator

This was covered in the Overview, we now do the analysis more carefully. After establishing some properties for $\Delta_0$ and $T_N$, it will be shown the $H_o = \overline{T}$ is self adjoint. Claim $\Delta_0$ ($\equiv \Delta_0^\alpha$) is the one-parameter family of self-adjoint operators for $\Delta_0 = (d/dr)^2$ with domain parameterized by $\alpha$:

$$D(\Delta_0^\alpha) = \{f \in L^2, \Delta_0 f \in L^2 \,|\, f(0^+)\sin(a) + f'(0^+)\cos(\alpha) \equiv F_\alpha(0^+) = 0, \alpha \in [0,\pi]\} \tag{13}$$

$$= \{f \in H^2 \,|\, F_\alpha(0^+) = 0\}, \tag{14}$$

With notation: $L^2 \equiv L^2(0,\infty), H^2 \equiv H^2(0,\infty)$

The equivalence on the domains is shown first. For this it is sufficient to show that $f, f'' \in L^2 \Rightarrow f' \in L^2$. So consider $f \in L^2 and\ f'' \in L^2$

$$(f, f'') = \int_0^\infty \bar{f} f'' \lim_{\substack{C\to 0 \\ R\to\infty}} \int_C^R \bar{f}(r) \frac{d^2 f(r)}{dr^2} dr \tag{15}$$

$$= \lim_{\substack{C\to 0 \\ R\to\infty}} \left\{\bar{f} f' \big|_\epsilon^R - \int_\epsilon^R |f'|^2 \, dr\right\} \tag{16}$$

Since $f'' \in L^2$ implies that $f' and\ f$ are continuous, $\lim_{\epsilon\to 0} \bar{f}(\epsilon) f'(\epsilon)$ is finite. If the integral

$$\int_{\epsilon}^{R} |f'|^2 \, dr \tag{17}$$

Diverges, then $R(\bar{f}f') \rightarrow \infty \; as \; R \rightarrow \infty$ . this implies $\lim_{R\rightarrow\infty} \frac{d|f|^2}{dx} \rightarrow \infty$ , which contradicts quadratic integrability of $(x)$ since:

$$\lim_{R\rightarrow\infty} \frac{d}{dx} |f(x)|^2 = \infty \tag{18}$$

$$\Leftrightarrow \forall k > 0 \exists R(k) > 0 \text{ such that } \frac{d}{dr} |f(R)|^2 \geq k \text{ for } R \geq R(k) \tag{19}$$

$$\Rightarrow |f(R)|^2 - |f(R_0)|^2 = \int_{R_0 \equiv R(k)}^{R} \frac{d}{dr} |f(R)|^2 dr \geq k(R - R_0) \tag{20}$$

$$\Rightarrow \lim_{R\rightarrow\infty} |f(R)| = \infty \tag{21}$$

Which contradicts $\int_0^{\infty} |f|^2 < \infty$ . thus, the limit as $\epsilon \rightarrow 0, R \rightarrow \infty$ of the integral in equation (17) converges for all $f$ , indicating that $f' \epsilon \, L^2 \, (0, \infty)$ if $f, f'' \epsilon \, L^2$

Note: no boundary conditions were needed to obtain this result, and a generalization to be used in what follows is:

$$\{f, f'' \epsilon \, L^2 \Rightarrow f', \epsilon \, L^2\} \Longrightarrow \{f^{(n)}, f^{(n+2)} \epsilon \, L^2 \; \Rightarrow f^{(n+2)} \epsilon \, L^2\} \tag{22}$$

The operator $\Delta_0^{\alpha}$ is symmetric with the minimal set of boundary condition required.

This result follows from consideration of the following inner products, $f \epsilon \, D(\Delta_0^{\alpha})$:

$$(f, \Delta_0^{\alpha} f) = \int_0^{\infty} \bar{f} f'' dr = \lim_{\substack{\epsilon\rightarrow 0 \\ R\rightarrow\infty}} \int_{\epsilon}^{R} \bar{f}(r) f''(r) dr, \tag{23}$$

$$\lim_{\substack{\epsilon\rightarrow 0 \\ R\rightarrow\infty}} \left[\bar{f} f' - \bar{f}' f\right] + \lim_{\substack{\epsilon\rightarrow 0 \\ R\rightarrow\infty}} \int_{\epsilon}^{R} \bar{f}'' f \, dr \tag{24}$$

Since $\bar{f}$and $f'$are in $L^2$ , then $\bar{f} f' \epsilon \, L^1$ , and the Riemann-Lebesgue Lemma for $L^1$ functions leads to $\lim_{R\rightarrow\infty} \bar{f} f' \rightarrow 0$ . Thus

$$(f, \Delta_0^{\alpha} f) = \lim_{\epsilon\rightarrow 0} \left[\bar{f} f' - \bar{f}' f\right] + (\Delta_0^{\alpha} f, f), \tag{25}$$

And symmetry is obtained in those instances where

$$\bar{f}(0^+) f'(0^+) - \bar{f}'(0^+) f(0^+) = 0 \tag{26}$$

The minimal condition for Equation (26) to be true leads to a one parameter family of nontrival solutions (parameterized be a ):

$$f(0^+) \sin(\alpha) + f'^{(0^+)} \cos(\alpha) = 0, \qquad \alpha \text{ real.} \tag{27}$$

Thus, $D(\Delta_0^{\alpha})$ impose the minimal boundary conditions such that $\Delta_0^{\alpha}$ is symmetric – a necessary and, as will be shown, sufficient condition for $\Delta_0^{\alpha}$ to be self-adjoint.

In the proof of the claim concerning the self-adjointness of $\Delta_0^{\alpha}$ , the adjoint remains to be determined. To obtain the adjoint we seek all possible pairs of elements $\{g, h\}$ in $L^2 \; (0, \infty)$ such that

$$(g, \Delta_0^\alpha f) = (h, f) \quad \forall f \in D(\Delta_0^\alpha). \tag{28}$$

The set of all $g\ is\ D(\Delta_0^{\alpha t})$, and if $\{g, h\}$ is any pair then $\Delta_0^{\alpha t}\ g = h$. So, for a pair $\{g, h\}$ consider test function $\varphi \epsilon C_0^\infty\ (0, \infty)$:

$$(g, \Delta_0^\alpha \varphi) = (H, \varphi) \quad \forall \varphi \epsilon\ C_0^\infty\ (0, \infty) \tag{29}$$

In terms of the distribution inner products this is written:

$$\langle \Delta_0^\alpha \overline{\varphi}, g \rangle = (\overline{\varphi}, \Delta_0^\alpha g) \tag{30}$$

However, $\langle \Delta_0^\alpha \overline{\varphi}, g \rangle = \langle \overline{\varphi}, \Delta_0^\alpha g \rangle$ by definition of the distribution derivative. Thus, $\langle \Delta_0^\alpha \overline{\varphi}, g \rangle = \langle \overline{\varphi},\ h \rangle, \forall \varphi,\ and\ h = \Delta_0^\alpha g$ result as a necessary condition for the pair $\{g, h\}$ to be of the required kind. (For $(\Delta_0^\alpha)^N$ , similar arguments using test functions may be used.

When $T \equiv \lim_{N \to \infty} T_N$ is considered, however, the implicit arguments involving a suitable choice of core become nontrivial.)

Suppose $h = \Delta_0^\alpha g\ \in\ L^2$, as shown earlier this implies $g' \in\ L^2$. So, for any $f\ \in\ D(\Delta_0^\alpha)$:

$$(g, \Delta_0^\alpha f) = \lim_{\substack{c \to 0 \\ R \to \infty}} \begin{matrix} \left\{ (\bar{g} f' - \bar{g}' f)|_c^R + \int_\epsilon^R (\Delta_0^\alpha \bar{g} f) dr \right\} \\ = \lim_{\epsilon \to 0} (\bar{g} f' - \bar{g}' f) + (\Delta_0^\alpha g, f), \end{matrix} \tag{31}$$

$$= -f(0^+)\big(\bar{g}(0^+)\tan(\alpha) + \bar{g}'(0^+) + (\Delta_0^\alpha g, f)\big) \quad (\alpha \neq \pi\ assumed).$$
(32)

In order that $(g, \Delta_0^\alpha f) = (h, f), \forall\ f\ \in\ D(\Delta_0^\alpha)$, it is sufficient that

$$g(0^+)\sin(\alpha) + \ g'(0^+)\cos(\alpha) = 0 \tag{33}$$

If the domain of $\Delta_0^{\alpha t}$ is restricted accordingly, then $D(\Delta_0^{\alpha t}) = D(\Delta_0^\alpha)\ and\ \Delta_0^\alpha$ is thus shown to be self-adjoint. (The $\alpha = \pi$ case is treated similarly.)

Claim: $T_N$ is self-adjoint

Recall that

$$D(T_N) = D(\Delta_0^{\alpha N}) = \{f\ \in\ D(\Delta_0^{\alpha N-1}) | \Delta_0^{\alpha N-1} \in\ D(\Delta_0^\alpha)\} \tag{34}$$

Thus,

$$D(T_N) = \{f \epsilon H^{2N}\ | f^{(2n-2)}(0^+)\sin(\alpha) + \ f^{(2n-2)}(0^+)\cos(\alpha) = 0, n \leq N, \alpha \epsilon [0, \pi]\} \tag{35}$$

Ignore the boundary restriction in the domain $D(T_N)$ for the moment and consider the inner product

$$(f, T_N f) = \lim_{\substack{\epsilon \to 0 \\ R \to \infty}} \Big[ -\tfrac{1}{2!}\left(\bar{f} f' - \bar{f}' f\right) + \tfrac{1}{4!}\left(\bar{f} f^{(3)} - \bar{f}^{(1)} f^{(2)}\right) +$$
$$\left(\bar{f}^{(2)} f^{(1)} - \bar{f}^{(3)} f\right) + \cdots$$
$$+ \left( \tfrac{(-1)^N}{(2N)!} \left(\bar{f} f^{(2N-1)} - \bar{f}^{(1)} f^{(2N-2)}\right) + \left(\bar{f}^{((2N-2))} f^{(1)} - \bar{f}^{(2N-1)} f\right) \right)$$
$$+ \int_e^R \left( \textstyle\sum_{n=0}^N \tfrac{(-1)^n}{(2n)!}\ \bar{f}^{(2n)} \right) f\ dr \Big]. \tag{36}$$

Since $f^{(n)} \in L^2, n \leq 2N$, the $R \to \infty$ boundary terms contribute zero. The $\epsilon \to 0$ boundary terms are zero as well – if the boundary conditions in $D(T_N)$ are satisfied, and this can be shown by substituting for even or odd-order derivative terms in $f$. So, for $f \in D(T_N)$: $(f, T_N f) = (T_N f, f)$, and $T_N$ is symmetric.

As in (28), self-adjointness is obtained by seeking pairs $\{g, h\}$ in $L^2(0, \infty)$ such that equation (28) is satisfied with $T_N$ replacing $\Delta_0^\alpha$ . The set of all such $g$ is $D(T^t{}_N)$, and for any pair $\{g, h\}$: $T^t{}_N g = h$ . The method is as before – test functions are introduced to obtain a necessary condition for the pair $\{g, h\}$. (The arguments pertaining to a choice of core are standard.) There then results the necessary condition on $\{g, h\}$ that $h = T_N g$, and $h \in L^2 \Rightarrow g \in H^{2N}$. So, for all $f \in D(T_N)$ it follows that

$$(g, T_N) = \lim_{\substack{\epsilon \to 0 \\ R \to \infty}} \Big[ -\tfrac{1}{2!} \left(\bar{g} f^{(1)} - \bar{g}^{(1)} f\right) + \tfrac{1}{4!} \left(\bar{g} f^{(3)} - \bar{g}^{(1)} f^{(2)}\right) + \left(\bar{g}^{(2)} f^{(1)} - \bar{g}^{(2)} f\right) + \cdots + \frac{(-1^N)}{(2N)!} \left( \left(\bar{g} f^{(2N-2)} - \bar{g}^{(1)} f^{(2N-1)}\right) + \ldots + \left(\bar{g}^{(2N-2)} f^{(1)} - \bar{g}^{(2N-1)} f\right) \right) + \int_\epsilon^R \left( \sum_{n=0}^{N} \frac{(-1)^n}{(2n)!} \bar{g}^{(2n)} \right) f dr \Big] \tag{37}$$

Since $f^{(n)}, g^{(n)} \in L^2, n \leq 2N$, the $R \to \infty$ boundary terms contribute zero. If $\alpha \neq \pi/2$ is taken, and the $F\ (0^+)$ boundary condition is substituted into all odd-order derivative factors in the boundary terms:

$$(g, T_N f) = \lim_{\epsilon \to 0} \Big[ -\tfrac{1}{2!} \left(\bar{g}(-\tan\alpha) - \bar{g}^{(1)}\right) f + \ldots + \frac{(-1)^N}{(2N)!} \left( \left(\bar{g}(-\tan\alpha) - \bar{g}^{(1)}\right) f^{(2N-2)} + \ldots + \left(\bar{g}^{(2N-2)}(-\tan\alpha) - \bar{g}^{(2N-1)}\right) f \right) \Big] + (T_N g, f) \tag{38}$$

For the sufficient condition on the domain of $g$, transfinite induction then establishes that $g^{(n)} \in L^2$ satisfies the same boundary condition as $f$ in the domain $D(T_N)$ (see Equation 35) – from this it follows that $T_N$ is self=adjoint.

The operator $T \equiv \lim_{N \to \infty} T_N$ is understood by first establishing a suitable core (where semi-group techniques are used). With this approach a proof is made of the following:

Claim: The operator $\bar{T}$ is self-adjoint.

Some supporting lemmas will be used in the proof.

Lemma 1: $-\Delta_0^\alpha$ is non-negative

$$(f, -\Delta_0^\alpha f) = \int_0^\infty |f'|^2 dr \geq 0 \tag{39}$$

Remark: since $-\Delta_0^\alpha$ is non-negative and self-adjoint it can be used to generate a smegroup $\{e^{t\Delta_0^\alpha} | t \geq 0\} \subset \mathcal{L}(L^2)$, where $\mathcal{L}^\epsilon(L^2)$ is the set of linear functional on $L^2$ .

Lemma 2: $T_N$ is non-negative.

$$(T_N f, f) = \sum_{n=0}^{N} \frac{(-1)^n}{(2n)!} ((\Delta_0^\alpha)^n f, f),$$

$$= \sum_{n=0}^{N} \frac{1}{(2n)!} |\nabla^n f|^2 \geq |f|^2 \geq 0, \tag{40}$$

And also establishes the relations

$$|T_N f||f| \geq |f|^2 \geq 0$$

$$|T_N f| \geq |f| \geq 0 \tag{41}$$

The limit of the operator sequence $T_N$ need only be weakly defined to then obtain

$$|Tf| \geq |f| \geq 0 \tag{42}$$

So, a corallary of the second lemma is:

Lemma 3: $T$ is non-negative

A crucial step in the analysis of $T$ is the determination of a suitable core, and for this purpose the following Claim will be proved:

Claim:

$$\mathcal{A}(\Delta_0^\alpha) = \{e^{s\Delta_0^\alpha} f \,| s \in (0,1|, f \in L^2\} \tag{43}$$

Is a core for $T$

Consider $e^{s\Delta_0^\alpha} f \ : f \in L^2$

$$T_N(e^{s\Delta_0^\alpha} f) = \sum_{n=0}^{N} \frac{(-1)^n}{(2n)!} (\Delta_0^\alpha)^n \, (e^{s\Delta_0^\alpha} f), \tag{44}$$

$$|(\Delta_0^\alpha)^n e^{s\Delta_0^\alpha} f| = \left|\left(\Delta_0^\alpha \, e^{\frac{s}{n}\Delta_0^\alpha}\right)^n f\right| \tag{45}$$

Since $e^{s\Delta_0^\alpha}$ is a semigroup (an analytic semigroup in fact), the theorem by Hille and Yosida [??] is applicable:

$$|\Delta_0^\alpha \, e^{s\Delta_0^\alpha}| \leq \frac{M}{s} \, |f|, \tag{46}$$

Where $f \in D(e^{s\Delta_0^\alpha})$ and M is an absolute constant. (Note, the same property, in this instance, can be obtained using spectral theory) Since $\left|\left(\Delta_0^\alpha \, e^{\frac{s}{n}\Delta_0^\alpha}\right)^n f\right| \leq M^n \left(\frac{n}{s}\right)^n |f|$, it is possible to write.

$$|T_N(e^{s\Delta_0^\alpha} f)| \leq \sum_{n=0}^{N} \frac{1}{(2n)!} \frac{M^n n^n}{s^n} |f| < \sum_{n=0}^{N} \frac{1}{n!} \left(\frac{M}{s}\right)^n |f|. \tag{47}$$

With either the strong or weak limit defining the operator sequence, it then follows that

$$|T(\, e^{s\Delta_0^\alpha} f)| \leq \sum_{n=0}^{\infty} \left|\frac{(-1)^n}{(2n)!} (\Delta_0^\alpha)^n \, e^{s\Delta_0^\alpha} f\right| < \sum_{n=0}^{\infty} \frac{1}{n!} \left(\frac{M}{s}\right)^n |f|$$

$< \exp\left(\frac{M}{s}\right)|f|$ (48)

So $e^{s\Delta_0^\alpha} f \in D_0$, i.e. $e^{s\Delta_0^\alpha} : L^2 \to D_0$. From the well established semi-group properties, $\mathcal{A}(\Delta_0^\alpha)$is clearly dense in $L^2$: $e^{s\Delta_0^\alpha} f \to f$ as $s \downarrow 0$. So, $f \in D_0$ :

$$e^{s\Delta_0^\alpha} f \to f, \tag{49}$$

$$Te^{s\Delta_0^\alpha} f = e^{s\Delta_0^\alpha} Tf \to Tf \tag{50}$$

The properties in equations (49) and (50) indicate that $\mathcal{A}(\Delta_0^\alpha)$ is a core. Since $\mathcal{A}(\Delta_0^\alpha)$ is dense in $L^2$ and $\mathcal{A}(\Delta_0^\alpha) \subset D_0$, this also proves that $D_0$ is dense.

Lemma 4: The closure of the range of $T$ $is$ $L^2$, i.e., $\overline{R(T)} = L^2$.

Definition: $R(T) = \{g \in L^2 \mid g = T f \text{ for some } f\}$

Suppose $L^2 = \left(\overline{R(T)}\right)^\perp \oplus \overline{R(T)}$ and $D_0 \cap \left(\overline{R(T)}\right)^\perp \neq \{0\}$, then $\exists f \neq 0$ in the intersection (by the denseness of $D_0$) which satisfies

$$0 = (f, T f) \geq |f| > 0, \tag{51}$$

Which is a contradiction. So $\left(\overline{R(T)}\right)^\perp = \{0\}$, i.e. $\overline{R(T)} = L^2$.

Lemma 5: $T^{-1}$ is non-expansive

This follows from

$$|Tf| \geq |f| \Rightarrow |T^{-1}g| \leq |g| \text{ for } g \in R(T) \tag{52}$$

Since $\overline{R(T)} = L^2$, the limit

$$g = \lim g_n, g_n \in D_0 \tag{53}$$

Exists. Furthermore,

Lemma 6: Then $n \to \infty$ limit of $T^{-1}$ $g_n$ exists

Proof:

$$|T^{-1} g_m - T^{-1} g_n| = |T^{-1}(g_m - g_n)| \leq |g_m - g_n|, \tag{54}$$

Where the nonexpansive property is used, and $|g_m - g_n| \to 0$ as a Cauchy sequence (implied by $\{g_n\} \in L^2$ and $L^2$ compact). Thus, $T^{-1}$ $g_n$ is also a Cauchy sequence, lim $T^{-1} g_n$ exists, and an extension of $T^{-1}$ is thereby defined:

$$\lim_{n\to\infty} T^{-1} g_n \equiv T^{-1} g \tag{55}$$

Lemma 7:$\overline{T}^{-1}$ exists

$$\begin{aligned} |T^{-1} g - T_N^{-1} g| &= |T^{-1} g - T_N^{-1} g| = |f - T_N^{-1}Tf|, \\ &= |T_N^{-1}T_N f - T_N^{-1}Tf| \leq |T_N f - Tf| \\ &\to 0 \text{ as } N \to \infty, \end{aligned} \tag{56}$$

Where the concluding relation follows from the definition of $T$. So $\overline{T}^{-1} g = lim_{N\to\infty} T_N^{-1} g$.
Since $T_N$ is self-adjoint the symmetry of $T$ is already established:
$(T\,f, g) = (f, Tg)\; f, g \in D_0$. (57)
This property allows the symmetry of $\overline{T}^{-1}$ to be established:
$T$ symmetric $\Rightarrow \overline{T}$ symmetric $\Rightarrow \overline{T}^{-1}$ symmetric (58)
The existence and symmetric of $\overline{T}^{-1}$ is established. Furthermore,
$D\left(\overline{T}^{-1}\right) = R(\overline{T}) = \overline{R}(T) = L^2$. (59)
Since $D\left(\overline{T}^{-1}\right) = L^2, D\left(\left(\overline{T}^{-1}\right)^t\right) = L^2$ also, and this established the self-adjointness of $\overline{T}^{-1}$.

Lemma 8: $\overline{T}^{-1}$ self-adjoint $\Rightarrow \overline{T}$ self-adjoint

$$
\begin{aligned}
(g, \overline{T}f) &= (h, f) \text{ for } f \in D(\overline{T}), \\
(g, \hat{f}) &= \left(h, \overline{T}^{-1}\hat{f}\right) \text{ for } \hat{f} = \overline{T}\, f \in L^2, \\
&= \left(\overline{T}^{-1} h, \hat{f}\right), \\
&\Rightarrow g = \overline{T}^{-1} h, h = \overline{T} g, \\
&\Rightarrow \overline{T}^{t} = \overline{T}.
\end{aligned}
\quad (60)
$$

So, $\overline{T}$ is self adjoint. This establishes the self-adjointness of this representation of the free shell operator.
Remark: $\overline{T} f = \lim_{N\to\infty} T_N\, f_N, f_N \in D(T_N), f_N \to f$, provides an indication of how much larger $D(\overline{T})$ is than $D(T)$ .

**C.2.4 Analysis of the Interaction Operator**

The shell Hamiltonian $H_0 = \cosh(p) \equiv \overline{T},$ is self-adjoint. Consider now the interaction contribution to the Hamiltonian Operator given by $C = r^{-1} e^{-p}$. Considered as an operator in its own right, lets study the properties of a representation for C like that used for $H_0$.

There are two fundamental complications in describing a suitable representation for C over the description presented for $H_0$: (1) there is a singular factor of $r^{-1}$ , and (2) there are odd order $(d/dr)$ terms in explansions of $e^{-p}$. The later difficulty motivates expressing

$$e^{-p} = \frac{1}{2}(\cosh(p) - \sinh(p))$$

and working with a somewhat familiar $\cosh(p)$ term first.

# Appendix D. Emanator Theory Overview

## D.1 Introduction

In quantum physics unitary propagation is a standard part of the description. Efforts to move to algebras to describe such propagation leads to formulations based on the normed division algebras (real, complex, quaternion, and octonion). In an effort to achieve maximal information propagation we relax the unitarity condition and show that multiplication (right) on a unit norm trigintaduonion base by a unit norm chiral trigintaduonions emanator results in a new unit norm product [104]. A path is comprised of repeated (right) multiplications. Each step of the 'emanation' arrived at is a multiplication by a chiral trigintaduonion. Use of methods from noise budget analysis, a constructive perturbation analysis, as well as analysis relating to maximal perturbation according to the Kato Rellich theorem, show that the chiral trigintaduonion with maximal perturbation has magnitude $\alpha$, precisely the fine structure constant. A relation between $\alpha$ and $\pi$ results. Suppose repeated achiral emanation steps can be described as an iterative mapping, with unit-norm constraint resulting in a quadratic relation on components, we then expect the Feigenbaum universal bifurcation parameter, $C_\infty$, to appear according to the number of independent dimensions in a chiral trigintaduonion emanation step and the precise form of the "emanator" construction. The number of effective dimensions is shown to be 29 plus a little more, and a relation between $\alpha$, $\pi$ and $C_\infty$results that is in agreement with the choice of emanator examined in computational studies shown here. The computational studies with the emanator have also been explored via "random walks" in the trigintaduonion space during emanation and to explore noise additivity effects. Component-level evolution is seen to behave like a random walk, with random walk asymptotics (established computationally). This helps to establish that the Emanation process is Martingale, since random walk processes are Martingale.

Just from the propagation structure on one path we already see core emergent structure that results in a universal emanation with structural parameters 10,22,78,137 and perturbation maximum $\alpha$=~1/137. The central notion in the universal emanation hypothesis is that there should be *maximal information flow*, where this is accomplished by finding the highest theoretical dimensionality of unit-norm 'propagation', here called an emanation, which turns out to be 10, then add the maximal perturbation that still allows unit-norm propagation, where that perturbation is into the space the 10D motion is embedded in, here a 32 dimensional (trigintaduonion algebra) space.

The existing Standard Model, and reasonable extensions for the massive neutrino, cannot explain the parameters of the model. Why they are 19, or so, in number, and why the local gauge structure exists with the odd-looking product form: $U(1) \otimes SU(2)_L \otimes SU(3)$.

At the heart of the quantum formulation underlying the standard model is the fundamental theoretical element known as the quantum propagator (corresponds to a complex unitary matrix). Most notably, if the wavefunction has unit norm at the outset, after unitary propagation it remains unit norm.

Efforts to generalize the quantum formulation by considering hypercomplex but non-unitary matrices have been stymied for a variety of reasons [95,99]. Given the existing close agreement of experimental observation and quantum predictions, it is hard to work within the unitary propagator-type theoretical framework and arrive at anything other than an equivalent quantum formulation. Instead of working within the existing theory, in Emanator Theory the objective is to obtain a generalized theory that can project the quantized Lagrangian and Standard Model (or closely related extended Standard Model). To seek a larger theory that projects the existing theory is an excellent way to break free of the Godel-Incompleteness Trap that occurs when working within the existing standard physics and standard model. The drawback, of course, is it's likely to be a far-fetched idea, whatever it is, so it will face a higher bar to be seen as even interesting, not to mention valid. In what follows a synopsis of emanator theory will be given, as well as a few of the latest results, that will hopefully be convincing on both accounts.

If trying to generalize to a theory that can project to a renormalizable quantum field theory (with the Standard Model parameters), there is the issue of why it should project the particular formulation described. Of all the possible projections, how do you argue why it should be the unitary propagator-type physics seen? Here the aforementioned difficulty in generalizing the existing propagator theory (from the inside out) is a strength as it explains why the projection should be the standard physics and standard model as seen. That being the case, there is still the matter of finding the missing structure elsewhere (the Path Integral Action-

Lagrangian formulation itself, for example, and the Standard Model parameters) and then projecting that as the renormalizable quantum field theory with Standard Model parameters as seen experimentally.

**D.1.1 Dirac used Lorentz Invariance**

If trying to guess the mathematical basis of a generalized quantum theory, that would project the existing Lorentz Invariant quantum field theory, then Dirac provides a powerful lesson. Recall that Dirac's guess and derivation for the relativistic wave equation was purely based on seeking a representation that was Lorentz invariant. In doing this, Dirac discovered the Dirac Equation that describes spin-1/2 fermionic matter (all the fundamental particles).

In Dirac's approach 4-vectors, $V^a$, were used to describe the spinors (spin ½ fermions), where the length of the 4-vector is constant under Lorentz Transformation $L$:

$$L\left\{\frac{1}{2}\eta_{ab}V^aV^b\right\} = \frac{1}{2}\eta_{ab}V^aV^b.$$

In Penrose's books on spinors [103,158] we see how to write a 4-vector as a 2x2 Hermitian matrix, $\psi(V^a)$, where the length of the four vector is equal to the determinant of the equivalent 2x2 matrix:

$$det[\psi(V^a)] = \frac{1}{2}\eta_{ab}V^aV^b.$$

Suppose the length of the 4-Vector is 1 (it's a spinor probability amplitude), then the associated 2x2 Hermitian matrix will have determinant equal to 1, thus $SL(2,\mathbb{C})$, and we have a direct representation of the Lorentz Group.

Let's now consider a similar process involving transformational invariance but instead of encoding the Lorentz transform in the form of matrix transformation invariance let's use elements of the Cayley algebras instead. Specifically, consider the following transformation:

$$q' = aqa_c^*, \;\; where \;\;\; aa_c = 1,$$

where $a$ is a (unitary) complex bi-quaternion: $H(\mathbb{C}) \times H(\mathbb{C})$, and $q = (ct, ix, iy, iz)$ (note this notation is for $q = (ReH_1, iImH_1, iReH_2, iImH_2)$ ). The $q' = (ct', ix', iy', iz')$ that results will correspond to a proper orthochronous Lorentz transform [100-102].

This is a remarkable result, but is there better (higher dimensionality/complexity)? Consider that a complex bi-quaternion is isomorphic to a complex octonion which is isomorphic to an 'achiral' sedenion, thus a theory for achiral sedenion emanation is indicated from this result as far back as 1917. The 'halting condition' on the generalization in 1917 seems to be that octonions are the highest-order of the division algebras that have inverses defined (which is necessary to have $aa_c = 1$ be defined). But they have already extended past octonions since these are complex octonions ≅ sedenions, so how are they guaranteed to have $aa_c = 1$ be defined? This is possible for the *chiral* sedenions, as shown in [104,21], if restricted to be unit norm. So now we have our answer based on the results from [104,21] – we can go one complexation order higher:

$$Q' = AQA_c^*, \ \ where \ \ AA_c = 1,$$

where $A$ is a (unitary) bi-complex bi-quaternion: $H(\mathbb{C} \times \mathbb{C}) \times H(\mathbb{C} \times \mathbb{C})$, which is isomorphic to a unit norm quaternionic bi-quaternion, $H(\mathbb{H}) \times H(\mathbb{H})$ , and $Q = (ct, ix, iy, iz)$.
where $Q = (ReH_1, iImH_1, iReH_2, iImH_2)$ as before, except $H_1 = H \times \mathbb{H}$ not $H_1 = H \times \mathbb{C}$
The $Q' = (ct', ix', iy', iz')$ that results will again correspond to a proper orthochronous Lorentz transform. Again, how do we know the critical operation $AA_c = 1$ can always be satisfied? Previously we saw that a complex bi-quaternion was equivalent to a 'chiral' sedenion (in the sense described in [104,21]), thus a bi-complex bi-quaternion is isomorphic to a complex chiral sedenion, which is isomorphic to a 'doubly chiral' trigintaduonion. This is precisely the construct examined in [104,21], so a generalization of the 1917 result to unitary quaternionic bi-quaternions appears possible. As shown in [104,21], however, there is no higher order construct. This latter form (on doubly-chiral trigintaduonions) establishes the Emanator as Lorentz Invariant.

Thus, in terms of Cayley algebras, we can generalize to encoding the Lorentz transformation into split-Cayley algebras as long as the Cayley transform has an inverse, and doing this for the highest order Cayley algebra possible. The Cayley algebras with inverses, the division algebras, consist of the first four Cayley Algebras: reals (1 parameter), complex numbers (2 parameter), quaternions (4 parameters), and octonions (8 parameters). Another aspect of the division algebras is that they have norm, and this then means that their unitary evolution can be described as unit norm propagation. Suppose we focus on this latter feature, unit norm propagation alone. Can we extend to even higher order Cayley Algebras in the sense of starting with unit norm and for a subset of that higher order algebra still effect a unit norm, invertible, propagation? In other words for the next higher algebras, sedenions (16 parameter) and trigintaduonion (32 parameter), etc., does there exist a subspace allowing unit norm propagation? This is equivalent to considering the maximal Cayley subalgebra order for which unit norms exist. This is the actual starting point of the Maximal Information Emanation Hypothesis

### D.1.2 The Maximum Information Emanation (MIE) Hypothesis and Prior Results

Emanator theory stems from the Maximum Information Emanation (MIE) Hypothesis. The definition of information is context dependent, so how the MIE hypothesis will manifest depends on circumstance. We start with the fundamental notion of the quantum propagator, for which mathematical 'propagators' satisfy unitarity. We seek to extend this foundational element so start by asking what is the highest Cayley algebra that can remain unitary – where the answer comes down to the highest order division algebra, which are the octonions. What if we change the desired property from unitary to unit-norm preserving? Then we can extend 2 more dimensions beyond the 8D octonion algebra to a chiral subspace of the trigintaduonions that is 10D [104,21].

The 10D chiral trigintaduonions are then identified as the maximal information 'carriers' or emanators, operationally like the quantum propagators in a larger theory, where evolution of the system will be shown to result from sums on paths of emanators (similar to quantum evolution in terms of a path integral on propagators 'steps'). Having posited this maximal construct we see the classic signs of "asking the right question" since we get a variety of clear results:

(1) the chiral emanator is manifestly Lorentz Invariant as indicated above, and the emanation step or propagation is simply multiplication by a unit-norm 'emanator' according to the Cayley algebra multiplication rules.

(2) chiral emanation involves a 10D element in a 32D space (trigintaduonions) for which maximum perturbation is determined computationally (still permitting unit-norm transmission) to be ~1/137, e.g., 'alpha' – we therefore have the mysterious alpha by a computational definition.

(3) The emanation process involves multiplication on the current unit norm trigintaduonion base element by a unit norm chiral trigintaduonion emanator to arrive at a new, unit norm, trigintaduonion base (and then it

repeats infinitely). When the trigintaduonion multiplication steps are expanded to octonionic level, we find that there are 137 independent tri-octonionic terms that occur in the emanation process. Due to complex noise contributions, the effective number is 137*, which is slightly greater than 137. Thus, theoretically, the maximum perturbation that is allowed is 1/137* which happens to be exactly 'alpha'. We, thus, have a theoretical derivation for alpha, referred to as the $\{\alpha, \pi\}$ relation:

$$\alpha^{-1} = \frac{137}{\cos\beta}\cos\theta\frac{\theta}{\sin\theta},$$

where

$$\beta = \frac{\pi}{137} \; and \; \theta = \frac{\pi}{137x29}$$

and

$$\alpha^{-1} = 137.03599978669910\,,$$

which agrees with 2002 [159] experimental observation:

$$\alpha^{-1} = 137.03599976(50).$$

(4) Chiral trigintaduonion (32D) emanation with perturbation does not have effective dimension 32D due to chiral and other constraints -- noise budget analysis or (equivalently) Kato Rellich operator analysis, both indicate effective dimension slightly greater than 29 referred to as "29*". It is hypothesized that the maximal level of information flow dimensionally should relate to the Universal fractal constant $C_\infty$ according to:

$$\alpha^{-1} = (C_\infty)^\gamma, \;\; \gamma \equiv 29^*,$$

where $\gamma$is estimated by [21]:

$$\gamma \cong \frac{1}{2}\left(29 + \left(\frac{4\pi}{72}\right)\left[1 + \left(\frac{\pi}{137 \times 29}\right)\left\{\frac{\pi}{72} + \frac{3}{72}\right\}\right]\right)$$

thus

$$\alpha^{-1} \cong 137.035999206 \ldots$$

in agreement with the exact alpha with nine digit accuracy.

In the process of estimating 29* we must explore the definition of the emanation process in the sense of is it one step then normalization, or a chain of steps (analogous to hands greater than one in size being dealt) then normalization, with infinite repeat. The answer appears to be two steps (since this suffices to 'flood' or max-out the noise channels, so no further steps needed) and working within this construct we obtain the $\gamma \cong$ 29* estimate mentioned above. It's not an exact match to 16 decimal places and known measurement, but at a 9-decimal place match its pretty good, certainly indicating that there may well be a relation $\alpha^{-1} =$

$(C_{\infty})^{29^*}$ as hypothesized (sometimes called the $\{\alpha, \pi, C_{\infty}\}$ relation). And, if there is such a relation, it would indication that not only is evolution at maximum perturbation in the quantum sense, but it is also evolution at the edge-of-chaos in the thermal/statistical mechanical sense.

(5) the chiral emanator indicates 'motion' in a 10D subspace of 32D, suggesting 22 constants of the motion. This is a naïve analysis, but it turns out to be true upon deeper analysis within the emanator formalism as there are 22 types of emanation that result in no change to the base trigintaduonion describing the system. This, in turn, shows that the emanator theory will have 22 fixed parameters.

(6) By using the split form of the trigintaduonions, we not only have manifest Lorentz invariance, we also have an exact algebraic split to a space that is simply the direct product of 29* real dimensions (not a local approximation to such). This is important because the fundamental existence of a complex structure means that we trivially have the extension $\mathbb{R}^{29^*} \rightarrow \mathbb{C}^{29^*}$. Suppose we have point-like singular elements in $\mathbb{C}^{29^*}$, such will occur due to zero-divisors (zd's) in the 32D trigintaduonion space. To achieve a maximal domain of analyticity (an application of MIE), we must remove the zd-singularities. In doing so we obtain point-like matter in the theory and a small-h constant that enters the sum on emanator paths just as Planck's constant in the sum on propagator paths in the quantum formulation – suggesting that these small-h numbers are related.

(7) Three derivations of alpha are thus obtained: (i) Computational: $\{\alpha\}$ based on the maximal perturbation for which chiral emanation retains the unit-norm property; (ii) Theoretical: $\{\alpha, \pi\}$ based on the maximal noise transmission on a chiral emanation path; and (iii) Approximate: $\{\alpha, \pi, C_{\infty}\}$ based on the maximal noise transmission on an achiral emanation path (where maximal emanation is at "the edge of chaos" which is defined according to Feigenbaum Universality [160]).

(8) At component level in the emanation product, using 100's of millions of computational steps, we see an excellent asymptotic fit to random walk behavior. Since a random walk is a Martingale process, this strongly suggests that the achiral emanation process is Martingale (between normalization steps involved with zero crossings in their values). In turn, the projected quantum process (standard theory) would retain the imprint of that Martingale process.

(9) The achiral emanator, a sum of achiral emanation paths, can be shown to have the mathematical form $\sum \exp(i\mathbb{H} \times \mathbb{O}) \rightarrow \mathbb{C} \times \mathbb{H} \times \mathbb{O}$, which can be shown to give the gauge theory of the standard model: $U(1) \times SU(2)_L \times SU(3)$. Thus, there is no grand unified theory in terms of gauges that is fundamental (although a GUT may approximately occur at early times cosmologically, at high temperature, where there is conformal flatness). The 'ugly' product gauge that is observed is precisely what is predicted by emanator theory.

(10) Universal thermality is indicated by application of the MIE hypothesis to the choice of whether 'effective' achiral emanation is associative or not, at least at the sedenion-level of propagation. In effect, the mechanism to extend unit norm (chiral) propagation on a 10D subspace of the 32D trigintaduonions can be repeated for other mathematical properties from the lower-level Cayley algebras. Thus, the mechanism identified for extending 'nice' properties of lower-order Cayley algebras to higher order (no more than two orders higher to be precise), can be used for other than existence of a norm. By the chiral extension mechanism, associativity can be extended to the chiral octonions and chiral sedenions. This provides the basis for the universal thermality relation that appears to be ubiquitous (e.g., analytic time). It also provides the basis for an associative matching of terms for cancellation, e.g., renormalization. Thus provides the mechanism to correct the projected quantum field theory such that it has the renormalization counter-terms needed.

(11) In the emanation process there is a clear separation between spinorial elements and manifold elements. Manifold elements include geometry and thermality, have no alpha-perturbation effects, and appear to be part of the 'apparatus' from the perspective of the quantum theory.

## D.2 Synopsis of Methods and Prior Results

### D.2.1 The Cayley Algebras

The list representation for hypercomplex numbers will make things clearer in what follows so will be introduced here for the first seven Cayley algebras:

Reals: $X_0 \rightarrow (X_0)$ .
Complex: $(X_0 + X_1\, \boldsymbol{i}) \rightarrow (X_0, X_1)$ with one imaginary number.

Quaternions: $(X_0 + X_1 \boldsymbol{i} + X_2 \boldsymbol{j} + X_3 \boldsymbol{k}) \rightarrow (X_0, X_1, X_2, X_3)$ with three imaginary numbers.
Octonions: $(X_0, \ldots, X_7)$ with seven imaginary numbers.
Sedenions: $(X_0, \ldots, X_{15})$ with fifteen imaginary numbers.
Trigintaduonions (Bi-Sedenions): $(X_0, \ldots, X_{31})$ with 31 imaginary numbers.
Bi-Trigintaduonions: $(X_0, \ldots, X_{63})$ with 63 types of imaginary number.

Consider how the familiar complex numbers can be generated from two real numbers with the introduction of a single imaginary number '$\boldsymbol{i}$', $\{X_0, X_1\} \rightarrow (X_0 + X_1 \boldsymbol{i})$. This construction process can be iterated, using two complex numbers, $\{Z_0, Z_1\}$, and a new imaginary number '$\boldsymbol{j}$':

$$(Z_0 + Z_1 \boldsymbol{j}) = (A+B\boldsymbol{i}) + (C+D\boldsymbol{i})\boldsymbol{j} = A+B\boldsymbol{i} + C\boldsymbol{j} + D\boldsymbol{ij} = A+B\boldsymbol{i} + C\boldsymbol{j} + D\boldsymbol{k},$$

where we have introduced a third imaginary number '$\boldsymbol{k}$' where '$\boldsymbol{ij}=\boldsymbol{k}$'. In list notation this appears as the simple rule ((A,B),(C,D)) = (A,B,C,D). This iterative construction process can be repeated, generating algebras doubling in dimensionality at each iteration, to generate the 1,2,4,8,16, 32, and 64 dimensional algebras listed above. The process continues indefinitely to higher orders beyond that, doubling in dimension at each iteration, but we will see that the main algebras of interest for physics are those with dimension 1,2,4,and 8, and sub-spaces of those with dimension 16 and 32 dimensional algebras.

Addition of hypercomplex numbers is done component-wise, so is straightforward. For hypercomplex multiplication, list notation makes the freedom for group splittings more apparent, where any hypercomplex product ZxQ to be expressed as (U,V)x(R,S) by splitting Z=(U,V) and Q=(R,S). This is important because the product rule, generalized by Cayley, uses the splitting capability. The Cayley algebra multiplication rule is:

$$\mathbf{(A,B)(C,D) = ([AC-D^*B],[BC^*+DA]),}$$

where conjugation of a hypercomplex number flips the signs of all of its imaginary components:

$$(A,B)^* = \mathrm{Conj}(A,B) = (A^*,-B)$$

The specification of new algebras, with addition and multiplication rules as indicated by the constructive process above, is known as the Cayley-Dickson construction, and this gives rise to what is referred to as the Cayley algebras in what follows.

If a Split Cayley algebra is used, then the multiplication rule has a single sign difference:

$$\mathbf{(A,B)(C,D) = ([AC+D^*B],[BC^*+DA]).}$$

### D.2.2 Unit-norm propagation

For a physical system, a unit norm object can be used to represent a system, and by repeated transformation to other unit norm objects, it thereby evolves. Mathematical objects that can effect this 'transformation' simply by the rule of multiplication would be objects like division algebras, ideals, and what I'll simply call projections or emanations. In the universal propagator we have a unit norm trigintaduonion (32D) and perform a right multiplication with a chiral (10D) unit norm 'alpha-step' (defined by a max perturbation $\alpha$ into the 29 free dimensions, given by 32 minus one for each chiral choice, and one for the unit normalization overall). Consider multiplication of a given (starting) trigintaduonion from the right with a chiral bi-sedenion as a 'projection' through the (chiral) step indicated. The repeated application and repeated 'chiral steps' thereby arriving at a path describing a chiral propagation. The resulting universal propagation consists of a 32D unit norm trigintaduonion with propagation via right multiplication using a unit-norm, chiral bi-sedenion, with max-$\alpha$ perturbation.

We thereby arrive at a 'Universe Propagator' that takes on the physics parameters desired (notably the fine-structure constant) and imprints them onto the evolution as seen from the 'internal reference frame' where we reference an object in the 4D spacetime with Standard Model gauge field, and where the standard Lagrangian emerges as the necessary 'propagate-able' structure (where Hilbert space must be complex, not real, quaternionic or octonionic, etc. [99]). From maximum information flow with the constructs, and the required emergent complex Hilbert space (thus complex path integral, thus standard quantum operator formalism) we arrive back at the familiar results with justification of their core mathematical representations (e.g., complex Hilbert space), and now with justification of all parameters, all from the emanation hypothesis.

***Emanator Theory uses unit-norm propagation***

Unit-norm right product propagation is trivial for the division algebras since norm(XY) = norm(X) × norm(Y). From this it is apparent that we have an automorphism group given by the norm itself (since an automorphism if A(XY)=A(X)A(Y) ), and in the case of the octonions this automorphism group is G2 [161]. It can be shown that SU(3) is in G2 [161]. Let's now consider the situation with a higher-order Cayley algebra, the Sedenions, 'S'. We obviously don't have norm($S_1S_2$) = norm($S_1$) × norm($S_2$) in general, as this would then allow S to join the ranks of the division algebras, and it is proven that such don't exist above the Octonions [162]. Can we still have a propagation structure? Is it possible to have a 'base' sedenion for which norm($S_{base}$)=1, and to have a right propagator (product) sedenion also norm($S_{right}$)=1, such that norm($S_{base}$ x $S_{right}$) =1? The answer is yes (see [104,21]), when the sedenion has the (chiral) form of an octonion crossed with a real octonion: $S_{chiral}$ = (O,$O_{real}$) or $S_{chiral}$ = ($O_{real}$,O). Can we continue this to arrive at a propagation structure on the Trigintaduonions? Again the answer is yes, with the chiral form generalizing off the chiral Sedenion as might be expected: $T_{chiral}$ = ($S_{chiral}$, $S_{real}$) or ($S_{real}$, $S_{chiral}$) [104,21]. It is proven that this extension process will go no further [104,21]. What happens is that due to the chiral form we are still able to re-express all T products (or S) as collections of terms involving tri-octonionic products (which have nice properties as described in [104,21]), and this can no longer occur above the (chiral) trigintaduonion level.

Thus, we have achiral emanation, and to get achiral there must be a way to sum over all chiral to get an achiral result to arrive at the full emanator process. These details are described next.

### D.2.3 Chiral T-emanation has 78 generators of change and 137 independent octonion terms

We begin with constructing the theoretical expression for a general element of the trigintaduonion algebra after two chiral trigintaduonion multiplicative propagation steps. A simple analysis of the number of terms in this expression, when reduced to three-element algebraic 'braid-level', results in a count on algebraic braids of 137, plus a little extra (e.g. some lagniappe for the best 'cooking') of a contribution towards a 138$^{th}$ braid when the "noise analysis" is done [21].

Consider a general Norm=1 (32D) Trigintaduonion (Bi-Sedenion): (A,B), where A and B are sedenions (16D). Then have (A,B) = ( (a,b), (c,d) ), where {a,b,c,d} are octonions. Slightly different than a propagator, we have an 'emanator' with the following notation and properties, where the emanator describes a 10D multiplicative step. The emanator is a chiral bi-sedenion: a trigintaduonion whose first sedenion half is itself a chiral bi-octonion, and the second sedenion half is a pure real (as is the second octonion half): $(\tilde{A},\beta)$, $\tilde{A} = (\tilde{a},\alpha)$, where the norm is 1, $\alpha$ is a real octonion, and $\beta$ is a real sedenion. Thus:

Emanator: $(\tilde{A},\beta) = ( (\tilde{a},\alpha), \beta)$.
Note: $\tilde{A}^* = (\tilde{a}^*,-\alpha)$.

Let's set up a description of the Universal 'Emanation' along a 'chiral path' resulting from a few emanation steps. To begin, suppose we have already arrived at, or received, a unit norm trigintaduonion (32D) state 'T', and suppose our emanations are the result of right multiplication with a chiral trigintaduonion (bi-sedenion) 'step', and suppose we consider one such path after just a few steps. Here's the notation to begin:

$\mathbf{T} = (A,B)$, a unit norm trigintaduonion.
$\boldsymbol{\tau} = (\tilde{A},\beta) = ( (\tilde{a},\alpha), \beta)$, the 'emanator' above (so named to distinguish from a 'propagator').

Universal Emanation from T on single path with three steps: $( (\mathbf{T} \bullet \boldsymbol{\tau}_1) \bullet \boldsymbol{\tau}_2) \bullet \boldsymbol{\tau}_3)$ …

Consider the first emanation step:
$\mathbf{T} \bullet \boldsymbol{\tau}_1 = (A,B) \bullet (\tilde{A},\beta) = ( [A\bullet\tilde{A}-\beta^*\bullet B] , [B\bullet\tilde{A}^*+\beta\bullet A] )$. (Standard Cayley algebra multiplication rules.)
$A\bullet\tilde{A} = (a,b) \bullet (\tilde{a},\alpha) = ( [a\bullet\tilde{a}-\alpha^*\bullet b] , [b\bullet\tilde{a}^*+\alpha\bullet a] )$
$B\bullet\tilde{A}^* = (c,d) \bullet (\tilde{a}^*,-\alpha) = ( [c\bullet\tilde{a}^*+\alpha^*\bullet d] , [d\bullet\tilde{a}-\alpha\bullet c] )$
Thus,
$\mathbf{T} \bullet \boldsymbol{\tau}_1 = (A,B) \bullet (\tilde{A},\beta) = ( [ (a\bullet\tilde{a}-\alpha^*\bullet b-\beta c) , (b\bullet\tilde{a}^*+\alpha\bullet a-\beta d) ] , [ (c\bullet\tilde{a}^*+\alpha^*\bullet d+\beta a) , (d\bullet\tilde{a}-\alpha\bullet c+\beta b) ] )$.

At the lowest octonion level, that covers the pure real trigintaduonion, we have:

$(a\bullet\tilde{a}-\alpha^*\bullet b-\beta c) \rightarrow 8x8 + 8 + 8 - 2 = 64+14 = 78$ independent octonion terms (78 independent generators of motion). The –2 comes from the unit norm constraints on T and $\tau$.

Now consider the second propagation step:
$(\mathbf{T} \bullet \boldsymbol{\tau}_1) \bullet \boldsymbol{\tau}_2 = ( [ (a\bullet\tilde{a}-\alpha^*\bullet b-\beta c) , (b\bullet\tilde{a}^*+\alpha\bullet a-\beta d) ] , [ (c\bullet\tilde{a}^*+\alpha^*\bullet d+\beta a) , (d\bullet\tilde{a}-\alpha\bullet c+\beta b) ] ) \bullet (\tilde{A},\beta)$,

where $\boldsymbol{\tau_2} = (\tilde{A}',\beta') = ( (\tilde{a}',\alpha'), \beta')$.

Let $(\mathbf{T} \bullet \boldsymbol{\tau_1}) \bullet \boldsymbol{\tau_2} = ( [Z_{11},Z_{12}] , [Z_{21}, Z_{22}] )$.
$Z_{11} = (a\bullet\tilde{a} - b\alpha - c\beta)\bullet\tilde{a}' - (b\bullet\tilde{a}^* + \alpha a - \beta d)\, \alpha' - (c\bullet\tilde{a}^* + d\alpha + a\beta)\beta'$.

In $Z_{11}$ we can replace the octonions with their unit component forms:

$$a = a_1e_1 + a_2e_2 + \ldots + a_8e_8 ,$$

where $\{e_1, e_2, \ldots, e_8\}$ are the unit octonions (one real, seven imaginary), while '$\alpha$'$=\alpha e_9$ and '$\beta$'$=\beta e_{17}$, originally, but in expressions, are reduced to just their real part. All expressions, thus, involve 10 components: $\{e_1, e_2, \ldots, e_8, e_9, e_{17}\}$, and as the equations for $Z_{11}$ shows, grouped in factors of three (three-element octonionic 'braids'). We don't have associativity but we do have alternativity and the braid rules on three-element octonionic products that allows their regrouping. Applying these rules to have only ordered $e_i\bullet e_j\bullet e_k$ products in a simplified expression, we will then have 10x9x8/3! = 120 independent terms when the products involve different components. We have 8 independent terms when the first product are on the same component (equals 1), have 8 independent terms when the second product involves the same component, and have 1 independent term when the three-way product equals 1 (further details on this and the properties of the exponentiation map on hypercomplex numbers is given in the next section. There are, thus, 137 independent terms in $Z_{11}$, where each term has norm less than unity (since each octonionic component has norm less than one and the norm of a product of octonions is the product of their norms). The terms involving products with the same component, or with the components three-way product equal unity, correspond to the 'telescoping terms' in what follows.

When $\mathbf{T=((a,b),(c,d))} \rightarrow ((\mathbf{T} \bullet \boldsymbol{\tau_1}) \bullet \boldsymbol{\tau_2})=((Z_{11},Z_{12}),(Z_{21}, Z_{22}))$. we have $a\rightarrow Z_{11}$ and the terms involving 'a' in $Z_{11}$ are referred to as 'telescoping' due to their simple math properties with further emanation steps. In particular, the terms involving 'a' are:

$Z_{11}[\text{a terms}] = a\bullet\tilde{a}\bullet\tilde{a}' - a\alpha\alpha' - a\beta\beta'$.

We can see that the original 'a' information is passed along three (telescoping) channels, one involving repeated full octonionic factors $\tilde{a}$, one involving repeated real-octonion $\alpha$ factors, and one involving repeated real-octonion $\beta$ factors:

(1) a → $(a \bullet \tilde{a}) \bullet \tilde{a}'$ , if this product is continued indefinitely, then we have ***the random product of a collection of octonions***, all of which have norm less than one (although their norms can be quite close to one). If their norms were perfectly equal to one, then the addition of their random 'phases' would tend to cancel to zero, giving only a real octonionic component (same argument for phase cancelation on S1 as on S7 or S15). What results is a 'mostly' real octonion, having some imaginary part. A more precise, and lengthy, derivation is given in the next section.

(2) a → $a\alpha\alpha'$ , if this product is continued indefinitely, 'telescoped' with repeated $\alpha$ products, we see that the original 8 independent terms arising from 'a' are passed forward with an overall real octonion product, giving rise to 8 independent terms.

(3) a → $a\beta\beta'$ , as with (2), we have 8 independent terms.

From the above, we see an alternative accounting of the extra 17 independent terms to go with the 120 for a total of 137 independent terms in the propagation of the octonionic sectors of the universal emanation. A benefit of the telescoping analysis is it clarifies how in (1) an imaginary component may arise, and in perturbation expansions it will then be natural to refer to an overall imaginary component.

There are 137 terms in the dually chiral 'emanation', each with norm bounded by unity, with total bi-sedenion norm equal to unity. In the analysis that led to the computational discovery of $\alpha$ (see [21] and references cited there), an imaginary (non 10D) component was added of growing magnitude until unit-norm propagation failed. In essence, a maximum perturbation, from propagation strictly in the 10D subspace of the 32D trigintaduonions, was sought.

We identify maximal perturbation by doing an independent term analysis, and by adding a maximum perturbation term that implicitly identifies a definition of maximum antiphase. From this definition of maximum antiphase, there results the parameter $\pi$.

**D.2.3.1 Exponential Map Properties when using hypercomplex numbers**

For what follows, it helps to recall some important properties of the exponential, particularly its well-defined properties with hypercomplex numbers [21]. Important map relations:

(1) exponential map on Im(T) gives unit norm object: $\exp(\mathrm{Im}(T)\theta) = \cos\theta + \mathrm{Im}(T)\sin\theta$.

(2) exponential map on iT gives $C \times T$:
$\exp(iT) = \exp(i\mathrm{Re}(T)) \times \exp(i\mathrm{Im}(T)) = (\cos\theta + i\mathrm{Re}(T)\sin\theta) \times (\cos\varphi + i\mathrm{Im}(T)\sin\varphi) = C \times T$

Use (1) to focus on fluctuations in imaginary parameters free of normalization concerns.

Use (2) to get complex structure $C \times$ (object). Note that exponentiation into phase terms is precisely what occurs in the path integral propagator formalism, and will occur here as well for the emanator formalism, thus the "$C \times$" complex factor. When drawn upon in the emanator formalism, this method of achieving additional "$C \times$" complex structure will be forced by the zero-divisor handling (that will give rise to point-like matter with very small phase coupling, thus a highly oscillatory integral, and ties over to foundational aspects of the path integral formalism).

**D.2.3.2 Alternate 137$^{th}$ count using Exponential Map**

The derivation below follows [104], but with a more succinct accounting of the independent terms.
Consider a general norm=1 bisedenion in list notation: (A,B), where A and B are sedenions. Consider a propagator bisedenion $(C,\beta)$, $C = (c,\alpha)$, where c is an octonion and $\alpha$ is shorthand for the real octonion $(\alpha,0,0,0,0,0,0,0)$, where $\alpha$ is a real number, and $\beta$ is shorthand for the real sedenion $(\beta,0,0,0,0,0,0,0,0,0,0,0,0,0,0,0)$, where $\beta$ is a real number. Using A=(a,b), B=(u,v), and the multiplication rule from Section 2, we have:
$(A,B)(C,\beta) = ([AC-\beta^*B], [BC^*+\beta A])$, where
$AC = (a,b)(c,\alpha) = ([ac-\alpha^*b],[bc^*+\alpha a])$; $BC^*=(u,v)(c^*,-\alpha) = ([uc^*+\alpha^*v],[vc^*-\alpha u])$.
Thus, we have:
$(A,B)(C,\beta) = ([(ac-\alpha^*b\ ,\ bc^*+\alpha a)-\beta^*(u,v)]\ ,\ [(uc^*+\alpha^*v\ ,\ vc-\alpha u)+\beta(a,b)])$, so,

$(A,B)(C,\beta) = ([ac-\alpha^*b-\beta^*u\ ,\ bc^*+\alpha a-\beta^*v]\ ,\ [uc^*+\alpha^*v+\beta a\ ,\ vc-\alpha u+\beta b])$.
Now consider another propagator bisedenion $(C',\beta')$, $C' = (c',\alpha')$, and form the product corresponding to the next multiplicative step:
$( (A,B)(C,\beta) )\ (C',\beta') = ( [(ac)c' - \alpha^*bc' - \beta^*uc' - \alpha'^*(bc^*+\alpha a-\beta^*v)\ , \ldots]\ , [\ldots\ , \ldots] )$, where only the first expression at octonionic level ( $T=(O_1,O_2,O_3,O_4)$ ) is shown:

$$O_1 = (ac)c' - \alpha^*bc' - \beta^*uc' - \alpha'^*(bc^*+\alpha a-\beta^*v).$$

At octonionic-level there are 10x9x8/3x2=120 independent terms for 8 octonionic components (labeled a, b, c) plus a separate octonion component ($\alpha$) and one sedenion component ($\beta$), e.g., have 10 choose 3. Also have telescoping terms with repeated real octonion factors, such as with the $a\alpha\alpha'^*$ term (think $a\alpha(\alpha'^*)^n$ ), which gives an additional 8 independent terms. Also have telescoping terms with alternating real octonion factors and real sedenion factors, such as with the $v\beta^*\alpha'^*$ term (think $v(\beta^*\alpha'^*)^n$ ), which gives another 8 independent terms. There is one other 'telescoping' term due to repeated octonion right products seen in $(ac)c'$ (now think $((ac)c')c'\ldots..c')$ ). The change in this term corresponds to an element of the automorphism group on octonions, G2, and as such provides one last independent term, for a total of 137 independent terms at octonion level.

All of the octonion products involve octonions with norms at most unity, and by the normed division algebra rules on octonions, their norm is simply the norm of the individual octonions multiplied together, all of which are bounded by unity, thus their product is bounded by unity. The overall bound for the expression, each individual term being bounded by unity, is therefore simply the counting on the independent terms.

The maximum magnitude of each component of the octonion in the product term is given with a 'channel multiplier' of 137. Also, in seeking the maximum information propagation we require that the real chiral component never cross zero (e.g., stay in its connected $\{\alpha,\beta\}$ quadrant), thus the strictest condition on evaluating evolution might be intuited to be when the imaginary components combine to have real component contribution that is antiphase, e.g., the total imaginary angle is $\pi$. The choice of antiphase will used in what follows and will be justified when "C ×" allows the antiphase to be understood in the context of the Universal Mandelbrot set [163] position on the negative real axis that gives the maximal magnitude of displacement from the origin: $C_\infty$. We limit the maximum perturbation allowable by the antiphase worst case. At octonionic-level there is thus the channel multiplier: $137 + i\pi$.

#### D.2.3.3 The {α, π} relation using Exponential Map

The maximum perturbation, referred to as maximum noise in what follows, is first evaluated for a chiral emanation where we take a norm=1 $T_{base}$=(A,B) and take the right product with $T_{chiral}$ in the form $T_{chiral}$=(C,β), with product (A,B)(C,β) proven to be unit norm above [104]. In the prior section we saw that there are 137 independent octonion terms at the octonion sub-level of the new unit norm trigintaduonion that results, which leads to 137 independent terms at component level. In order to use the map rules mentioned in the previous section, it is necessary to move from the trigintaduonion, T, space, to the C × T space. This is done in later sections anyway where we consider sums on exp(iT). The exponential function (map) provides a well-defined 'lift' of a hypercomplex (Cayley) algebra from T to C × T. The exponential map also provides a very useful maneuver when working with unit-norm hypercomplex numbers via the generalized deMoivre theorem exp(Im(T))=cos(θ)+Im(T)sin(θ), with the real part recoverable from cos(θ). More details on this follow later but for now, in evaluating the maximum noise allowed we have three structures to adopt: (1) the noise is generalized to be complex (as will be the case for the components themselves once the T→C × T structure is adopted). (2) At component-level, the noise (for maximum noise) is equipartitioned in both real and imaginary parts. (3) Total imaginary noise magnitude is π for maximal antiphase (to be justified later).

(I) Chiral emanation noise: have 137 terms with max unit norm each, for the real part, and for the imaginary part have a "phase angle" β such that 137β=π (here referred to as a phase angle in the sense that the exp(Im(T)) map is being used). The noise magnitude at octonionic-component level is then given by the right triangle with real part = 137 and angle β=π/137, thus maximum chiral emanation noise magnitude is:

$$H = 137/\cos(\pi/137)$$

(II) Achiral emanation noise: now have 29 "free" components, each with 137 independent terms. For maximum achiral emanation we thus have 137 x 29 independent terms that are built from the aforementioned chiral emanation terms (to make achiral). If we equipartition as before, with noise magnitude Hc, we have a "noise triangle" with magnitude (hypotenuse) Hc and with angle $\theta = \pi/137x29$. The imaginary part is then (Hc)sin(θ). As regards the H magnitude separated form (separating out the 'H' factor for now), we have for the imaginary part sin(θ)c. As before, we take maximal noise transmission when all the imaginary parts add to maximal antiphase. Given the equipartitioning assumption, we then simply have the factor 137x29:

$$\sin(\theta)\, c\ (137x29) = \pi \rightarrow c = \theta/\sin\theta.$$

The maximum real noise perturbation that the system can have is then $\alpha$, where:

$$\alpha^{-1} = \frac{137}{\cos\beta}\cos\theta\frac{\theta}{\sin\theta}, \qquad where\ \beta = \frac{\pi}{137}\ and\ \theta = \frac{\pi}{137x29}$$

$$\alpha^{-1} = 137.03599978669910\,,$$

where the evaluation was done at WolframAlpha to high precision [164] (e.g., higher precision than that reported in earlier work [104]). This matches the experimentally observed value to all 11 decimal places currently known. As of 2002 [159], the measured value of α is:

$$\alpha^{-1} = 137.03599976(50).$$

Note that in quantum field theory the parameters are renormalized at a particular energy scale. Thus choice of energy scale impacts the value of α (as a coupling constant in the classical theory or a perturbation expansion factor in the quantum theory). At 0K we have the extreme low-energy end of the renormalization group (with the largest α value). We are at the 2.7K CMBR, so we have the max α to very high precision. (In studies at high energy scale at LEP, at the energy scale of the Z-boson (91GeV), we get the renormalized value to be [165,166]: $\alpha^{-1}[M_Z] \cong 127.5$. Note that 91GeV is way above the energy scale of the familiar Hagedorn temperature at ~pion mass=150MeV or 1.7x10^12 K) [167], where hadronic matter 'evaporates' into quark matter.)

**D.2.4 Trigintaduonion Emanation: achirality from chirality**

There are four chiralities, and for a given chirality (with unit norm) there are 29 dimensions of freedom (10D + 19D of chiraly consistent perturbation). When analytic extension is taken to give maximal information flow, the effective dimension for each of the four chiralities is 29* (detailed in [168]). This clear decomposition into 29* independent effective dimensions is then revealed in the {α,π,$C_\infty$} relation in [168].

The Mandelbrot Set (see Appendix) is one of many that encounter the universal constant $C_\infty$. The Mandelbrot set also describes a boundary with 2D fractal dimension [163] at its "edge of chaos" [168]. If driven to similar optimality in approaching a zero-value (a zero-divider issue), we see a two-value zero-crossing specification effectively like a double zero. The parameterization of the zeros of the Emanator at chiral zero-divisor points will thus be as double-zeros.

For what follows we use the simple description of the emanator:

$$T_{chiral}^{(k)} = \begin{cases} ((0,\alpha),\beta) \\ ((\alpha,0),\beta) \\ (\beta,(0,\alpha)) \\ (\beta,(\alpha,0)) \end{cases}.$$

$$\text{Emanation}(\mathbf{T}) = \frac{1}{N} \sum_{k\in\{4x72\}^n} \mathbf{T} \bullet \boldsymbol{T}_{chiral}^{(k)} = \frac{1}{N} \sum_{K\in 4\ chiralities} \mathbf{T} \bullet \overline{\boldsymbol{T}}_{chiral}^{(K)}$$

If working with non-split T's, then we restrict to emanations that are perturbations of unity:

$$T_{chiral}^{(k)} = \begin{cases} ((0,\alpha),\beta) \\ ((\alpha,0),\beta) \\ (\beta,(0,\alpha)) \\ (\beta,(\alpha,0)) \end{cases}, where\ T_{chiral}^{(k)} = \mathbf{1} + i\boldsymbol{\delta}.$$

From unit norm we have $\alpha^2 = 1 - 0^2 - \beta^2$, with $\pm$ sign choice on $\alpha$, similarly for $\beta$.

If working with split T's (bi-sedenions, etc.), then we have manifest Lorentz Invariance (shown in Chapter 7). So, it is often convenient to work with split-T' since this is manifest from the outset.

***Issues with zero-divisors***

Suppose we add the rule that emanation may not proceed when a particular chirality is zeroed-out, in other words:

$$\mathbf{T} \bullet \overline{\boldsymbol{T}}_{chiral}^{(K)} \neq \mathbf{0}.$$

For 'normal' numbers this goes without saying, since for real numbers if we have $r_1 \times r_2 = r_3$ then $r_3 \neq 0\ if\ neither\ r_1 = 0\ or\ r_2 = 0$. This holds true for the Real, Complex, Quaternion, and Octonion numbers. This does not hold true for Sedenions or higher. For sedenions the dimensionality of the zero-divisor event is mostly constrained, while for trigintaduonions it is significant. If such zeros were eliminated from the emanator description by using analytic extension component-wise (on 29* effective components) we see how a description devoid of matter (pure static field with no source or sink) might acquire matter by way of extending to a maximal domain of analyticity be removing zero-divisor events (a Wick transformation from real dimensionless action to pure imaginary action that is dimensionless but consisting of a dimensionful ratio). For further discussion along these lines, see Book 7 [21].

### D.2.4.1 Achiral T-emanation has 29* effective dimensions

Let's estimate of the effective dimension of information transmission in an achiral T-emanation process. There are 4 chiralities, so to get an achiral emanator candidate, minimally need a "4-card deck" to emanate in the four chiralities, with emanator equal to normalized sum. The actual deck appears to require a normalized sum over sub-chiralities, as will be explicitly enumerated in what follows. Here are the four chiralities with real fluctuation noise shown:

$$
\begin{array}{c}
(\,(\,(O[0]\pm\delta,\dots),\alpha\pm\delta),\beta\pm\delta)\\
(\,(\,\alpha\pm\delta,(O[0]\pm\delta,\dots)),\beta\pm\delta)\\
(\,\beta\pm\delta,(\,(O[0]\pm\delta,\dots),\alpha\pm\delta)\,)\\
(\,\beta\pm\delta,(\,\alpha\pm\delta,(O[0]\pm\delta,\dots)\,)\,)
\end{array}
$$

where $\alpha$ is a real octonion and $\beta$ is a real sedenion, and Tem is an equal weight sum of the action of each of the sub-chiral propagations on the base T, with the fluctuations indicated each done separately. We have the constraints $\alpha \neq 0$, $\beta \neq 0$, and common octonion O not pure real.

Each of the δ's is an independent fluctuation corresponding to its own sub-chiral emanation, but no subscripting on δ's is used or shown. There are thus 9x2x4=72 independent *imaginary* noise fluctuations to consider in the exp(Im(Tem)) evaluation (that automatically provides unit-norm). The real noise fluctuations in the real (first) component are, thus, not counted. If our definition for Tem entails only one card being dealt, then the sum over those possibilities is the sum

$$\mathbf{T}\bullet \mathrm{T}_{\mathrm{em}} \equiv \mathrm{Emanation}(\mathbf{T}) = \frac{1}{72}\sum_{k\in\{72\}} \mathbf{T}\bullet \boldsymbol{T}_{chiral}^{(k)}$$

For one-card, or a one-step, emanation, with real components and real noise, this makes sense from the counting shown, and it's what we use going forward. Using this will allow an entirely separate method for evaluating α (here at the one-card hand approximation). This will be done by determining the effective dimension 29*>29 of maximal information propagation (or maximal noise fluctuation). Before moving on, however, let's examine what happens when we allow complex noise fluctuations as this will trivially be allowed when we consider C × T via exp(iT) in later discussion anyway.

**D.2.4.2 Effective Deck size is 72, which is consistent with the {α,π,$C_\infty$} relation**

Maximum information transmission involves a complex extension to the T components and their noise fluctuations, but in doing this it must retain emanation structures such as the octonionic triple that occurs in previous expressions (starting with the proof of the $T_{chiral}$ solution itself), which leads to the counting that gives 137 independent terms, etc. Thus, the maximal complex extension on the noise is that it remain real in the octonion components:

$$((\,(O[0] \pm \delta, \dots), \alpha \pm i\delta), \beta \pm i\delta)$$
$$((\,\alpha \pm i\delta, (O[0] \pm \delta, \dots)), \beta \pm i\delta)$$
$$(\beta \pm i\delta, (\,(O[0] \pm \delta, \dots), \alpha \pm i\delta)\,)$$
$$(\beta \pm i\delta, (\,\alpha \pm i\delta, (O[0] \pm \delta, \dots)\,)\,)$$

The first chiral T component is where new imaginary terms might arise (the others are already counted since in imaginary components). We see there are six more, so the deck is now78. In application, as we will see, those added six are precluded due to constraints such that the effective deck size (impacting the sums and numbers of independent terms in the emanator definition) remains 72.

All noise terms will be treated additively, including terms in different imaginary components as well as imaginary noise terms in the real component. The criterion for max noise (in-phase constructive interference) gives the extreme of linear additivity. (Not like Gaussian statistical noise that adds in quadrature.) Also note that the discussion in terms of "noise transmission" and "information transmission" will be used almost interchangeably, whenever one description or the other best suits the analysis it will be used. Note that with this kind of noise analysis we can effectively shift around T noise terms associatively. Also note that application of the Kato-Rellich theorem [168,169] is related to the noise budget analysis done here focusing on first order terms.

There are 137 independent tri-octonionic terms in each of 29 free components indicated by a particular chirality (within the 32 components of a general trigintaduonion). This is a nontrivial result since ($T_{chiral} \bullet T_{chiral}$) is no longer $T_{chiral}$ type (but still $T_{norm1}$ type), so direct expansions are needed to identify the number of independent terms and this is briefly described below, with more detail in [170].

Obtaining an achiral emanation from a collection of chiral emanations requires that all chiralities be summed over (there are four) as well as sub-chiralities (there are 72). Noise analysis requires collecting of first-order terms. Analysis of noise transmission indicates 29* dimensions, where:

$$29^* \cong 29 + \left(\frac{4\pi}{72}\right)\left[1 + \left(\frac{\pi}{137 \cdot 29}\right)\left(\left(\frac{\pi}{72}\right) + \left(\frac{3}{72}\right)\right)\right]$$

The above result was obtained in [168] to describe the 72-card chiral 'deck' of chiral emanation products for a "single-step" emanation. In the new Results to follow this is reviewed and elaborated further.

**D.2.4.3 'Edge of chaos' maximal perturbation hypothesis [168]**
Consider the 'edge of chaos' maximal perturbation in each of the 29* dimensions to be at position $C_\infty$ (see Appendix for background on Mandelbrot Set), which is on the negative real axis, i.e., at π rotation to have −1 factor, ***thus at maximal antiphase***. This results in the relation for maximal perturbation at maximal antiphase (maximum reference angle with sign chosen positive by convention) has a lower bound on α given by:

$$\alpha_{\square}^{-1} = \left(\sqrt{C_\infty}\right)^{29^*}.$$

where

$$C_\infty = 1.4011551890920506004\ldots$$

This ties $1/\alpha$ to the second Feigenbaum constant $C_\infty$ in the context of the Mandelbrot set. It is well known that the Feigenbaum constants are universal, and part of a description of a universal transition to chaos regime. The Mandelbrot set is also universal [163], and maximal in that its fractal boundary has maximal fractal dimension of 2 [163], a detail that will be important in the meromorphic matter description given later.

For $C_\infty$, most references only provide $C_\infty = 1.401155189$ ..., and a higher precision tabulation is not readily found, so use is made of the relation

$$C_n = a_n(a_n - 2)/4,$$

together with the tabulation on $a_\infty$ [171]:

$$a_\infty = 3.56994567187094490 18 \ldots$$

The resulting $C_\infty$ is:

$$C_\infty = 1.4011551890920506004 \ldots$$

The resulting $\alpha_{\square}^{-1}$ is:

$$\alpha_{\square}^{-1} = 137.03599933370198263 \ldots$$

## D.3 Implication of MIE

### D.3.1 MIE requires a complex Hilbert Space

As mentioned previously, according to [99], a complex Hilbert space is selected by the quantum deFinetti theorem, since it is required for information propagation (and thereby consistent with the maximum information propagation concept in its selection). Because it's a complex Hilbert space, this explains why the path integral operates in a complex space, even though the underlying universal algebraic construct from which it is emergent is hypercomplex to the level of the trigintaduonions.

From [99], a simple derivation shows why the quantum deFinetti Theorem requires amplitudes to be complex. Suppose f(n) is the number of real parameters to specify an n-dimensional mixed state. For real amplitudes f(n)=n(n+1)/2, for complex amplitudes f(n)=n^2, and for quaternionic f(n) = n(2n-1). For propagation, etc., need f(n1n2)=f(n1)f(n2), which only works for complex amplitudes.

### D.3.2 Time is analytic and Matter is meromorphic

A variety of efforts have been made to find a definition of time that is somehow implicit to the main quantum field theory and General Relativity formalisms, whether it be a choice of vacuum for quantum field theory in curved spacetime (and even if the spacetime is not curved [172]) which is indirectly a choice of time. Or seeking an internal time-reference in a full-General Relativity quantum minisuperspace analysis of dust-shell collapse [173]. Or in seeking a notion of time in full general relativistic models, in the equilibrium sense, with an assumption of euclideanizability [125,126]. For the latter, the self-consistent stable solutions that were indicated showed the general utility of the euclideanizability hypothesis on emanation/propagation solutions in general (that is especially relevant, or interpretable, when the system is in equilibrium). In none of these efforts, however, was there success in identifying some internal notion of time, time, it seems, is an added construction, and this is consistent with the results shown here, where we find that time is likely analytic and an emergent construct.

In Book 7 [21], we see matter as meromorphic residue precipitation, in amounts of one quantum given by a precursor to Planck's constant h*. The meromorphic residue winding number is also notable in that it gives an integer that stays constant in the meromorphic region. This raises the possibility that elementary particle attributes might encode by way of different winding numbers, with reference to their different winding numbers at residues, but this will not be discussed further here.

**D.3.3 Emergent Evolution and Emergent Universal Learning**
We see that the definition of the emanator process is not known, but that consistency arguments (such as achirality constructed from a collection of chiral emanations) lead to a certain set of forms. And that consistency with the $\{\alpha, \pi\ C_{\infty}\}$ relation imposes further constraints on the form of the emanator. What is hypothesized is that the emanator is selected for maximal information transmission, thus emergent itself under that criterion. Let's now consider the maximal information transmission idea from the receiving end, e.g., maximal information receiving, or learning, in this context. If we turn to the information geometry analysis of learning in neural nets [174-178] (which uses differential geometry) we obtain a fundamental origin for statistical entropy (Shannon Entropy), and we identify optimal learning processes, based on expectation/maximization, that involves two-steps (as the name suggests) that may be done according to two fundamentally different conventions, e.g. the optimal learning involves four types of step, or is doubly chiral, consistent with the emanator 4-chiral processes described [179]. The potential applications of these results to Trigintaduonion encoded neuromanifolds is beyond the scope of this paper.

**D.3.4 Objective Reduction, Zero-Divisors, and possible origins of Planck 's constant**
A new mechanism for objective reduction [180,181] is also indicated by the way π enters the theory as a maximum anti-phase amount comprising part of the maximal perturbation propagation. Consider in the context where there is a 'classical' trigintaduonion path in a congruence of paths (a flow-line description). On the classical path in the congruences, we

have $\alpha$ calculated using a $+\pi$ maximal anti-phase, but this could also occur with $-\pi$ maximal anti-phase as well, thus we could have a $\pm\pi$ phase toggle when a zero divisor is encountered in the 32D propagation (given the perturbations extending outside the 10D somewhat into the entire 32D). The zero-divisor discontinuity requires the field to reformulate a new 'consistency' with the 32D algebraic propagation (and 64D and higher, as well), which could have the result that since the prior $\pi$ phase had the discontinuity, then it must toggle to the other, negative, phase, e.g., objective reduction may occur as a zero-divisor phase-toggle event.

### D.3.5 Where's the geometry?

The geometry side of emanation theory does not result from the action of the repeated emanator product directly, but from the accumulated product in the T base that results. Geometry is, in effect, emergent (projected) on the $T_{base}$ 'space' of the $T_{em}$ product action. Geometry appears as a manifold construct in both space-time curvature (where it is locally given with the standard model action) and as an intrinsic entropic property, via neuromanifold 'geodesic' motion being equivalent to, and possibly the origin of, the minimization of the relative entropy (and maximization of entropy, the 2nd Law) [179]. Setting aside thermodynamic issues in this discussion, this puts the Lagrangian formulation with standard model terms, and Hilbert action for General Relativity, into better perspective. The representation of the geometry via the Hilbert action for General Relativity suffices with maximal extension in whatever causally connected domain of interest. So, we've got the existing quantum field theory in CST space-time formulations in the black hole exterior, for example. We may have the resolution at the black hole horizon causal boundary via String Theory on the surface (using Ads/CFT relation and related holographic hypothesis [182,183]).

We describe repeated chiral product action on the trigintaduonion spinor space. The emanation process, consisting of a chain of chiral trigintaduonion products, leads to a Lagrangian variational formalism *with the standard model.* The origins of the parameters of the model are beginning to be understood as well. Apparently state information memory/inertia is carried via the manifold curvature response to the matter density, where 'G' is the linkage for the balance on this 'learning' process. Presumably the G learning rate is set for optimal learning, e.g., maximal information flow, and as such its value may eventually be clarified theoretically.

## D.4 Results

### D.4.1 Two-card Hand consistent with 137 and Neuromanifold Two-step Learning

MIE may guide our choice of 'hand' size dealt. Consider that one chiral emanation step (multiplication) off a base trigintaduonion will only have 78 independent tri-octonionic noise terms introduced. With a second step we 'flood the channels' and get the full assortment of 137 tri-octonionic noise terms (as shown in the derivation that obtained the 137 terms in [21]). Thus, a minimum of a two card hand must be considered. Having achieved a maximal perturbation scenario, what need for further cards in the hand, i.e., or further multiplicative factors in the pre-achiral normalization step?

Previously, when speaking of "Single-step achiral" we presumed initial conditions for the noise that effectively performed an average two-card 'hand' emanation analysis. If not in such an approximate form the ***{α, π}*** relation is recovered. The approximate form, however, expresses a new hypothesized relationship: $\alpha_{\square}^{-1} = \left(\sqrt{C_{\infty}}\right)^{29^*}$, that is itself verified to many decimal places.

Two-card-hand and two-step emanation jibes with two-step neuromanifold 'learning via 'em' and indicating relative entropy as optimal in this process (similar to choice of Euclidean distance in Riemann spatial change, here we have relative entropy as local entropic measure (with fixed reference, this then gives Boltzmann's entropy.

### D.4.2 The Chiral-Extension Cayley-Family relation

Recall that the unit-norm propagation was extended from octonion-level to trigintaduonion-level by introduction of chiral trigintaduonion emanators. The mathematical construct at the trigintaduonion-level that results can be represented in terms of tri-octonion products, thus all of the octonion properties are thereby inherited with such a construct since it is represented in terms of octonions, most notably a norm (here unit norm is of interest physically, for repetitive 'propagation'). The same double-chiral extension can be done at any Cayley level, and we thereby have a Cayley-Family extension property, via use of chiral Cayley emanation only, that is general.

***Associativity Extension most likely Candidate for Renormalization and Thermality reification***

Application of the Chiral-Extension Cayley-Family relation is first considered for associativity. In the Cayley Algebras, the highest algebra that retains the property of associativity is the quaternions. If we consider extending this associativity to higher -order algebras in the context of chiral propagation we can do this up to algebras at the level of the octonions and the sedenions. Consider that ***unit-norm*** trigintaduonion emanation will reduce to sedenion propagation in a general sense, and it is in this space that we can now add another sedenion element that is associative.

### D.4.2.1 Sedenion Associative Element and the Axiom of (Renormalization) Choice

The description of the renormalization process involves counter-terms. The emanation/selection of the standard gauge field group, and other elements, seems consistent with the Standard Model (and certain extended versions), but makes no mention of counterterms. Rather than add counter-terms as an external regularization element to the theory, here we can consider them as a more complete manifestation of the existing Emanator theory (again the MIE hypothesis justifies this). So, let's suppose there really is chiral trigintaduonion emanation, we can now see where a side-version of the theory presents that is sufficient to provide a (reified) version of the renormalization terms needed to renormalize the theory. Note that the application of the renormalization 'cancellation' is a fundamental application of the Axiom of Choice. More accurately, the universe that is hypothesized to be MIE, implicitly selects for the existence of the Axiom of Choice (and well-ordering foundation of mathematics) in its choice of renormalization cancellation.

### D.4.2.2 Octonion Associative Element and Complex Periodic Time

If extension of associativity on a sub-class of chiral-trigintaduonion emanations was possible, via the aforementioned effective sedenion associativity on chiral propagation, what of the indicated octonion-associativity extension? At the level of the octonion we can see a more direct manifestation of Lorentz invariance, including terms that are spatial or temporal (see generalized Lorentz invariance representations in Section 7.2), so suppose it is here that the theory creates ties between the real time quantum dynamics on a manifold and the complex-time statistical dynamics on a neuromanifold. These ties are shown to exist during equilibrium analyses in a number of applications (see Book 6

[20]), and in thermal quantum field theory applications (see Book 5 [15]), but here we speak to this being a more fundamental attribute, revealed in the indicated circumstances, yes, but also present in general (albeit perhaps not in a useful form).In essence, we use our second wish/demand (we only get two) of the associative-extension genie that it explain the recurrence of the appearance of a fundamental thermality by way of time having a complex periodicity proportional to inverse system temperature.

'Contact' Associativity from Higher-order Cayley Algebras (64 and 128 orders for example) was suggested previously, involving use of limit processes and infinitesimal processes on the non-propagatable orders (even chiraly) above 32-order trigintaduonion, to be at zero displacement from the 32-algebra (thus 'contact'), but still imparting possible nontrivial contributions, such that the above renormalizability and complex-time relations might be obtained. Here we find a better route to this result via the chiral-extension property.

### D.4.2.3 Octonion Commutative Element and Analytic Time

If we are considering operations reduced to octonion-level, there are also chiral extensions on the commutative complex numbers to arrive as a subset of commutative octonion-level operations in the unit-norm chiral emanation process. This allows further modification to the octonions as regards the internal time parameter analyticity, not just its imaginary-time periodicity. Together these chiral extensions may provide for the analytic time that is observed in the many successful applications of such a notion (required in path integral descriptions of bound states, for example).

The multiplication properties of the Cayley algebras are: Commutativity, Associativity, Alternativity, and Power Associativity. As you go to higher order algebras you lose properties. The Complex numbers are the last to have commutativity, the quaternions the last to have associativity, and the octonions the last to have alternativity. All the algebras have power associativity. Thus, the chiral extensions described are now exhausted since the interesting cases to extend, from complex, quaternionic, and octonionic, have now been considered.

### D.4.3 The Extended Standard Model predicted by Emanator Theory

According to the Standard model of Particle Physics, the Universe consists of point-like spin-1/2 fermions that reside in two families: the Leptons and the Quarks. These particles exist in a geometry with a gauge field, where they interact through that gauge field according to local gauge invariance and indirectly through the geometry according to general relativity. Even more indirectly, the gravitational interaction via mass has mass itself determined through exchange of spin-0 Higgs particles.

The local gauge invariance of the Standard model is:

$$U(1)_Y \otimes SU(2)_L \otimes SU(3)_C.$$

Let's consider each part separately:

$U(1)_Y$→ gives rise to the electromagnetic force between charged particles.

$SU(2)_L$→ gives rise to the force between left-handed particles.

$SU(3)_C$→ gives rise to the strong force that operates on particles (hadrons) that carry color charge (three types).

In Book 5 [15] we see that each gauge group listed above derives from the general local gauge invariant group form:

$$SU(N)_L \otimes SU(N)_R.$$

For $N = 1$, we have the trivial case (no left or right): $U(1)$.
For $N = 2$, we have the asymmetric case of only left: $SU(2)_L$.
For $N = 3$, we have the symmetric case of left-right, or simply a fixed ratio: $SU(3)$.

In Book 5 [15] we see that there is mixing between the $U(1)$ and $SU(2)$ parts to give the actual electroweak theory observed:

$$U(1) \otimes SU(2) \rightarrow \; U(1)_Y \otimes SU(2)_L.$$

The mixing introduces universal (globally gauge invariant) 'mixing angles' that number among the fundamental parameters of the theory (along with particle masses and a few other constants).

In Book 7 [21] we see that Emanator Theory predicts the form of the local gauge invariance as a general form, to be exactly that observed: $U(1)_Y \otimes SU(2)_L \otimes SU(3)_C$. Emanator theory also predicts a theory governed by 22 parameters, not 19 as currently listed for the Standard Model.

As strange as the form $U(1)_Y \otimes SU(2)_L \otimes SU(3)_C$ might seem, it is so far only applied to the interaction element of te theory. Let's now consider an interaction with what? The answer resides in the representations of the indicated groups (the left-handed particles interacting according to $SU(2)_L$ will be seen as interacting visa the weak force, etc.). So, saying there is a local gauge invariance is to say that there will be particles according to the irreducible, independent, representations of these groups. Thus, for any chosen representation there will be a collection of particles predicted. This is precisely what is observed experimentally. Thus, we not only know the groups $SU(2)_L$ and $SU(3)_C$, say, we also know their specific representations to obtain particle numbers and groupings observed experimentally.

So, the Standard Model is actually indicting the Model pair of Local Gauge Field and Representation $\mathcal{R}$:

$$\{U(1)_Y \otimes SU(2)_L \otimes SU(3)_C;\ \mathcal{R}\},$$

and it (the Standard Model) then says "times three", where the representation has there generations or copies:

$$\{U(1)_Y \otimes SU(2)_L \otimes SU(3)_C;\ 3 \times \mathcal{R}\}.$$

For the fundamental Leptons and Quarks this is shown in tabular form with the symbols for the particles as:

| | electroweak left-handed Leptons | electro-only right-handed Leptons | Electro-only Quarks |
|---|---|---|---|
| 1st Gen. | $[e^-{}_L, \nu_{e,L}(m=0)]$ | $[e^-{}_R, ---]$ | $d, u$ |
| 2nd Gen. | $[\mu^-{}_L, \nu_{\mu,L}(m=0)]$ | $[\mu^-{}_R, ---]$ | $s, c$ |
| 3rd Gen. | $[\tau^-{}_L, \nu_{\tau,L}(m=0)]$ | $[\tau^-{}_R, ---]$ | $b, t$ |

Note that there are no right-handed neutrinos in the Standard Model and the left-handed neutrinos are massless. We already have ample evidence that the left-handed neutrinos are massive, so we know the Standard model will extend in this regard, which leaves the supposedly non-existent right-handed neutrinos. There probably are right-handed neutrino but if "electro-only" and having no charge, what results is a particle completely decoupled from the other matter, except gravitationally, precisely what describes dark matter. More on this will follow once we go from the 19 parameter model to the hypothesized 22 parameter model, as this suggests limits on extensions to the theory, as does cosmological data, which suggests a possible 4th neutrino and no more scenario consistent with the cosmological evolution, unless the other neutrinos are very massive (all probably the case).

With the Standard Model quantum field theory structure, with scatter results according to the indicated local gauge field and particle representation, we find additional conservation rules. One of these rules describes the conservation of Lepton number at each generation, in other words:

$$L_e = N(e^-) + N(\nu_e) - N(e^+) + N(\bar{\nu}_e)$$

is a constant, and the same for constants for the muon and tau generations. There is also conservation of baryon number and the familiar conservation of charge.

As described in Section X, the Standard Model is a renormalizable/renormalized quantum field theory, and while this poses no difficulty with the $SU(2)$ elements alone or $SU(3)$ alone, here we have a product gauge and, what's more, it has handedness with use of $SU(2)_L$ not $SU(2)$. Not surprisingly, anomalous terms arise in he renormalization effort and for the renormalization to work, additional constraints are imposed on the structure of the theory. Most notable is that at a generational level across both leptons and baryons there should be a charge sum of zero. This is observed and was used to predict missing quarks (charm) in the early discovery process.

The structure identified by the standard model and the high decimal number agreement on key results from quantum field theory together provide a theory that can't be displaced directly by anything better, and only minimally augmented (with addition of right-handed neutrino), so trying to find a unified theory is impossible from 'within' the theory, and Godel's incompleteness theorem also suggests the hopelessness of such a task. This, then leaves no choice but to look for an encompassing theory that projects the quantum theory with the Standard Model structure in its entirety. And this is what is attempted with emanator theory (Book 7 [21]).

The Standard Model can't explain the three generations of matter, the origin of the local gauge group with its odd product form, the dimensionless constant alpha, and certainly can't explain the extension to massive(light) neutrinos, or the possible extension to massive (dark) right-handed neutrinos. In Emanator Theory (Book 7 [21]) we have:
(1) The local gauge $U(1) \otimes SU(2) \otimes SU(3)$ is projected by the theory. Why it should have three generations and why it should by asymmetric with use of $SU(2)_L$ is due to those choices providing the maximal packing of the Light matter sector (for maximal complexity information flow, an aspect of the MIE hypothesis), within he constraint of a 22-parameter emanation process.

(2) The 22 parameter, constants of the motion (emanation), result helps to constrain the particle representations such as to stay at 22-parameters, yet have maximally complex interaction.
(3) The MIE Hypothesis indicates the given generational structure given the constraint of working with the indicated product algebra. It also suggests any ‘fine-tuning’ on gravitational constant G might be dominated by the dark matter sector and its contribution to the evolution of the universe. Such fine-tuning would be most powerful if it indicated a fractal scale invariance property (see fractal G discussion in [21]).

Before considering the 22-parameter Emanator theory prediction of an extended Standard Model, let’s first recount the 19-parameter Standard Model, which I’ll separate into four groups:

> (I) 9 Yukawa coupling constants (masses) for the charged fermions
> (II) 5 constants for Weinberg Angle and the CKM matrix (with three mixing angles and CP-violating phase)
> (III) 3 Constants for electromagnetic coupling (alpha), for strong interaction (g3), and strong CP-violating phase ($\theta_3 \approx 0$).
> (IV) 2 Higgs parameters: Mass and Vacuum Expectation

If we allow for the left-handed neutrinos to have mass, then we get 3 more masses and another 4 constants for the PMNS matrix (three mixing angles and a CP-violating phase):

> (V) Extended model: 7 more constants → We, thus, have 26 parameters.

Let’s update out table with this extended version of the theory:

| | electroweak left-handed Leptons | electro-only right-handed Leptons | Electro-only Quarks |
|---|---|---|---|
| 1st Gen. | $[e^-{}_L, \nu_{e,L}(m \neq 0)]$ | $[e^-{}_R, ---]$ | $d, u$ |
| 2nd Gen. | $[\mu^-{}_L, \nu_{\mu,L}(m \neq 0)]$ | $[\mu^-{}_R, ---]$ | $s, c$ |
| 3rd Gen. | $[\tau^-{}_L, \nu_{\tau,L}(m \neq 0)]$ | $[\tau^-{}_R, ---]$ | $b, t$ |

There is now a problem. In order to maintain renormalizability, if there are left-handed neutrinos with mass there must be right handed neutrinos with mass [184] So, just how sure are we about neutrinos having mass? There is strong evidence not only for neutrino mass, but for neutrino family dynamics (here seen as oscillation between neutrino mass states). We'll see more evidence of generational family kinematics/dynamics in a later section. Neutrino family dynamics was first seen in measurements indicating that neutrinos spontaneously changed flavor (the Solar neutrino experiments [185]), which not only indicates they have mass, but allows us to determine the mass differences.

Clearly the Extended Standard Model needs to be extended further, to allow for the massive right-handed neutrino that is hypothesized to have no weak interaction like its right-handed electron cousin (and no electric interaction since no electric charge and no strong interaction since no color charge, thus 'dark' matter). Such a right-handed neutrinos (with no charge) can act as their own antiparticle (which is consistent with formation of "Bright" supermassive Black Holes in the early universe as seen with Webb). Furthermore, in addition to the Dirac mass relation to the left-handed neutrino it also has a Majorana mass term not tied to the Higgs mechanism [186]. What results is that instead of mass ${e^-}_L$ equal to mass ${e^-}_R$ here we have

$$m_{\nu_{e,L}} \propto \frac{1}{m_{\nu_{e,R}}},$$

which is known as the see-saw mechanism [187]. Thus the very low-mass left-handed neutrinos indicate very large mass right-handed neutrinos. Large mass neutrinos are an excellent candidate for cold dark matter, precisely what is needed to complete the Cosmological Standard Model.

Given the renormalization constraints and the observation of neutrino mass we arrive at the following updated Model:

| | electroweak left-handed Leptons | electro-only right-handed Leptons | Electro-Quarks |
|---|---|---|---|
| 1st Gen. | $[{e^-}_L, \nu_{e,L}(m \ll m_e)]$ | $[{e^-}_R, \nu_{e,R}(m \gg m_e)]$ | $d, u$ |
| 2nd Gen. | $[{\mu^-}_L, \nu_{\mu,L}(m \ll m_e)]$ | $[{\mu^-}_R, \nu_{\mu,R}(m \gg m_e)]$ | $s, c$ |
| 3rd Gen. | $[{\tau^-}_L, \nu_{\tau,L}(m \ll m_e)]$ | $[{\tau^-}_R, \nu_{\tau,R}(m \gg m_e)]$ | $b, t$ |

which is described by the 26 parameter theory indicated for massive left-handed neutrinos (if we assume that the right handed neutrino masses can be determined from the left-handed neutrino masses).

The standard counting to arrive at the 19 parameters (or, now, 26) includes $\theta_{QCD} \cong 0$ as a parameter of theory and its nearness to exactly zero is often referred to as the Strong CP problem. Going forward this is removed as a concern. The 'fine-tuning' that selects $\theta_{QCD} = 0$ is considered to be the same as that which selects the generational number, to arrive at 22 parameters while respecting the local gauge field constraint. Thus t is part of the MIE optimal selection.

Also included in the 19 parameter count is the U(1) gauge coupling $g_1$ and the SU(2) gauge coupling $g_1$, and these coupling are related to alpha by:

$$\alpha = \frac{1}{4\pi}\frac{{g_1}^2{g_2}^2}{{g_1}^2 + {g_2}^2}$$

Now, the emanation's projection of the quantizable Lagrangian theory involves various constants (22) in the resulting quantization. In that process, however, alpha already exists, in fact, from the derivation of alpha, we see that it establishes maximal perturbation starting at the level of the emanation process itself. Thus, alpha doesn't number among the 22, and the relation between coupling constants involving alpha means only one of those coupling constants is independently specifiable, so only one should be counted. At tis juncture we've reduced the count on 26 parameters to 24.

Notice in the prior counting, to arrive at the count of 26, we extended the count to be the same with dark neutrinos included since the dark neutrino mass was assumed to have some fixed (inverse) relation to the light neutrino mass. Let's not make this assumption, but retain the same mixing angle matrix, to now have 3 more masses in the theory (adding to the 24 count we now have 27).

In order to reduce the 27 parameter count to 22 we need to remove an 'overcount' of 5. Such an overcount would occur if there was a family-relationship generationally, for all masses. Consider such a mass relationship for the electron-muon-tau family, it is already known to exist and is known as the Koide relation. The Koide relation [188] was first observed for the three massive leptons currently known:

$$\frac{m_e + m_\mu + m_\tau}{\left(\sqrt{m_e} + \sqrt{m_\mu} + \sqrt{m_\tau}\right)^2} = \frac{2}{3}$$

To a lesser extent this relation is satisfied for the quarks as well, particularly for the three most massive, where the value is 0.6695. The problem with a simple application to the quark masses is that they are dependent on energy scale. A theoretical explanation for the Koide relation describes how this relation might exist for the masses of a given generation (or family group) [189]. As mentioned with neutrino oscillations, their oscillations involve the existence of such a family relationship in the quantum field theory implementation, so the hypothesis that it exists for each column in the table is not such a stretch. There are five columns of (independent) masses (the right handed electrons not independent of the left handed electrons removes them from the count). Thus, the five groups of three represent five groups with only two independent mass parameters, and the overall parameter count is thereby reduced by 5 from 27 to 22, as needed to be in agreement with Emanator theory. Thus Emanator theory, with the MIE hypothesis, would select for family group dynamics as observed with the neutrinos since it is what allows the maximal particle set within the 22-parameter constraint.

### D.5 Conclusion

Maximal information propagation as an emergent construct appears to require two forms of propagation, an early hypercomplex 'emanation' that involves a chiral 10D propagation in a 32D trigintaduonion space, and standard propagation with complex propagators (consistent with the quantum deFinetti relation) operating inside the geometry and gauge field that is projected. From the 'emanation' stage we see the maximum dimensionality and fractal limits provide the fundamental constants that then imprint upon the emergent geometry and gauge field, including giving rise to the constants $\alpha$. The origin of $\alpha$ has been a long-standing mystery. So much so that the central role of $\alpha$ in modern physics is literally engraved in stone, the tombstones of Sommerfeld (which displays $e^2/\hbar c$, which is $\alpha$) and Schwinger (which displays $\alpha/2\pi$) for example. Its origin has eluded physics for over a century, and appears to reside in the algebra of trigintaduonions.

Emanator Theory results from a hypothesized maximal information propagation and this means maximal analyticity, maximal domain, etc. As a process, Emanator theory is also hypothesized to operate up to "the edge of chaos" to permit maximal perturbation (noise) domain. When taken with the results showing that Emanator theory is Martingale, thus has well-defined limits, we then must wonder if there are well-defined multi-scale (fractal) limits. In other words, is there a relation that would tie the micro scale constant to fundamental constants (as they are counted in the 22) with the cosmological scale 'constants' that have settled out, at macro scale, in the current evolution of the Universe? In this context, the

Gravitational constant G is hypothesized to be a multiscale fractal coupling parameter.

In seeking a deeper theory we build on the sum-on-paths with propagator formulation to arrive at a sum-on-emanations with emanator formulation. Propagation in a complex Hilbert space, however, in a standard Quantum Mechanics/quantum field theory formulation, requires the propagator function to be a complex number (not real or quaternionic, etc., [99]). This prohibits what would otherwise be an obvious generalization to hypercomplex algebras. In order to achieve this generalization, we have to introduce a new layer to the theory, one with universal emanation involving hypercomplex algebras (trigintaduonions) that is hypothesized to project to the familiar complex Hilbert space propagation with associated fixed elements (e.g., the emanator formalism projects out the observed constants and group structure of the standard model). The 'projection' is an induced mathematical construct, like having SU(3) on products of octonions, but here it we be the standard model U(1)xSU(2)xSU(3) on products of emanator trigintaduonions. Thus, a unified variational formulation is posed, one that arrives at alpha as a natural structural element, among other things, uniquely specified by the condition of maximal information emanation.

A 'deeper' phase of universal evolution is described by a theory of emanations, where mathematically invariant emergent structures appear:

$$\boldsymbol{emanation \rightarrow propagation \rightarrow trajectory}$$

At the emanation to propagation emergence, one of the emergent constructs is the familiar path integral based on standard (unitary) propagators in a complex Hilbert space.

We have $\alpha$, 10,22,78,137 as parameters resulting from analysis on a single path maximal information flow construct, where the number 22 corresponds to the number of emergent parameters in the description of the propagating construct (exact derivation of the 22-parameter in [21]). In addition, the time choice is emergent via a multi-path construct, along with the propagator construct, and is coupled in both time step and imaginary time increment. The formulation is inherently embedded in a higher dimensional complex space, thus all of the quantum field theory complex analysis analyticity methods are valid as the assumptions made are now part of the maximal information flow emergent construct.

## References

[1] Liboff, Richard L. Introductory Quantum Mechanics. Longman Higher Education (1987).

[2] Yariv, Amnon. An Introduction to Theory and Applications of Quantum Mechanics. John Wiley; First Edition (1982).

[3] Baym, Gordon. Lectures On Quantum Mechanics (Lecture Notes and Supplements in Physics). The Benjamin / Cummings Publishing Company; 1st edition (1969).

[4] Gasiorowicz, Stephen. Quantum Physics. John Wiley & Sons; 3rd Edition (2003).

[5] Dirac, P. A. M. . The Principles of Quantum Mechanics. Oxford: Clarendon Press. (1930).

[6] Neumann, John von. The Mathematical Foundations of Quantum Mechanics. Princeton University Press; First Edition (1971).

[7] Feynman, R.P. and Hibbs, A.R. Quantum Mechanics and Path-Integral. McGraw-Hill, New York. (1965).

[8] Schulman, L.S. Techniques and Application of Path Integration. Dover Publications (2005).

[9] Chaichian, M.; Demichev, A. P. Introduction: Path Integrals in Physics Volume 1: Stochastic Process & Quantum Mechanics. Taylor & Francis. (2001).

[10] Richtmyer, R. D. Principles of Advanced Mathematical Physics. Springer, 1978.

[11] Hermann Weyl. The Theory of Groups and Quantum Mechanics. Dover Publications (1950).

[12] Kleinert, H. Path Integrals in Quantum Mechanics, Statistics and Polymer Physics. World Scientific, Singapore. (1995).

[13] Glimm, J. and Jaffe. Quantum Physics: A Functional Integral Point of View. Springer; 2nd edition (1987)

[14] Zinn-Justin, Jean. Path Integrals in Quantum Mechanics. Oxford University Press; 1st edition (2010)

[15] Winters-Hilt, S. Quantum Field Theory and the Standard Model. 2024. (Physics Series: "Physics from Maximal Information Emanation" Book 5.)

[16] Winters-Hilt, S. Classical Mechanics and Chaos. 2023. (Physics Series: "Physics from Maximal Information Emanation" Book 1.)

[17] Winters-Hilt, S. The Dynamics of Fields, Fluids, and Gauges. 2023. (Physics Series: "Physics from Maximal Information Emanation" Book 2.)

[18] Winters-Hilt, S. The Dynamics of Manifolds. 2023. (Physics Series: "Physics from Maximal Information Emanation" Book 3.)

[19] Winters-Hilt, S. Quantum Mechanics, Path Integrals, and Algebraic Reality. 2024. (Physics Series: "Physics from Maximal Information Emanation" Book 4.)

[20] Winters-Hilt, S. Thermal & Statistical Mechanics, and Black Hole Thermodynamics. 2024. (Physics Series: "Physics from Maximal Information Emanation" Book 6.)

[21] Winters-Hilt, S. Emanation, Emergence, and Eucatastrophe. 2023. (Physics Series: "Physics from Maximal Information Emanation" Book 7.)

[22] D'Alembert, Jean Le Rond (1743). Traité de dynamique.

[23] Laplace, P S (1774), "Mémoires de Mathématique et de Physique, Tome Sixième" [Memoir on the probability of causes of events.], Statistical Science, 1 (3): 366–367.

[24] Benedicks, M. "On Fourier transforms of functions supported on sets of finite Lebesgue measure", *J. Math. Anal. Appl.*, **106** (1): 180–183. (1985).

[25] Heisenberg, Werner. "Über quantentheoretische Umdeutung kinematischer und mechanischer Beziehungen". Zeitschrift für Physik (in German). 33 (1): 879–893. ("Quantum theoretical re-interpretation of kinematic and mechanical relations"). (1925).

[26] Schrödinger, E. (1926). "An Undulatory Theory of the Mechanics of Atoms and Molecules". Physical Review. 28 (6): 1049–1070.

[27] Heisenberg, W. (1927) [1927-03-01]. "Über den anschaulichen Inhalt der quantentheoretischen Kinematik und Mechanik". Zeitschrift für Physik (in German). **43** (3): 172–198. Heisenberg, W (1983) [1927]. "The actual content of quantum theoretical kinematics and mechanics". No. NAS 1.15: 77379. 1983. **43** (3–4): 172.

[28] Born, Max; J. Robert Oppenheimer. "Zur Quantentheorie der Molekeln" [On the Quantum Theory of Molecules]. Annalen der Physik (in German). 389 (20): 457–484. (1927).

[29] Dirac, P. A. M. "The Quantum Theory of the Electron". Proceedings of the Royal Society A: Mathematical, Physical and Engineering Sciences. 117 (778): 610–624. (1928).

[30] Popov, V.S. Positron Production in a Coulomb Field with Z > 137. Soviet Physics JETP. Volume32, Number 3 (1971).

[31] Balmer, J. J. "Notiz über die Spectrallinien des Wasserstoffs" [Note on the spectral lines of hydrogen]. Annalen der Physik und Chemie. 3rd series (in German). 25: 80–87. (1885).

[32] Maxwell, James Clerk. "On Faraday's Lines of Force". Transactions of the Cambridge Philosophical Society. (1855).

[33] "1861: James Clerk Maxwell's greatest year". King's College London. (2011).

[34] Hertz, H.R. "Ueber einen Einfluss des ultravioletten Lichtes auf die electrische Entladung", Annalen der Physik, vol. 267, no. 8, p. 983–1000. 1887.

[35] Michelson, A. A. and E. W. Morley, Amer. J. Sci. 34, 427 (1887); Phil Mag. 24, 463 (1887).

[36] Sommerfeld, A., Atombau und Spektrallinien. Friedrich Vieweg und Sohn, Braunschweig, 1919.

[37] Thomson, J.J. Cathode rays, Philosophical Magazine, 44, 293. (1897).

[38] Thomson, J.J. "On the Structure of the Atom: an Investigation of the Stability and Periods of Oscillation of a number of Corpuscles arranged at equal intervals around the Circumference of a Circle; with Application of the Results to the Theory of Atomic Structure," Philosophical Magazine Series 6, Volume 7, Number 39, pp. 237–265. (1904).

[39] Geiger, Hans. "On the Scattering of α-Particles by Matter". Proceedings of the Royal Society of London A. 81 (546): 174–177. (1908).

[40] Geiger, Hans, and Ernest Marsden (1909). "On a Diffuse Reflection of the α-Particles". Proceedings of the Royal Society of London A. 82 (557): 495–500.

[41] Rutherford, E., "The Scattering of α and β Particles by Matter and the Structure of the Atom", Philosophical Magazine. Series 6, vol. 21. May 1911

[42] Rydberg, J.R. "Recherches sur la constitution des spectres d'émission des éléments chimiques" [Investigations of the composition of the emission spectra of chemical elements]. Kongliga Svenska Vetenskaps-Akademiens Handlingar [Proceedings of the Royal Swedish Academy of Science]. 2nd series (in French). 23 (11): 1–177.

[43] Rydberg, J.R. On the structure of the line-spectra of the chemical elements. Philosophical Magazine. 5th series. 29: 331–337, 1890.

[44] Berestetskii, V. B.; E. M. Lifshitz; L. P. Pitaevskii (1982). *Quantum electrodynamics*. Butterworth-Heinemann.

[45] Winters-Hilt, S. The Mystery of Alpha. 2024.

[46] Wolfgang Pauli – Nobel Lecture. NobelPrize.org. Nobel Media AB 2021. Tue. 2 Mar 2021.
<https://www.nobelprize.org/prizes/physics/1945/pauli/lecture/>

[47] Pauli, Wolfgang; Jung, C. G. (2001). C. A. Meier (ed.). Atom and Archetype, The Pauli/Jung Letters, 1932–1958. Princeton, NJ: Princeton University Press.

[48] Arthur Miller, I. Arthur. 137: Jung, Pauli, and the Pursuit of a Scientific Obsession. W. W. Norton & Company, 2010.

[49] Feynman, R.P. QED: The Strange Theory of Light and Matter. Princeton University Press. p. 129. (1985) ISBN 978-0-691-08388-9.

[50] Kirchhoff, Gustav. "IV. Ueber das Verhältniß zwischen dem Emissionsvermögen und dem Absorptionsvermögen der Körper für Wärme und Licht," Annalen der Physik 185(2), 275-301.

[51] Einstein, A. "On a heuristic point of view concerning the production and transformation of light" . Ann. Phys., Lpz 17 132-148.

[52] Wheaton, Bruce R. (1978). "Philipp Lenard and the Photoelectric Effect, 1889-1911". Historical Studies in the Physical Sciences. 9: 299–322.

[53] deBroglie, Louis. Recherches sur la théorie des quanta (Researches on the quantum theory), Thesis, Paris, 1924, Ann. de Physique (10) **3**, 22 (1925)

[54] Compton, Arthur H. "A Quantum Theory of the Scattering of X-Rays by Light Elements". *Physical Review*. **21** (5): 483–502. (1923).

[55] Bohr, Niels. "On the Constitution of Atoms and Molecules, Part II Systems Containing Only a Single Nucleus". Philosophical Magazine. 26 (153): 476–502. (1913).

[56] Sommerfeld, A. (1916). "Zur Quantentheorie der Spektrallinien". Annalen der Physik (in German). 51 (17): 1–94.

[57] Wilson, W. (1915). "The quantum theory of radiation and line spectra". Philosophical Magazine. 29 (174): 795–802.

[58] Kennard, E. H. "Zur Quantenmechanik einfacher Bewegungstypen". *Zeitschrift für Physik.* **44** (4–5): 326–352. (1927).

[59] Gerlach, W.; Stern, O. "Der experimentelle Nachweis der Richtungsquantelung im Magnetfeld". *Zeitschrift für Physik.* **9** (1): 349–352. (1922).

[60] Dirac, Paul A. M. (1933). "The Lagrangian in Quantum Mechanics" (PDF). Physikalische Zeitschrift der Sowjetunion. 3: 64–72.

[61] Feynman, Richard P. (1942). The Principle of Least Action in Quantum Mechanics (PhD). Princeton University.

[62] Feynman, R.P. Space-Time Approach to Non-Relativistic Quantum Mechanics. Rev. Mod. Phys. 20, 367 – Published 1 April 1948.

[63] Erdeyli, A. Asymptotic Expansions. 1956 Dover.

[64] Erdeyli, A. Asymptotic Expansions of differential equations with turning points. Review of the Literature. Technical Report 1, Contract Nonr-220(11). Reference no. NR 043-121. Department of Mathematics, California Institute of Technology, 1953.

[65] Carrier, G.F, M. Crook and C.E. Pearson. Functions of a complex variable. 1983 Hod Books.

[66] Cameron, R.H. A family of integrals serving to connect he Wiener and Feynman integrals. J. Math. and Phys. 39, 126-140. (1960).

[67] Chung, K.L. and Varadhan, S.R.S. Kac functional and Schrodinger Equations. Studia Math. 68, 249-260.

[68] Szabados, Tamas. Pathwise approximation of Feynman Path Integrals using simple random walks. arXiv: 1803.07681v1.

[69] Pauli, W. (1925). "Über den Zusammenhang des Abschlusses der Elektronengruppen im Atom mit der Komplexstruktur der Spektren". Zeitschrift für Physik. 31 (1): 765–783.

[70] Wolfgang Pauli (15 October 1940). "The Connection Between Spin and Statistics". Physical Review. 58 (8): 716–722.

[71] positive energy condition by its existence [Witten ref]

[72] Lüders, Gerhart; Zumino, Bruno (1958-06-15). "Connection between Spin and Statistics" (htt ps://link.aps.org/doi/10.1103/PhysRev.110.1450). Physical Review. 110 (6): 1450–1453.

[73] Duck, Ian; Sudarshan, Ennackel Chandy George; Sudarshan, E. C. G. (1998). Pauli and the spin-statistics theorem (1. reprint ed.). Singapore: World Scientific. Pg. 393.

[74] Duck, Ian; Sudarshan, Ennackel Chandy George; Sudarshan, E. C. G. (1998). Pauli and the spin-statistics theorem (1. reprint ed.). Singapore: World Scientific. Pg. 425

.

[75] Aharonov, Y; Bohm, D (1959). "Significance of electromagnetic potentials in quantum theory". *Physical Review*. **115** (3): 485–491.

[76] Born, Max (1926). "Quantenmechanik der Stossvorgänge". Zeitschrift für Physik. 38 (11–12): 803–827.

[77] Brillouin, Léon (1926). "La mécanique ondulatoire de Schrödinger: une méthode générale de resolution par approximations successives". Comptes Rendus de l'Académie des Sciences. **183**: 24–26.

[78] Kato T. Fundamental Properties of Hamiltonian Operators of Schrodinger Type, Transactions of the American Mathematical Society, 1951, Pg. 195-211.

[79] Kato, T. Perturbation theory for linear operators. Springer 1980.

[80] Ghirardi, G.C., Rimini, A., and Weber, T. (1986). "Unified dynamics for microscopic and macroscopic systems". *Physical Review D*. **34** (2): 470–491.

[81] Penrose, Roger (May 1996). "On Gravity's role in Quantum State Reduction". *General Relativity and Gravitation*. **28** (5): 581–600.

[82] Albeverio, Sergio and Sonia Mazzucchi (2011) Path integral: mathematical aspects. Scholarpedia, 6(1):8832.

[83] Kumano-go, N and Fujiwara, D (2008). Feynman path integrals and semiclassical approximation. RIMS Kokyuroku Bessatsu B5: 241--263.

[84] Albeverio, S and Mazzucchi, S (2005). Feynman path integrals for polynomially growing potentials. J. Funct. Anal. 221(1): 83--121.

[85] Nelson, E (1964). Feynman integrals and the Schrödinger equation. J. Math. and Phys. 5: 332-343.

[86] Klauder, J R (2000). Beyond conventional quantization. Cambridge University Press, Cambridge

[87] Hida, T; Kuo, H H; Potthoff, J and Streit, L (1993). White noise. An infinite-dimensional calculus. Mathematics and its Applications, 253. Kluwer Academic Publishers Group, Dordrecht.

[88] Itô, K (1961). Wiener integral and Feynman integral. Proc. Fourth Berkeley Symposium on Mathematical Statistics and Probability. California Univ. Press, Berkeley 2: 227-238.

[89] Albeverio, S; Høegh-Krohn, R and Mazzucchi, S (2008). Mathematical theory of Feynman path integrals. An Introduction. 2nd and enlarged edition. Lecture Notes in Mathematics 523. Springer-Verlag, Berlin.

[90] Elworthy, D and Truman, A (1984). Feynman maps, Cameron-Martin formulae and anharmonic oscillators. Ann. Inst. H. Poincaré Phys. Théor. 41(2): 115--142.

[91] Albeverio, S and Mazzucchi, S (2005). Generalized Fresnel integrals. Bull. Sci. Math. 129(1): 1--23. doi:10.1016/j.bulsci.2004.05.005.

[92] Ryder, Lewis H. (1985). *Quantum Field Theory*. Cambridge University Press.

[93] Feynman, R.P. (1950) Mathematical Formulation of the Quantum Theory of Electromagnetic Interaction. Physical Review, 80, 440-457.

[94] Feynman, R.P. Space-Time Approach to Quantum Electrodynamics. Phys. Rev 76, pg. 769. (1949).

[95] Aaronson, Scott. Is Quantum Mechanics an Island in Theoryspace? arXiv: quant-ph/040.1062v2.

[96] Gleason, A.M. Measures on the closed subspaces of a Hilbert space, J. Math. Mech. 6:885-893, 1957.

[97] Deutch, D. Quantum Theory of probability and decisions. arXiv:quant-ph/9906015.

[98] Zurek, W.H.. Environment-assisted invariance, causality, and probabilities in quantum physics. Phys. Rev. Lett. **90**, 120404

[99] Caves, C.M., C.A., Fuchs, R. Schack. Unknown quantum states: The Quantum de Finetti Representation. J. Math. Phys. 43, 4537 (2002).

[100] Cailler, C. 1917. Archs. Sci. Phys. Nat. ser. 4, 44 p. 237.

[101] Girard, P.R.. The Quaternion group and modern physics. Eur. J. Phys. 5 (1984): 25-32.

[102] Synge, J.L. Quaternions, Lorentz Transformations and the Conway-Dirac-Eddington Matrices.

[103] Penrose, R., W. Rindler (1984) Volume 1: Two-Spinor Calculus and Relativistic Fields, Cambridge University Press, United Kingdom.

[104] Winters-Hilt, S. Feynman-Cayley Path Integrals select Chiral Bi-Sedenions with 10-dimensional space-time propagation. Advanced Studies in Theoretical Physics, Vol. 9, 2015, no. 14, 667 – 683. dx.doi.org/10.12988/astp.2015.5881.

[105] Gibbons, G.W. and M.J. Perry. Black Holes and Thermal Green's Functions. Proc. R. Soc. Lond. A, 358, 467-494 (1978).

[106] Popov, V.S. and V. D. Mur. Quasistationary levels in the lower continuum. Yad. Fiz. 18, 684-698 (1973).

[107] Duru, İ. H.; Kleinert, Hagen (1979). "Solution of the path integral for the H-atom" (PDF). *Physics Letters*. **84B** (2): 185–188. (1979).

[108] B. K. Berger, D. M. Chitre, V. E. Moncrief, and Y. Nutku, Phys. Rev. D 8, 3247 (1973).

[109] W. G. Unruh, Phys. Rev. D 14, 870 (1976).

[110] T. Thiemann and H. A. Kastrup, Nucl. Phys. B399, 211 (1993). (grqc/ 9310012)

[111] H. A. Kastrup and T. Thiemann, Nucl. Phys. B425, 665 (1994). (grqc/ 9401032)

[112] K. V. Kuchar, Phys. Rev. D 50, 3961 (1994). (gr-qc/9403003)

[113] J. Louko and B. F. Whiting, Phys. Rev. D 51, 5583 (1995). (gr-qc/9411017)

[114] M. Cavaglia, V. de Alfaro, and A. T. Filippov, Int. J. Mod. Phys. D 4, 661 (1995).

[115] M. Cavaglia, V. de Alfaro, and A. T. Filippov, "Quantization of the Schwarzschild Black Hole," Report DFTT 50/95, gr-qc/9508062.

[116] S. R. Lau, Class. Quantum Grav. 13, 1541 (1996). (gr-qc/9508028)

[117] J. Louko and J. Makela, Phys. Rev. D 54, 4982 (1996). (gr-qc/9605058)

[118] T. Thiemann, Int. J. Mod. Phys. D 3, 293 (1994).

[119] T. Thiemann, Nucl. Phys. B436, 681 (1995).

[120] J. Gegenberg and G. Kunstatter, Phys. Rev. D 47, R4192 (1993). (grqc/ 9302006)

[121] J. Gegenberg, G. Kunstatter, and D. Louis-Martinez, Phys. Rev. D 51, 1781 (1995).

[122] D. Louis-Martinez and G. Kunstatter, Phys. Rev. D 52, 3494 (1995). (grqc/ 9503016)

[123] M. Varadarajan, Phys. Rev. D 52, 7080 (1995). (gr-qc/9508039)

[124] S. Bose, J. Louko, L. Parker, and Y. Peleg, Phys. Rev. D 53, 5708 (1996). (gr-qc/9510048)

[125] J. Louko and S. N. Winters-Hilt, Phys. Rev. D 54, 2647 (1996). (grqc/ 9602003)

[126] J. Louko, J. Z. Simon, and S. N. Winters-Hilt, Phys. Rev. D, 55, 3525 (1997).

[127] R. LaFlamme, in: Origin and Early History of the Universe: Proceedings of the 26th Liege International Astrophysical Colloquium (1986), ed. J. Demaret (Universite de Liege, Institut d'Astrophysique, 1987); Ph.D. thesis (University of Cambridge, 1988).

[128] B. F. Whiting and J. W. York, Phys. Rev. Lett. 61, 1336 (1988).

[129] H. W. Braden, J. D. Brown, B. F. Whiting and J. W. York, Phys. Rev. D 42, 3376 (1990).

[130] J. J. Halliwell and J. Louko, Phys. Rev. D 42, 3397 (1990).

[131] G. Hayward and J. Louko, Phys. Rev. D 42, 4032 (1990).

[132] J. Louko and B. F. Whiting, Class. Quantum Grav. 9, 457 (1992).

[133] J. Melmed and B. F. Whiting, Phys. Rev. D 49, 907 (1994).

[134] S. Carlip and C. Teitelboim, Class. Quantum Grav. 12, 1699 (1995). (grqc/ 9312002)

[135] S. Carlip and C. Teitelboim, Phys. Rev. D 51, 622 (1995). (gr-qc/9405070)

[136] G. Oliveira-Neto, Phys. Rev. D 53, 1977 (1996).

[137] P. Kraus and F. Wilczek, Nucl. Phys. B433, 403 (1995). (gr-qc/9408003)

[138] P. Kraus and F. Wilczek, Nucl. Phys. B437, 231 (1995). (hep-th/9411219)

[139] W. Fischler, D. Morgan, and J. Polchinski, Phys. Rev. D 42, 4042 (1990).

[140] V. A. Berezin, N. G. Kozmirov, V. A. Kuzmin, and I. I. Tkachev, Phys. Lett. B 212, 415 (1988).

[141] V. P. Frolov, Zh. Eksp. Teor. Fiz. 66, 813 (1974) [Sov. Phys. JETP 39, 393 (1974)].

[142] P. Hajcek, Commun. Math. Phys. 34, 37-52. (1973).

[143] P. Hajcek, B. S. Kay, and K. V. Kuchar, Phys. Rev. D 46, 5439 (1992).

[144] J. D. Romano, "Spherically Symmetric Scalar Field Collapse: An Example of the Spacetime Problem of Time," Report UU-REL-95/1/13, gr-qc/9501015.

[145] J. D. Romano, Phys. Rev. D 55, 1112 (1997).

[146] K. V. Kuchar, J. D. Romano, and M. Varadarajan, Phys. Rev. D 55, 795 (1997). (gr-qc/9608011).

[147] Winters-Hilt S. Topics in Quantum Gravity and Quantum field Theory in Curved Spacetime. UWM PhD Dissertation, 1997.

[148] Kuchař, Karel. (2011). Time and Interpretations of Quantum Gravity 10.1142/S0218271811019347. International Journal of Modern Physics D.

[149] Kuchar, Karel. (1999). Black Hole Formation by Canonical Dynamics of Gravitating Shells: An Equatorial View. 10.1023/A:1026694027074. International Journal of Theoretical Physics.

[150] Kuchař, Karel. (1994). Geometrodynamics of Schwarzschild Black Holes. 10.1103/PhysRevD.50.3961. Physical Review D.

[151] Regge, Tullio and Claudio Teitelboim. Role of surface integrals in the Hamiltonian formulation of general relativity. Annals of Physics, Volume 88, Issue 1, 1974, 286-318.

[152] Straumann, Norbert. General Relativity. Springer 2013.

[153] Weinberg, S. (2002) [1995], *Foundations*, The Quantum Theory of Fields, vol. 1, Cambridge: Cambridge University Press,

[154] Misner, Charles W., Thorne, K. S., & Wheeler, J. A. Gravitation. Princeton University Press, 2017.

[155] Constantinescu, F. and E. Magyari. Problems in Quantum Mechanics. Pergamon Press 1972.

[156] Cohen-Tannoudji, Claude et al.. Quantum Mechanics, Volume 1. Wiley-VCH, 1991.

[157] Cohen-Tannoudji, Claude et al.. Quantum Mechanics, Volume 2. Wiley-VCH, 1991.

[158] Penrose, R., W. Rindler (1984) Volume 2: Two-Spinor Calculus and Relativistic Fields, Cambridge University Press, United Kingdom.

[159] Fritzsch, Harald (2002). "Fundamental Constants at High Energy" . Fortschritte der Physik. 50 (5–7): 518–524. arXiv: hep-ph/0201198 . (https://arxiv.org/abs/hep-ph/ 0201198) (https://arxi v.org/abs/hep-ph/0201198).

[160] Feigenbaum, M. J. (1976) "Universality in complex discrete dynamics", Los Alamos Theoretical Division Annual Report 1975-1976 .

[161] Gunaydin, M. and F. Gursey. Quark structure and the octonions. J. Math. Phys., 14, 1973.

[162] Hurwitz, A. (1923), "Über die Komposition der quadratischen Formen", Math. Ann., 88 (1–2): 1–25, doi:10.1007/bf01448439, S2CID 122147399.

[163] McMullen, Curtis T. 2000. The Mandelbrot set is universal. In The Mandelbrot Set, Theme and Variations, ed. T. Lei, 1–18. Cambridge U.K.: Cambridge Univ. Press. Revised 2007.

[164] https://www.wolframalpha.com

[165] Collins, J. Renormalization: An Introduction to Renormalization, the Renormalization Group and the Operator-Product Expansion. ISBN-13: 9780521311779.

[166] Bogoliubov, N.N.; Shirkov, D.V. (1959). The Theory of Quantized Fields. New York, NY: Interscience.

[167] Gaździcki, Marek; Gorenstein, Mark I. (2016), Rafelski, Johann (ed.), "Hagedorn's Hadron Mass Spectrum and the Onset of Deconfinement", Melting Hadrons, Boiling Quarks – From Hagedorn Temperature to Ultra-Relativistic Heavy-Ion Collisions at CERN, Springer International Publishing, pp. 87–92.

[168] Winters-Hilt, S. Fiat Numero: Trigintaduonion Emanation Theory and its Relation to the Fine-Structure Constant $\alpha$, the Feigenbaum Constant $C\infty$, and $\pi$. Advanced Studies in Theoretical Physics Vol. 15, 2021, no. 2, 71 - 98 HIKARI Ltd, www.m-hikari.com https://doi.org/10.12988/astp.2021.91517.

[169] Winters-Hilt, S. The 22 letters of reality: chiral bisedenion properties for maximal information propagation. Advanced Studies in Theoretical Physics, Vol. 12, 2018, no. 7, 301-318. https://doi.org/10.12988/astp.2018.8832.

[170] Winters-Hilt, S. Unified propagator theory and a non-experimental derivation for the fine-structure constant. Advanced Studies in Theoretical Physics, Vol. 12, 2018, no. 5, 243-255. https://doi.org/10.12988/astp.2018.8626.

[171] Briggs, K. A precise calculation of the Feigenbaum constants. Mathematics of Computation, Vol. 57, Num.195, July 1991, pages 435-439.

[172] Winters-Hilt S, I. H. Redmount, and L. Parker, "Physical distinction among alternative vacuum states in flat spacetime geometries," Phys. Rev. D 60, 124017 (1999).

[173] Friedman, J.L., J. Louko, and S. Winters-Hilt. Reduced phase space formalism for spherically symmetric geometry with a massive dust shell. Phys Rev. D Vol. 56, Num 12 (1997).

[174] Nielsen, Frank (2022). "The Many Faces of Information Geometry". Notices of the AMS. 69 (1). American Mathematical Society: 36-45.

[175] Nielsen, Frank (2018). "An Elementary Introduction to Information Geometry". Entropy. 22 (10).

[176] Amari S; Dualistic Geometry of the Manifold of Higher-Order Neurons. Neural Networks, Vol. 4(4), 1991:443-451.

[177] Amari S: Information Geometry of the EM and em Algorithms for Neural Networks. Neural Networks, Vol. 8(9), 1995:1379-1408.

[178] Amari S and Nagaoka H: Methods of Information Geometry. 2000. Translations of Mathematical Monographs Vol. 191.

[179] Winters-Hilt, S. Informatics and Machine Learning. Wiley Publishing. 9781119716747, Sept. 2021.

[180] Diósi, L. (1989). "Models for universal reduction of macroscopic quantum fluctuations". Physical Review A. 40 (3): 1165–1174.

[181] Penrose, Roger (1996). "On Gravity's role in Quantum State Reduction". General Relativity and Gravitation. 28 (5): 581–600.

[182] Maldacena, J.. The Large N limit of superconformal field theories and supergravity. Advances in Theoretical and Mathematical Physics. 2 (4): 231–252.

[183] Susskind, L. "The World as a Hologram". Journal of Mathematical Physics. 36 (11): 6377–6396. arXiv:hep-th/9409089.

[184] Peskin, M.E., Schroeder, D.V. (1995). An Introduction to Quantum Field Theory.

[185] Haxton, W.C.; Hamish Robertson, R.G.; Serenelli, Aldo M. (18 August 2013). "Solar Neutrinos: Status and Prospects". *Annual Review of Astronomy and Astrophysics*. **51** (1): 21–61.

[186] Sonjanovic, G. 2011. Probing the origin of neutrino mass: from GUT to LHC.

[187] Grossman, Y. 2003. TAST 2002 lectures on neutrinos.

[188] Koide, Y., Nuovo Cim. A 70 (1982) 411 [Erratum-ibid. A 73 (1983) 327].

[189] Sumino, Y. (2009). "Family Gauge Symmetry as an Origin of Koide's Mass Formula and Charged Lepton Spectrum". Journal of High Energy Physics. 2009 (5): 75. arXiv:0812.2103.

[190] Jackson, J.D. Classical Electrodynamics, 2nd Edition. Wiley 1975.

[191] Lorentz, Hendrik Antoon (1899), "Simplified Theory of Electrical and Optical Phenomena in Moving Systems" , Proceedings of the Royal Netherlands Academy of Arts and Sciences, 1: 427–442.

[192] Winters-Hilt, S. Theory of Trigintaduonion Emanation and Origins of $\alpha$ and $\pi$. Researchgate 05/24/20.

[193] Winters-Hilt, S. Chiral Trigintaduonion Emanation Leads to the Standard Model of Particle Physics and to Quantum Matter. Advanced Studies in Theoretical Physics, Vol. 16, 2022, no. 3, 83-113.

[194] Winters-Hilt, S. Meromorphic precipitation of quantum matter with dimensionful action. May 2021. DOI:10.13140/RG.2.2.32294.24640.

[195] Winters-Hilt, S. Emanator Theory using split octonions is Manifestly Lorentz Invariant and reveals why the fundamental constant $\hbar$ should be so small. Advanced Studies in Theoretical Physics, 2023.

[196] Winters-Hilt S. Emanator Theory is shown to be an optimal Martingale process at the fractal edge of chaos, where the Gravitational constant G is hypothesized to be a multiscale fractal coupling parameter. Advanced Studies in Theoretical Physics, 2023.

[197] Landau, Lev D.; Lifshitz, Evgeny M. (1969). Mechanics. Vol. 1 (2nd ed.). Pergamon Press.

[198] Goldstein, Herbert (1980). Classical Mechanics (2nd ed.). Addison-Wesley.

[199] Fetter, A.L and J.D Walecka, Theoretical Mechanics of Particles and Continua, Dover (2003).

[200] Percival, I.C. and D. Richards. Introduction to Dynamics. (1983) Cambridge University Press.

[201] Arnold, V.I. Ordinary Differential Equations. MIT Press. (1978).

[202] Arnold, Vladimir I. (1989). Mathematical Methods of Classical Mechanics (2nd ed.). New York: Springer.

[203] Woodhouse, N.M.J. Introduction to Analytical Dynamics. Springer, 2nd Edition. 2009.

[204] Bender, C.M. and S.A. Orszag. Advanced Mathematical Methods for Scientists and Engineers: Asymptotic Methods and Perturbation Theory. Springer. 1999.

[205] Robert L. Devaney. An Introduction to Chaotic Dynamical Systems. Addison -Wesley.

[206] Landau, Lev D.; Lifshitz, Evgeny M. (1971). The Classical Theory of Fields. Vol. 2 (3rd ed.). Pergamon Press.

[207] Penrose, Roger (1965), "Gravitational collapse and space-time singularities", Phys. Rev. Lett., 14 (3): 57.

[208] Hawking, Stephen & Ellis, G. F. R. (1973). The Large Scale Structure of Space-Time. Cambridge: Cambridge University Press.

[209] Peebles, P. J. E. (1980). Large-Scale Structure of the Universe. Princeton University Press.

[210] B. Abi et al. Measurement of the Positive Muon Anomalous Magnetic Moment to 0.46 ppm. Phys. Rev. Lett. 126, 141801 (2021).

[211] Hawking, S. W. (1974-03-01). "Black hole explosions?". Nature. 248 (5443): 30–31.

[212] Birrell, N.D. and Davies, P.C.W. (1982) Quantum Fields in Curved Space. Cambridge Monographs on Mathematical Physics. Cambridge University Press, Cambridge.

[213] Witten, Edward (1998). "Anti-de Sitter space and holography". Advances in Theoretical and Mathematical Physics. 2 (2): 253–291.

[214] Maldacena, J.. The Large N limit of superconformal field theories and supergravity. Advances in Theoretical and Mathematical Physics. 2 (4): 231–252.

[215] Tolkien, J.R.R. (1990). The Monsters and the Critics and Other Essays. London: HarperCollinsPublishers.

# Index

## A

# B

# C

# D

# E

# F

# G

## J



## K



## L



## M

## N

# O



# P

# Q



# R

# S

## T

## U



## V



## W

## Y



## Z